DIFFRACTION GRATINGS AND APPLICATIONS

OPTICAL ENGINEERING

Series Editor

Brian J. Thompson

Distinguished University Professor
Professor of Optics
Provost Emeritus
University of Rochester
Rochester, New York

1. Electron and Ion Microscopy and Microanalysis: Principles and Applications, *Lawrence E. Murr*
2. Acousto-Optic Signal Processing: Theory and Implementation, *edited by Norman J. Berg and John N. Lee*
3. Electro-Optic and Acousto-Optic Scanning and Deflection, *Milton Gottlieb, Clive L. M. Ireland, and John Martin Ley*
4. Single-Mode Fiber Optics: Principles and Applications, *Luc B. Jeunhomme*
5. Pulse Code Formats for Fiber Optical Data Communication: Basic Principles and Applications, *David J. Morris*
6. Optical Materials: An Introduction to Selection and Application, *Solomon Musikant*
7. Infrared Methods for Gaseous Measurements: Theory and Practice, *edited by Joda Wormhoudt*
8. Laser Beam Scanning: Opto-Mechanical Devices, Systems, and Data Storage Optics, *edited by Gerald F. Marshall*
9. Opto-Mechanical Systems Design, *Paul R. Yoder, Jr.*
10. Optical Fiber Splices and Connectors: Theory and Methods, *Calvin M. Miller with Stephen C. Mettler and Ian A. White*
11. Laser Spectroscopy and Its Applications, *edited by Leon J. Radziemski, Richard W. Solarz, and Jeffrey A. Paisner*
12. Infrared Optoelectronics: Devices and Applications, *William Nunley and J. Scott Bechtel*
13. Integrated Optical Circuits and Components: Design and Applications, *edited by Lynn D. Hutcheson*
14. Handbook of Molecular Lasers, *edited by Peter K. Cheo*

15. Handbook of Optical Fibers and Cables, *Hiroshi Murata*
16. Acousto-Optics, *Adrian Korpel*
17. Procedures in Applied Optics, *John Strong*
18. Handbook of Solid-State Lasers, *edited by Peter K. Cheo*
19. Optical Computing: Digital and Symbolic, *edited by Raymond Arrathoon*
20. Laser Applications in Physical Chemistry, *edited by D. K. Evans*
21. Laser-Induced Plasmas and Applications, *edited by Leon J. Radziemski and David A. Cremers*
22. Infrared Technology Fundamentals, *Irving J. Spiro and Monroe Schlessinger*
23. Single-Mode Fiber Optics: Principles and Applications, Second Edition, Revised and Expanded, *Luc B. Jeunhomme*
24. Image Analysis Applications, *edited by Rangachar Kasturi and Mohan M. Trivedi*
25. Photoconductivity: Art, Science, and Technology, *N. V. Joshi*
26. Principles of Optical Circuit Engineering, *Mark A. Mentzer*
27. Lens Design, *Milton Laikin*
28. Optical Components, Systems, and Measurement Techniques, *Rajpal S. Sirohi and M. P. Kothiyal*
29. Electron and Ion Microscopy and Microanalysis: Principles and Applications, Second Edition, Revised and Expanded, *Lawrence E. Murr*
30. Handbook of Infrared Optical Materials, *edited by Paul Klocek*
31. Optical Scanning, *edited by Gerald F. Marshall*
32. Polymers for Lightwave and Integrated Optics: Technology and Applications, *edited by Lawrence A. Hornak*
33. Electro-Optical Displays, *edited by Mohammad A. Karim*
34. Mathematical Morphology in Image Processing, *edited by Edward R. Dougherty*
35. Opto-Mechanical Systems Design: Second Edition, Revised and Expanded, *Paul R. Yoder, Jr.*
36. Polarized Light: Fundamentals and Applications, *Edward Collett*
37. Rare Earth Doped Fiber Lasers and Amplifiers, *edited by Michel J. F. Digonnet*
38. Speckle Metrology, *edited by Rajpal S. Sirohi*
39. Organic Photoreceptors for Imaging Systems, *Paul M. Borsenberger and David S. Weiss*
40. Photonic Switching and Interconnects, *edited by Abdellatif Marrakchi*
41. Design and Fabrication of Acousto-Optic Devices, *edited by Akis P. Goutzoulis and Dennis R. Pape*
42. Digital Image Processing Methods, *edited by Edward R. Dougherty*
43. Visual Science and Engineering: Models and Applications, *edited by D. H. Kelly*
44. Handbook of Lens Design, *Daniel Malacara and Zacarias Malacara*
45. Photonic Devices and Systems, *edited by Robert G. Hunsperger*

46. Infrared Technology Fundamentals: Second Edition, Revised and Expanded, *edited by Monroe Schlessinger*
47. Spatial Light Modulator Technology: Materials, Devices, and Applications, *edited by Uzi Efron*
48. Lens Design: Second Edition, Revised and Expanded, *Milton Laikin*
49. Thin Films for Optical Systems, *edited by François R. Flory*
50. Tunable Laser Applications, *edited by F. J. Duarte*
51. Acousto-Optic Signal Processing: Theory and Implementation, Second Edition, *edited by Norman J. Berg and John M. Pellegrino*
52. Handbook of Nonlinear Optics, *Richard L. Sutherland*
53. Handbook of Optical Fibers and Cables: Second Edition, *Hiroshi Murata*
54. Optical Storage and Retrieval: Memory, Neural Networks, and Fractals, *edited by Francis T. S. Yu and Suganda Jutamulia*
55. Devices for Optoelectronics, *Wallace B. Leigh*
56. Practical Design and Production of Optical Thin Films, *Ronald R. Willey*
57. Acousto-Optics: Second Edition, *Adrian Korpel*
58. Diffraction Gratings and Applications, *Erwin G. Loewen and Evgeny Popov*

Additional Volumes in Preparation

Characterization Techniques and Tabulations for Organic Nonlinear Optical Materials, *edited by Mark Kuzyk and Carl Dirk*

DIFFRACTION GRATINGS AND APPLICATIONS

ERWIN G. LOEWEN

Spectronic Instruments, Inc.
Rochester, New York

EVGENY POPOV

Institute of Solid State Physics
Sofia, Bulgaria

NEW YORK • BASEL

CRC is an imprint of the Taylor & Francis Group,
an informa business

Library of Congress Cataloging-in-Publication Data

Loewen, E. G. (Erwin G.)
Diffraction gratings and applications / Erwin G. Loewen, Evgeny Popov.
p. cm. — (Optical engineering ; 58)
Includes bibliographical references and index.

ISBN 0-8247-9923-2 (hardcover : alk. paper)
1. Diffraction gratings. 2. Diffraction gratings—Industrial applications.
I. Popov, Evgeny. II. Title. III. Series: Optical engineering
(Marcel Dekker, Inc.); v.58.
QC417.L64 1997
621.36'1—dc21

97-2659
CIP

The publisher offers discounts on this book when ordered in bulk quantities. For more information, write to Special Sales/Professional Marketing at the address below.

This book is printed on acid-free paper.

MARCEL DEKKER, INC.
270 Madison Avenue, New York, New York 10016

Current printing (last digit):
10 9

To the Memory

of

Lyuben Mashev

From the Series Editor

Our series of books on optical engineering continues to grow in number and in scope. It is a particular pleasure to be able to add this current work to the series because it represents a very important fundamental tool (the diffraction grating) based on the cornerstone of wave optics (diffraction). Since the diffraction grating is a device or subsystem, it is incorporated into systems and instruments that allow its versatility to be expressed in a wide range of applications. Of course, I also admit to my own biases, having spent many years being intrigued and amazed by the beauty and diverse manifestations of the diffraction of light whether by one-, two-, or three-dimensional structures.

The topic of diffraction gratings and their applications seems forever new in spite of (or because of) its venerable history. The process of diffraction was first observed and recorded by Francesco Maria Grimaldi and published in 1665. The diffraction grating didn't arrive on the scene until over one hundred years later in 1785; it was discovered by the American astronomer David Rittenhouse. This fact is usually only a footnote in many technical books and historical treatises because it didn't have much impact. Joseph von Fraunhofer rediscovered the diffraction grating somewhat by chance. Let me quote from one of my favorite early texts on diffraction (*The Diffraction of Light, X-Rays, and Material Particles* by Charles F. Meyer, University of Chicago Press, 1934):

> Fraunhofer, in studying the pattern due to a slit, sought to obtain this pattern with greater intensity. In order to achieve the desired result he made a series of slits close together by winding a hair or wire upon a frame. What he then found was not at all what he had expected. The pattern was *not,* as he expected, that due to a single slit only more intense; it was that due to a *diffraction grating*--he had discovered the diffraction grating. His discovery of the grating was thus a large extent accidental, but he showed great genius in the manner in which he followed up on this discovery as well as in the manner in which he followed up his discoveries of the sodium line and of the dark lines in the solar spectrum which were also to a certain degree accidental.

And as that great commentator Walter Cronkite often says, " . . . and the rest is history." An important history, a viable present, and a strong future for diffraction gratings and applications are detailed in this volume.

Brian J. Thompson

Preface

The importance of diffraction gratings in the field of science has never been expressed with more feeling as well as accuracy than by George Harrison some 50 years ago when he wrote:

"It is difficult to point to another single device that has brought more important experimental information to every field of science than the diffraction grating. The physicist, the astronomer, the chemist, the biologist, the metallurgist, all use it as a routine tool of unsurpassed accuracy and precision, as a detector of atomic species, to determine the characteristics of heavenly bodies, the presence of atmospheres in the planets, to study the structures of molecules and atoms, and to obtain a thousand and one items of information without which modern science would be greatly handicapped."

Today we could add to this list the important symbiosis that exists between gratings and lasers, which ranges from miniature couplers in integrated optics to giant gratings for laser pulse compression, not to mention the millions of tiny transmission gratings that are found in almost every CD-player, where they serve as beam splitters required for keeping reading heads in focus and on track. The recent discovery that optical fibers can have diffraction grating structures superimposed has great potential impact on the efficiency and capacity of fiber optics networks.

There exists an enormous literature on gratings and their many applications, spread over dozens of journals and chapters in textbooks, but only a few monographs, such as *Electromagnetic Theory of Gratings*, edited by Petit (1980) and *Diffraction Gratings* by Hutley (1982), and *Le Multiplexage de Longuers d'Onde* by Laude (1992).

The aim of this book is to provide an overview of the field of diffraction gratings and their applications in a single volume. To maintain a reasonable length many details must be left out, but an attempt is made to provide a bibliography extensive enough so that anybody who wants to follow a more detailed trail can either find it directly or be led to it. Our aim has been to reach the many users of gratings rather than specialists in their production.

The book is made up of three parts: I. Properties of diffraction gratings, discussed in Chapters 2 to 9; II. Diffraction grating treatment, Chapters 10 to 13; and III. Diffraction grating manufacture, Chapters 14 to 17. The boundaries between these topics are not always rigid. For example efficiency behavior depends on groove profile quality, which in turn depends on manufacturing. Echelles are reflection gratings, but of a special type, and concave gratings can be in bulk form or as waveguides, and some transmission gratings can reflect totally, etc.

The historical review of Chapter 1 covers the early work in spectroscopy, mainly in the 19th century, and other historical aspects are found in the respective chapters.

Each chapter is designed to be as self-consistent as possible, but a reader without experience in this field should start by reading Chapter 2. Some topics are discussed throughout several chapters, which leads to repetitions necessary for better understanding and clarity.

No book is conceived in a vacuum. This one began a long time ago with E. Loewen joining the grating group at Bausch & Lomb, which was started by George Harrison as consultant, David Richardson as the coordinator and Robert Wiley the mechanical engineer who not only made the ruling engines work but took a leading part in the many aspects of ruling, replication and testing of gratings. This included a collaboration of many years with the ruling development at MIT, where echelles were the main goal, under George Harrison's leadership that ended with his death in 1979. In 1974 there began a long period of collaboration with the Laboratoire d'Optique Electromagnetique, which was based on their pioneering effort to establish accurate solutions to the problem of energy partition at the surface of a grating. The key developments were the integral code of Daniel Maystre and the differential method of Michel Nevière, and their continued help and interest ever since is gratefully acknowledged here. The results are especially visible in Chapters 4 to 6.

The collaboration that led to this book owes its origin to Lyuben Mashev of the Institute of Solid State Physics in Sofia, where he created a laboratory for holographic gratings. He spent a post-doctoral fellowship with Bausch & Lomb in Rochester, N.Y., in 1986. The idea of joining in the writing of a book was first broached by him on a ski lift at nearby Bristol Mountain. He was the tutor of Evgeny Popov as well as a great friend. Unfortunately his untimely death in Sofia in 1988 forced a passing of the torch. We acknowledge here his many contributions and dedicate the book to his memory.

A decade of collaboration with and within the Laboratoire d'Optique Electromagnetique in Marseille, where E. Popov has worked since 1993, not only provided scientific survival, but permitted deeper understanding of many problems and their detailed analysis.

An important participant in Sofia was Lyubomir Tsonev, who since 1988 contributed his experimental and analytical skills to grating studies. He had a great role in preparation of Chapter 7. Special thanks are devoted to Evguenia Anachkova-Scharf, who contributed by locating some rare old papers in Munich. We also want to acknowledge the staff of the Richardson Grating Laboratory of Spectronic Instruments, Inc., formerly Bausch & Lomb, who have given so much of their experience of many years, and helped in the preparation of numerous figures. We could single out a few: Robert Wiley, Tom Blasiak, John Hoose, Chris Palmer, and Sam Zhelesnyak.

Garry Blough devoted many hours to reviewing the entire text, for which we are most grateful.

Erwin G. Loewen
Evgeny Popov

Contents

DIFFRACTION GRATINGS AND APPLICATIONS

Chapter 1

A Brief History of Spectral Analysis

No history of gratings would be complete without taking a look at the history of spectral analysis. Much of the early history can be found in a monumental 10-year compilation on spectroscopy published by Heinrich Kayser in 1900 [1.1]. Not much of the work of the 19th century appears to be missing from this 780 page compendium.

1.1 Work Before the Year 1800

The first serious study of solar radiation was conducted by Isaac Newton in 1672 and described in his famous *Treatise on Optics*. His principal observation was that separation into what he felt were an infinite number of colors occurred due to the differences in refractive index of the prism as a function of color. He noted that when his "entrance slit" was reduced to 1/2 mm, the colors appeared with greater purity than with the circular entrance apertures he had been using. His most important finding was that the basis of light lies in individual colors whose eventual mixture is perceived as white. One may wonder why, given slits that were sufficiently small, as well as adequate linear dispersion, he failed to see any of the solar absorption lines. The explanation lies in the poor quality of the glass available to him, full of inclusions and inhomogeneities, as well as the low quality of the polished faces. It is an early example of how important the quality of instrumentation can be to obtaining important results.

No additional work in this direction was published until 1800, when W. Herschel discovered with the help of a sensitive thermometer that the spectrum extends beyond the visible red, and is perceived as heat. He noted that while maximum perceived brightness is in the yellow, the maximum energy of solar radiation occurs just beyond the visible red, and also that what we now call the infrared can be reflected and focussed by mirrors, exactly like the visible [1.1]. Only a few years later J.Ritter found that just beyond the violet end of the spectrum there was radiation whose existence was verified by its blackening effect on silver chloride or what we might be tempted to call photography [1.2].

In the meantime Thomas Young, father of the wave theory of light, published a paper in 1803 describing what happened when he passed light through a transmission grating in the form of a glass stage micrometer that had been ruled with 500 lines/inch [1.3]. He noted not only that red light could be observed in 4 different directions but that the sine of their respective angles varied as the ratio 1:2:3:4. He was clearly the first to use this finding to identify different colors by their wavelength, with results translated into nm that we would accept today:

Start	red	yellow	green	blue	violet	end
675	650	576	536	498	442	424

With his discovery of the sine relationship, Young ought to be given more credit than has been customary for being the first to do scientific studies with diffraction gratings. He was, however, not the first to have taken a look at this phenomenon. In 1785 Francis Hopkinson, who was one of the signers of the Declaration of Independence (and George Washington's first Secretary of the Navy), was one night observing a distant street lamp through a fine French silk handkerchief. He noticed that this produced multiple images, which to his astonishment did not change location with motion of the handkerchief. He passed on this discovery to his friend the astronomer David Rittenhouse, who recognized it as a diffraction effect, and promptly made himself a 1/2 inch diffraction grating by wrapping fine wire around the threads of a pair of fine pitch screws, ending up with 53 apertures. Knowing the pitch of his screws in terms of the Paris inch, he determined the approximate wavelength of light [1.4]. This appears to be the end of his investigations, and he went on to become the first Director of the U.S. Mint. Rittenhouse used wires of different diameter in the same screw threads and was able to observe that this changed the relative amount of light going into different orders. Since it is highly unlikely that any of the European investigators bothered to read American journals, we can safely assume that Fraunhofer's development of a similar wire grating three decades later was a rediscovery.

1.2 The Early Work in Gratings

Fraunhofer was without doubt the first scientist to take gratings seriously, as described in Chapters 12 and 14. He built the first ruling engine, details of which are lost to posterity because he and his optical company in Munich regarded it as an important and proprietary secret. He studied the ruling process and discovered the need for extraordinary accuracy (1% of the groove spacing), observed the behavior of the various higher orders, the presence of occasional "very peculiar" polarization effects, and even had the insight to associate the difference in efficiencies with variations in the shape of the grooves. He did all this with gratings that never exceeded one inch in size (25 mm), and groove spacings that were never less than 3 μm. The cosine relationship for skew rays was clear to him, as was the earlier work of Thomas Young.

On the instrumental side he was one of the first to appreciate operating in collimated light, using telescopes to do so, and by using theodolites was able to make more accurate angular measurements than anyone before. Not so well known is that he was also the first to use cross dispersion by means of a small angle prism attached to the entrance slit. The purpose was to allow him to see the higher orders separately, avoiding the normal overlap. All of this was written up in great detail, so that we are entirely justified to call him the father of grating technology [1.5, 6].

The initial impetus to Fraunhofer's work was not just pure science. As manager of the leading optical shop of its time, he was greatly concerned with accurate measurement of the refractive index of his glass blanks so as to make better achromatic lenses. In order to do this he needed accurate determination of the wavelengths at which the index was measured, for which the solar absorption lines, which he discovered, were excellent markers. Even though he had no idea what caused them, it was the start of the field of spectral analysis.

Interesting also is his insight into the complexity of the diffraction process, with his correct prediction that the laws of its (efficiency) behavior would "strain even the cleverest of physicists", which it did for the next 150 years.

1.3 The Beginnings of Spectral Analysis

W.H.Wallaston, best known for the polarizing prisms named after him, was actually the first to have seen the solar absorption lines in 1802, with a high quality flint glass prism held near his eye while standing some 3 m from a 1 mm wide slit illuminated by the sun [1.7]. Nobody paid much attention to this discovery, nor to his description of seeing different colored images when looking at the lower (blue) part of candle light. He also saw the sodium lines, without appreciating their origin. Thomas Young quickly confirmed these findings, but also failed to follow up.

In his 1823 paper, Fraunhofer paid more attention to the possibilities of spectroscopy than is usually assumed [1.5]. He noted that in the bright spectrum of a flame there are two yellow lines that occur at exactly the same wavelength as the double D lines in the solar spectrum and that a glowing piece of glass produces a spectral continuum. Given his company's commitment to building telescopes, we should not be surprised that Fraunhofer spent considerable effort examining stellar spectra, building a special 100 mm telescope for the purpose (and equipped with an objective prism). To better observe the spectra, he provided a 50 mm telescope mounted at an angle of 26° to the first, grumbling about the need for a second observer. He looked at all the bright stars and planets, noting that the latter, as well as Capella, had spectra just like that of the sun. Sirius, on the other hand had no absorption lines, showing a band in the green and two in the blue. Of course, while these differences must have been fascinating there was no hint of an explanation.

J.F.W. Herschel, son of the discoverer of infrared, narrowly missed becoming the father of spectral analysis in 1823. He investigated the spectra of many different substances, noting amongst other things the pale violet of potassium hydroxide injected into a flame, even when the quantity was minute [1.8]. He also pointed out absorption lines in various colored glasses. He was fooled by the ubiquitous presence of sodium into believing that differences in emission somehow were related to temperature and were not intrinsic to the elements.

David Brewster expressed considerable interest in this field, but never could give up the false notion that absorption lines where somehow the property of light rather than of the substances.

Henri Fox Talbot, another one of the many scientists who acquired fame in other directions, photography in this instance, did useful but not defining work in emission spectra in 1834 [1.9]. To quote him: "*The strontium flame exhibits a great number of red rays well separated from each other, not to mention an orange and a very bright blue one. Lithium on the other hand exhibits a single red ray. I hesitate not to say that optical analysis can distinguish the minutest portion of these two substances from each other, with as much certainty, if not more, than any other known method.*" What Talbot missed was the recognition that these lines are emitted by the substances involved, and function (not just by coloring the flame) without being consumed.

Charles Wheatstone, famous for the electrical bridge circuit bearing his name, was another scientist interested in observing spectra, this time excited by electric sparks [1.10]. He reported all the now well-known mercury lines, as well as spectra of metals such as Zn, Cd, Bi, Sn, Pb. His comment is prophetic: "*The number, position and color of these lines differ in each of the metals employed. These differences are so obvious that any metal may instantly be distinguished from the others by the appearance of its spark and we have a mode of discriminating metallic bodies more readily even than chemical examinations, and which hereafter may be employed for useful purposes.*" Wheatstone proved that the spectra were intrinsic to the metals and felt that it had to do with their 'molecular structure', and thought that this had the possibility of being the clue to their study, a remarkably prescient idea.

The fame of A.J.Ångström, in contrast to the others, rests almost solely on his spectroscopic research, and he was honored by having a unit of wavelength named after him. The choice of 10^{-10} m was not arbitrary, but was picked because it represented the smallest significant figure that instruments of the time could measure. It has proved so convenient that it is still much used, even though absent from the SI system. The first of his many publications dates to 1855 [1.11]. Ångström was the first to describe the fundamental difference between solid and gaseous bodies, and began to speculate on resonance effects to explain the relationship of emission and absorption spectra. He missed the need to make comparisons at equivalent temperatures, a failure he was to greatly regret. He was also under the impression that metallic lines obtained from alloys were slightly displaced in wavelength from the corresponding pure metals, which gives an indication of the limitations of his equipment. Remarkable, from a current perspective, was that Ångström covered up a calibration error of his reference scale that he used to determine the absolute groove spacing of his gratings, and which he discovered two years after his first publication. However, it was not until 1884, a decade after Ångström's death, that Thalén revealed that every one of these published wavelengths was too short by 130 parts per million [1.12].

Not as well known is the work of Alter, who not only studied metallic spark spectra but also those of gases [1.13]. He speculated that the change from white to red in the color of lightning propagating through water was due to the strong red line of hydrogen.

Another outstanding member of the early group of this period was J. Plücker, who spent much effort examining gas emission lines produced by electric discharge (in glass tubes made for him by his glass blower Geissler) [1.14]. History has given Geissler's name to such tubes, while Plücker has been almost forgotten. He determined that the three hydrogen lines, which he termed H_α, H_β and H_γ, coincided exactly with the absorption lines that Fraunhofer had designated F, C, and G. In looking at the spectra of several tubes containing arsenic Plücker kept seeing new lines that were absent in others and suspected that they came from a new element. Luckily he was cautious about publishing this finding because he later found they were merely nitrogen lines from gas that had crept in.

1.4 Nobert

F. Nobert played an important role in 19th century spectrometry, because for 30 years he was the world's only source of diffraction gratings, from 1850 until his death in 1881 (i.e., almost until Rowland came upon the scene a few years later). He had started earlier (1833) to make circular rulings, and then spent much effort to rule microscope resolution targets, which sold for £15 each in London. Four of his gratings were used by Ångström, others by Quincke, Rayleigh and many others. His ruling engine is now in the storage area of the Smithsonian Institution in Washington, D.C. [1.15, 16].

1.5 Kirchhoff and Bunsen

There is no doubt that the science of spectrometry owes its firm foundation to Gustav Kirchhoff, professor of physics in Heidelberg, and his friend and colleague Rudolph Bunsen, professor of chemistry. In 1859 Kirchhoff's landmark paper announced the general law that connected emission and absorption of light, and clearly pointed out the significance of the unique spectra emitted by different elements [1.17]. Kirchhoff's law states that "*the relation between the powers of emission and the powers of absorption for rays of the same wavelength is constant for all bodies at the same temperature.*" It is clear from this that a gas that radiates a line spectrum must, if at the same temperature, absorb the line that it radiates. Kirchoff showed that the Fraunhofer D lines were identical to the yellow lines of sodium, and that a sodium flame absorbs the same yellow light from a stronger source behind it. He announced with fine insight that the dark Fraunhofer lines were due to absorption by their corresponding elements located in the cooler parts of the solar atmosphere, while the continuum came from the sun's interior.

Kirchhoff and Bunsen started a thorough analysis of every pure element they could get their hands on. Bunsen acquired fame in his own right through the invention of the gas burner that will forever carry his name, which was useful for spectral analysis because it was hot and nearly colorless, unlike previously used candles or oil burners. Kirchhoff's name is also for ever linked to electric network analysis and diffraction theory of optics. In their spectral study of the alkali metals, Kirchhoff and Bunsen discovered in 1860 a fourth hitherto unknown member of the family, which they named caesium, and shortly thereafter a fifth named rubidium. Once identified they were soon isolated. This work was done with surprisingly crude spectrometric equipment, using a hollow glass prism filled with CS_2, with no attempt to obtain absolute wavelengths [1.18]. The spectral light source was a loop of platinum wire coated with a salt of the compound under study and heated in a Bunsen burner.

These and other activities were a scientific sensation of the time. Not only were scientists from all over the world drawn to this new field, very much like what happened when lasers were discovered 100 years later, but, in another parallel, there was also a great deal of popular interest.

One effect of so much publicity was to arouse envy in some of their predecessors, most of whom are mentioned above (or in some instances their nationalistic partisans), because they had been so close to making the same discovery. It may have galled them to realize that by missing crucial insights, major fame had just eluded them. However, Kirchhoff's vision was to state the general laws so clearly and convincingly that it attracted the attention of the whole scientific world. He felt so secure in this that he never worried about petty sniping from abroad. One is reminded of the remarkably parallel events a century later that took place in the field of lasers.

1.6 Georg Quincke

Quincke appeared in the field of optics, and especially gratings, somewhat like a nova, shining brightly and then departing the scene. As a result he is relatively little known, despite some real contributions. A former graduate student of both Kirchhoff and Bunsen, he become the first professor of experimental physics in Berlin, but later was appointed to Kirchhoff's chair in physics at the University of Heidelberg, where he died in 1924 [1.19]. He introduced the first practical laboratory course in physics in a German university; evidently the ideal person for the job. He spent several years studying the behavior of gratings, all of them obtained from Nobert, publishing the results in 1872 in a 65-page paper, after which he dropped the subject [1.20]. Quincke started with experiments in which transmission gratings were immersed in various liquids (with the aid of a cover glass), and noted that the diffracted beam did not change direction. Hardly a surprising conclusion.

Not well known are Quincke's experiments with laminar gratings, which he derived from what we would call Ronchi rulings in silver. Exposing them to iodine vapor he converted the silver into silver iodide, which is transparent. If made to the correct thickness to get half-wave retardation he could reduce to near zero the transmission of a small wavelength band. He even made one such grating with tapering thickness, so that the wavelength of extinction would shift progressively. He found the resulting color sequence to be identical to that of Newton's fringes.

Interested in reflection gratings, he produced them by silver coating the Nobert glass rulings. In order to observe what would happen if the groove geometry were inverted, he produced the world's first grating replicas, using a cleverly contrived galvanoplastic procedure. Building a dam of Guttapercha around a silver grating, he devised a plating cell that would generate a copper replica that could be pealed off. Of course he had no thoughts of deriving a business from this. He noted that giving his silver gratings a light polish, so that the grooves were only partly filled with silver, would change the intensities of the spectral orders, but never their direction.

Quincke's careful observations led him to discover secondary images in all of his gratings that he was unable to explain. They were, of course, what we now call Rowland ghosts, the result of Noberts considerable periodic errors. His paper [1.20] was well known to R.W. Wood, who some 30 years later was the first to address some of the same topics.

1.7 Progress in Solar Spectroscopy

The ready availability of sunlight, together with the richness of its spectrum, gave impetus to many studies of its spectrum. One of the best known investigators was Lockyer. Once he obtained an instrument of high resolution he devoted many years to his studies, and began a custom followed ever since, which was to combine work in the field with work in the laboratory. He discovered a prominent line which he first suspected to be hydrogen, but then realized it was from an element hitherto unknown [1.21]. This turned out to be helium and his name is in every physics book for his discovery, and the fact that it was first found in the sun. He also coined the term chromosphere for the surface of the sun, as distinguished from the corona above it.

Lockyer tried to test for the effects of gas pressure and sparks of different strengths, imaging them onto his entrance slit. From the length of line image he could distinguish between lines that were formed over a wider temperature range (long line vs. short). He was also the first to make extensive use of photography, a truly vital addition to the field, even today. Lockyer was the first also to properly study the Doppler effect [1.22].

To improve on this study, Zöllner built a special reversion spectroscope to observe the small shift in wavelength of solar prominences from opposite sides of the sun [1.23], to the red on one side and blue on the other. The difference confirmed the known rotation speed of the sun. Huggins appears to be the first to look for Doppler shifts in stellar spectroscopy in 1872 [1.24], and it is interesting that this is still a major field of research in modern astronomy, because it allows accurate velocity measurements.

Soret added an unusual piece of technology in search for spectral lines at wavelengths below the visible. To do this he designed a fluorescent eyepiece. He spent years using this new tool to investigate UV absorption spectra, including rare earths [1.25].

A short time later Cornu undertook a careful study of solar lines in the UV, naturally using photography, publishing results in a series of atlases over the years 1872-1880. He was concerned with the nature of the UV cut-off, and observed that moving his equipment from sea level to 800 m moved the cut-off wavelength by only 1 nm, and correctly identified absorption by air as responsible [1.26].

An important event was the first grating-based photograph of the solar spectrum, obtained in 1873 by H. Draper, with a Rutherfurd grating [1.27]. He set in motion a permanent trend to photography as a basic tool for stellar work, especially spectroscopy. It gave access to weak stars through long integrating exposures, and allowed data reduction of increased accuracy and away from tedious observations at telescope eyepieces in cold observatories.

Somewhat later Mouton began studies in the solar infrared. He was able to go as far as 1.85 μm, using a thermopile as a detector [1.28]. In order to calibrate the wavelengths he adopted the idea of Fizeau and Foucault of interposing in front of the entrance slit a birefringent plate. This generates a pattern of interference fringes that serve as a calibration marker. Much better results were obtained by Langley in 1881, due to his invention of the bolometer, for which he received wide and well deserved credit [1.29]. However, in another of history's oddities, the concept had been described in great detail over 30 years earlier by Svanberg, but unused the idea was forgotten [1.30]. History is often unkind to concepts described before their time.

Abney's unique achievement was to make photographic plates sensitive up to 1 μm. He used them to make studies of the many near IR Fraunhofer solar lines [1.31]. The procedure was so complex that nobody successfully followed up. However, Lommel got around this by taking advantage of a technique developed earlier by Becquerel. This was to use phosphorescent surfaces which lose their properties temporarily following radiation by IR light. This resulted in negative pictures but allowed him to go up to 1.8 μm [1.32].

Huggins had gone in the opposite direction with observations of UV spectra in stars, in which he was able to discern hydrogen lines that were progresively closer together, later also found in the lab [1.33].

The discovery of new spectral lines and their subsequent identification in the lab is another endeavor that was carried on for another century, as technology allowed exploration of new spectral regions along with gradually improving accuracy.

1.8 The Era of Rowland

A whole new era of spectral analysis opened up with Rowland's famous 1882 paper [1.34]. The world was presented, as if by magic, with gratings that were much larger and much more accurate than anything available before. Probably an even greater impact was generated by his almost simultaneous invention of the concave grating. The absorption losses and wavelength limitation of collimating lenses vanished, and resolution increased to where it exceeded that of a large array of prisms with a base length exceeding one meter. In addition wavelength accuracy greatly increased. Over his lifetime he supplied the world with about 100 of his master gratings, charging only for his expenses.

Higher orders used to be a nuisance, and thanks to chromatic aberrations, usually in poor focus. Now they suddenly became useful adjuncts to determining wavelengths by the methods of coincidences that Rowland developed. Not content with making gratings and supplying them to colleagues around the world, he began a long cycle of experimental work of the highest order, beginning with the solar spectrum [1.35]. He published a detailed photographic map, beginning in 1887, achieving what he felt was an accuracy of 0.01 Å [1.36], at least an order more accurate than anything done before. As so often in the history of science even Rowland succumbed to the lure of underestimating systematic errors, which were about 35 PPM, or twice his estimate. He identified many of the solar lines by comparison with arc spectra, and found several for which the element had not yet been discovered on earth. It represented a monumental effort, for which he was the ideal person. The work was continued both by his pupils in the USA and in Europe, and contributed greatly to his fame.

Despite his acclaim in other aspects of physics, for example the definitive work in measuring the mechanical equivalent of heat, Rowland was most proud of his achievements in the field of spectroscopy and diffraction gratings. He died at age 53 in 1901, and his ashes are interred in the wall of his ruling laboratory.

1.9 Origin of Spectral Lines

The period from 1860 to 1885 was one during which publications on spectroscopy jumped from about 20 per year worldwide to about 200, all traceable to the influence of Kirchhoff. Beginning in 1885 the rate increased quite sharply again, reaching about 400 by the turn of the century. What brought about this interest was the discovery that there was some mathematical order to the location of spectral lines. Specifically it was Balmer who discovered that the wavelengths of hydrogen could be represented by a simple mathematical formula [1.37]. This naturally gave rise to a search for similar rules for other elements, especially amongst groupings of families from the periodic table. Rydberg found regular grouping of lines in the alkali metals, and noted that they followed a progression that was tied to their atomic weight [1.38]. Similar series were found in heavier metals such as copper–silver–gold, by Kayser and Runge [1.39], while Runge and Pashen studied the helium lines [1.40].

Well known today is how this background information led Bohr in 1913 to announce his theory of the hydrogen atom, which was the beginning of atomic physics, culminating in quantum mechanics that could explain nearly all spectral lines on the basis of electrons changing their orbital states.

1.10 The Vacuum UV

It is rare to find any scientific endeavor of major interest in which a single person was able to dominate the entire basic development. In vacuum

UV spectrometry we find such a case in the person of Victor Schumann. Working as an amateur scientist in Vienna, he decided to attack spectrometry at wavelengths below the air cut-off wavelength at 1855 Å. He found this such a challenge and faced so many difficulties that he decided he either had to drop it or give up his business and devote full time to it. He chose the latter course. Vacuum pumps, for example, were not exactly an article of commerce. The simple tools we use to seal vacuum boxes did not exist. Electric sparks had to be mounted inside the shielded instrument. Fluorite (CaF_2) was known to transmit, but nobody knew how far, nor was there a way to measure index as function of wavelength and therefore there was no calibration. Finally, what was he to use as a detector? Unknown was the fluorescence of salycilic acid, taken for granted today. A major hurdle was Schumann's discovery that photographic plates were useless because the gelatin in which the sensitive silver halide crystals are imbedded was totally opaque to the UV radiation, even in very thin layers. He solved this key problem by preparing what we now call Schumann plates. These are made by allowing silver bromide to form in solution and deposit slowly onto a glass plate in the bottom of the dish, which calls for extremely careful handling in near darkness. He published a large number of papers beginning in 1890. The wavelength calibration problem was solved when he was able to acquire a small Rowland concave grating, leading to an 1893 paper in which he claimed to reach 1000 Å [1.41]. The collection of problems, which he carefully described, were so great that it was a long time before anybody else developed enough courage to pick up this field.

1.11 Some Special Effects

The high resolution possible with Rowland's gratings opened up some new experimental avenues. One was Zeeman's great discovery that if the spectral source is placed in a sufficiently strong magnetic field, most lines will double [1.42]. Eventually this provided great insight into atomic processes, and quickly earned its discoverer the Nobel prize. Interesting is that his first attempt to observe the effect was a failure. However, discovering that the famous Michael Faraday had made a similar abortive attempt in 1862 (one of his last experiments), he felt it was worth one more try. This time he succeeded simply because he had access to a new 10 ft radius Rowland grating of 600 gr/mm in Kammerlingh Onnes lab. As an interesting aside, he found that this delicate work in Amsterdam was severely impeded by the traffic, even when he worked in the middle of the night [1.43].

An entirely different discovery, also based on superior instrumentation, was that there were often very slight differences between Fraunhofer lines in the sun and their corresponding laboratory equivalent. This was examined by Humphrey and Mohler, and traced to pressure effects [1.44]. The effect differed considerably by element but was never seen with band spectra.

1.12 Some Historical Aspects of Ruled Gratings

A few aspects of ruled grating history are presented here, because they still have some interest today.

1.12.1 Blazing and Efficiency

All the early workers in the field of gratings, from Fraunhofer on, were well aware that the ability of gratings to distribute energy into various orders or directions was hardly ever the same twice. They were aware that in some mysterious fashion this was connected with the shape of the groove, but knew that there was little they could do to control it. The reason was simple enough: an inability to shape the diamond tools, that were largely picked by guess from a collection of splinters. One finds in the literature occasional expressions of delight that a certain grating was *"unusually bright in the second order green,"* but there was surprisingly little grumbling. Even Rowland, who understood the game well, was relatively unconcerned. Presumably in a more leisurely age it did not matter if photographic exposures were long ones.

The first to point out what the ideal grating groove shape should look like was Lord Rayleigh in 1874 [1.45]. He writes:

To obtain a diffraction spectrum in the ordinary sense, containing all the light, it would be necessary that the retardation should gradually alter by a wavelength in passing over each element of the grating and then fall back to its previous value, thus springing suddenly over a wavelength.

He was not exactly encouraging about achieving such a geometry, because he adds:

It is not likely that such a result will ever be obtained in practice; but the case is worth stating, in order to show that there is no theoretical limit to the concentration of light of assigned wavelength in one spectrum, and as illustrating the frequently observed unsymmetrical character of the spectra on either side of the central image.

One is left perhaps to speculate on the meaning of *"will ever."* It was not until about 40 years later that Wood produced the first grating that we would call blazed, and that with a tool of carborundum, ruled into copper, for use in the infrared [1.46].

The missing insight, that we now take for granted, was provided by John Anderson (in 1916, while working at the Mt. Wilson Observatory), who not only showed how one can shape diamonds into the so-called canoe form, but also that much better results were obtained through generating grooves by burnishing rather than by cutting, as reported by Babcock [1.47] (see Ch.14).

The final crucial development in this chain was the discovery by Strong that vacuum deposited aluminum on glass is a far superior medium into which to rule than Speculum metal, which had reigned supreme (see Ch.14).

1.12.2 Defects of Grating Ruling

That gratings usually contain ruling deficiencies capable of influencing results was already well understood by Fraunhofer. For example there are the "secondary spectra" mentioned by Quincke. The first published analysis of their cause (i.e., periodic errors of the lead screw), is by Peirce, who may not have been acquainted with Nobert or Quincke, but knew all about Rutherfurd's gratings [1.48]. The paper contains no hint as to the source of his insight. He must have been a good experimenter as well as mathematician, because he also noted that Rutherfurd's gratings focused closer or further from the central image, depending on whether the first or final sections were illuminated. He immediately deduced that this was caused by error of run (i.e., a progressive change in the pitch of the screw). The same observation had actually been made earlier in France by Mascart in 1864, who later became a senior statesman of science. It was picked up by Cornu, who worked on this effect for many years [1.49]. What makes this interesting today is that it constitutes the beginning of the currently active field of diffraction optics, where diffractive patterns are applied to refractive elements to provide special optical behavior.

1.13 Spectrographs and Spectrophotometers

Over the century and 1/3 that have elapsed since Kirchhoff pointed the way to spectrometric measurements, there have always been new approaches and new applications. The basic instrumental developments are described in Chapter 12.

In general the path has been from spectroscopes, with the human eye as the somewhat limited detector, to spectrographs where photographic film provided not only wider wavelength coverage, but also enormous capacity for parallel recording.

As electronic detectors and amplifiers were developed, the nature of recording changed again to take advantage of higher speed and sensitivity. The final stage of electronic detection was the introduction of array detectors that have made film almost obsolete, not only because of high detectivity but the direct links to subsequent data processing, now nearly always digital in form.

Spectrophotometers, in their many guises, have been developed in specialized forms for a host of different fields and are briefly described below.

1.13.1 Infrared Spectrometry

For the study of a large number of organic molecules, the infrared absorption lines are invaluable for identifying and investigating structure by their molecular vibrations and rotations. The basis for this had already been noted in the late 19th century by people like Ångström and Abney, even though they could not progress too far due to instrumental limitations. Both gratings and prisms have been used as the dispersing medium for many decades in instruments that started out strictly manual and gradually progressed to more and more automation. Prisms have the advantage of having no overlapping orders and high efficiency, but all the available materials have limited bands of transmission and are expensive in the larger sizes needed for good throughput. As a result they were often replaced by echelette gratings, which could be made in larger sizes, but suffered from the need for filtering out higher orders. One solution is a small low dispersion prism fore-monochromator in front of a grating spectrometer, but is limited to wavelengths < 40 μm by the availability of transmissive prism materials. A good review of near IR instruments (to 3.5 μm) is found in reference [1.50].

One of the first IR spectrophotometers to use the double beam approach was introduced by Perkin-Elmer in 1950, later followed by a whole series [1.51]. However, about 20 years later the picture started to change, and today analog IR recording spectrophotometers have a much reduced role. The reason is that the advent of high speed computers changed the ground rules and allowed the development of a whole new class of instruments, Fourier transform instruments. As pointed out by Gebbie, a spectrum is the plot of energy as a function of frequency. The necessary wavelength separation is derived from a phase delay, and the instruments can be thought of as interferometers that differ chiefly by the number of beams involved [1.52]. Prisms represent an infinite number of beams, and at the other end of the scale is the Michelson interferometer with just two. In between are echelette gratings, in which each groove represents a beam, with a delay of one groove. If in a Michelson interferometer the intensity variation as a function of mirror displacement is recorded and then given a Fourier transform, it provides the input spectrum. Michelson himself built an analog Fourier transform device and used it to analyze the fine structure of all the important spectral lines available to him. He found just one, the cadmium red line, which was free of such structure, and thus became the reference for comparing its wavelength to the meter bar. He was not able to use the analyzer any further, because of its limited capabilities. What he missed was access to high-speed digital computers, as developed in the late 1950's. Also important was the invention of high speed algorithms in 1965, and about the same time the development of He-Ne lasers for accurately and simply monitoring the mirror travel of the interferometer. What provided the incentive to adapt the old concept and put it to commercial use, starting in the 1970's, was that infrared detectors are noise limited, so that high resolution data over any significant range could be obtained much faster through the Felgett (multiplexing) advantage. This derives from passing all the input light to the detector, rather than 'squeezing' it through an entrance slit of a monochromator. There is also gain derived from using a circular aperture rather than a slit, plus the Jacquinot advantage of increased optical acceptance angle. Resolution is limited mainly by the maximum path traveled by the mirror, and wavelength calibration is directly traceable to the wavelength of the He-Ne laser. As a result very high resolution is available when needed. However, in most applications it is more important to take advantage of the increased speed with which data can be accumulated, a rate that can be as much as 100 times that of an equivalent grating instrument. This explains why Fourier transform instruments (FTR) have come to dominate the IR field. Included is the ease with which computers can add the results of a large number of scans and present the average as the output.

At wavelengths $< 1\mu m$ the advantage of FTR quickly vanishes, as described in a detailed survey by Kneubühl [1.53], first because detectors are no longer noise limited, and secondly because of the greater accuracy required. For example at 500 nm the steps must be no greater than 1/4 μm. A serious concern is the transmission properties of the interferometer beam splitters, which may limit the spectral region that can be covered. How important this type of instrument has become can be judged by a world market that at this writing exceeds $200 M per annum.

1.13.2 Raman Spectrometry

The idea that light impacting on a liquid would generate scattered light at wavelengths specific to molecular vibrations as the result of inelastic molecular scattering was predicted by Smekal in 1923 [1.54], and also considered by Kramers and Heisenberg. It was demonstrated experimentally first by Raman in 1928, with liquids such as CCl_4 and benzene [1.55]. Although the scattering is always weak (typical scattering efficiency being about 1 part in 10^7), and can be judged by photographic exposures that typically lasted 24 hours, Raman not only observed polarization effects, but also the so-called anti-Stokes lines that occur at wavelengths less than the excitation. The importance to the field of physics can be judged by the award of a Nobel prize to Raman just two years later. For the next 40 years it remained an important but strictly research endeavor. However, as soon as intense monochromatic light sources became available in the form of lasers, it became practical to use the technique more widely in industry [1.56], first with ruby lasers but quickly switching to argon ion lasers when they became available. Since then Nd:YAG and diode lasers have been added as alternate sources. Longer wavelengths have an advantage in not exciting fluorescence that would disturb the reading of Raman lines.

The approach is especially useful in the study of symmetrical molecules, but being complementary never displaced the important role that infrared spectrometry had by then established for itself in molecular analysis. From an instrumental point of view the problem is always to extract weak signals that are not too far removed from the highly intense exciting line. The classical approach has been to use double monochromators. The demand for exceptionally low stray light levels, has led to general use of interference (holographic) gratings, usually with 1800 grooves per mm to obtain the desired high dispersion together with high efficiency. If data is required within 10 wavenumbers of the exciting line a double monochromator no longer suffices, and it becomes necessary to adopt a triple monochromator design to obtain the required isolation [1.57]. However, if a gap of 100 wavenumbers is acceptable, the instrumentation can be simplified and reduced in cost by the use of a holographic notch filter, produced in a thick film of dichromated gelatin, which is capable of filtering out the exciting line by a factor of 10^6, and therefore requires only a single monochromator and CCD detector to quickly read the spectrum. The monochromator suggested differs from the usual Czerny-Turner configuration in two respects. Taking advantage of a relatively short spectral interval, it becomes safe to use suitable lenses in place of the focusing mirror. In addition, the reflection grating can be replaced by a Bragg transmission grating. It is also made in dichromated gelatin. [1.58].

1.13.3 Atomic Absorption Spectrometry

While the principles of atomic absorption have their roots in Fraunhofer's work of the 1820's, and more fully established by Kirchhoff some 35 years later, it took another 100 years before it found a place in routine laboratory instrumentation, gradually replacing many of the emission spectrographs. In 1952 Walsh in Australia patented the principle of atomic absorption analysis (generally known by its initials AA). The idea is that instead of narrow peaks being supplied by high resolution instruments, they are derived from narrow band spectral lamps, one for each element, so that analysis can be performed by absorption which requires no more than a small low resolution monochromator [1.59]. The lamps are in the form of hollow cathode discharge devices, whose output light passes through the flame of a specially designed burner, usually using gas or nitrous oxide fuel, and then enters the monochromator with a sensitive detector at its output. Injected into the flame is a fine mist prepared from a solution of the substance to be analyzed. The purpose of the flame is to dissociate the atoms from the compound, without ionizing them. The result is much improved sensitivity from very small samples, and systems are more compact and easier and faster in operation than the previously used spectrographs. As result there was quick worldwide acceptance, as can be judged from the instrument's first appearance in 1963, and the annual production of 5,000 units just 12 years later.

Competition to AA was eventually provided by the development of the inductively coupled plasma (ICP) system, largely due to Wendt and Fassel [1.60]. It operates by injecting an aqueous aerosol of the analyte into a very hot plasma of Argon, the radiation of which is picked up by a high-resolution monochromator and detector. A major advantage is that there is no longer a need for a family of expensive cold cathode lamps. A considerable number of such instruments have been built and used as process monitors for a large number of materials.

1.13.4 Fluorescence Spectrometry

For compounds that show fluorescence it has long been practice to take advantage of the exceptional sensitivity with which they can be detected. This derives from the low background that is associated with illuminating the sample with monochromatic exciting radiation and detecting the resulting fluorescence with a second monochromator set to the longer wavelength radiated.

However, in recent years a much more sophisticated application has been developed for the life sciences, which involves measuring short fluorescent life times (i.e., the time elapsed after light has been absorbed before re-emission begins). In particular there are a large number of nucleotides which can be differentiated on this basis, effectively adding another degree of freedom to spectral analysis, which is important in biologic research and enzyme analysis. The instrumental problems are severe because the lifetimes in question are measured in nsec, so that a precision of the order of 30 psec is required. While such measurements can be performed with high-speed pulsed lasers, a more versatile approach is to modulate the input light with a Pockel cell, at about 500 Mhz, and determine the relative phase lag from the PMT (photomultiplier tube) detector at the output end of the second monochromator. The concept was first described by Gaviola [1.61], who used Kerr cells and polarizing prisms, with mirror separation as a timing device to measure lifetimes as small as 4 nsec. A more recent review of lifetime fluorimeters can be found in reference [1.62]. Additional useful information is obtainable by tracking the degree of depolarization of the emission from fluorescing molecules because that due to torsional vibrations is instantaneous, while that due to Brownian motion is time dependent [1.63].

1.13.5 Colorimetry

A great deal of effort has been devoted for a long time, going back to Newton and Goethe, to not only define the visual aspects of color, but also to making accurate measurement of colors and colored objects. The instruments for this purpose, called colorimeters (and for purposes of this section) are considered as instruments that measure the color of objects by establishing their color coordinates. This has always been considered a difficult assignment because accuracy depends on illumination, its color temperature and spatial distribution, and even temperature. Simple instruments have been built that use three-color filters, as are many used today, but better results are obtained with specialized spectrophotometers. They are often combined with integrating spheres for uniformity of illumination, but they add to bulk and cost. Optical fibers are especially useful for transporting light to the correct places, because this conserves space and can distribute light evenly enough to do without integrating spheres. Modern instruments all use microprocessors for data reduction, and readily derive color coordinates from reflectance values taken in 10 or 20 nm increments. If array detectors are used for the latter we obtain instruments with no moving parts.

Of great historical interest is the Hardy color measuring recording spectrophotometer, first described in 1929 and improved in 1935 [1.64]. An abbreviated description of its operation follows: Beams of white light are made to fall alternately onto the sample and a reference, and the reflected light from each is passed through a double monochromator and then to a photodetector. The intensity of the standard beam is continually adjusted by means of a cam-controlled shutter so that it matches that of the sample, and its position fed to a recording pen whose position represents the spectral reflection as a function of wavelength. The wavelength scan takes place by motor-driven cam control in such a way that the rotation of a recording drum is a linear function of the wavelength to which the monochromator is adjusted. Additional cams adjust the entrance and exit slits to maintain constant bandpass. The results, which take from 1/2 to 3 minutes to obtain, are recorded on special paper. Not content with this, Hardy added an analog integrating system from which tri-stimulus values could be read out. For many years these large and expensive instruments were made by the General Electric Co., and were considered the standard against which all others were judged [1.65]. The difficult task of dealing with accurate cams has, in today's world, been completely replaced by simple digital equivalents.

1.14 Transformation of the Field to the Present Day

The 20th century, especially the second half, has seen great expansion in general spectrometric instrumentation, and diffraction gratings in particular. In the manufacture of gratings the advances in the technical infrastructure have led to the ruling of bigger and better gratings, largely because of the accuracy derived from interferometric feedback control of ruling engines. For many applications a key step has been the routine replication of masters, with no loss in wavefront quality.

A second revolution, due to the outside development of ion lasers and photoresists, made it possible to create high-grade gratings by recording interference wavefronts. The technology of vacuum coating, of both metals and dielectrics, has proved to be an important tool.

Lasers for testing are a great asset, and digital electronics makes possible tricks that could only be dreamed of in earlier days, for example ruling variable spacing to high accuracy.

On the instrumental side there have also been important advances. New light sources, such as deuterium lamps for the UV, play a useful role, as do lasers for certain applications. The ability to direct light to any point desired, even over lengthy distances, with the aid of fiber optic bundles, has turned out to be highly useful.

Of enormous influence has been the development of solid-state detectors with not only greatly enhanced sensitivity over wide regions of the spectrum, but also could be made in the form of large arrays. This made it possible to build high-resolution instruments that not only eliminated moving parts, but in addition could be connected to computers that rapidly manipulated data in any way desired.

References

1.1 H. Kayser: *Handbuch der Spectroscopie*, Vol.**1**, (Hirtzel, Leipzig, 1900).

1.2 J. Ritter: "Versuche über das Sonnenlicht," Gilberts Ann. **12**, 409-415 (1803).

1.3 T. Young: "On the theory of light and colors," Phil. Trans. **II**, 399-408 (1803).

1.4 D. Rittenhouse "An optical problem proposed by F. Hopkinson and solved," J. Am. Phil. Soc. **201**, 202-206 (1786).

1.5 J. Fraunhofer: "Kurtzer Bericht von the Resultaten neuerer Versuche über die Gesetze des Lichtes, und die Theorie derselbem," Gilberts Ann. Phys. **74**, 337-378

(1823).

1.6 J. Fraunhofer: "Über die Brechbarkeit des electrishen Lichts," K. Acad. d. Wiss. zu München, April-June 1824, pp.61-62.

1.7 W. Wallaston: "A method for examining refractive and dispersive powers, by prismatic reflection," Phil. Trans. **II**, 365-380 (1802).

1.8 J. F. W. Herschel: "On the absorption of light by colored media, and on the color of certain flames," Edinb. Trans. **9 II**, 445-460 (1823).

1.9 H. F. Talbot: "Facts relating to optical science," Phil Mag. **4**, 112-114 (1834).

1.10 Ch. Wheatstone: "On the prismatic decomposition of the electric, voltaic, and electro-magnetic sparks," Chem. News, **3**, 198-201 (1861).

1.11 A. J. Ångström: "Optical investigations," Phil. Mag. **9**, 327-342 (1855), "Optische Untersuchungen,"Pogg. Ann. **94**, 141-165 (1855).

1.12 R. Thalén: "Sur le spectre de fer obtenue à l'aide de l'arc électriqe," Nova acta Upsala **12**, 1-49, (1884).

1.13 D. Alter: "On certain physical properties of the light of the electric spark," Am. J. **18**, 213-214 (1855).

1.14 J. Plücker: "Fortgesetzte Beobachtungen über die electrishe Entladung," Pogg. Ann. **104**, 113-128 (1858), Pogg. Ann., **105**, 67-84 (1858).

1.15 F.A.Nobert: "Ueber eine Glassplatte mit Theilungen zur Bestimmung der Wellenlänge und relativen Geschwindigkeit des Lichtes in der Luft und im Glase," Annal.der Physik, **85**, 83-92 (1851).

1.16 *Dictionary of Scientific Biographies*, **X,** (Scribners, N.Y. 1972), p.133.

1.17 G. Kirchhoff: "Über den Zusammenhang zwischen Emission und der Absorption von Licht und Wärme," Monatsber. d. Berlin. Akad., pp.783–787 (1859). Also Pogg. Ann. **109**, 275-301 (1860).

1.18 G. Kirchhoff and R. Bunsen: "Chemische Analyse durch Spectralbeobachtungen," Pogg. Ann. **110**, 160-189 (1860).

1.19 *Dictionary of Scientific Biographies*, **XI,** (Scribners, N.Y. 1972), p.241.

1.20 G. Quincke: "Optische Experimentaluntersuchungen: XV. On diffraction gratings." Ann. der Physik, **146**, 1-65 (1872).

1.21 J. N. Lockyer: "Preliminary note of researches on gaseous spectra in relation to the physical constitution of the sun," Proc. Royal Soc. **17,** 288-291, 453-454 (1869).

1.22 J. N. Lockyer: "Spectroscopic notes, I to III," Proc. Royal Soc. **22**, 371-380 (1874).

1.23 F. Zöllner: "Ueber die spectroscopische Beobachtung der Rotation der Sonne und ein neues Reversionsspectroscop," Pogg. Ann. **144**, 449-456 (1871).

1.24 W. Huggins: "On the spectrum of the great nebula in Orion and on the motions of some starts towards and away from the sun," Proc. Royal Soc. **20**, 379-394 (1872).

1.25 J. L. Soret: "On harmonic ratios in spectra," Phil. Mag. **42**, 464-465 (1871).

1.26 A. Cornu: "Sur le spectre normal de soleil, partie ultraviolet," Ann. Scientific de l'Ecole Norm. Supér. **3**, 421-434 (1874).

1.27 H. Draper: "On diffraction spectrum photography," Phil. Mag. **46**, 417-425 (1873).

1.28 L. Mouton: "Sur la determination des longueurs d'onde calorifique (On the determination of infrared wavelengths)," Compt. Rendue, **88**, 1078-1082 (1879).

1.29 S. P. Langley: "The actinic balance," Amer. J. **21**, 187-198 (1881).

1.30 A. F. Svanberg: "Om uppmating of lednings mot ständet for electriska strömmer," Pogg. Ann. **84**, 411-417, (1951).

1.31 W. de W. Abney: "On the photographic method of mapping the long wavelength end of the spectrum," Phil. Trans. Royal Soc. **171**, II, 653–667 (1880).

1.32 E. Lommel: "Phosphoro-Photographie des Ultraroten Spectrums," München Sitzber. **18**, 397-403 (1888).

1.33 W. Huggins: "On the photographic spectra of stars," Phil. Trans. **171**, 669-690 (1880).

1.34 H. Rowland: "Preliminary notice of results accomplished on the manufacture and theory of gratings for optical purposes," Phil. Mag. Suppl. to v.**13**, 469-474 (1882).

1.35 H. Rowland: "On the relative wavelength of the lines in the solar spectrum," Phil. Mag. **23**, 257-286 (1887).

1.36 H. Rowland: *Preliminary table of the solar spectrum*, Johns Hopkins Univ. Press, Vols.1 to 6. (1895 to 1898).

1.37 J. Balmer: "Notiz über die Spectrallinien des Wasserstoffs," Wied. Ann. **25**, 80-87 (1885).

1.38 J. Rydberg: "Recherche sur le constitution des spectres d'émission des éléments chimiques," Compt. Rend. **110,** 394-400 (1890).

1.39 H. Kayser and C. Runge: "Über die Spectren der Elemente: I - VII Abschnitt," Physik. Abh. d. Königlichen Akad. der Wiss. zu Berlin, S. 22 (1888), S. 1, pp. 1-16 (1890), S. III, pp. 1-20 (1893).

1.40 C. Runge and F. Pashen: "Ueber das Spectrum des Heliums," Astrophys. Jl. **3**, 4-28 (1895).

1.41 V. Schumann: "Ueber die Photographie des Gitterspectrums bis zur Wellenlänge 1000 Å im luftlehren Raum," Photogr. Rundshaue, Wien. Ber. **102,** 415-475, 625-

694, 944-1024 (1893).

1.42 P. Zeeman: "On the influence of magnetism on the nature of light emitted by a substance," Phil. Mag., fifth series, **43**, 226-239 (1897).

1.43 *Dictionary of Scientific Biographies*, **XV,** (Scribners, N.Y. 1972), p.497.

1.44 W. Humphrey and J. Mohler: "A study of the effect of pressure on the wavelength of arc spectra of certain elements,"Astrophys. Jl. **3**, 114-137 (1896).

1.45 J. W. Strutt (Lord Rayleigh): "On the manufacture and theory of diffraction gratings," Phil.Mag. **XLVII**, 193-205 (1874).

1.46 R. Wood: "The echelette grating for the infra-red," Phil. Mag **XX** (Series **6**), 770-778 (1910).

1.47 H. D. Babcock: "Bright diffraction gratings," J. Opt. Soc. Am. **34**, 1-5 (1944).

1.48 C. S. Peirce: "On the ghost in Rutherfurd's diffraction spectra," Am. Jl. of Mathem. **2**, 330-347 (1879).

1.49 A. Cornu: "Études sur les réseaux diffringents. Anomalies focales," Comptes Rendu, **116**, 1215-1222, 1421-1428, **117**, 1032-1039, 1455-1461 (1893).

1.50 W. Kaye: "Near IR spectroscopy: Instrumentation and technique (A review)," Spectrochimica Acta, **7**, 181-204 (1955).

1.51 J. U. White and M. D. Liston: "Construction of a double beam recording infrared spectrophotometer," J. Opt. Soc. Am. **40**, 29-35 (1950).

1.52 H. A. Gebbie: "Fourier transform versus grating spectroscopy," Appl. Opt. **8**, 501-504 (1969).

1.53 F. Kneubühl: "Diffraction grating spectroscopy," Appl. Opt. **8,** 505-519 (1969).

1.54 A. Smekal: "Zur Quantentheorie der Dispersion," Die Naturwissenschaften, **11**, 873-875 (1923).

1.55 C. V. Raman and K. S. Krishnan: "The production of new radiations by light scattering," Proc. Royal Soc. (London) **122a**, 23-35 (1928).

1.56 S. P. Porto and D. L. Wood: "Ruby optical maser as a Raman source," J. Opt. Soc. Am., **52**, 251-252 (1962).

1.57 V. L. Chupp and P. C. Granz: "Coma cancelling monochromator with no slit mismatch," Appl.Opt. **8**, 925-929 (1969).

1.58 H. Owen, D. E. Battley, M. J. Pelletier and J. B. Slater: "New spectroscopic instrument based on volume holographic elements," S.P.I.E. **2406**, (*Practical Holography* **IX**), 260-267 (1955).

1.59 A. Walsh: "The application of atomic absorption spectra to chemical analysis," Spectrochimica Acta, **7**, 108-117 (1955).

1.60 R. H. Wendt and V. Fassel: "Induction-coupled plasma spectrometric excitation source," Analyt. Chem., **37**, 920-922 (1965).

1.61 E. Gaviola: "Die Abklingungszeit der Fluoreszenz von Farbstoff Lösungen," Z. Physik, **35**, 748-756 (1926).

1.62 J. B. Birks and I. H. Monroe: "The fluorescence lifetimes of aromatic molecules," Progr. Reaction Kinetics, **4**, 239-249 (1967).

1.63 Chester O'Konski, Ed: *Electro-optics,* (Marcel Dekker, Inc., NY, 1976), ch.16.

1.64 A. C. Hardy: "A new recording spectrophotometer," J. Opt. Soc. Am. **25**, 305-311 (1935).

1.65 F. W. Billmeyer, Jr: "Comparative performance of color measuring instruments," Appl. Opt. **8**, 775-783 (1969).

Chapter 2

Fundamental Properties of Gratings

2.1 The Grating Equation

When light is incident on a grating surface it is diffracted from the grooves. In effect, each groove becomes a very small source of reflected and/or transmitted light. The usefulness of gratings is derived from the fact that there exists a unique set of angles where the light scattered from all facets is in phase. This can be visualized in Fig.2.1 which shows a plane wavefront, incident at an angle θ_i with respect to the grating normal. It is easy to see that the geometrical path difference between the light diffracted by successive grooves in a direction θ_d is simply d $\sin\theta_i$ - d $\sin\theta_d$, where d denotes the groove spacing. The principle of interference dictates that only when this difference equals the wavelength of light, or a simple integral multiple thereof, the light will be in phase (i.e., reinforce itself). At all other angles there will be destructive interference between the wavelets originating at successive grooves.

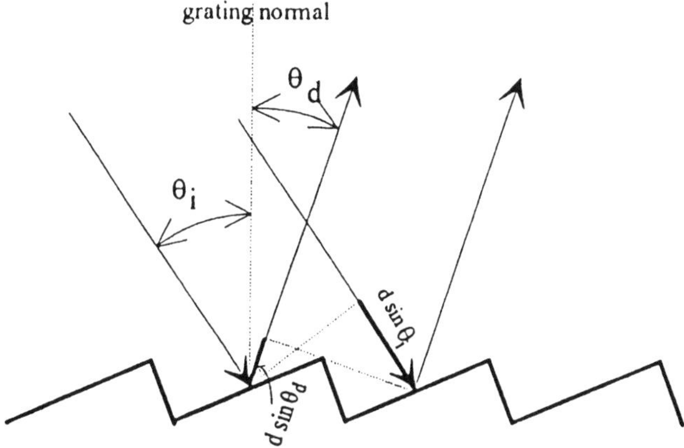

Fig.2.1 Diagram for phase relation between the rays diffracted from adjacent grooves.

The famous property of gratings to diffract incident light into clearly distinguished directions is expressed in a simple equation, called *the grating equation*:

$$\sin\theta_m = \sin\theta_i + m\frac{\lambda}{d}, \quad m = 0, \pm 1, \pm 2, \cdots , \qquad (2.1)$$

where θ_i and θ_m are the angles between the incident (and the diffracted) wave directions and the normal to the grating surface, λ is the wavelength and d is the grating period (Fig.2.2). m is an integer, numbering the orders so that the specular reflected one is numbered as 0. The grating period d is usually measured in μm, but its inverse, called groove (or grating) frequency is in common use, given in the number of grooves per mm (gr/mm), so that d = 0.8333 μm will correspond to 1200 gr/mm, etc. The order number m represents the number of wavelengths between light reflected from successive grooves. It is assumed that the incident wave is monochromatic and perfectly collimated and that the plane of incidence is perpendicular to the grooves. For linear media with a non-monochromatic or/and non-collimated incident beam, the grating response is also linear (i.e., its diffraction can be expressed as a superposition

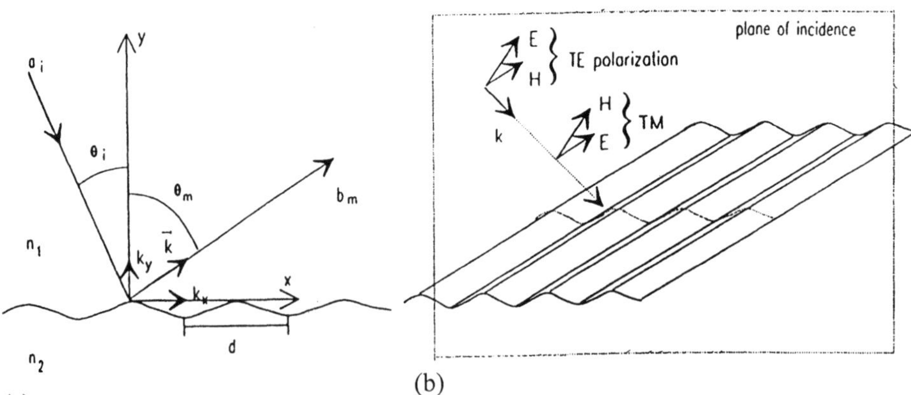

Fig.2.2 Schematic representation of light diffraction by relief grating in a classical diffraction mounting. a) Cross-section view, introducing the coordinate system axis, angles of incidence and diffraction, and wavevector and its components. b) TE and TM fundamental cases of polarization with the electrical **E** and magnetic **H** vectors.

of the diffraction of all its plane-wave components). Of course, as far as the superposition is carried out over the sine of the angles of incidence and diffraction (eq.2.1), the diffracted beam may be shifted, widened or narrowed when compared to the incident one and can become quite asymmetrical, especially at high angles of incidence and diffraction.

When m is equal to zero, the grating acts as a mirror, all wavelengths being superimposed. For non-specular orders ($m \neq 0$) the angle of diffraction depends on the wavelength value so that wavelengths are separated angularly.

If the plane of incidence is not perpendicular to the grating, eq.2.1 is transformed into a more general law:

$$\begin{vmatrix} k_{m_x} = k_{i_x} + m\dfrac{2\pi}{d} \\ k_{m_z} \equiv k_{i_z} \end{vmatrix}, \qquad (2.2)$$

where k_x, k_y, and k_z are the wavevector components (see Fig.2.2). In the grating theory instead of k_x and k_y another set of notations is often used, namely α and β, equal to k_x and k_y, normalized by the modulus of the wavevector k, so that for the propagating diffraction orders, $\alpha = n \sin\theta$ and $\beta = n\cos\theta$, where n is the refractive index.

The quantity

$$K = \frac{2\pi}{d} \qquad (2.3)$$

is called *grating wavenumber* (or *grating vector*). It can be shown that if $k_z \neq 0$ the directions of the diffracted orders, determined by (2.2) lie on a cone. That is why this case is usually called *conical diffraction mounting*, whereas the more simple case when $k_z = 0$ - *classical diffraction mounting*. Further on we shall pay attention predominantly to the classical mounting.

It is necessary also to distinguish between the two cases of polarization. If the incident wave is linearly polarized and the electric field vector is perpendicular to the plane of incidence, all the diffracted orders have the same polarization. It is called s, or P, or TE polarization. The other case, when the electric field lies in the plane of incidence, also preserves the polarization direction and is called p, or S, or TM case. Any other polarization state can be represented as a linear combination of the two *fundamental* cases, so, luckily, it is necessary to investigate the grating response only for these polarizations.

Considering a transmission grating, the direction of propagation of the transmission orders can be determined by an equation, similar to (2.1):

$$n_2 \sin\theta_{2_m} = n_1 \sin\theta_i + m\frac{\lambda}{d} , \tag{2.4}$$

where the subscript 1 denotes the cladding and 2 - the substrate. In fact, this equation is a direct consequence of the wavenumber summation law (2.2) because of the relation:

$$k_x = n\frac{2\pi}{\lambda}\sin\theta . \tag{2.5}$$

2.2 Propagating and Evanescent Orders

Let us return back to the grating equation, something that will be done quite often in this book. For a given set of incident angles, groove spacing and wavelength values, the grating equation can be satisfied for more than one value of m. It is obvious that there is a solution only when

$$\left|\sin\theta_m\right| < 1 . \tag{2.6}$$

Diffraction orders with number m such that condition (2.6) is fulfilled are called *propagating* orders. The vertical wavevector component can be easily found from the wave equation:

$$k_x^2 + k_y^2 = \left(\frac{2\pi}{\lambda}n\right)^2 , \tag{2.7}$$

so that

$$k_{m_y} = \sqrt{\left(\frac{2\pi}{\lambda}n\right)^2 - k_{m_x}^2} = \frac{2\pi}{\lambda} n\cos\theta_m , \tag{2.8}$$

and the x and y variation of the propagating orders represent a plane wave, propagating in a direction $\vec{\mathbf{k}}_m$:

$$\exp(ik_{m_x}x + ik_{m_y}y) \equiv \exp\left[in\frac{2\pi}{\lambda}(x\sin\theta_m + y\cos\theta_m)\right] . \tag{2.9}$$

For other orders having $\left|\sin\theta_m\right| > 1$, we have to look at eq.2.2 instead of 2.1. Wave equation (2.7) implies then that the vertical component of the wavevector is imaginary (i.e., these orders decrease exponentially with the distance from the grating surface). Their amplitudes are proportional to:

$$\exp(ik_{m_x}x - k_{m_y}y) \tag{2.10}$$

with

$$k_y = \sqrt{k_x - \left(\frac{2\pi}{\lambda}n\right)^2} \quad . \tag{2.11}$$

These orders are called *evanescent orders.* They can not be detected at a distance greater than a few wavelengths from the grating surface, but can play an important role in some surface-enhanced grating properties and must be taken into account in any electromagnetic theory of gratings. Evanescent orders are essential in some special applications: waveguide and fiber gratings.

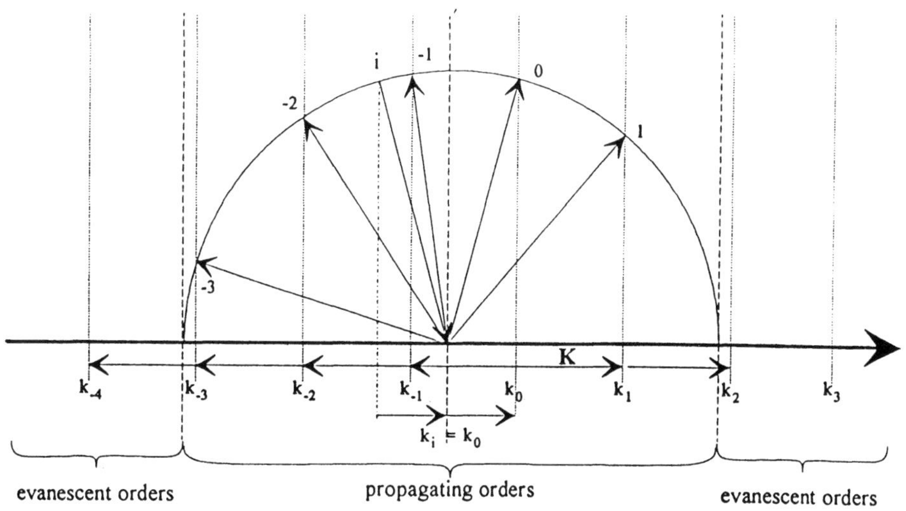

Fig.2.3 Schematic representation of grating orders wavevectors. Incident wave **i** has a horizontal component of the wavevector $k_i = k_0$. A grating vector K is added or substracted from k_0 to form the diffracted order horizontal wavevector component. As the length of the propagating order wavevectors are equal and limited, only a limited number of orders propagate (namely, from -3 to +1), and the others are evanescent.

The integer number of the diffraction orders can have both negative and positive values. There are several conventions, but the most common implies that positive orders are those where the angle of diffraction exceeds the angle of incidence and lie on opposite sides of the grating normal. Fig.2.3 represents the directions of diffraction orders formed by adding or substracting grating wavevector from the zeroth reflected wavevector k_{0_x}. The wavevectors of the propagating orders have moduli equal to $2\pi/\lambda$ so that they are limited in number. Evanescent orders lie outside this region. When a grating vector is added to form a diffraction order its number is considered positive, whereas subtraction of a grating vector gives negative diffraction orders. Throughout the book, except for Chapter 6, we use this convention. The other commonly used convention names the orders in an opposite sense, so that the sign plus in eq.2.1 must be replaced by minus. This choice comes from the fact that in most cases the grating is utilized in its "negative" mode, according to the first convention for orders but to avoid the minus sign, one can choose the opposite sign convention. Fortunately, this rarely becomes a problem.

There is also another important aspect of the grating equation. For a fixed angle of incidence and groove spacing, there is an infinite set of wavelength values that can diffract in the same direction. It is evident that for any particular grating instrument configuration, the spectral slit image corresponding to wavelength λ will coincide with that of the second order image of $\lambda/2$, the third order image of $\lambda/3$, etc. This results in overlapping of successive orders and a detector with a broader range will see several wavelengths simultaneously, unless prevented from doing so by suitable filtering at either source or detector. The higher the spectral orders, the shorter the wavelength range where successive orders fail to overlap.

2.3 Dispersion

The angular separation $d\theta_d$ of two different wavelengths of light differing by $d\lambda$ can be obtained by differentiating the grating equation, assuming the angle of incidence to be fixed:

$$\frac{d\theta_d}{d\lambda} = \frac{m}{d\cos\theta_d} \quad . \tag{2.12}$$

The ratio $d\theta_d/d\lambda$ is known as *angular dispersion*. The *linear dispersion* of a grating system is simply a product of this and the effective focal length. The usual instrument design calls for as much linear dispersion as possible in an instrument whose compactness limits focal length. Hence the desire for relatively large angular dispersion. It is important to realize that the ratio m/d in eq.2.12 is not the independent variable it is frequently taken to be. When this ratio is derived from the grating equation we obtain the general equation for angular dispersion:

$$\frac{d\theta_d}{d\lambda} = \frac{1}{\lambda}\frac{(\sin\theta_i \pm \sin\theta_d)}{\cos\theta_d} . \tag{2.13}$$

The important conclusion is that, for a given wavelength, angular dispersion is purely a function of the angles of incidence and diffraction. This becomes more obvious when we consider the Littrow case, defined by $\theta_d = \theta_i$ (see later Section 2.9). Then eq.2.13 reduces to:

$$\frac{d\theta_d}{d\lambda} = \frac{2}{\lambda}\,\mathrm{tg}\theta_d . \tag{2.14}$$

It is evident that when θ_i increases from 10° to 63° in Littrow mount, the angular dispersion increases by a factor of 10. Once θ_i has been determined, the designer must choose between working in a low order of a fine pitch grating, or a higher order of a coarse grating.

In the grazing incidence configuration, where θ_d and θ_i are both large but on opposite sides of the grating normal, the expression $(\sin\theta_i - \sin\theta_d)/\cos\theta_d$ comes out numerically much less than $\mathrm{tg}\theta_d$. This is also true for grazing incidence and diffraction direction close to the grating normal. However, such configurations are likely to be used only at shorter wavelengths, so that reasonable values of dispersion are still obtained.

From the foregoing it is clear that high dispersion is associated with large angles of diffraction, and these in turn are associated with relatively steep groove angles (see Chs. 4 and 6). For a given wavelength and order, steep blaze angles lead to finely spaced rulings, which explains the frequent request for such gratings. However, there are natural limitations in this direction: In the first order, the finest groove spacing theoretically possible is when $\theta_i = \theta_d =$ 90°, in which case d = $\lambda/2$, showing that a grating cannot diffract at wavelengths greater than 2d. Alternative solutions for increasing dispersion is to work in high orders, as it is done with echelles, but then the free spectral range becomes narrow due to the large number of orders that can overlap.

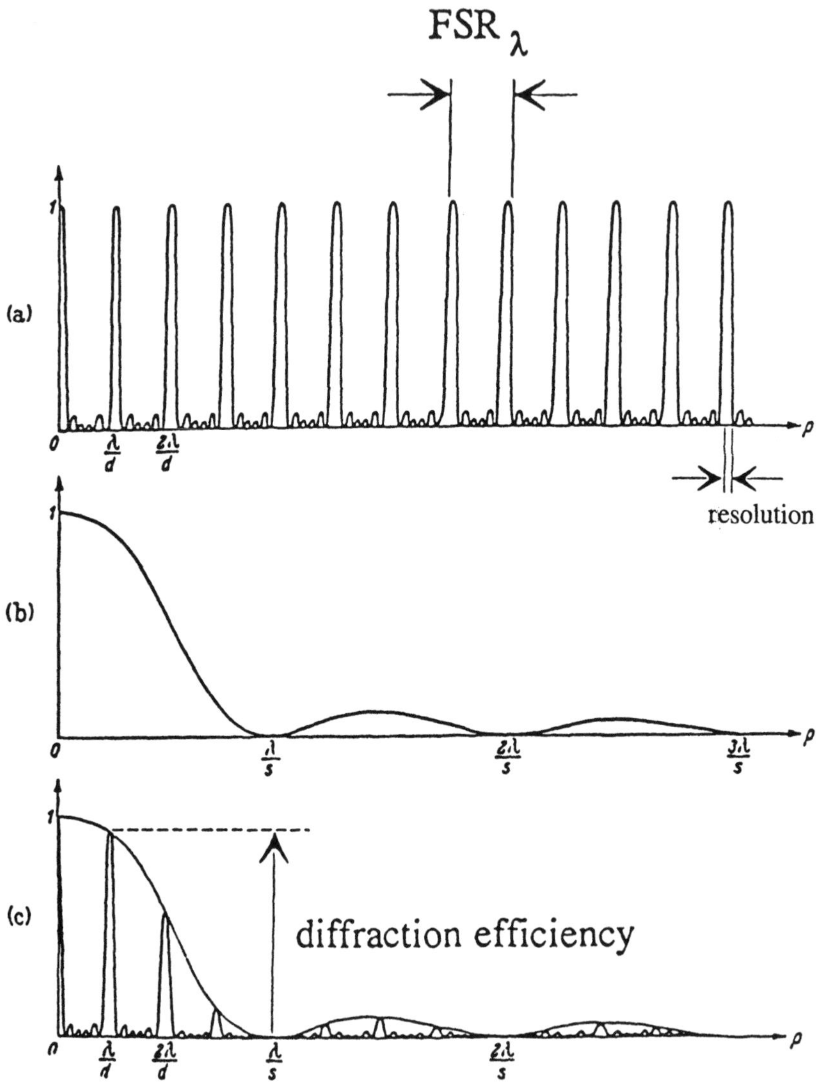

Fig.2.4 (a) The normalized interference function H; (b) the normalized slit intensity function I_s; and (c) the normalized grating intensity function I_g (after [2.1]).

2.4 Free Spectral Range

The range of wavelengths, for which overlapping from adjacent orders does not occur, is called the free spectral range FSR_λ. This means the range of wavelengths $\Delta\lambda = \lambda_{2,m+1} - \lambda_{1,m}$ for which the m-th order of the wavelength λ_2 coincides with the (m+1)st order of wavelength λ_1 (see Fig.2.4). The concept applies to all gratings in a spectral range where more than one order propagates, but is particularly important in the case of echelles because they operate in high orders with correspondingly short free spectral range.

The free spectral range can be calculated directly from its definition

$$m(\lambda_1 + \Delta\lambda) = (m+1)\lambda_1 \ , \tag{2.15}$$

from which

$$FSR_\lambda = \Delta\lambda = \frac{\lambda}{m} \ . \tag{2.16}$$

It is evident that the free spectral range is directly proportional to the wavelength and inversely proportional to the order. In terms of wavenumbers, $\xi = 1/\lambda$, the free spectral range is defined as follows:

$$FSR_\xi = \frac{1}{\lambda_1} - \frac{1}{\lambda_2} = \frac{\lambda_2 - \lambda_1}{\lambda_1 \lambda_2} \ . \tag{2.17}$$

Since for practical purposes the product of wavelengths, when compared to their difference, can be substituted by their mean value λ^2, eqs.2.16 and 17 result in

$$FSR_\xi = \frac{\xi}{m} \ . \tag{2.18}$$

2.5 Passing-Off of Orders

Varying the angle of incidence and/or the wavelength, it can happen that when some of the diffraction orders have a gradually increasing angle of diffraction, they become parallel to the grating surface, and instantly disappear. Or just the opposite - some orders appear "from nowhere," from a direction parallel to the surface and then can be detected. During such appearance and disappearance, called *cut-off of orders*, one can observe sudden changes in the diffraction efficiency of the other propagating orders. These changes are called *threshold anomalies*, but for highly conducting bare metallic gratings in TM polarized light these anomalies coincide with a *resonance anomaly* (see Chapter 8) and can be quite strong.

The new orders appear from the "pool" of evanescent orders, which is infinitely large. The conditions for the cut-off of the m-th order are rather simple, and can be derived easily from the condition $|n \sin\theta_m| = 1$:

$$\left| n \sin\theta_i + m\frac{\lambda}{d} \right| = 1 \, . \tag{2.19}$$

2.6 Guided Waves

The property of gratings to couple multiple (in fact, infinite number) of electromagnetic waves plays an important role in integrated and fiber optics. The property of waveguide modes to propagate long distances without substantial scattering losses is essential and is only possible when the electromagnetic field is evanescent in the regions outside to the waveguide layer and fiber core. In the core the mode must propagate, so that its field is characterized by eqs. (2.8 and 9), but in the cladding eqs. (2.10, 11) are valid. The property of the grating to couple evanescent to propagating orders (in the cladding) is widely used to couple light into and out of the waveguide. The mode is characterized by its propagation constant, the phase velocity k_G in the propagation direction (see Chapter 8), which takes discrete values depending on the waveguide optogeometrical properties, the wavelength, and polarization. This constant is always greater than the modulus of the wavevector in the cladding:

$$k_G > |\mathbf{k}| \equiv \frac{2\pi}{\lambda} n_{\text{cladding}} \quad , \tag{2.20}$$

so that the radiated field in the cladding is evanescent, according to eqs.(2.10 and 11). The grating can couple this evanescent field to a propagating order, say the m-th one, under specific conditions called *phase-matching condition*:

$$k_G = \frac{2\pi}{\lambda}\left(n_1 \sin\theta_m + m\frac{\lambda}{d} \right), \tag{2.21}$$

which is another form of the grating equation (2.1). Strictly speaking, when the guided mode is coupled to a propagation diffraction order, it no longer remains bound to the core, because it is radiated into the outer region. However, then another more general approach to the guided waves can be used, defining them as a solution of the homogeneous problem (having a scattered field without waves incident from outside).

Another application of gratings in waveguides and fibers is to mutually couple two (or very rarely more) guided modes, or one and the same mode propagating in two different directions. The phase-matching condition is then called *Bragg condition* and it again represents another form of the grating equation:

$$k_{G_1} = k_{G_2} + m\frac{2\pi}{\lambda} \tag{2.22}$$

with indices 1 and 2 used to distinguish between the propagation constants of the two modes.

2.7 Diffraction Efficiency

2.7.1 Definition

The grating equation determines where light goes but says nothing about how much goes where. This "how much" has been a question of great interest since the first grating was made. The physical quantity that characterizes how the incident field power is distributed between the different orders is called *diffraction efficiency*. It is defined as the ratio between the energy flow of a particular order in a direction perpendicular to the grating surface (i.e., parallel to the y-axis in Fig.2.2) and the corresponding flow of the incident wave through the same surface. We have already observed that the spatial variation of the propagating diffracted orders is given by eq.2.9. The propagating exponential term is multiplied by a constant b_m, different for each order and called *diffraction order amplitude*. In fact, the total electromagnetic field component parallel to the grooves can be represented as a sum of the incident wave (having an amplitude a_i) and all the diffracted orders. For TE polarization this component is the electric field E_z and for TM polarization it is the magnetic vector H_z (see Fig.2.2b):

$$\begin{pmatrix} E_z \\ \text{or} \\ H_z \end{pmatrix} = \sum_m b_m \, e^{ik_{m_x}x + ik_{m_y}y} + a_i \, e^{ik_{i_x}x - ik_{i_y}} \tag{2.23}$$

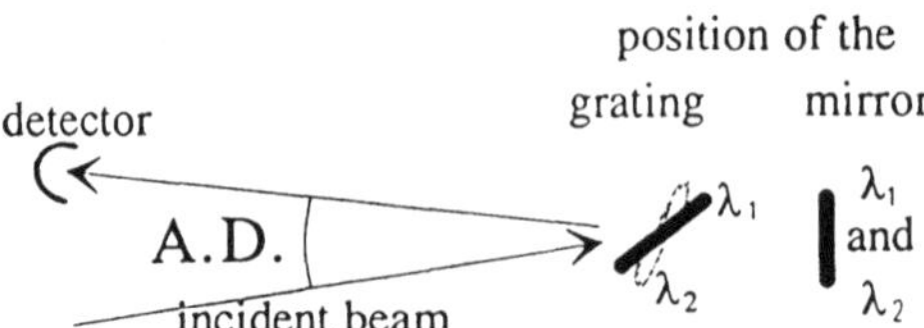

Fig.2.5 A mounting to measure relative efficiencies. The angular deviation (A.D.) between the diffracted and the incident beam is shown. In order to ensure constant A.D., the grating rotates with varying wavelength.

and the sum is carried over all possible orders. However, only the propagating orders can carry energy away from the grating, because the evanescent orders have an imaginary wavevector component in the vertical direction. The *diffraction efficiencies* η_m are simply connected to the diffraction order amplitudes:

$$\eta_m = \frac{|b_m|^2}{|a_i|^2} \frac{k_{m_y}}{k_{i_y}} . \qquad (2.24)$$

Curiously, it is easier to measure the grating efficiencies, except in some special cases, than to define them theoretically. Usually it is enough just to measure the incident beam intensity and the diffracted order intensities and to take the ratio. However, if the incident intensity is unstable, the two measurements must be performed simultaneously with electronic division. Other difficulties arise from the size of the incident and diffracted beams and the size of the available detectors, more important near grazing incidence.

For historical reasons, instead of definition (2.24), which provides the values of the *absolute efficiency*, the so called *relative efficiency* is sometimes used to characterize the properties of reflection gratings. This is the ratio between the intensity of light diffracted into a given order and the reflectivity of a plane mirror of the same material used under the same conditions, i.e., measured at the same angles of diffraction (Fig.2.5).

2.7.2 Classical Model of Grating Efficiency

Unfortunately there is no simple way to describe grating efficiency behavior, at least in the most interesting and widely used cases. However, for the physical understanding of grating properties it is useful to return to the classical optical textbooks. Kirchhoff's diffraction theory, in the Fraunhofer approximation, expresses the grating scattering of incident light as a product of two terms, the *interference function* H and the *intensity function of a single slit* I_s so that the normalized *intensity function* of the grating consisting of M identical slits I_g is given by:

$$\frac{1}{M^2} I_g(p) \equiv H\, I_s = \left[\frac{\sin\left(M\frac{kdp}{2}\right)}{M\sin\left(\frac{kdp}{2}\right)}\right]^2 \left[\frac{\sin\left(\frac{ksp}{2}\right)}{\frac{ksp}{2}}\right]^2 , \quad (2.25)$$

where the first term represents the normalized interference function and the second the slit intensity function, s denoting the slit width, and $p = \sin\theta_d - \sin\theta_i$. The interference function H has maxima when $p = m\lambda/d$, i.e., in directions, given by the grating equation. Between them there are weak secondary maxima (Fig.2.4a). For large values of M (the number of the illuminated grooves) the secondary maxima are very weak. They are separated by points of zero intensity in directions given by

$$p \equiv \sin\theta_d - \sin\theta_i = \frac{m}{M}\frac{\lambda}{d} \ . \quad (2.26)$$

The slit intensity function depends on the form of the grooves. It has a maximum in some direction, called *blazed direction*, or *blazed wavelength*, if considered as a function of the wavelength. When compared with the interference function for large values of M the slit (groove) intensity function falls off slowly on both sides of its maximum so that the grating response (the Intensity function of the grating) consists of sharp peaks (determined by H), modulated by the slit intensity function (Fig.2.4c).

The grating response, determined by the interference function, can be easily predicted. The influence of the intermediate secondary maxima can be important in certain spectroscopic applications, thus the desire for gratings of large size. The deviations from the regular alternation of diffraction orders, separated by weak maxima, comes from grating imperfections: large-scale non-periodical variations of the period *deform the wavefront* of the diffracted orders; periodical variations of the spacing give rise to *ghosts*; if only a small number of grooves are displaced, the ghosts lie close to the strong parent line and are called *satellites*. The role of these errors and the technique for measuring them are described in more detail in Chapter 11.

The slit (groove) intensity function determines the distribution of the diffracted light among the diffraction orders. The simple formulas where the Fraunhofer approximation is valid no longer hold when the spacing is reduced to near wave length values. Moreover, even for echelles which one can consider as the purest "scalar limit" devices, there are noticeable deviations from the simple expectations, given by eq.2.25, even if the groove intensity function is evaluated correctly (see Chapter 6).

The challenging task of developing an appropriate theory for grating efficiency has lead to numerous approximate and rigorous theories (see Chapter 10) so that finally the performance of any existing or imagined grating can be predicted, given the correct groove and material parameters. It is impossible to summarize in a simple way the variety of grating properties. However, several general rules do exist that can serve to eliminate some theoretical and experimental errors. The first rule comes from the basic laws of physics and is called *energy balance criterion*. It states that the sum of efficiencies of all the propagating orders must equal the intensity of the incident light minus the losses. More important practically are the following two properties.

2.7.3 Reciprocity Theorem and Symmetry with Respect to Littrow Mount

If a grating is utilized under the same conditions as in Fig.2.2a, but with angles of incidence and diffraction exchanged, the *Reciprocity Theorem* states that the efficiency in the diffraction order under consideration remains the same[1]. A direct consequence of the Reciprocity theorem is that the efficiency of the m-th diffraction order, as a function of the sine of the angle of incidence, has to be symmetrical with respect to this order Littrow mounting. It is important to notice that the symmetry is valid with respect to $sin\theta_i$, rather than just the *angle* of incidence. This rule is easily forgotten, because for moderate angles of incidence the difference is not well-pronounced, but if one goes to high incidence or diffraction angles (gratings used in grazing incidence and echelles), the asymmetry with respect to θ_i does become significant.

The reciprocity theorem is a direct consequence of the periodicity of the grating and it is rigorously fulfilled for perfectly and highly conducting metallic substrates, and for lossy or lossless dielectric gratings, provided the incident wave is close to a plane wave. If these two conditions (periodicity and plane incident wave) are not fulfilled, the experiment can show noticeable deviations when measurements are performed at both sides of Littrow mount. Such cases can involve considerable surface roughness to spoil the periodicity, but this is never enough. Asymmetry with respect to Littrow is sometimes reported at high incident angles, but the difference between $\sin\theta_i$ and θ_i is usually enough to explain the discrepancy. Rarely has one to take into account the convolution

---

[1] Of course, the angle of incidence and, thus, the effective grating aperture is changed, which may be important at steep angles of incidence or diffraction.

between the incident beam divergence, that is represented as a function of θ_i and the grating response function, symmetrical with respect to $\sin\theta_i$. The influence of the other optical components response function can also be of some importance and in addition the surface roughness may not always be negligible.

2.7.4 Perfect Blazing - Does It Really Exist?

Probably since the first use of gratings the desire to force the entire incident light to diffract into a given order went hand in hand with the desire to maintain perfect wavefront. The property of gratings to concentrate the diffracted light into a specific order is called *blazing*. It is *perfect* when no light goes elsewhere, the absolute efficiency limited only by the absorption losses and diffuse scatter.

One of the rare mistakes of Lord Rayleigh lies in his rather off hand prediction that "*To obtain a diffraction spectrum containing all the light it would be necessary that the retardation gradually alter by a wavelength in passing over each element of the grating and then fall back to its previous value. However, it is not likely that such a result will ever be obtained in practice*" [2.2]. To make ironclad predictions is a dangerous thing to do, although it took 36 years for R. Wood [2.3] to make the first blazed grating, and that was limited to the infrared region. Moreover, despite the numerous arguments of scalar theory, in only rare cases is the blazing perfect. It was necessary to wait until 1980 when Marechal and Stroke formulated their theorem [2.4] to undestand that while perfect blazing is possible theoretically it is rarely seen in practice. It is important to distinguish between fine pitch gratings supporting only a few orders and coarse ones that have a large number. When the grating supports only two orders, namely the zero and minus first, one can find the optimum groove depth to suppress the zero order independent of the profile form. Examples are the cases of 40% modulated sinusoidal reflection gratings with 1800 or more grooves per mm (which have 85% efficiency in TM polarization, as discussed in Sections 4.6, 4.8, and 4.11). Moreover, when working in the total internal reflection regime, without metallic coatings to increase absorption, one can expect almost 100% efficiency in reflection (section 5.11), or 90% in transmission (section 5.10) by increasing the groove frequency and depth.

Although quite important in laser applications, gratings with only two diffracted orders give rise to several problems, limiting their use in other applications. In addition to the numerous technological problems, such as precise control of groove parameters (which is unfortunately typical of all high-efficiency gratings), these fine pitch gratings have some common disadvantages. First of all, perfect blazing occurs in the spectral region of resonance and cut-off anomalies, characterized by rapid variation of efficiency. Second, when blazing in one polarization, the efficiency in the other is typically low, a property undesirable in many applications. Third, the spectral interval of near perfect blazing is quite limited.

As discussed in Section 2.7.2 gratings with multiple diffraction orders can blaze if the profile of the groove is given a special form, typically a triangle with a 90° apex angle. This is the optimal geometry predicted by geometrical optics considerations, but Marechal and Stroke have shown that there is a much stronger electromagnetic basis of these expectations. The *only* condition is that the grating material is perfectly conducting and the polarization is TM. The arguments are so simple that they are worth repeating. Consider the geometry given in Fig.2.6. In the TM case the two waves, the incident and the backward diffracted, satisfy the boundary conditions at the second facet B, as the tangential components of their electric field vectors are null (Fig.2.6a). The boundary conditions at the "working" facet A are satisfied when the amplitudes of the incident and diffracted orders are equal but of opposite sign. Thus these two waves are the solution of the diffraction problem which means that all the other orders are null and the blazing is perfect.

The problem is that these arguments, and thus the theorem of Marechal and Stroke, lose their validity when the conditions are changed. In TE polarization, even for perfectly conducting echelettes, the incident wave and a single backward diffracted order cannot satisfy the boundary conditions simultaneously at both facet A and B, because the tangential components of electric field are not null automatically along the facet B (see Fig.2.6b), as happens in the TM case. Of course, this provides no information whether perfect blazing in the TE case might exist for other incident angles, but it does state clearly that when a perfectly conducting echelle blazes perfectly in TM

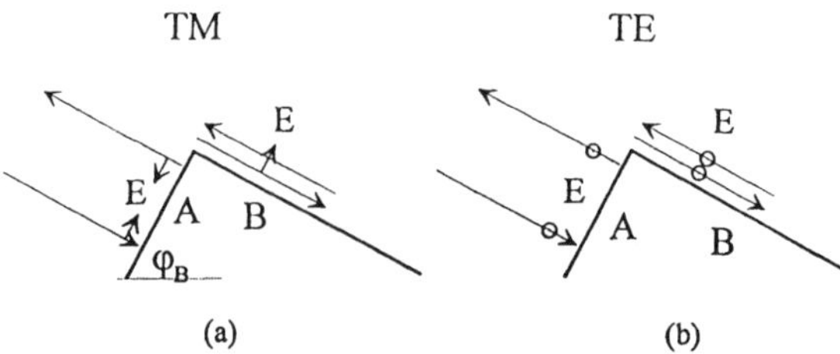

Fig.2.6 Schematic representation of the wave vector and electric field vectors of incident and diffracted-backward plane waves along the two facets of a triangular grating with a 90° apex angle. The two planes of polarization are TM and TE. E represents the electric field and the open circles indicate that the direction of the field is perpendicular to the paper in the TE case.

polarization, its efficiency in the TE case can never be 100%. How much the reduction is can be determined only by electromagnetic numerical simulation. Moreover, when going to real metals used as reflection coatings, even in TM polarization one observes a decrease of efficiency with increase of the groove angle, when compared to the reflectivity of a corresponding plane mirror. An intuitive argument is that when the groove angle (and thus the groove depth) increases so does the length and the influence of the "parasitic" facet B.

While this reduction is considered as an inevitable nuisance for classical echelettes (which rarely have groove angles higher than 22°), this is where the effect starts to become important (see later Section 4.3). It does play a rather important role in the case of echelles (see later, Section 6.4). The echelles work with groove angles of typically 63° and 76° and the "working" facet is shorter than the "parasitic" one.

Moreover, the effect can also be observed in transmission, a fact which some experimentalists find hard to accept and is usually attributed to technological difficulties. While blazed transmission gratings usually have groove angles less than 15° to 20°, recent Fresnel lenses and zone plates can have an aperture large enough to require 36° or higher groove angles, or equivalently, higher-than-the-first working diffraction order. As shown later, in sections 5.4 to 5.6, these extreme conditions lead to a drastic reduction of blazed order efficiency by tens of percents, even for a perfectly ruled triangular groove. The reasons are similar to the case of reflection gratings and have electromagnetic origin [2.5]. Consider a transmission grating with a triangular groove having a 90° apex. An artificial, infinitely thin but perfectly conducting layer, is deposited on the "parasitic" facet. This imposes artificial boundary conditions requiring that the tangential electric field be null there. It can be easily observed that if the optogeometrical parameters (period, wavelength and refractive indices) allow for two diffraction orders to propagate, one transmitted in the same direction as the incident wave and another one reflected

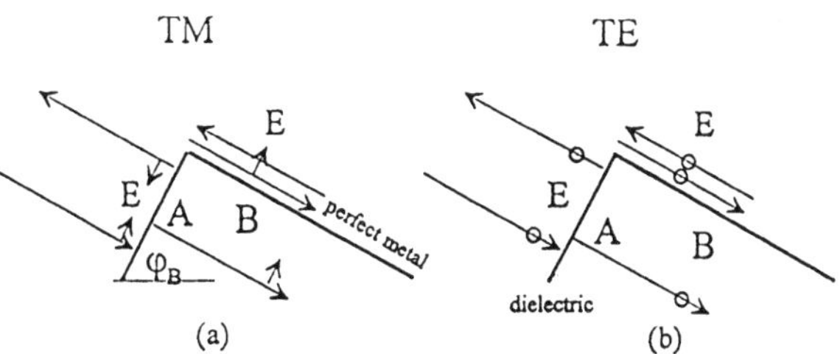

Fig.2.7 The same as in Fig.2.6 but in transmission.

backwards (Fig.2.7), the boundary conditions in TM polarization are satisfied and the amplitudes of the transmitted and reflected orders are determined by Fresnel's formulas [2.1]. This would ensure the desired 96% efficiency maximum, regardless of the groove angle or diffraction order number. While not perfect, this value can be considered more than enough, and we can refer to it using the practically acceptable term of *perfect blazing*.

However, life does not provide us with an infinitely thin perfectly conducting layer or with the means of its deposition. As for reflection gratings, the inability to correctly satisfy the boundary conditions along the "parasitic" groove facet for a purely dielectric grating, and/or TE polarizations, does not prove that perfect blazing does not exist for other mountings. But it proves that under the optimal blazing conditions determined by the geometrical optics considerations, i.e., with only two diffraction orders propagating according to Fig.2.7, these two orders are not enough to satisfy the boundary conditions. Other (usually an infinite number of) orders will be required, which inevitably reduces the efficiency of the only useful transmitted order. And it is. As shown numerically [2.6] and as observed experimentally, even with a perfect triangular profile, the maximum transmission efficiency can go down well below 80% as the groove angles reaches 23° and can drop below 60% with a groove angle reaching 45°. Further discussion is found in sections 5.4 to 5.6, but to the disgust of users there is no solution, unless the deposition of an infinitely thin highly conducting layer on the "parasitic" surface becomes possible.

2.8 Resolution

The *resolution* of the grating is a measure of its ability to separate adjacent spectrum lines. Quite often in the literature and in conversations, the term *resolving power* is used instead of resolution. We should avoid that, however, since the word *power* has a very specific meaning of energy per time and it has nothing to do with the grating (or, more general, optical system) resolution. Assuming for simplicity the Rayleigh criterion[1], the separation R between the primary maximum from neighbouring minimum in Fig.2.4a is given from eq.2.26:

$$R \equiv \frac{\lambda}{\Delta\lambda} = |m|\, M \quad . \tag{2.27}$$

[1] Two maxima of equal intensity are considered separated, when the minimum adjacent to one of the maxima coincides with the maximum of the other. This definition is somewhat arbitrary, but remains useful.

However, as M and m are not independent variables, a more meaningful expression is derived from the grating equation:

$$R = \frac{Md}{\lambda}\left|\sin\theta_d - \sin\theta_i\right| = \frac{W}{\lambda}\left|\sin\theta_d - \sin\theta_i\right| \quad , \qquad (2.28)$$

where W denotes the illuminated width of the grating. Citing Born and Wolf, "*the resolving power is equal to the number of wavelengths in the path difference between rays that are diffracted in the direction θ from the two extreme ends ... of the grating.*" As is obvious, resolution can be increased either by increasing the *illuminated* area of the grating and/or by increasing the corresponding optical path difference by going to steeper angles of incidence and diffraction. It is also evident that R is not dependent on the order or the number of grating grooves and, thus the pitch, but is a direct function of ruled width, wavelength and the angular configuration (mounting).

Since the maximum value of the angular part is equal to 2, the maximum attainable resolution of a given grating is simply equal to twice the number of wavelengths located in the grating width:

$$R_{max} = \frac{2W}{\lambda} \; . \qquad (2.29)$$

The corresponding resolution of a prism depends on the greatest optical thickness of the glass utilized in the beam, and thus gratings can have much greater resolution since nobody has been able to produce a high quality prism of 1/2 m thickness.

Fig.2.8 presents high-resolution spectra of mercury obtained in a 50 foot grating spectrograph. The theoretical resolution of the grating with 184 mm width at 63° incidence is 1,500,000 at 435.8 nm, and the experiment confirms its abilities, since the separation of the 202 and 200 lines requires 1,000,000 resolution for separate imaging.

The degree to which theoretical resolution (2.29) is attained depends not only on the diffraction angles used, but also on the optical quality of the grating surface, the uniformity of the groove spacing, and the quality of all the associated optics. Any departure greater than λ/4, or even λ/20, from the flatness of the plane grating, or from the sphericity of a concave one, will result in a loss of resolution. Grating groove spacing must be kept constant to within about λ/100. Experimental "details", such as slit width, air currents, vibrations, and temperature fluctuations, can seriously interfere with obtaining optimum results.

The practical resolution is of course limited also by the spectral width of the source lines and systems with resolution greater than 500,000 are usually

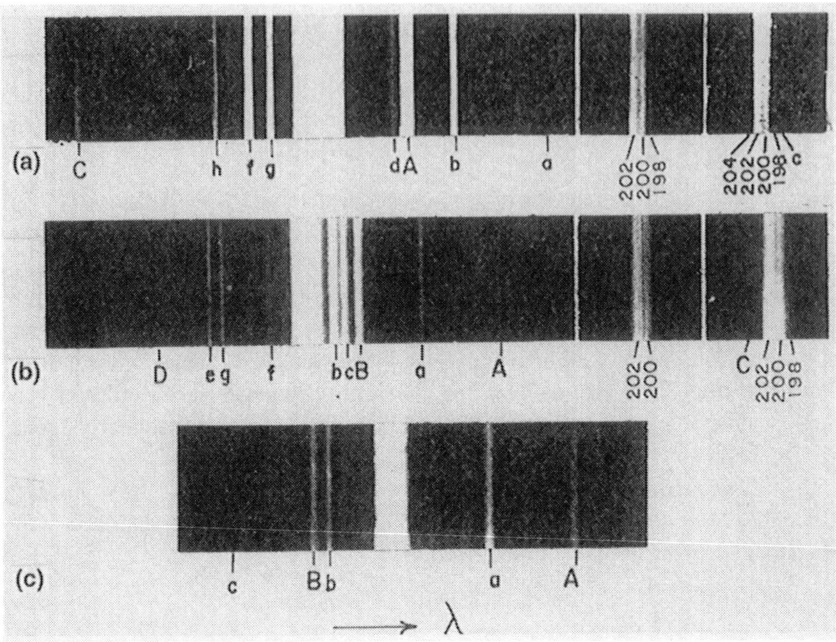

Fig.2.8 High resolution spectra of mercury obtained in a 50 foot spectrograph and using double passing for increased resolution. All gross hyperfine patterns were obtained in absorption and all central components in emission. The 546.1 nm lines were obtained from 240 mm wide echelle type grating with a maximum theoretical resolution of 1,585,000 in double pass. Resolution of 685,000 is required for the 200-198 pair and 910,000 for the 204-202 pair. The other two lines are resolved with a similar grating of 184 mm width. The separation of 202-200 pair at 435.8 nm is only 0.0057Å, but with a Doppler width of 0.0038Å a 1,000,000 resolution is required for separate imaging (after [2.7]).

required only in the study of spectral line shapes, Zeeman effects, line shifts (mainly in astronomy), and are not needed for separating individual lines.

2.9 Mountings

In spectroscopic devices the light path is usually fixed, because entrance and exit slits are (normally) fixed and the grating works in a so called "mounting" (i.e., it is mounted in some mechanical device that ensures that light from the entrance slit is focused at the exit slit). Even in non-classical systems that utilize CCD cameras or photodiode arrays, the grating is mounted in a specific way, i.e., works in a specific mounting. The term *mounting* has come to have a separate meaning, indicating more than just the technical details of how the grating is mounted. It determines the way the grating is utilized - what combination of parameters is varied and what is kept constant. An example of a specific mounting is the system for measuring the relative diffraction efficiencies (Fig.2.5). The mirror is mounted firmly, independently of the wavelength to ensure the required reflection angle. The grating has to be rotated according to a sine law, due to its dispersive properties, but the angle between the incident and the diffracted beams is kept constant, naming this configuration "constant angular deviation mounting".

There is a mounting that plays the greatest role both in grating experiments and theories. This is the famous *Littrow mounting*, when light diffracted in a given diffracted order (namely, the m-th) propagates backward toward the source. This mounting is also called *Autocollimation*. The link between the angle of incidence and the wavelength-to-period ratio in the Littrow mount is quite simple and is easily derived from the grating equation (2.1):

$$2\sin\theta_i = m\frac{\lambda}{d} . \tag{2.30}$$

This simple equation had been somewhat of a headache to generations of instrument designers before the stepping motor was invented, because the sine rule of rotation of the grating requires a sine drive to achieve a linear wavelength readout.

Littrow mount is considered important in utilizing gratings because it corresponds to maximum efficiency of diffraction. This assumption comes from the scalar theory of echelette gratings, when light diffracted along the angle of geometrical reflection by the plane facet is supposed to be maximal. Although not rigorous, this rule holds for surprisingly large class of gratings, even when there is some necessary departure from Littrow mount (see Chapters 4 to 7). It is not practical to use a system in which entrance and exit slits coincide, especially when the exit slit is exchanged for an array of detectors. It becomes necessary to change the angle of incidence so that the diffracted beam includes some angle with it. If this angle is kept constant the mounting is called *constant angular deviation mounting* and the angle between the diffracted and incident beams is called *angular deviation (A. D.)*. Littrow mount is a particular case with zero A. D. Of course, in order to ensure constant A. D. (or Littrow mount), it is necessary to rotate the grating while changing the wavelength (Fig.2.5). In spectrographs the detection at different wavelengths is separated in space. Such are the classical Rowland circle spectrographs, where many detectors (often photomultipliers), each with its own exit slit, are located on the focal curve of a concave grating or an imaging optical system that includes a plane grating. The other typical example is the so called flat-field spectrometer, where detection is carried out by an array of photodetectors. These devices utilize gratings in a mounting, where the *angle of incidence is constant* and the wavelength scan is performed using *different diffraction angles*.

Several other terms are used by grating specialists. The first one determines if the incident (and thus the diffracted) light plane is perpendicular to the grooves and is called *classical diffraction mounting*, in opposition to the *conical one*, where the direction of incidence is not perpendicular to the grooves. In the latter case the diffracted orders form a cone, naming the mounting correspondingly. Most gratings are used in the classical diffraction mounting. This is because of simpler mechanical requirements. Otherwise instead of the sine law of rotation one needs two-dimensional rotation. However, in many devices the grating is slightly inclined from the classical mounting. This is necessary when working near the Littrow mount in order to avoid the overlapping of exit and entrance areas. Even when the inclination is small it may require special attention.

Grazing incidence mounting is well known to laser system designers. In fact the term includes two types of mountings. In the first one the grating is used in very steep angles of incidence and diffraction in order to increase dispersion or reflectance in the X-ray domain. The second, often termed *Littman dye laser tuning*, is shown in Fig.2.9. A tuning cavity needs to fill the aperture of a relatively large grating to get sufficient resolution. At grazing

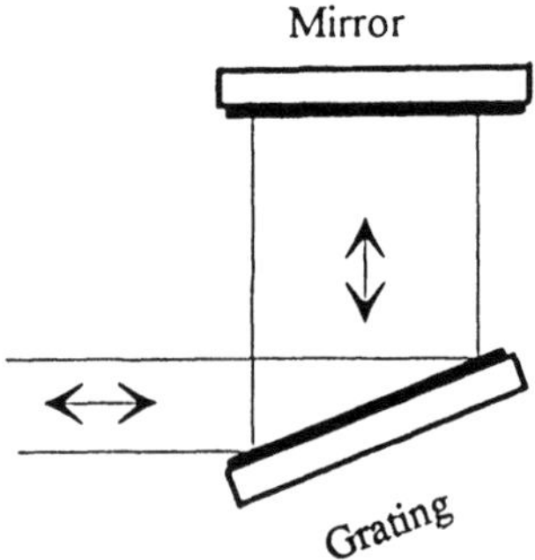

Fig.2.9 Littman dye laser tuning with grating utilized in grazing incidence.

incidence even a small beam can fill the grating and tuning is obtained by rotating the retro-reflecting mirror.

2.10 Some Electromagnetic Characteristics

It is necessary to introduce several general characteristics of the electromagnetic field, required to understand some of the grating properties.

2.10.1 Energy Flow (Poynting) Vector

This characteristic has a clear intuitive meaning, although in some cases the intuition can fail. There is a simple definition, at least for homogeneous isotropic lossless media: The *energy flow vector* is locally tangent to the direction of energy transfer (flow) and its amplitude is proportional to the quantity of the electromagnetic energy transferred through a unit surface. It can be shown that this vector is usually equal to:

$$\mathbf{P} \propto \mathrm{Re}(\mathbf{E} \times \overline{\mathbf{H}}) \ , \tag{2.31}$$

where the overbar means complex conjugation. Far from the grating surface, where the diffraction orders can be considered independent (and spatially separated), the direction of **P** coincides with the direction of the wavevector. And indeed, taking into account the transverse character of the plane wave (**E** is perpendicular to **H**, both of them perpendicular to **k** and thus to the direction of propagation), it immediately follows from the definition of vector multiplication that **P** and **k** are parallel (see Fig.2.2b).

In the vicinity of the grating surface all the orders are mixed together (including the evanescent orders) and the direction of the Poynting vector can not be determined by such a simple consideration. A guiding example is reflection by a perfectly conducting plane mirror: While far from the surface, where the incident and reflected beams are separated, energy flow follows the direction of the beams (more or less, depending on their width and divergence); near the mirror **P** is parallel to the surface, because there is no flow through the mirror face. In the grating case complicated pictures can be observed (see Chapter 8). These pictures can serve as a proper explanation for some general properties of grating efficiencies, although it is very difficult, if not impossible, to measure directly the direction of the energy flow near the surface. The main difficulty comes from the fact that any measuring device will be large enough to drastically modify the flow direction. Its magnitude is measured more easily, because detector response is usually proportional to the energy flow through its working surface.

2.10.2 Electromagnetic Energy Density

The definition of the density of energy is quite straightforward:

$$\mathcal{E} \propto \varepsilon_0 n^2 |E|^2 + \mu_0 |H|^2, \tag{2.32}$$

the coefficients of proportionality in 2.31 and 2.32 depend on the system of units. The energy density is equal to the magnitude of the Poynting vector. For a plane wave it is constant over the space. In some cases (connected, generally with the excitation of guided waves near the grating surface), the density of the electromagnetic field energy can grow significantly (several orders of magnitude) near the grating surface and can be indicated using some substances and phenomena that are field-sensitive (have a nonlinear response). Such field enhancement is discussed in detail in Chapter 8.

2.11 Two Simple Methods of Determining the Grating Frequency

People always ask simple questions that require complex answers. Fortunately, there are a few exceptions and to roughly determine the grating period is one of them. To presuppose that everybody will have a small laser pointer at hand is not necessary. A *small* flashlight is sufficient. The smaller the source, the cleaner will be the spectrum. Clear bulbs, because their filament acts as an effective entrance slit, are always better than ceiling lights. The first method works for coarse gratings. You hold the flashlight close to the eye pointing to the grating. Starting as close as possible to the eye, the grating is moved away since both the zero order and the first order green are to be observed. Next, adjust the position of the grating until the distance between the images is a known one. This could be 1 or 2 cm as measured with a ruler, or it could be the width of the grating, call it w. If the distance from the eye to the grating, as measured with a ruler or tape is q, then the angle subtended is 2arcsin(w/2q), or simply w/q radians for coarse gratings. Since the conditions involved are close to normal incidence, the grating equation (2.1) is λ/d = sin(w/q), and with λ known at 0.54 μm, it is not hard to find a rough value of d.

This method does not work as well with finer pitch gratings, because the zero and the first order images are too far apart to be seen simultaneously. However, an even simpler method becomes available. Again you start with a small flashlight next to your eye (take off any glasses). Preferably in a darkened room you look at the grating and see the zero order reflection. It can be easily distinguished from the others due to lack of dispersion. Then rotate the grating, preferably in the blaze direction, if blazed, and note the orders as they go by. With a 1200 gr/mm you can see the first two orders quite well, but when you start to see the third order colors, the grating will be near grazing incidence and you cannot get beyond the green. Since you are operating in Littrow you can use eqs.(2.19) and (2.30), or $d = m\lambda/2$. For green (540 nm) that makes $d = 0.81$ µm, or 1235 gr/mm, good enough for identification. For a 600 gr/mm, grazing incidence occurs in the 5th order, just obtainable.

For transmission gratings you look at a flashlight through the grating, with the eye acting as an imaging system. The rest is the same.

Why the choice of green? Because the eye is most sensitive and the flashlight provides enough intensity. Because green band is relatively narrower and wavelength shorter than red, accuracy is better and there are a greater number of diffraction orders. Otherwise, a red laser pointer suffices, unless the groove frequency exceeds 3000 gr/mm.

2.12 Pulse Compression by Diffraction Gratings

There are many applications where the ability of lasers to deliver short pulses, particularly in the psec and fsec range, is inadequate for experiments because the energy levels are too low. In order to amplify them at least two critical requirements must be fulfilled by the amplifying medium. First the bandwidth of the amplifier must be capable of accommodating the full spectrum of the short pulse. Secondly the intensity within the amplifier must stay below the level at which non-linear effects start distorting the spatial and temporal profiles of the pulse. Several amplifying materials have proven effective in the near IR region ($\lambda \sim 1$ µm) which represents a center of interest, such as Ti:saphire, Nd:glass, and alexandrite, and which can operate at average power levels of 100 W in compact systems.

A basic problem is that as pulses become shorter the ability of the amplifying medium to accept peak intensities beyond a certain level becomes the limiting factor. The solution is based on stretching out or chirping the initial input pulse in such a way that peak powers are greatly reduced (so they can safely be amplified), the long, high power pulses are recompressed after amplification back to the original width, or possibly even shorter. It is possible to increase energy levels in this fashion by ratios as high as 10^9, into the TW domain for energy with pulses as short as 80 fs [2.8].

One of the methods for pulse stretching uses the self-phase modulation and group velocity dispersion (GVD) of single-mode optical fibers typically 1 km long, Fig.2.10. In this example the initial pulse is stretched from 55 to 300 ps. The Nd:glass regenerative amplifier boosts energy to 2 mJ, which is further amplified by a factor of 50 in a 4-pass Nd:glass amplifier. The final stage of Nd:glass amplification increases energy by a similar ratio to 1 J. Finally the 300 ps pulse is sent through a double passed two-grating compressor, which in this instance compresses the pulse to 1 ps [2.8]. Since each of the matched gratings

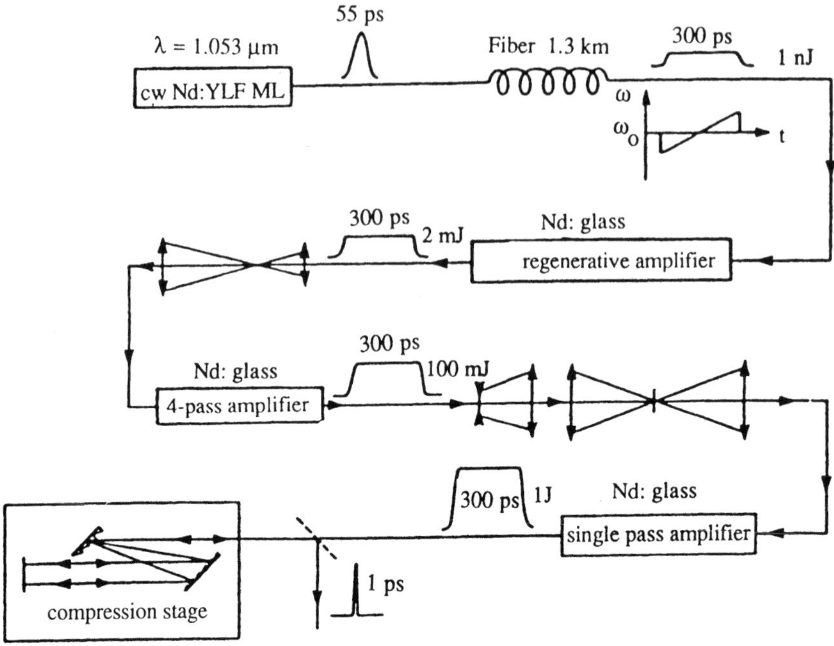

Fig.2.10 Diagram of multistage pulse amplification system for transforming a low power 55 ps pulse to 1 ps 1 J pulse. Fiber pulse stretching and grating pulse compression (after [2.8]).

is double passed, their efficiency is critical, entering as the fourth power. Even with 90% efficiency, which is attainable with a gold coated 1700 groove per mm grating (at 1.06 μm wavelength) in the TM plane of polarization, 35% of the energy is lost, mostly to the zero orders. In many instances the maximum output is limited by the ability of the gratings to survive high flux densities (see Ch.13). With 20 mJ/cm^2 a typical safe figure, a grating area of 50 cm^2 area would be required in the above example.

Another possible limitation of such systems is the mismatch between the imperfect linearity of the dispersive properties of the optical fiber pulse stretcher and the linear ones of the diffraction grating pulse compressor. A possibly obvious solution is to replace the fiber with another grating pair to perform the the pulse stretching function, as shown in Fig.2.11, where it is advantageous to locate the stretching gratings inside the focal points of a telescope system [2.9], but are otherwise identical to the compression gratings.

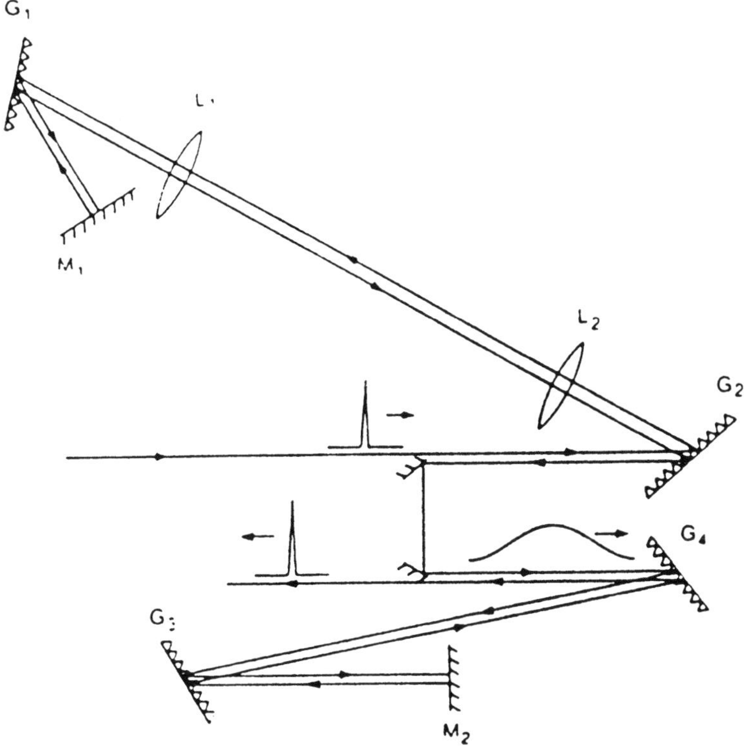

Fig.2.11 Experimental arrangement of grating pulse expansion and compression system, with four 1700 gr/mm gratings labeled G_1, G_2, G_3, G_4. Lenses L_1 and L_2 have 500 mm focal length and mirrors M_1 and M_2 allow double passing (after [2.10]).

A major advance of such four-grating systems is the large stretching and compression ratios that can be attained, with ratios easily exceeding 1000 [2.10].

A practical problem in such systems is that the alignment of all the gratings have to be readjusted if the wavelength is changed. One suggested solution is to use a single grating together with retroreflecting mirrors, although the grating has to be twice as large, Fig.2.12 [2.11].

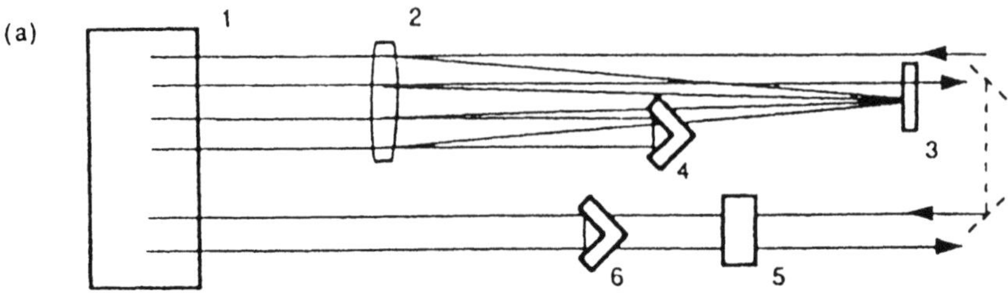

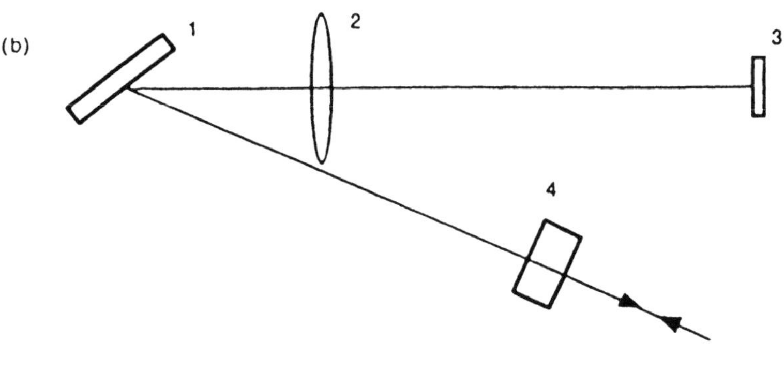

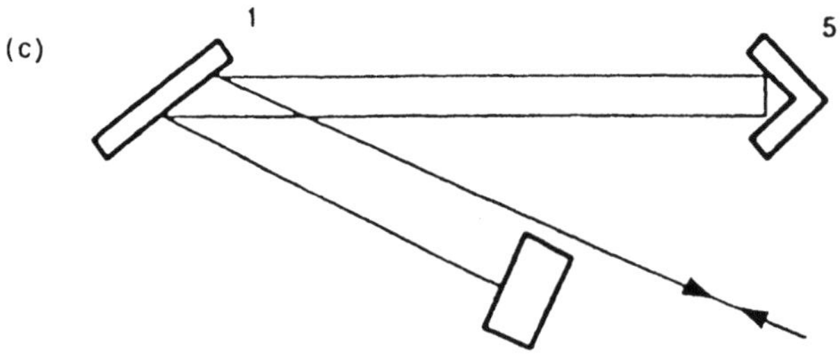

Fig.2.12 a) Schematic diagram of a single grating two-level laser pulse stretcher-compressor. 1 grating, 2 lens, 3 mirror, 4 and 6 roof mirror reflectors for vertical displacement, and 5 roof mirror for horizontal displacement. b) Top view of stretcher and c) Top view of compressor (after [2.11]).

References

2.1 M. Born and E. Wolf, *Principles of Optics*, 4th edition, (Pergamon Press, Oxford 1968).

2.2 Lord Rayleigh: "On the manufacture and theory of diffraction gratings," PhiL Mag. Series 4, **47**, 193-205 (1874).

2.3 R. Wood: "The echelette grating for the infra-red," Phil. Mag **XX** (Series **6**), 770-778 (1910).

2.4 A. Marechal and G. W. Stroke: "Sur l'origine des effets de polarisation et de diffraction dan les réseaux optiques," C. R. Ac. SC **249**, 2042 – 2044 (1980).

2.5 M. Neviere, D. Maystre, and J-P. Laude: "Perfect blazing for transmission gratings," J. Opt. Soc. Am. A **7**, 1736-1739 (1990).

2.6 M. Neviere: "Electromagnetic study of transmission gratings," Appl. Opt. 30, 4540-4547 (1991).

2.7 D. H. Rank, G. Skorinko, D. P. Eastman, G. D. Saksena, T. K. McCubbin Jr., and T. A. Wiggins: "Hyperfine structure of some HgI lines," J. Opt. Soc. Am. **50**, 1045–1052 (1960).

2.8 P. Maine, D. Strickland, P. Bado,M. Pessot, and G. Moourou: "Generation of ultrahigh peak power pulses by chirped pulse amplification," IEEE J. Quantum Electr. **QE-24**, 398-403 (1988).

2.9 O. Martinez: "3000 times grating compressor with positive grooup velocity dispersion: Application to fiber compensation in 1.3 – 1.6 μm region," IEEE J. Quant. Electr. **QE-23**, 59-64 (1987).

2.10 M. Pessot, P Maine, and G. Mourou, "1000 times expansion/compression of optical pulses for chirped pulse amplification," Opt. Comm. **62**, 419-421 (1987).

2.11 M. Lai, C. Lai, and C. Swinger: "Single-grating laser pulse stretcher and compressor," Appl. Opt. **33**, 6985-6987 (1993).

Additional Reading

J. A. Anderson and C. M. Sparrow: "On the effect of the groove form on the distribution of light by a grating," Astroph. Jl. **XXXIII**, 338-352 (1911).

P. Beckmann: "Scattering of light by rough surfaces," E. Wolf, ed. *Progress in Optics* (Elsevier, North-Holland, Amsterdam, 1967) v. **VI**, pp.53-69.

O. Bryngdahl: "Evanescent waves in optical imaging," E. Wolf ed. *Progress in Optics* (North Holland, Amsterdam, 1973), v. **XXI**, pp.167-221.

J. A. De Santo and G. S. Brown: "Analytical techniques for multiple scattering from rough surfaces," E. Wolf, ed. *Progress in Optics* (Elsevier, North-Holland, Amsterdam, 1986) v. **XXIII**, pp.1-62.

Diffraction Gratings, Special issues of J. Opt. Soc. Am. A, v.**8**, no.8 and 9 (1990).

T. K. Gaylord and M. G. Moharam: "Analysis and applications of optical diffraction by gratings," Proc. IEEE **73**, 894-937 (1985).

C. Hafner: *Generalized multipole technique for computational electromagnetics*, (Artech, Boston, 1990).

D. G. Hall: "Optical waveguide diffraction gratings: coupling between guided modes," E. Wolf, ed. *Progress in Optics* (Elsevier, North-Holland, Amsterdam, 1991) v. **XXIX**, pp.1-63.

G. R. Harrison: "The diffraction grating - an opinionated appraisal," Appl. Opt. **12**, 2039-2049 (1973).

W. R. Hunter: "Diffraction gratings and mountings for vacuum ultraviolet spectral region," ch.2, *Spectrometric Techniques,* v. **IV**, G. Vanasse, ed. (Academic, London, 1985).

M. C. Hutley: *Diffraction Gratings*, (Academic, London, 1982).

M. Gale and K. Knop: "Surface relief images for color reproduction," in *Progress Reports in Imaging Science 2*, (Focal Press, London, 1980).

J. M. Lerner, ed. *International conference on the applications, theory, and fabrication of periodic structures, diffraction gratings, and Moire phenomena*, II, SPIE, v.**503** (1984).

J. M. Lerner, ed. *International conference on the applications, theory, and fabrication of periodic structures, diffraction gratings, and Moire phenomena*, III SPIE, v.**815** (1987).

E. G. Loewen: "Diffraction gratings, ruled and holographic," ch.2, Appl. Opt. and Opt. Engineer., v. IX, R. R. Shannon and J. C. Wyant, eds. (Academic, London, 1983).

D. Marcuse: "Light transmission optics," *Bell Laboratories Series*, Van N. Reinhold, ed. (New York, 1972).

A. Maréchal: "Optique géometrique générale," in *Handbook of Physics,* v.**24**, *Foundations of Optics*, S. Flugge, ed. (Springer, Berlin, 1956).

D. Maystre: "General study of grating anomalies from electromagnetic surface modes," in *Electromagnetic Surface Modes*, A. D. Boardman, ed. (John Wiley, 1982), ch.17.

D. Maystre: "Rigorous vector theories of diffraction gratings," E. Wolf, ed., *Progress in Optics* (Elsevier, North-Holland, Amsterdam, 1984) v. **XXI**, pp.2-67.

D. Maystre, ed. "Selected papers on diffraction gratings," *SPIE Milestone Series*, v. **MS 83**, (1993).

D. Mendlovic and O. Melamed: "Grating triplet," Appl. Opt. **34**, 7807-7814 (1995).

W. McKinney: "Diffraction gratings: manufacture, specialization, and application," SPIE 28th Annual Intern. Techn. Symp., Tutorial 25, San Diego 1984.

T. Namioka, T. Harada, and K. Yasuura: "Diffraction gratings in Japan," Opt. Acta **26**, 1021-1034 (1979).

E. W. Palmer, M. C. Hutley, A. Franks, J. F. Verrill, and B. Gale: "Diffraction Gratings," Rep. Progr. Phys. **38**, 975-1048 (1975).

R. Petit and D. Maystre: "Application des Lois de l'Electromagnetisme a l'Etude des Reseaux," Rev. de Phys. Appliquée, **7**, 427-441 (1972).

R. Petit, ed. *Electromagnetic Theory of Gratings*, (Springer-Verlag, Berlin, 1980).

E. Popov: "Light diffraction by relief gratings: a microscopic and microscopic view," E. Wolf, ed. *Progress in Optics* (Elsevier, North-Holland, Amsterdam, 1993) v. **XXXI**, pp.139-187.

G. Schmahl and D. Rudolph: "Holographic diffraction gratings," E. Wolf, ed. *Progress in Optics* (Elsevier, North-Holland, Amsterdam, 1977) v. **XIV**, pp.195-244.

A. E. Siegman and P. M. Fauchet: "Stimulated Wood's anomalies on laser-illuminated surfaces," IEEE J. Quant. Electron. **QE-22**, 1384-1402 (1986).

G. I. Stegeman and D. G. Hall: "Modulated index structures," J. Opt. Soc. Am. A **7**, 1387-1398 (1990).

G. W. Stroke: "Diffraction gratings" in *Handbook of Physics*,v.29, *Optical Instruments*, ed. S. Flugge (Springer, Berlin, 1967).

G. W. Stroke: "Ruling, testing and use of optical gratings for high-resolution spectroscopy," E. Wolf, ed. *Progress in Optics* (Elsevier, North-Holland, Amsterdam, 1963) v. **II**, pp.1-72

P. M. Van den Berg: "Diffraction theory of a reflection grating," Appl. Sci. Res. **24**, 261-293 (1971).

W. T. Welford: "Abberation theory of gratings and grating mountings," E. Wolf, ed. *Progress in Optics* (Elsevier, North-Holland, Amsterdam, 1965) v. **IV**, pp.241-280.

Chapter 3

The Types of Diffraction Gratings

3.1 Introduction

Diffraction gratings are usually divided according to several criteria: their geometry, their material, their efficiency behaviour, the method of their manufacturing, according to the working spectral interval, or their usage. A few examples are cited to show the complexity and lack of clarity: amplitude - phase, phase - relief, ruled - holographic - lithographic, symmetrical - blazed, transmission - reflection, concave - plane, flat-field spectrographic - operating on the Rowland circle - those for Seya-Namioka monochromators, echelettes - echelles - echelons, those for integrated optics - those for distributed feedback (DFB), lamellar - triangular - sinusoidal - trapezoidal groove shape, dielectric - metallic, masters - replicas, etc. A good example could be: Special Type of Grating Ordered for OUR Experiment and Destroyed After It.

The separation of gratings into such doublets (triplets, etc.) are sometimes due to historical reasons, others have a strong background (difference in properties) and others are due to difference in applications. Several classes are now considered old-fashioned and used mainly by non-specialists in grating theory and manufacturing. For example the division between amplitude and phase types and between Raman-Nath and Bragg type. Although these two classifications play a minor practical role, we discuss them in order to clear up some problems. Other types are more important, concerned with basic grating properties and they are discussed in detail within the following chapters. A key distinction is reflection versus transmission gratings, where the case of reflection gratings working in very high diffraction orders is discussed separately in Ch. 6. Second, there is an important difference between plane and concave gratings. The division into ruled and holographic can be disturbing as regards their properties, although there are well defined differences in the process of manufacturing. In some cases masters have better performance than replicas, especially regarding certain properties such as stray light, but not always. In the case of high angle echelettes and echelles odd-order generation replicas have better performance than even (the master can be considered as a zero-order replica).

3.2 Amplitude and Phase Gratings

This is probably the most confusing of all grating classifications. There are two complementary definitions. The first one draws the boundary according to the influence the gratings have on the incident light and the second definition concerns the way this influence is achieved. Phase gratings are supposed to change only the phase of the incident light in the different groove regions and amplitude gratings are presumed to change the amplitude. These changes are believed to be executed through groove geometry, consisting of variation of the real part of the refractive index of the grating material (phase gratings, Fig.3.1a) or, for the amplitude gratings, through varying the absorption along the grooves (in Fig.3.1b the imaginary part of the refractive index is varied). Typical examples of gratings considered to be phase or amplitude types are given in Fig.3.1a,c and b,d, respectively. When transmission in the absorbing part of the amplitude grating is zero and its thickness is much smaller than its width (Fig.3.1d), they are called Ronchi Rulings. The most common way to produce them is through photography (lithography) or mechanical ruling a thin metallic (Al) layer deposited on a glass substrate.

This classification is somewhat arbitrary and rather confusing, because the type of grating material does not determine uniquely its efficiency behaviour and because the "phase" gratings also change the amplitude of the incident light and vice versa. It is also very difficult, if not impossible, to change losses (i.e., absorption) without changing the real part of the refractive index (phases).

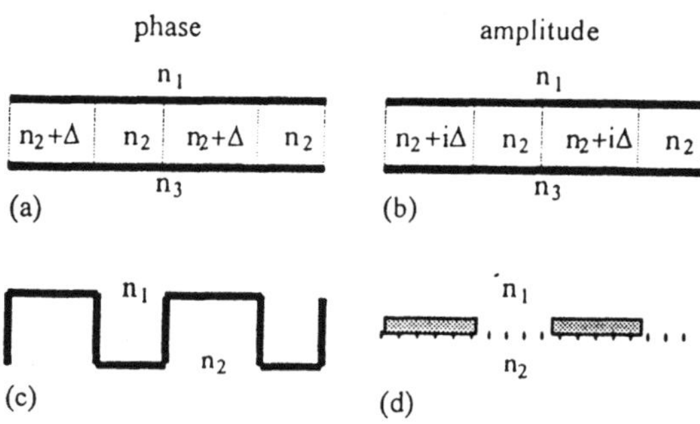

Fig.3.1 Schematic representation of phase (a, c) and amplitude (b, d) gratings, and of phase (a, b) and relief (c, d) gratings.

3.3 Phase and Relief Gratings

This separation becomes clearer by considering phase gratings to consist of refractive index variations and relief gratings a surface relief structure that separates two media of different optical properties (compare Fig.3.1a with c, and b with d). Nevertheless, many cases can fall outside this definition: for example the relief dielectric grating in Fig.3.1c can be considered not only as a relief, but as a phase grating as well. The border line is drawn by the magnitude of the refractive index changes: whereas in Fig.3.1c the change is great, in Fig.3.1a (e.g. gelatine grating) the refractive index varies in the 2nd or 3rd digit. However, this makes it difficult to distinguish clearly where Ronchi rulings belong.

Except for special situations grating applications center on relief gratings because of their greater efficiency. An important exception is holography, where the recording is made in the volume of the photosensitive material (even for thin holograms), but holography lies outside the scope of this book so we discuss only relief gratings.

3.4 Reflection and Transmission Gratings

Intuitively it is much easier to distinguish between reflection and transmission gratings, the former working in reflection regime and the latter in transmission. Usually reflection gratings are relief gratings, covered with some highly reflecting material. Depending on the spectral region reflectivity can vary significantly, so that the choice of material can be critical for grating performance (see Chapter 4). In particular, gratings that are reflective in one spectral region, can become transmitive ones in another. On the other hand, metallic strip gratings (e.g., Ronchi rulings) can have adequate efficiency in both transmission and reflection, at least in first order.

Transmission gratings are rarely used in standard spectrometric instruments because of inherent limitations, except for certain direct imaging spectrographs. However, they find wide applications as beam dividers and combiners, together with laser sources, with groove shapes that may be triangular, rectangular or trapezoidal. If only a few orders are required the period is comparable to the wavelength, but transmission gratings can also supply large number of equal intensity orders if the profile form is suitably chosen (fan-out gratings). A common configuration converts a camera into a simple spectrograph by inserting the grating in front of the lens so that the distant luminous spark of, say, falling meteors or the re-entry of space vehicles, acts as an entrance slit. Another application is the determination of spectral sensitivity of photographic emulsions, where a combination of high speed lenses and diffraction gratings is used.

Since a transmission blank is part of the imaging optics, specifications of its optical quality are tighter. The back face needs an anti-reflection coating to prevent light losses due to reflection and, even more important, to prevent multiple scattering effects inside the substrate. For wavelengths between 220 and 300 nm fused silica blanks are used together with a special resin, transparent at these short wavelengths.

Properties of reflection gratings are discussed in detail in Chs. 4, 6, 7, and of transmission gratings in Ch.5.

3.5 Symmetrical and Blazed Gratings

When the grating groove is symmetrical, normally incident light delivers equal efficiencies into the symmetrical orders (+1 and -1, +2 and -2, etc.). Blazed gratings are usually characterized by a triangular groove shape with 90° apex angle, so that when light is incident close to the direction normal to one of the facets, almost all of it is diffracted into the backward direction. When grating period and wavelength allow diffraction in that direction blazing occurs in the corresponding diffraction order. Blazed gratings can be either reflection or transmission.

Of course, while scalar expectations about blaze behaviour of echelette gratings are valid under most conditions they can change in the resonance domain (wavelength close in magnitude to the groove period). For example, sinusoidal groove gratings can have blaze behavior when only the specular and -1st order can propagate and in that case may even appear to blaze better than triangular-groove gratings. Thus, when a grating operates under these conditions, non-blazed gratings may be preferred. When large spectral intervals must be covered in one step without changing the grating only blazed gratings will suffice. Gratings with profile close to sinusoidal are easily obtained by the holographic manufacturing method.

Depending on groove form, gratings can be divided into lamellar (rectangular), triangular (blazed), sinusoidal, trapezoidal and undetermined gratings. Blazed gratings are often called *echelettes*, with working orders from -1st to -4, or -5th. *Echelles*, with working orders up to several hundred have light incident close to the normal of the small facet. They are becoming ever more popular tools due to the high angles of incidence and diffraction; their dispersion can be as high as that of very fine-pitch gratings, but because of their low λ/d ratio polarization effects are much less. They also cover a much wider range of wavelengths with good efficiency.

Lamellar gratings consist of ridges with rectangular cross section. Most often, the space between them is equal to their width; lamellar gratings are also called *laminar*. When the height of the ridges is such that the optical path difference between the rays reflected (or transmitted) at the top and on the bottom is $\lambda/2$, the zero order at this wavelength may be eliminated, while equally strong first orders appear at either side. Such gratings make useful beam dividers. Many compact disc heads have them working in transmission under conditions where both the zero and first orders are used. The zero order serves for data reading and orders +1 and -1 are used to maintain tracking and focus.

In reflection form as a stack of many layers, lamellar gratings have been used successfully for X-ray dispersion in the 0.5 - 10 Å region.

Detailed analysis of properties of blazed and symmetrical reflection gratings can be found further on in Ch.4, in Ch.5 for transmission gratings, and in Ch.6 for echelles.

3.6 Ruled, Holographic and Lithographic Gratings

As is obvious from their names, these grating classifications concern the process of manufacturing. *Mechanical ruling* is believed to produce only blazed gratings, but the profile can be triangular or trapezoidal. Holographic recording (of the interference pattern of two monochromatic coherent beams) usually leads to symmetrical profiles with smooth grooves, but, more rarely, can produce asymmetrical blazed gratings. With interferometric feedback, mechanical ruling is responsible for gratings of higher quality, as regards the groove spacing control over the entire blank, especially for large grating areas (a property of vital importance for high-resolution spectroscopy and astronomy). *Holographic gratings* are easily manufactured, at least when the grating dimensions do not exceed 100 mm.

Lithographic gratings are a product of single-beam mask transfer into a photosensitive layer, followed by etching of this layer and the underlying material (glass, semiconductor, or metal). Diffraction phenomena are responsible for the lower limit of the grating periods than can be copied, so that they are usually greater than 5 - 10 μm. A special solution is described in Ch.16, where a property of the mask acting also as a grating can reduce this limit significantly.

Electron-beam ruling is a flexible and powerful tool for drawing different patterns, including diffraction gratings with curved and chirped grooves. Multistep processes can lead to a partially blazed groove profile, although with severe difficulties and poor reproducibility. A key limitation for E-beam processes is the slow rate of writing. Limitations of linearity of electron beam position control dictate that gratings larger than few mm have to be made by a step and repeat process that may be adequate to generate mask patterns but not for high quality gratings. The basic processes of grating manufacturing are discussed in detail in Chs. 14 to 17.

3.7 Plane and Concave Gratings

The most commonly used gratings have a plane substrate, and straight and equidistant grooves. At least, that is what they should have. In rare instances (mainly in integrated optics), it is useful to bend the grooves slightly and to vary (chirp) their separation along the grating surface to provide some focusing properties. Much more common is to make gratings with curved and chirped grooves on a concave substrate, so that the grating can act as a single-element spectrograph or monochromator, combining the dispersion properties of plane gratings with the focusing properties of concave mirrors. There are several degrees of freedom, regarding the substrate curvature (spherical, toroidal, aspherical) and the groove form. Holographic recording using two point sources is the most common method of reducing aberrations, but the grooves in that case have almost symmetrical form that leads to modest diffraction efficiency in their usual working conditions. Additional ion-beam blazing is necessary if high efficiency is required, although that tends to increase stray light. Concave gratings always work in reflection. Typical examples of concave gratings and their applications are discussed in Ch.7.

3.8 Bragg Type and Raman-Nath Type Gratings

This is another example of a confusing classification that has little to do with recent grating concepts, but these terms, together with "Bragg type diffraction" are widely used in optical textbooks and integrated optics papers, so they deserve some attention. With a few exceptions, Bragg type gratings work at angles where only a few orders can propagate, usually 0th and 1st. Sometimes in corrugated waveguides there is not even a single order in the cladding and the substrate, all of them being evanescent and the gratings provide the phase vector for interaction (Bragg type phase matching, Bragg type diffraction) between the waveguide modes (Chapter 9). The equation always used to describe Bragg diffraction will be recognized by readers of this book as being identical to the Littrow formulation of the grating equation. This is also the case with Bragg transmission gratings (Chapter 5.10).

On the other hand, Raman-Nath regime of diffraction is characterized by many propagating orders so that no single one is predominant, as it is with sinusoidal gratings in the scalar region. The uselessness of this classification comes from the fact that in different spectral regions all the gratings (even the blazed ones) can work in one or the other regime, or in an intermediate one.

3.9 Waveguide Gratings

Diffraction gratings are used with planar optical waveguides as input and output couplers, beam-splitters, wavelength demultiplexers, and beam directional switchers and modulators, etc. (Chapter 9). Both relief and phase gratings are used (Fig.3.2). Holographic and, recently, lithographic techniques can be applied to form a surface relief pattern on the top (or bottom) of planar waveguides. The flexibility of optical or electron-beam lithography enables curved and chirped lines for additional beam shaping - focusing and waveform changing, but unfortunately the groove profile is not so easily controlled and

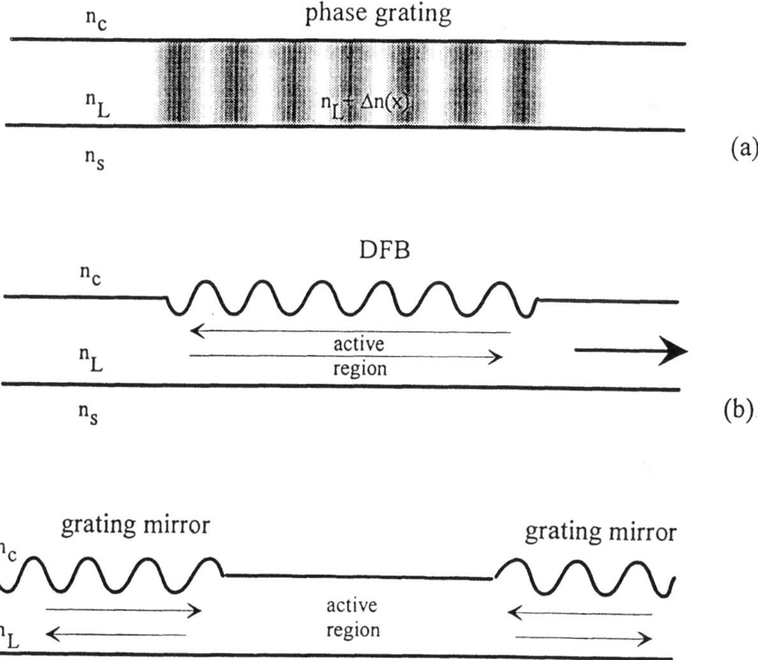

Fig.3.2 Waveguide gratings: phase (a) and surface relief (b,c) ones. The corrugated region coincides with the lasing region of semiconductor laser in so called distributed feedback (DFB) grating (b). The grating can lie outside the grating region (c), often called distributed Bragg reflector (DBR).

the recording areas are limited. Comparatively large period gratings made by lithography require blazing (i.e., control of finer structure of the profile), which is again limited by the manufacturing problems.

Phase gratings can be formed by classical holography or UV lasers. A special case is grating formation by electrooptical or acoustooptical effects, where the grating strength or even its period can be varied to modulate or switch beam direction and intensity.

Waveguide gratings find large application in semiconductor lasers for feedback of emission within the guiding layer. They are used in distributed feedback (DFB) geometry with a grating covering the active (lasing) region or in other configurations with the grating outside the lasing region (Fig.3.2c), often called distributed Bragg reffectors (DBR). The latter is more practical because the grating formation does not interfere with the active zone manufacturing by, for example, molecular beam epitaxy. Recently there are attempts to use the special properties of deeply modulated surface relief waveguide gratings to form forbidden regions for optical mode propagation (optical band-gaps) in order to direct the emission of laser diodes in a narrower cone.

3.10 Fiber Gratings

A special case of waveguide gratings are fiber gratings, usually phase gratings formed inside the core of optical fibers (Fig.3.3). Several methods can be used - holographic recording, lithography, self-interference method, laser chirping (see Chapters 9 and 16), resulting in straight or slanted grooves, with constant or chirped period. The applications vary from input and output coupling and wavelength division to polarization and dispersion compensation. A grating with larger period will couple the guided mode to a radiated wave. Straight grooves radiate a cylindrically symmetrical wave (Fig.3.3a). With slanted direction of the "grooves" the cylindrical symmetry can be broken so that the output is more or less directional (Fig.3.3b). If the period is shorter, the grating can couple modes inside the fiber without radiating in the cladding (Fig.3.3c). They can be different modes in multimode fibers or the same mode propagating in opposite direction for monomode fibers. The latter effect can be used for narrow-band (weak gratings) or broad-band (stronger gratings) mirroring and wavelength selection. Such fibers can serve as deformation sensors when, for example embedded inside large constructions: strain will stretch or compress the fiber so that the grating period deformations will result in modifying the reflected wavelength, which can easily be detected at a distance. Further details can be found in Chapter 9.

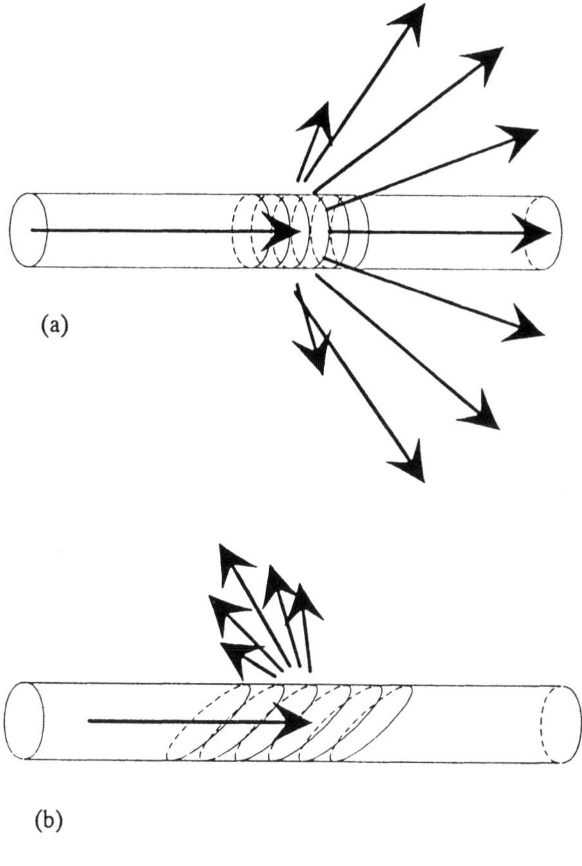

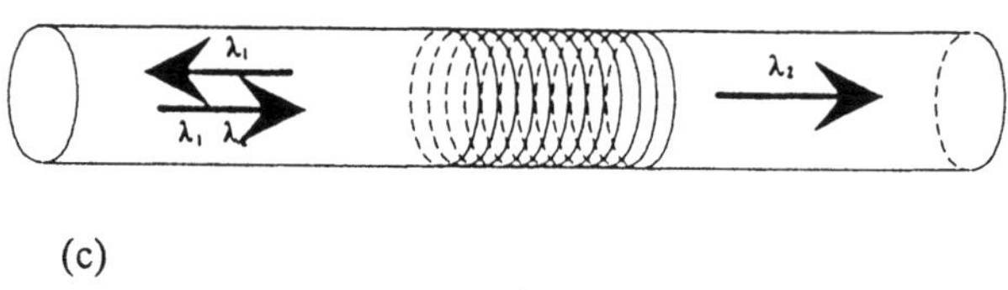

(c)

Fig.3.3 Fiber gratings used for input and output coupling with straight (a) and slanted (b) grooves, and for mode conversion (c).

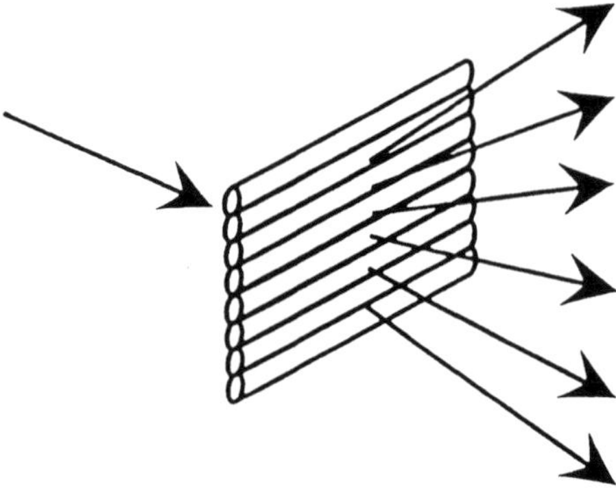

Fig.3.4 Grating consisting of fibers used for multiple beam sampling.

A peculiar application that points out the ambiguity of naming conventions was proposed a decade ago (see ref.5.17). It is also called 'fiber grating', but instead of having a grating in(side) the fiber, it consists of a grating assembly made of simple adjacent fibers used as transmission fanning gratings (Fig.3.4). Two sets of fibers crossed at 90° will serve to generate 2-D pattern (see Chapter 5).

3.11 Binary Gratings

Binary gratings are a special type of photolithographic gratings. They are made by a multistep process that consists of consecutive exposure through masks of reducing period, the ratio between two consecutive mask periods being two, from where the term *binary* appears. The aim is to obtain groove profiles with more complicated forms than the lamellar one given by the natural tendency of the photolithographic process. The lowest limit comes from the shortest period which is reproducible by photolithography.

3.12 Photonic Crystals

A multilayer plane mirror with layers of $\lambda/4$ optical thickness can be made to reflect all the incident light over a specific spectral region. The width of such a region increases with the difference between the optical index of consecutive layers. However, the effect has angular selectivity that may be undesirable. To improve on this, the multiple stack can be replaced by a volume grating consisting of periodically arranged particles within a transparent medium. The term *photonic crystal* is taken from solid state physics. These devices are supposed to totally reflect incident light over a large range of incident angles (including normal incidence) within a specified spectral range, called *band-gap* (*photonic stop-band*). Introducing some disorder in the arrangement of 'atoms' can allow light to pass the 'crystal' in a very narrow spectral interval inside the band-gap, forming an 'impurity level'. Such phenomena fall outside the topics of this book, although similar beavior in waveguide and fiber gratings is discussed in Chapter 9.

3.13 Gratings for Special Purposes

Experimental requirements sometimes call for gratings with special properties or extreme quality. They can be considered a piece of art in science and engineering and only few of them are mentioned here.

3.13.1 Filter Gratings

Gratings are sometimes used as reflectance filters when working in the far infrared, as a convenient tool for removing second and higher orders from the light incident on the grating. For this purpose, small plane gratings are used which are blazed for the wavelength of the unwanted radiation. The grating acts as a mirror, reflecting the wanted light into the instrument while diffracting shorter wavelengths out of the beam.

3.13.2 Gratings for Electron Microscope and Scanning Microscope Calibration

Lightly ruled masters with space left between the grooves can serve to produce carbon replicas that act as scales for calibrating the magnification of electron microscopes. Line frequencies of up to 10,000 gr/mm have been produced experimentally. Besides having a great variety of spacings, they can be ruled in two sets of grooves at right angles so as to form a grid that shows up distortion of the field.

3.13.3 Electron Interaction Gratings

Electron beams passing close to and across a metal grating will generate light from interaction with the grooves. The heat generated in this process makes it necessary to use original gratings which are typically ruled in silver deposited on stainless steel blanks. Small gratings can be directly ruled in polished stainless steel, but diamond tool wear limits diffraction efficiency attainable. A more suitable material is electroless deposited nickel.

Heat resistance is required also for gratings used in high-power lasers and which often use copper blanks for good heating sinking. For maximum power they are used in the form of master ruled into the metal, rather than as replicas. Such solutions depend strongly on the spectral region and experimental scheme (see Chapter 13).

3.13.4 Rocket and Satellite Spectroscopy

Much of the early work in space was done with original gratings (masters), but it has since been shown that cast plastic replicas do not suffer any degradation even over extended periods of time in space. The advantage of replicas lies not only in lower cost, but in availability of exact duplicates whenever needed.

3.13.5 Metrology

Standard high precision gratings are ruled with groove spacing precision close to 10^{-6}. Exact values of absolute spacing are of no consequence spectroscopically. In metrological applications there is need to maintain absolute spacing to less than 10^{-6}, and demands corresponding calibration and temperature control.

3.13.6 Synchrotron Monochromators

In synchrotrons, vacuum levels of 10^{-12} torr are aimed for. This is a problem because standard replica gratings cannot be subjected to the high temperature bake-out as with other components. This points to original gratings etched into fused silica.

3.13.7 X-Ray Gratings

The X-ray region imposes special requirements on diffraction gratings for spectrographs and monochromators. Most of the materials used in grating applications are practically transparent so that gratings are used in grazing incidence to enhance the reflectivity. Combined with the very small wavelength-to-period ratio this results in very low efficiency levels (typically few percents). Blazed profiles and multilayered reflection coatings are used to significantly improve the performance, although the surface roughness and the deviation from the blazed profile are critical at these short wavelengths.

3.13.8 Chemical/Biological Monitoring

Small changes in refractive index are the hallmark of certain chemical or biological process. These may be detected by placing a few drops of solution and generating a thin layer on a small disposable grating that is illuminated so that diffraction lies in the anomaly region. Even small index changes result in significant changes in the diffracted beam intensity, which are readily detected.

3.14 "Good" and "Bad" Gratings

Good gratings are those that satisfy the apparent needs of the user. This may seem obvious, but unfortunately it can lead to many misunderstandings. For example, the eye can easily pick up cosmetic defects that have no measurable influence on spectrometric performance, since the latter is an integrated effect. This is especially true of fine pitch gratings, where relatively trivial changes in local efficiency manifest themselves in a change of color appearance of the zero order to which the eye is very sensitive.

Sometimes high efficiency is stressed when wavefront quality or signal to noise ratio is more important, and vice versa.

Stray light effects are notoriously difficult to assess or quantify, because their influence depends so much on the nature of the light source, the detector, and even on the design of an instrument case and its baffles. Ghosts in modern gratings are so low that they rarely constitute a problem in instruments, as they were before the introduction of interferometric control.

References

N. S. N. Nath: "The diffraction of light by supersonic waves," Proc. Ind. Acad. Sci. A **4**, 222-242 (1936).

C. V. Raman and N. S. N. Nath: "The diffraction of light by high frequency sound waves." Part I: Proc. Ind. Acad. Sci. A v.**2**, 406-412 (1935), Part II: ibid, v.**2**, 413-420 (1935), Part III: ibid, v.**3**, 75-84 (1936), Part IV: ibid, v.**3**, 119-125 (1936), Part V: ibid, v.**3**, 459-465 (1936).

Chapter 4

Efficiency Behavior of Plane Reflection Gratings

4.1 Introduction

Grating efficiency is defined as the fraction of incident monochromatic light diffracted into a specific order. It is one of the most important and basic attributes of a grating — there are few applications where it is not a factor in the functioning of a spectrometric instrument. What generates so much concern is that efficiency is rarely constant, but varies considerably as a function of wavelength. Thus a typical UV–VIS monochromator could start at the short wavelength end with an efficiency of 50%, increase to a maximum of 75%, and gradually decrease to 10% at longer wavelengths, clearly variations that cannot be ignored. This behavior is not due to any lack of quality, but is the natural result of the interaction of light with a modulated metallic surface and the fact that light will always be diffracted into a minimum of two orders whenever there is useful dispersion. The number of orders can increase to as many as 100 or even more. One can imagine a silent competition between orders for photons. Just how this works out for a wide variety of practical gratings will be shown in this chapter.

At low angles of diffraction the division of energy between orders can be analyzed as a scalar phenomenon, but a large fraction of gratings operate at angles where electromagnetic theory is required to explain the behavior [4.1]. The most obvious evidence of electromagnetic behavior is that the diffracted light is partially polarized, because diffraction efficiency is different in the two planes of polarization, which are defined as having the electric vector parallel to the grooves (TE, P or s designations) or perpendicular to the grooves (TM, S, or p designation). The choice of designations, especially the S and P, is sometimes a problem, due to historical usage. The lower case s and p designations are commonly used to describe mirror reflections, where polarization is defined with respect to the plane of incidence. For gratings it was considered more descriptive to use the groove direction as the reference, which is at right angles to the plane of incidence. The choice of S, based on the word Senkrecht, German for perpendicular, was capitalized to make it different. For most reflection gratings the difference between the efficiency behavior between the two planes is so marked that rarely will one be mistaken for the other.

For some applications the polarization properties have a real advantage, in others it is a nuisance to be tolerated or suppressed. In addition there are angular regions where the efficiency, especially in the TM plane, undergoes such sharp changes that it is termed anomalous, even though there is no longer any theoretical mystery attached (see Chapter 8) [4.2, 3]. While usually considered annoying, there are special applications where they turn out to be an asset.

To understand how this non-monotonic behavior is reflected in designs we must appreciate what aspects of a grating dictate its performance, the purpose of this chapter. The picture is complicated because there are so many different factors which can interact, sometimes in complex ways. Each of the factors will be considered and enough data presented to obtain an insight into the interactions. This will be done with a set of over 40 families of efficiency curves chosen to serve as typical examples. To present a still more coherent picture would require an almost infinite amount of data which would be overwhelming rather than instructive.

The two single most important determining factors are the ratio of wavelength to groove spacing (λ/d) and the groove depth modulation (h/d). Next in importance is the groove geometry itself, followed by the nature of the metal surface and its optical properties (as a function of wavelength). There is an interaction with the optical system, or mount, which derives from the choice of incident angle (for spectrographs) and the angular deviation (A.D.), the angle between incident and diffracted beams, as used in monochromators.

In almost every case it will be important to distinguish between the two planes of polarization. To collect the required data experimentally would be an enormous undertaking, but we can take advantage of accurate theoretical calculations with which to explore the behavior of ideal gratings under a wide variety of conditions.

To make the resulting information as directly useful as possible, the bulk of the graphs are plotted as a function of wavelength for 1200 gr/mm frequency, which is by far the single most used, covering the UV-VIS region so well. Since other groove frequencies can be fairly well approximated by simple ratioing this seems more useful than plotting figures against the usual alternative in the form of the dimensionless ratio λ/d.

Because of wide spectral band coverage and minimal order overlap, 90% of all gratings are used in first order, and thus the bulk of the data presented is for this order. However, enough data will be shown for orders 2 to

4 to enable adequate judgments to be made. Gratings that are used in high orders only, i.e., *echelles*, are covered separately in Chapter 6. Transmission gratings are rarely used in standard instruments, and their role is covered in Chapter 5.

To distinguish low order gratings from echelles they have often been termed *echelettes*, from the French diminutive for staircase, the well known profile for blazed gratings. However, the chapter also includes a survey of sinusoidal grooves, as made by the interference (holographic) process.

Some 95% of all gratings have an aluminum surface, because no other metal combines so many useful properties, both optical and mechanical. For that reason the bulk of the data presented here is for this metal, rather than the infinite conductivity surface often used in the past (but included for reference). Absolute efficiencies are used, since that is the quantity needed for instrumental applications. In all cases both TE and TM efficiencies are reported. For unpolarized systems one simply takes the arithmetic average between the two.

Using Table 4.1 the reader can directly locate the figures corresponding to the grating of interest without following in detail the entire chapter.

Table 4.1 Number of the figures presenting efficiency in the first order.

Echelette gratings:

gr/mm	Perfectly conducting	Al	Au	Ag
1200	4.3, 4.16 - 4.21 (higher orders: 4.11 - 4.15)	4.4 - 4.10, 4.27 (apex angle)	4.23	4.23
1800	-	4.22	4.24	4.24
2400	-	4.22	4.25	4.25
3600	-	4.22	4.26	4.26

Sinusoidal gratings:

gr/mm	Perfectly conducting	Al	Au	Ag
1200	4.28 (higher orders: 4.36 - 4.38)	4.29 - 4.35	-	-
1800	-	4.39, 4.49, 4.50	4.45	4.42
2400	-	4.40	4.46	4.43
3600	-	4.41	4.47	4.44

4.2 General Rules

4.2.1 Reflection Coatings

The distinguishing boundary between reflection and transmission seems quite firm, although there are a few exceptions to prove the rule. One such case is with the grating that works in total internal reflection (see Chapter 5.11). Although transmission gratings can reach the highest absolute efficiency values, reflection gratings are far more common. This is because they have only one working surface and operate in a non-dispersive environment (air or vacuum). However, their properties are determined not only by groove shape, depth and period, but also by the optical properties of the substrate (its refractive index). Only rigorous electromagnetic theory can tell whether diffraction efficiency can be determined by simple multiplication of material reflectivity with the efficiency of an equivalent perfectly conducting grating. Although usually true, there are several important exceptions:

1. In the so-called resonance domain, where the wavelength is of the same order of magnitude as the period, guided wave excitation (plasmon wave along the metallic surface, or leaky wave in the covering dielectric layer) can significantly destroy performance. Many such examples, including grating induced total absorption of light by an otherwise highly reflecting metallic surface are discussed in Chapter 8.

2. Only rarely can efficiency exceed the reflectance of the corresponding plane mirror (see Fig.8.12), but this one example in no way contributes to the frequently expressed optimism that a combination of multilayer dielectric stacks on a highly efficient metal or dielectric grating can revolutionize grating performance. As in other instances nature is beyond such simple reasoning, and obeys more complex laws. We must realize that any multilayer coating thick enough to efficiently reflect light is also capable of supporting many waveguide modes, which are readily excited by the grating, unless the λ/d ratio is very low. As a result in the visible or near UV, multilayer coated gratings tend to be characterized by multiple resonance anomalies (Fig.8.12) which makes them useless, except perhaps over a highly restricted spectral interval. Fortunately, resonance phenomena play a minor if not negligible role in the X-ray domain where multicoated gratings are widely used.

3. Relatively deep corrugations can lead to significant non-resonant absorption of light when the reflectivity of the substrate is low. This can be observed when gold coated gratings are used in the 500 – 700 nm domain where they are not normally applied. In this instance the low reflectivity region is extended to longer wavelengths than expected from comparison with the reflectivity of a plane mirror.

4. Going from the visible to the X-Ray domain in grazing incidence, diffraction efficiency is always lower (sometimes one or two orders of magnitude) than expected from simple considerations of material reflectivity and grating efficiency.

Fig.4.1 (pp.76 and 77) shows the reflectivity of nearly all the materials used as reflection coatings, extracted from several references listed under [4.4]. At long wavelengths (IR to microwave) there is no problem in making a choice, except if *low* reflectivity is required, like suppressing radar response. Above 700 nm in the near IR gold is commonly used to obtain maximum efficiency, since silver is too fragile. Solid copper gratings are often used in high power IR laser applications because of its high thermal conductivity which increases damage resistance.

In the visible and near UV (above 200 nm) it is easy to see why aluminum is the most used metal. Below 200 nm the oxide layer gradually loses its transparency, but its formation can be prevented by overcoating the aluminum with a thin layer of MgF_2, which in turn remains transparent down to 120 nm. Replacing MgF_2 with LiF reduces the lower limit to 110 nm. Below this limit gold and other heavy metals become interesting, for example platinum and osmium. At wavelengths below about 25 nm virtually all materials are transparent at normal incidence, forcing reliance on grazing incidence, which leaves the question about the concept of refractive index for describing material optical properties. Fortunately, comparison between experiment and rigorous electromagnetic theory has shown the validity of this concept as low as 1 Å. Recently it has been found that multilayer structures, for example tungsten and carbon, are useful as grating coatings in this domain, which provides welcome relief from grazing incident mounts.

4.2.2 Scalar Behavior of Reflection Gratings

It is generally assumed that in the short–wavelength limit reflection gratings can safely be characterized by scalar theory, although on many occasions greater accuracy is needed. It turns out that there is no approximate method from which one can determine deviations. However, scalar theory predictions serve as a good rule of thumb, especially for echelette gratings.

As discussed in Chapter 10.4 the direct consequence of eq.(10.3) is that maximum efficiency in the N–th order of echelette gratings can be expected when it is diffracted as if being reflected by the facet. In Littrow mount this points out the simple relation between the facet angle φ_B and the blaze wavelength λ_B:

$$2\sin\varphi_B = \frac{\lambda_B}{d} \quad , \qquad (4.1)$$

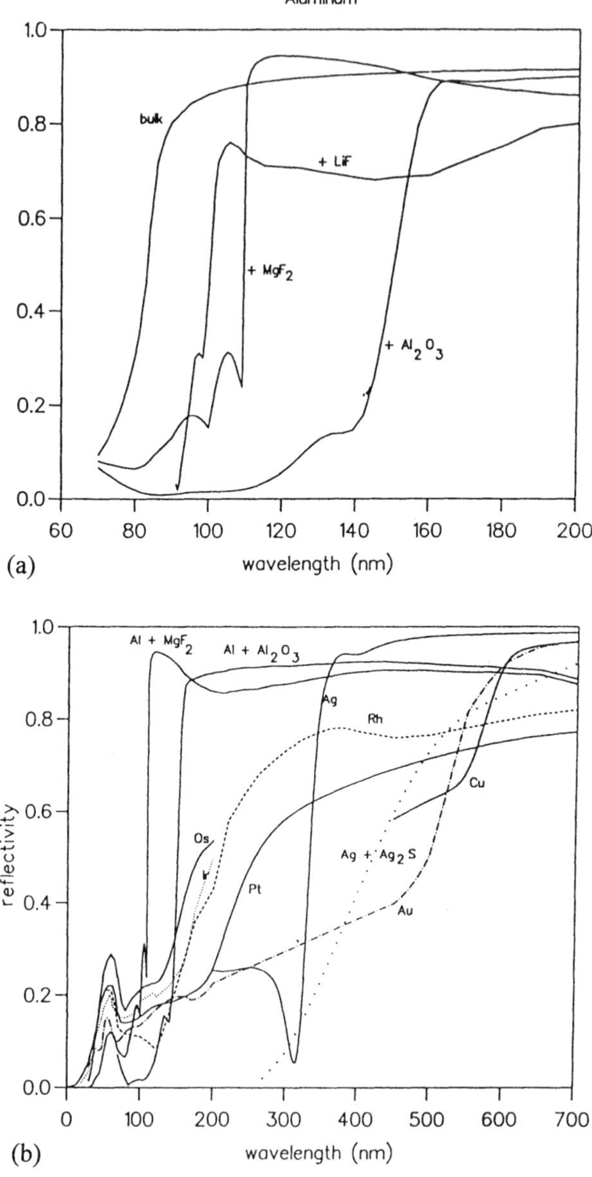

Fig.4.1 Reflectivity at normal incidence of metals useful for gratings as a function of wavelength: a) Reflectivity of Al and the useful metals over the region 20 to 150 nm. b) Similar to (a) except the wavelength scale is expanded to 700 nm.

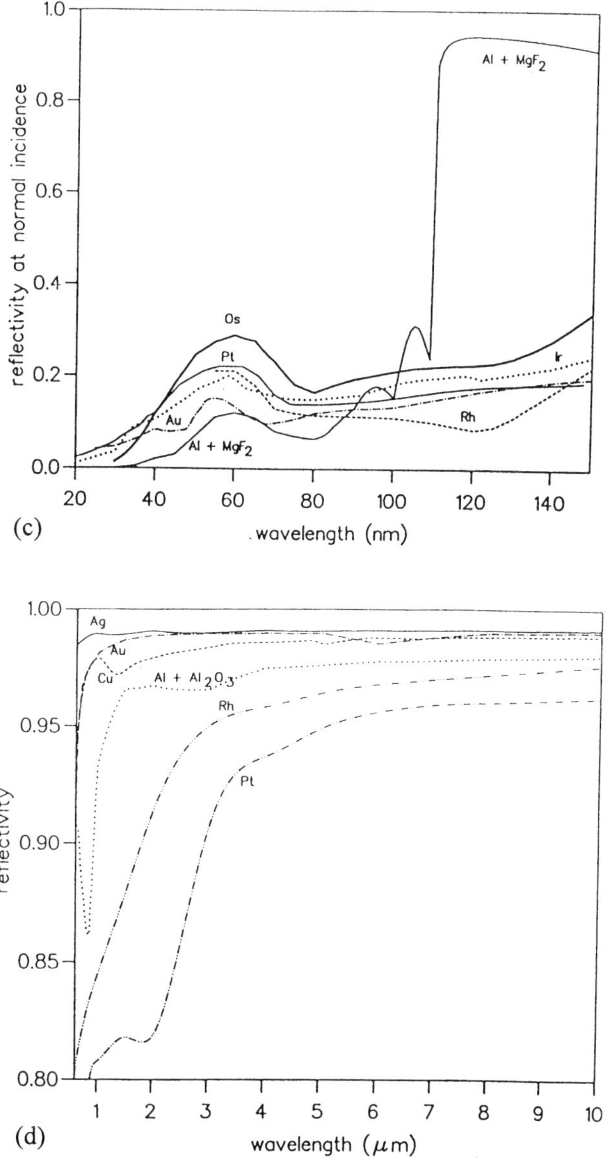

c) Reflectance of metals in the IR region from 0.7 to 10 μm. Note the expanded reflectivity scale. d) Reflectance of Aluminum, bulk and with normal oxide coat, as well as with 25 nm thick protective coats of MgF_2 and LiF.

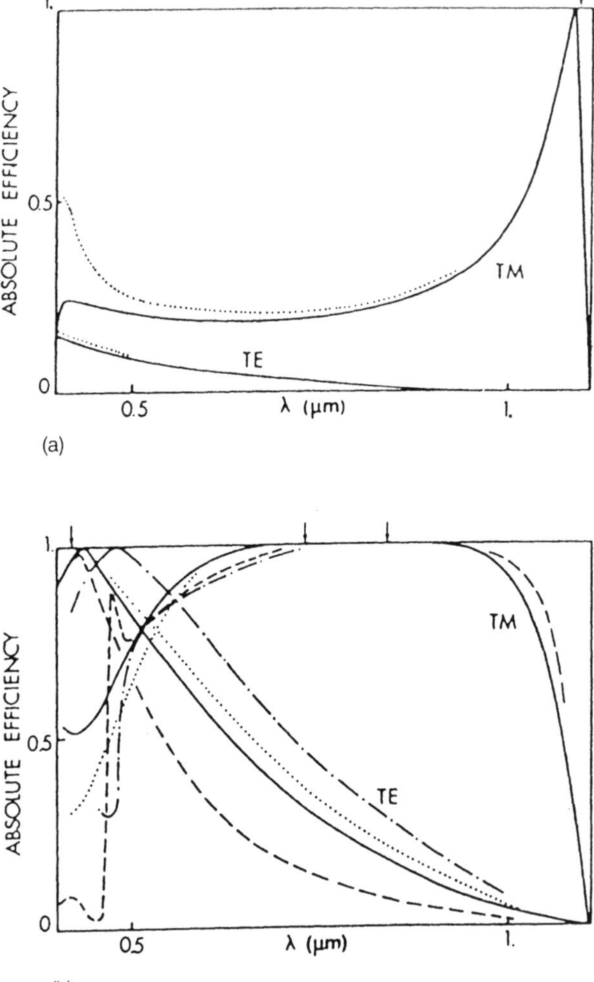

Fig.4.2 Absolute efficiency as a function of wavelength of 1800 gr/mm perfectly conducting grating with different groove profiles (after [4.6]):

a) ——— sinusoidal: h/d = 0.10,
- - - - - - lamellar: h/d = 0.07854,
·········· ruled: $\varphi_B = 9°$, apex 90°,
-·-·-·-·- ruled: $\varphi_B = 18.5°$, apex 150°.

b) ——— sinusoidal: h/d = 0.23,
- - - - - - lamellar: h/d = 0.1806,
·········· ruled: $\varphi_B = 69.5°$, apex 90°,
-·-·-·-·- ruled: $\varphi_B = 86°$, apex 74°,
oooooo ruled: $\varphi_B = 28°$, apex 120°.

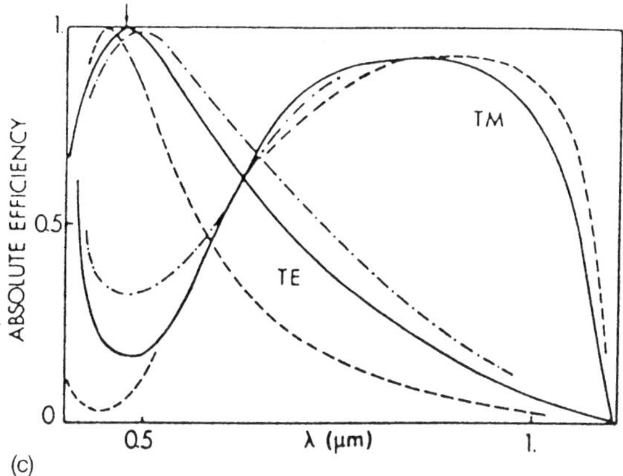

(c)

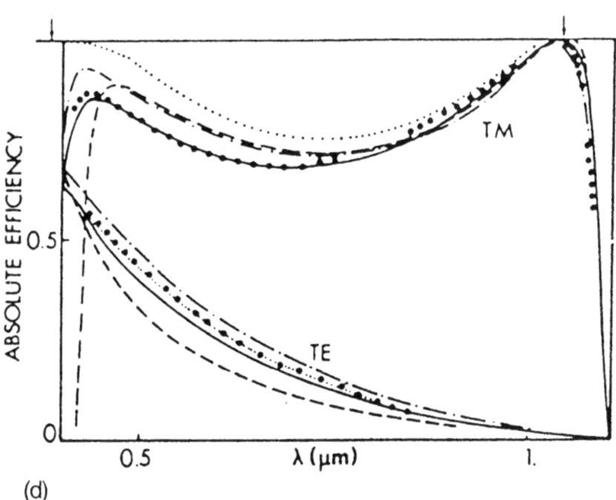

(d)

c) ——— sinusoidal: h/d = 0.40,
- - - - - lamellar: h/d = 0.3142,
········· ruled: $\varphi_B = 41°$, apex 90°,
-·-·-·-·- ruled: $\varphi_B = 32°$, apex 60°.

d) ——— sinusoidal: h/d = 0.50,
- - - - - lamellar: h/d = 0.3927,
······· ruled: $\varphi_B = 43°$, apex 74°.

and the angle of incidence is equal to the facet angle.

This simple rule holds rigorously for perfectly conducting gratings in the TM plane, according to the Marechal–Stroke theorem [4.5]. TE polarization, finite conductivity, and departure from Littrow have several effects: degradation of efficiency maximum of the blazed order, increase of energy leakage into other diffraction orders, and shift of the maximum position. However, for highly conducting echelette gratings with relatively low blaze angles and moderate angular deviation from Littrow, eq.(4.1) predicts the blaze wavelength quite well.

The scalar limit for sinusoidal gratings is given by eq.(10.5) and predicts a maximum of 33.4%, a value well know from experiments. Higher values can readily be obtained when the number of propagating orders is reduced, but then only electromagnetic theory can give reliable results.

4.2.3 Gratings Supporting Only Two Diffraction Orders: The Equivalence Rule

Gratings with a single dispersive order are of special interest because they are associated with high dispersion, wide free spectral range, and the ability to preserve high efficiency values over a large spectral interval. They rarely have scalar behavior, except when the grooves are very shallow, seldom a case of interest. While perhaps trivial it is worth recalling that the number of propagating orders is not a characteristic of the grating but depends on the wavelength–to–period ratio.

When comparing the efficiencies of highly reflecting gratings with only two propagating orders, the curves for different groove profiles have a common behavior, at least with commercially available modulation depths. This similarity is expressed in the so–called *equivalence rule* [4.6]. It holds for a variety of groove profiles, with the only limitation the requirement for a center of symmetry, so that the profile function f(x) can be expanded in Fourier sine series:

$$f(x) = \sum_{n=0}^{\infty} f_n \sin\left(n\frac{2\pi}{d}x\right). \tag{4.2}$$

Holographic gratings, ruled echelette gratings, and lamellar gratings with a 0.5 filling ratio all fit this limitation. According to the equivalence rule the efficiency of such gratings in the spectral region where only the 0–th and –1st orders propagate is determined mainly by the value of the first Fourier harmonic f_1 of the profile. Sinusoidal, echelette, laminar, etc. gratings with equal values of f_1 will have similar efficiencies. This is not a rigorous theorem but an empirical rule that has a strong argument in the perturbation theory for shallow gratings. Fortunately it holds more or less for commercial gratings, at least if an error of 10 – 20% is acceptable (Figs.4.2a – d). Due to the theorem of Marechal and Stroke the error in the TM plane, where the efficiency is higher, is much smaller and the rule can be used to provide an idea of grating behavior.

4.3 Absolute Efficiencies of 1200 gr/mm Aluminum Echelettes

Absolute efficiencies of 1200 gr/mm echelettes are shown in the figures as a function of wavelength from about 0.1 to about 1.67 μm, the maximum wavelength at which diffraction is possible under Littrow conditions, but reducing as A.D. increases, in Figs.4.3 to 4.15. Note that the abscissa does not always start exactly at zero and wavelengths will need to be extrapolated backwards if accurate scaling is attempted. Fig.4.3 is special in that the grating surface is considered infinitely conducting (perfectly reflecting) for reference purposes. It should be compared to Fig.4.5, which shares the same A.D., but accounts for the complex index of aluminum. Both of these figures differ from the rest in that they contain data for six blaze angles, from 5° to 48°, while all the remaining figures are limited to just four blaze angles, from 9° to 36° in logical steps that cover values most often used in practice. Each of the figures differs from the rest by a progressive change in A.D. from 0° (Littrow) to 90°.

4.3.1 Discussion of Efficiency Behavior of 1200 gr/mm Echelettes

All the efficiency curves demonstrate that values start low at short wavelengths, increase with varying degrees of smoothness to a peak, and then decrease. In the TE plane the decrease is always monotonic, reaching zero when the angle of diffraction becomes 90°. In the TM plane behavior is much more complex, due to its response to higher orders passing off, as well as resonance effects (see Chapter 8 for details).

Fig.4.3 and 4.5 display the classical behavior of echelette gratings, with 8° chosen for A.D. At low blaze angles, such as 5°, polarization effects are minimal, and the curve takes on the scalar shape, except for anomalous spikes in the TM plane. High efficiency is limited to a rather narrow band, say from 0.18 to 0.4 μm, or a ratio of just 2:1. The effect of aluminum is no more than to suppress efficiency values by roughly the reflectance of the metal.

The 9° groove angle curves are of special interest because they represent the single most frequently used blazed grating. The 1200 gr/mm frequency puts it in the most popular wavelength region and the 9° angle provides the widest usable wavelength band obtainable in one order, from 0.2 to 1.5 μm in

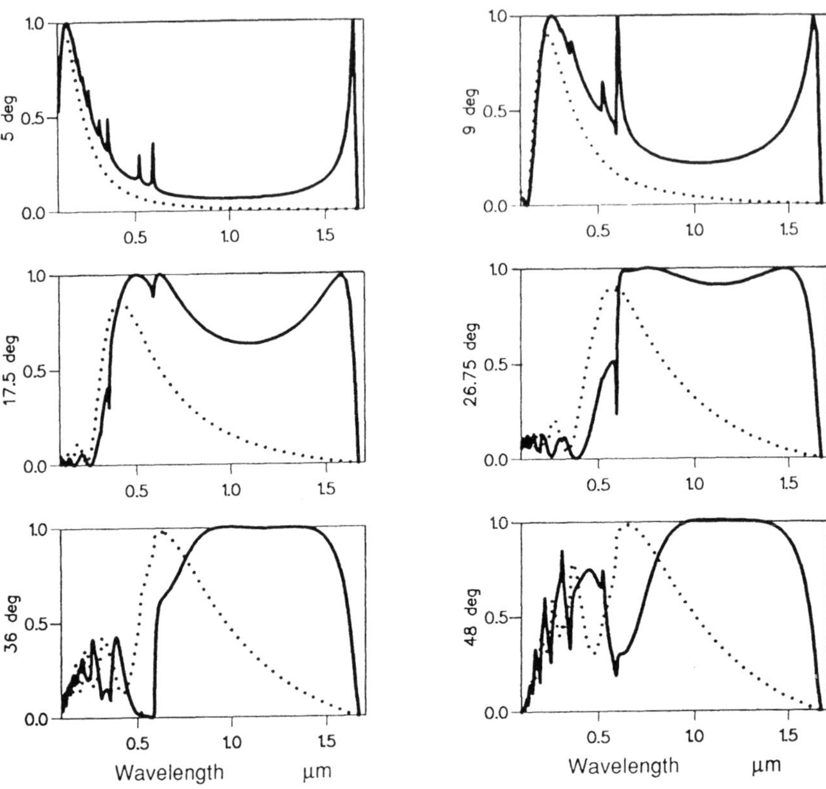

Fig.4.3 Absolute first order efficiencies of 1200 gr/mm echelette grating as a function of wavelength in µm. Numerical values are for perfect conductivity and angular deviation 8°. Solid lines for TM and dotted lines for TE polarization efficiencies as a function of wavelength in µm. Six blaze angles as marked.

unpolarized light. This 7.5 to 1 ratio is not attainable with any other groove angle or groove shape.

The 17.5° groove or blaze angle indicates a second interesting special quality, which is the relatively low contribution of the TM plane anomalies. It still has a respectably wide wavelength coverage of about 4:1, a number which naturally depends on the lowest efficiency that is considered acceptable. A small change to 20° (not shown) is enough to reduce the main anomaly still further, because in this case the peak of the TE plane coincides with the main anomaly spike near 0.6 µm.

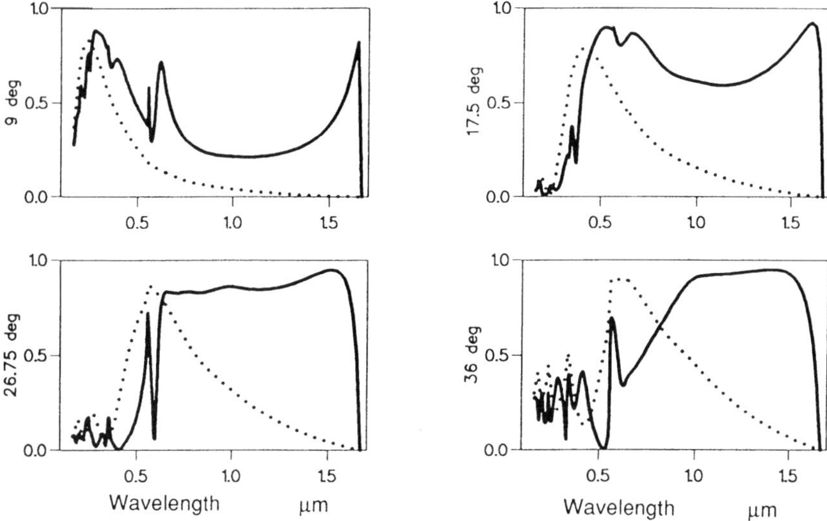

Fig.4.4 Same as Fig.4.3, except for aluminum surface and A.D. 0°, with four blaze angles, as marked.

When the blaze angle reaches 26.75° the character of the TM curve changes quite drastically. There is a sharp anomaly in TM plane before it comes to its peak, but it is followed by a long wavelength span (0.7 to 1.5 μm) over which the efficiency remains substantially constant (85 to 90%). This is a valuable attribute for applications that operate naturally in polarized light, such as laser wavelength tuning.

This last attribute continues at higher blaze angles, as can be judged from the 36° curve, although the wavelength range is somewhat reduced. If the groove angle is increased further still, for which the 48° curves are an example, the efficiency swings in first order are so great as to make them only marginally useful. In practice the unusual difficulty of attaining good groove geometry at these angles adds to the custom of restricting their use to higher orders.

The TE curves always have a much more monotonic behavior, except for the anomalous appearance at 48° blaze. The efficiency peaks in this plane always occur at wavelengths less than the peaks predicted by scalar considerations, while for TM they always occur later. Through an accident of nature the average of the two comes surprisingly close to the simple scalar prediction. In the TE plane we can expect virtually all the light diffracted to add up to 100%, reduced only by reflectance losses. When the diffraction angle exceeds about 20° only the zero and first orders can diffract, and their sum will remain constant. In the TM plane the situation is much more complex, due to resonance effects. A certain amount of light is likely to be absorbed by the grating and turned into heat, but in amounts that vary considerably with wavelength, or the λ/d ratio.

When TE and TM curves are arithmetically averaged the combined peak happens to remain close to the one predicted by scalar theory, or simply twice the groove depth, even well into the electromagnetic domain. In the past this sometimes masked the need to examine all gratings in both planes of polarization. The peak values in the TE plane for aluminum gratings differ little with groove angle (Fig.4.5), but for perfect conductivity (Fig.4.3) TE efficiencies reach 100% for both the smallest and largest groove angles, while for intermediate angles (9 to 26°) it does not exceed 90%.

The choice of angular deviation A.D. is an important one to instrument

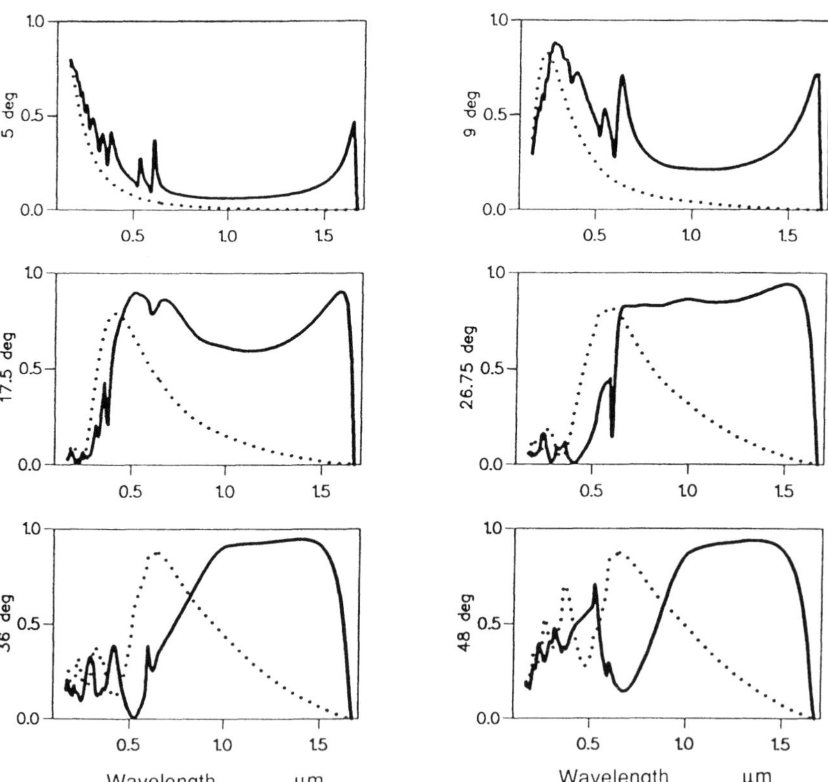

Fig.4.5 Same as Fig.4.4, except A.D. 8°.

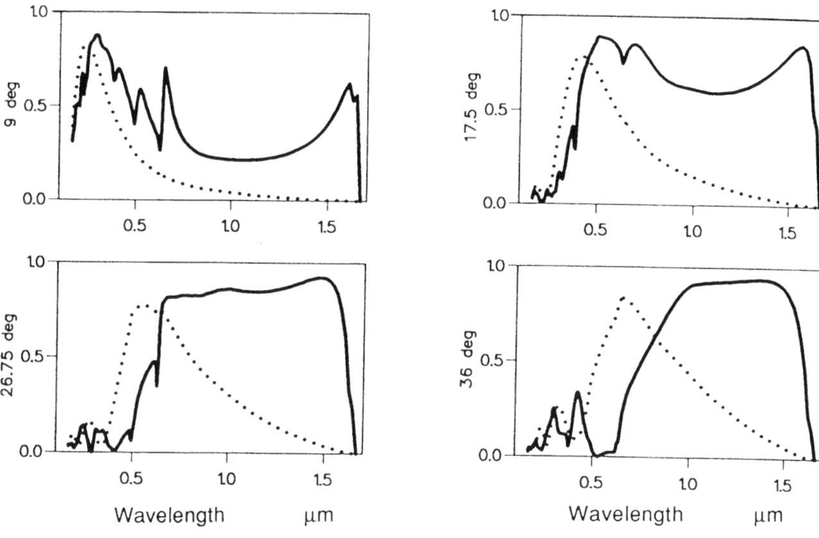

Fig.4.6 Same as Fig.4.4, except A.D. 15°.

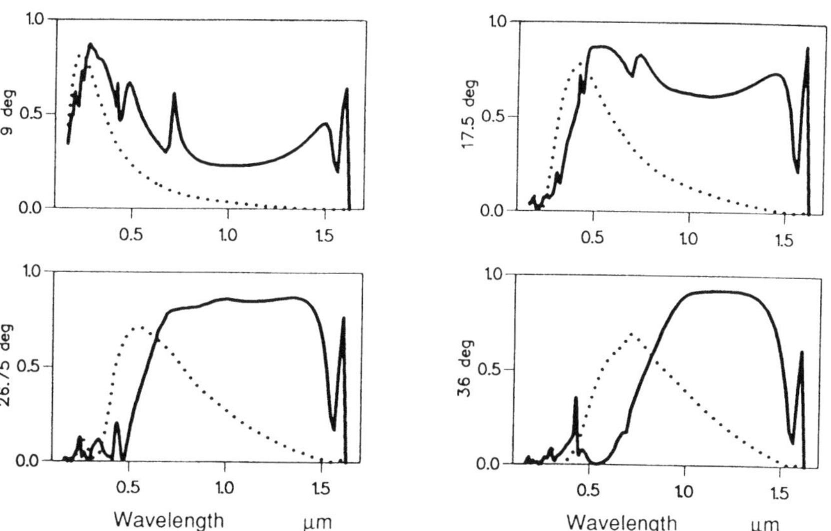

Fig.4.7 Same as Fig.4.4, except A.D. 30°.

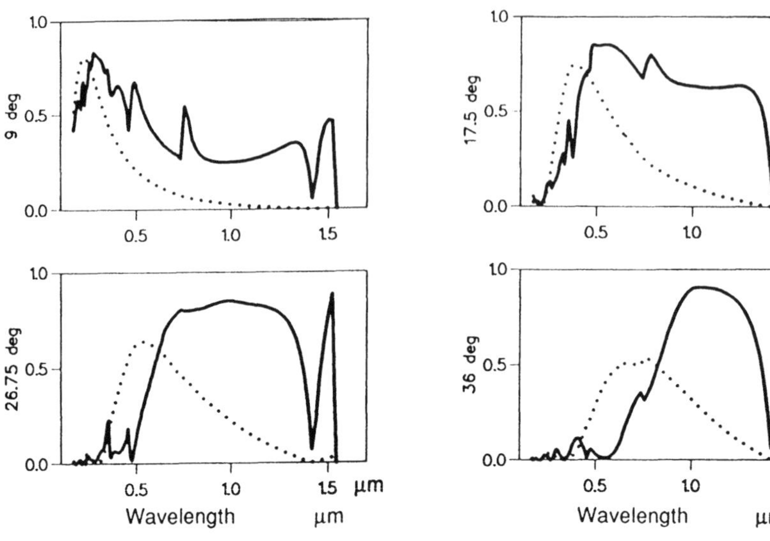

Fig.4.8 Same as Fig.4.4, except A.D. 45°

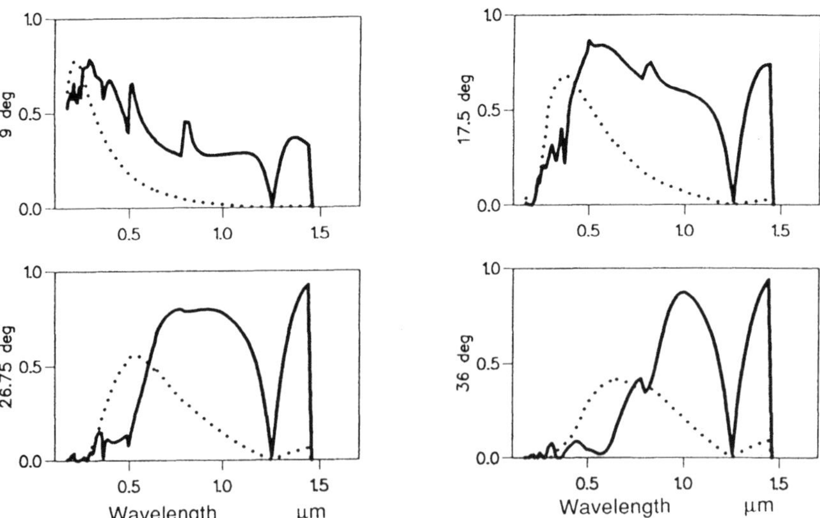

Fig.4.9 Same as Fig.4.4, except A.D. 60°.

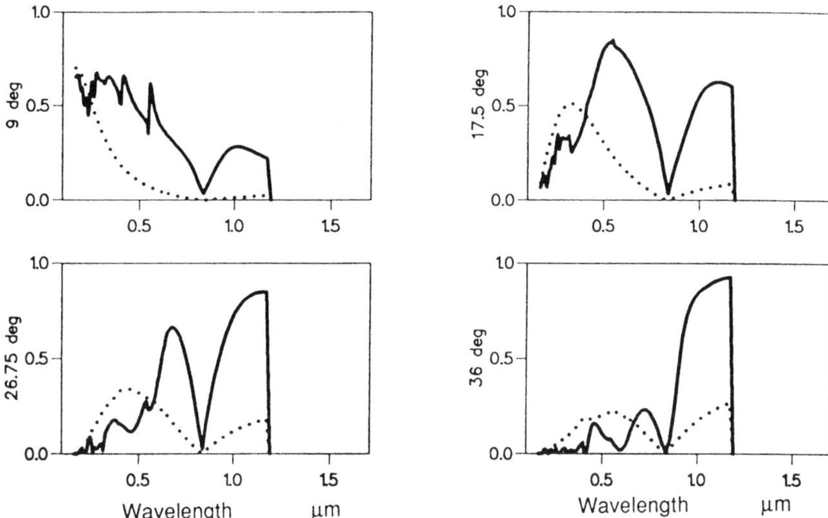

Fig.4.10 Same as Fig.4.4, except A.D. 90°.

designers. The angle must be large enough to enable proper beam separation, at least 8°, but not so large as to reduce attainable efficiencies. Comparing Figs.4.4 to 4.10 gives direct insight into what happens as A.D. increases from 0 to 90° for each of the four standard groove angles chosen. For 5° blaze the most prominent effect is to increasingly separate the anomaly spikes (a direct response to the grating equation), but otherwise leaving efficiency little changed, except for a sharp drop when A.D. goes to 90°. For the 17.5° angles the effect of increasing A.D. is small, again until the 90° angle is reached. The 26.75° grating behavior is interesting because at A.D. 30° anomalies disappear, at the expense of some efficiency reduction, the effect still stronger at 45°; as usual 90° appears to be useless. With the steeper blaze angles, like 36°, an A.D. of only 15° is already sufficient to suppress anomalies, by greatly lowering the TM efficiency in the wavelength region where they would occur. An increase in A.D. to 30° changes the picture only slightly, but beyond 45° efficiency is sharply reduced. In other words, the deeper the grooves the greater the influence of A.D.

One important aspect of going to larger angles of deviation, which is easily overlooked, is that the maximum wavelength at which diffraction can take place becomes progressively less. This follows directly from the grating equation. The picture becomes clearly visible by comparing Figs.4.9 and 4.10 with 4.4, or Figs.4.34 and 4.35 with 4.29.

4.3.2 *Reflection Efficiencies of 1200 gr/mm Echelettes in Orders 2, 3 and 4*

Reflection efficiencies in orders 2 to 4 are presented in Figures 4.11 to 4.15, again for 1200 gr/mm gratings. To maximize the amount of information in a given space the TE and TM curves are shown separately, which enables the three orders to be shown superimposed. One additional modification in the display is necessary to achieve an orderly arrangement, which is to plot the abscissa as a function of $m\lambda/d$, where m is the order, λ the wavelength in micrometers, and d the grating spacing. The effect is to superimpose the efficiencies, with the peaks sharing the same values of $m\lambda/d$. This would result in some confusion if it were not for the property of successively higher orders to cover a narrower angular range.

Each figure contains data from the same set of blaze angles as before, 5° to 48°, but calculated for infinite conductivity, because including the effects of complex index, which varies with λ, might interfere with the clear family relationships. The role of A.D. is covered for the same five angles as above, one in each figure. At the smaller blaze angles it is convenient to put two sets of curves on one graph, without problems of superposition, only the 38° and 48° needed to be separately displayed. As a result, 18 pairs of curves can be shown in one figure. Note that the abscissa scales are contracted for low blaze angles in order to show details.

Peak efficiencies are easily located by noting that they occur when $m\lambda/d=2\sin\theta_B$, for A.D. = 0. In the TE plane only at high blaze angles do peaks reach 100%, reducing to 90% at 26.75°, but increasing again to 95% at 5°. There is a tendency for higher order peaks to shift slightly to longer wavelengths, the more so with higher groove angles. Interesting are the efficiency swings below $m\lambda/d$ of 1.0 for the deep grooves. In TM plane efficiency peaks always reach 100%, and do not drift with groove angle, as a direct consequence of the Marechal – Stroke theorem. When compared to experimental data of relative efficiency, conformance is always good, except for TM values at 36° and 48° blaze. This is traceable to the high sensitivity at these steep angles to small groove shape departures from the ideal triangles assumed in calculations.

4.3.3 *Effect of A.D. on Peak Efficiency Values and Location in Orders Two to Four of 1200 gr/mm Echelletes*

In observing the figure families for the influence of A.D. we can see that for blaze angles of 5° to 26.75° raising A.D. from 0 to 30° has relatively minor influence, except for the anomaly reduction in TM plane for the two higher blaze angles. At 36° and 48° angles, the second order efficiency is seen visibly

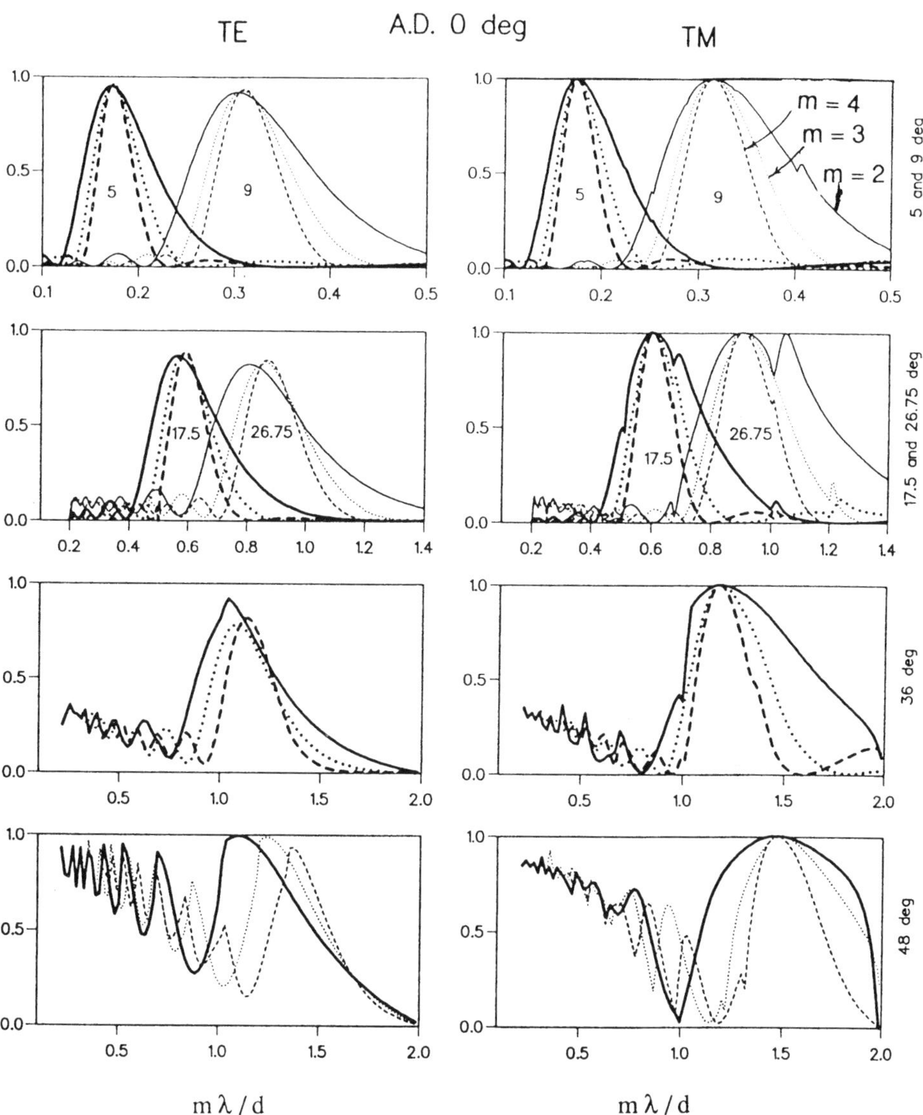

Fig.4.11 Absolute efficiency of 1200 gr/mm echelette in orders 2 to 4. TE on left, TM on the right. Wavelength in mλ/d, four blaze angles as marked. Order symbols marked on TM 9° groove angle grating. A.D. = 0°. Perfect conductivity assumed.

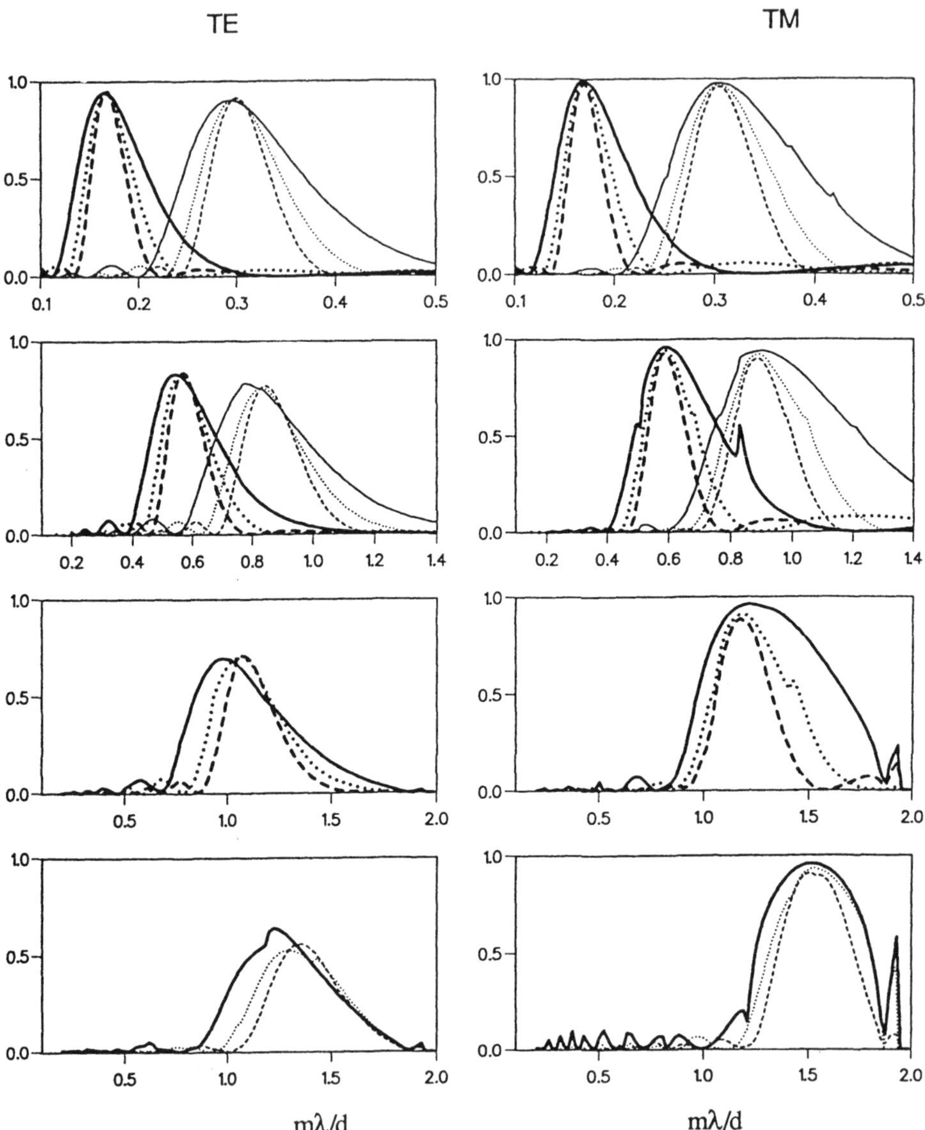

Fig.4.12 Same as Fig.4.11, except A.D. 30°. In sequence, top to bottom, blaze angles are [5 and 9°], [17.5 and 26.75°], [36°], [48°].

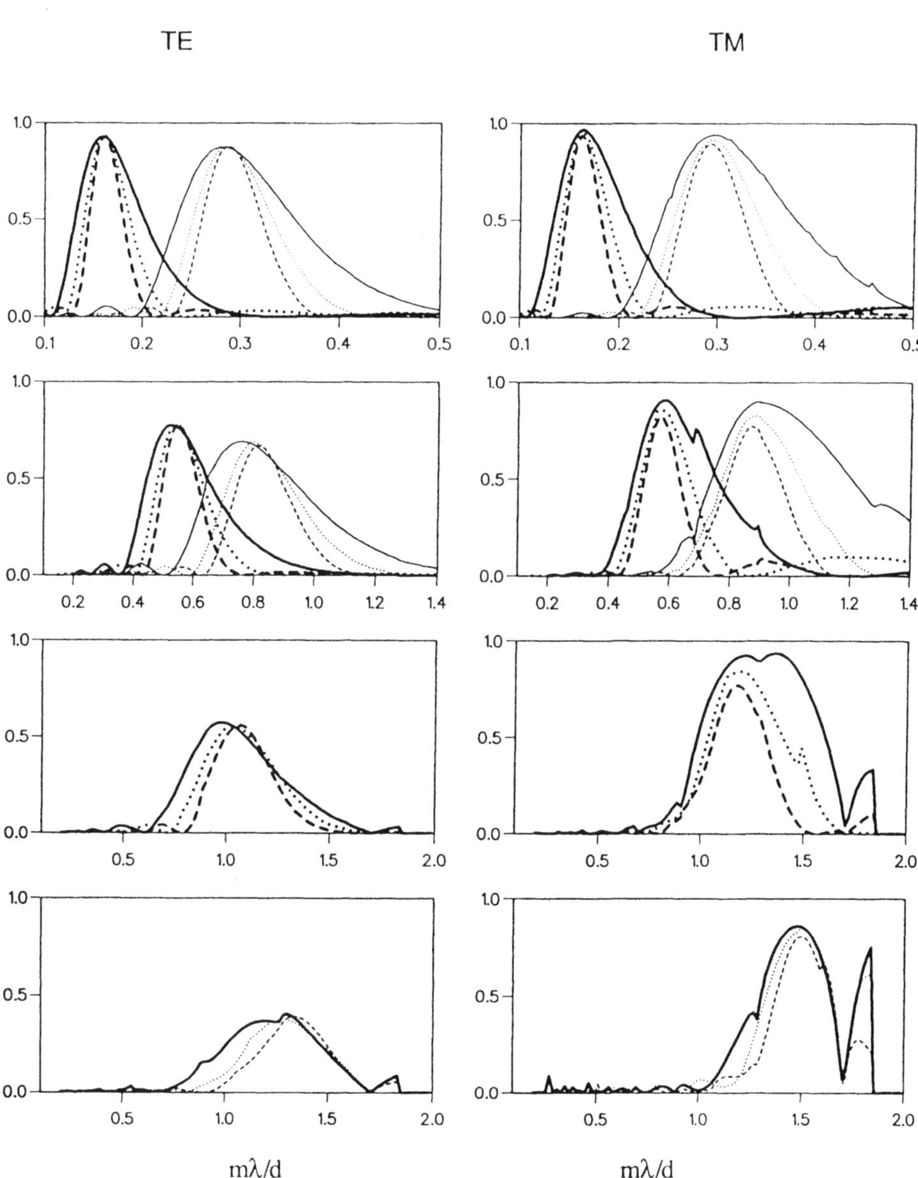

Fig.4.13 Same as Fig 4.11, except A.D. 45°. In sequence, top to bottom, blaze angles are [5 and 9°], [17.5 and 26.75°], [36°], [48°].

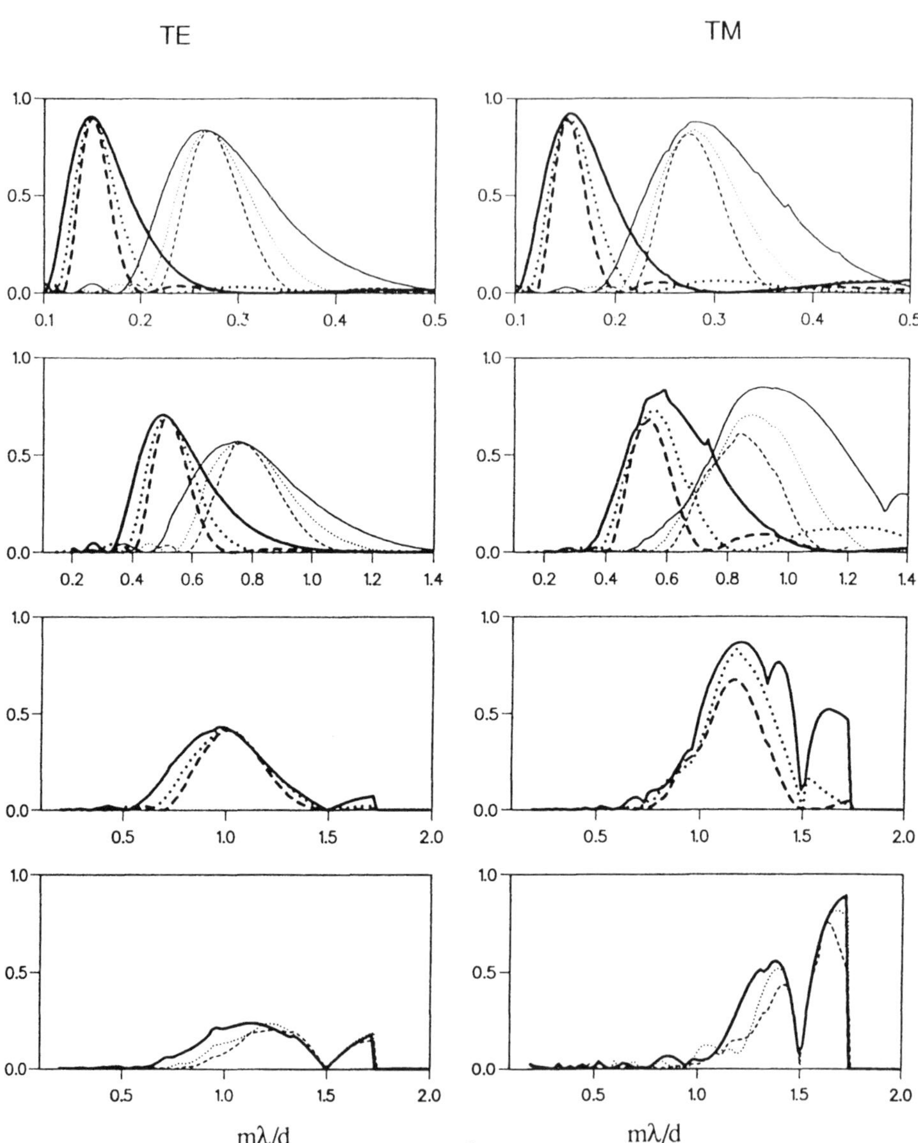

Fig.4.14 Same as Fig.4.11, except A.D. 60°. In sequence, top to bottom, blaze angles are [5 and 9°], [17.5 and 26.75°], [36°], [48°].

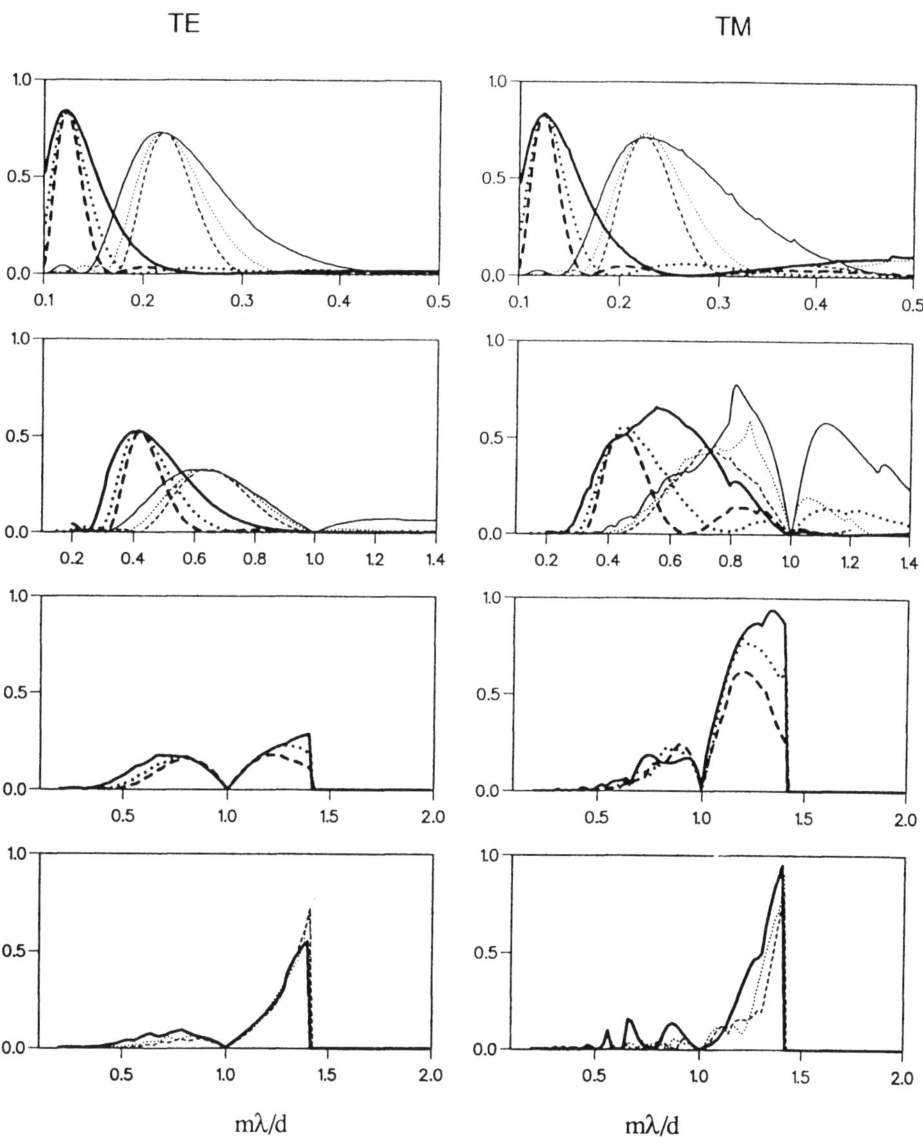

Fig.4.15 Same as Fig.4.11, except A.D. 90°. In sequence, top to bottom, blaze angles are [5 and 9°], [17.5 and 26.75°], [36°], [48°].

constricted, and in TE plane also reduced in value.

The same trends are seen progressing in Fig.4.13, where A.D. increases to 45° and still more at 60°, at which point the 48° blaze angle appears useless.

For the maximum A.D. of 90° Fig.4.15 indicates that once the blaze angle exceeds 9° efficiencies drop rapidly to levels that make them useless in practice. This information is of value especially in applications where extreme angles of incidence are desired for mixing beams from different wavelengths into a single exit beam.

As A.D. increases there will always be decreases in the efficiency peaks in both planes of polarization, as well as shifts in the wavelength at which they occur. In order to summarize this behavior the effects for orders 2 to 4 is shown in Figs.4.16 to 4.21. They include the simple scalar wavelength shift described by the cos(A.D./2) factor, which applies quite well in the first order. For better oversight the TE and TM data is plotted separately side by side. Blaze angles from 5° to 48° are covered.

In general we can see that for low blaze angle gratings such as 5° the efficiency peaks drop only about 15%, even for A.D. as large as 90°, Fig.4.16. The location of the wavelength peak drops as a function of cos(A.D./2), as predicted under scalar theory. When the blaze angle increases so does the effect of increasing A.D. The simple cos(A.D./2) ratio applies with less and less accuracy as the blaze angle increases, and for once in a favorable direction. Note should be taken of the progressively larger drop in efficiency that accompanies an A.D. increase when the blaze angle becomes larger.

4.4 Reflection Efficiencies of Echelettes at Higher Groove Frequencies and the Roles of Aluminum vs. Gold and Silver Coatings

There are many applications for groove frequencies exceeding the common 1200 gr/mm, especially for shorter wavelengths. Since the influence of the complex metal index of refraction increases inversely with wavelength the data are repeated for gold and silver coatings. The results are shown in Figs.4.22 to 4.26, where the value of A.D. was held constant at 8°, and blaze angles chosen from 17.5° to 36° to show up the differences.

The first Figure, 4.22, shows aluminum gratings at 1800, 2400, and 3600 gr/mm, whose character may be compared with Fig.4.5, which contains corresponding information for 1200 gr/mm. Note that the wavelength scales are chosen in proportion to the change in groove frequencies so that any changes in curve shape become evident at a glance. For the 17.5° blaze there is not much visible change until 3600 gr/mm is reached, where there are significant reductions in the UV region, as one would expect. The same general conclusion

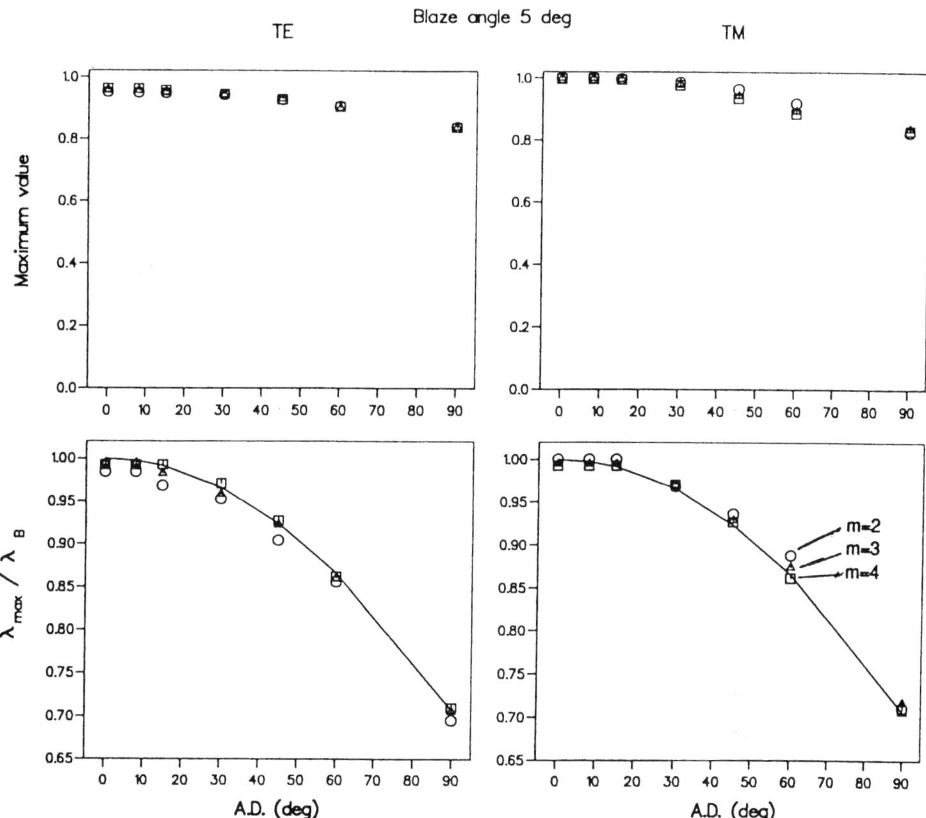

Fig.4.16 Effect of increased A.D. on efficiency peaks of 1200 gr/mm echelette gratings, in orders two, three and four, for 5° groove angle. TE values on the left and TM on the right. The order symbols as marked. Upper set of curves shows the effect on maximum efficiency values and lower curves the shift in location of the peak wavelength λ_{max} as a ratio to the blaze wavelength λ_B. Solid line is the cos (A.D./2) function.

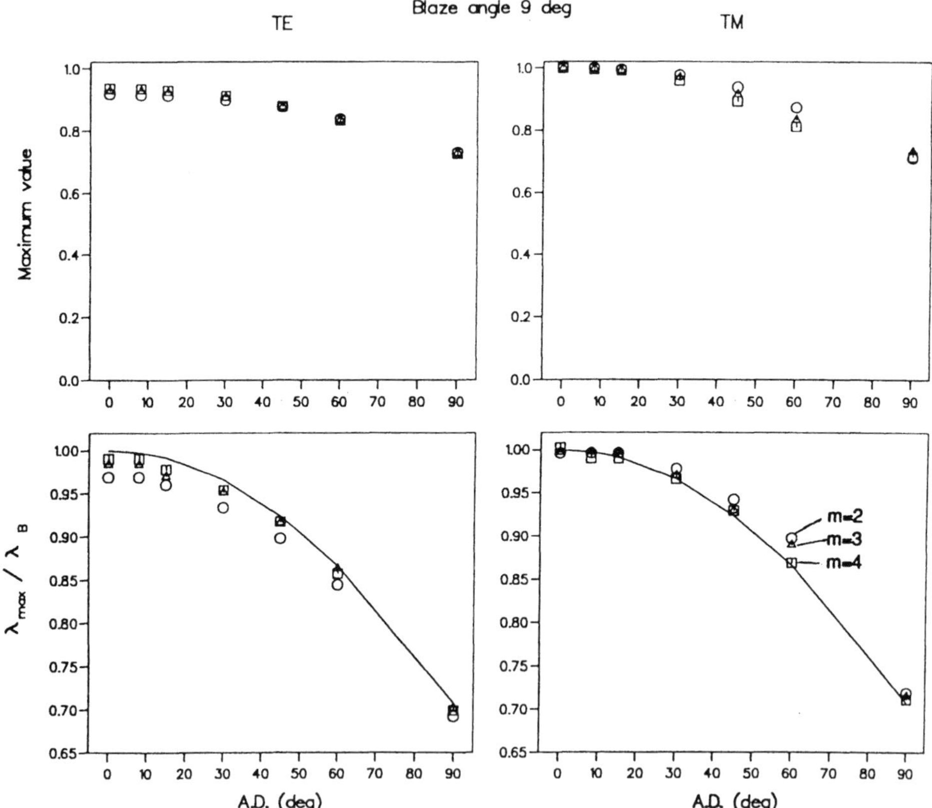

Fig.4.17 Same as Fig.4.16, except the blaze angle is 9°.

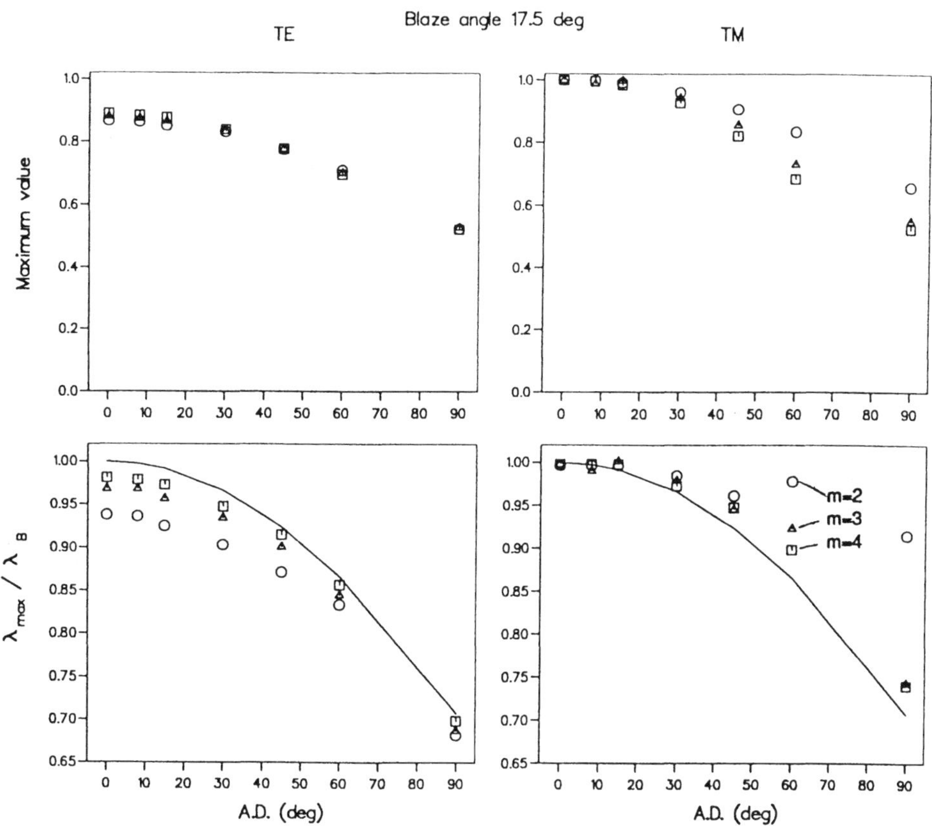

Fig.4.18 Same as Fig.4.16, except the blaze angle is 17.5°.

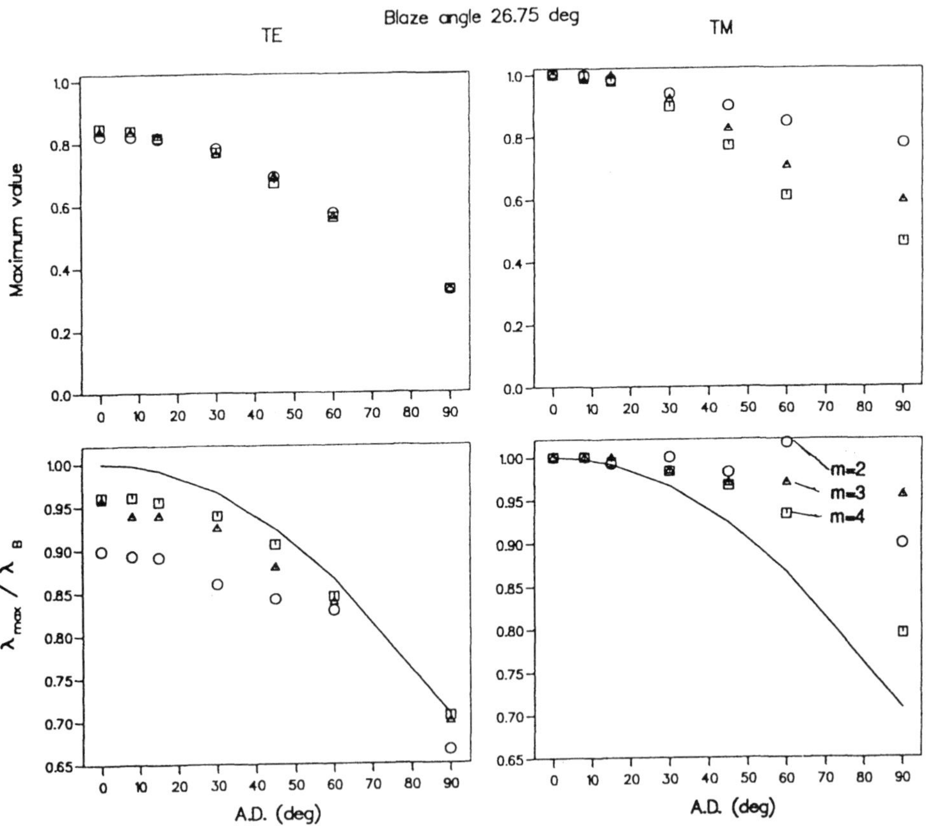

Fig.4.19 Same as Fig.4.16, except the blaze angle is 26.75°.

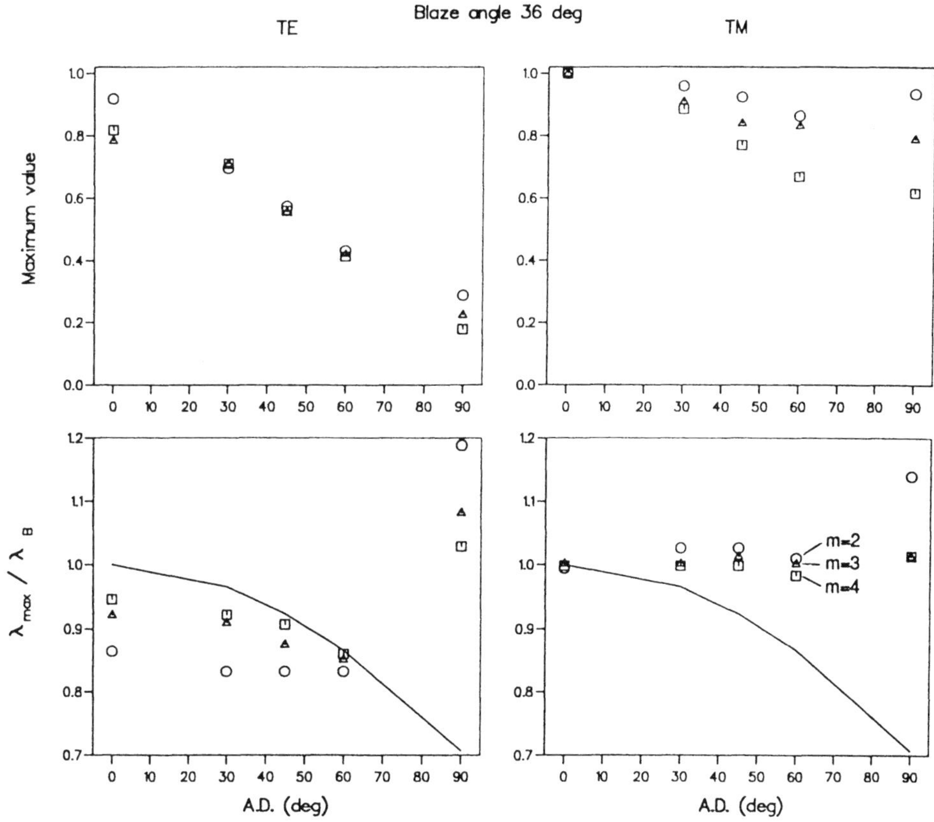

Fig.4.20 Same as Fig.4.16, except the blaze angle is 36°.

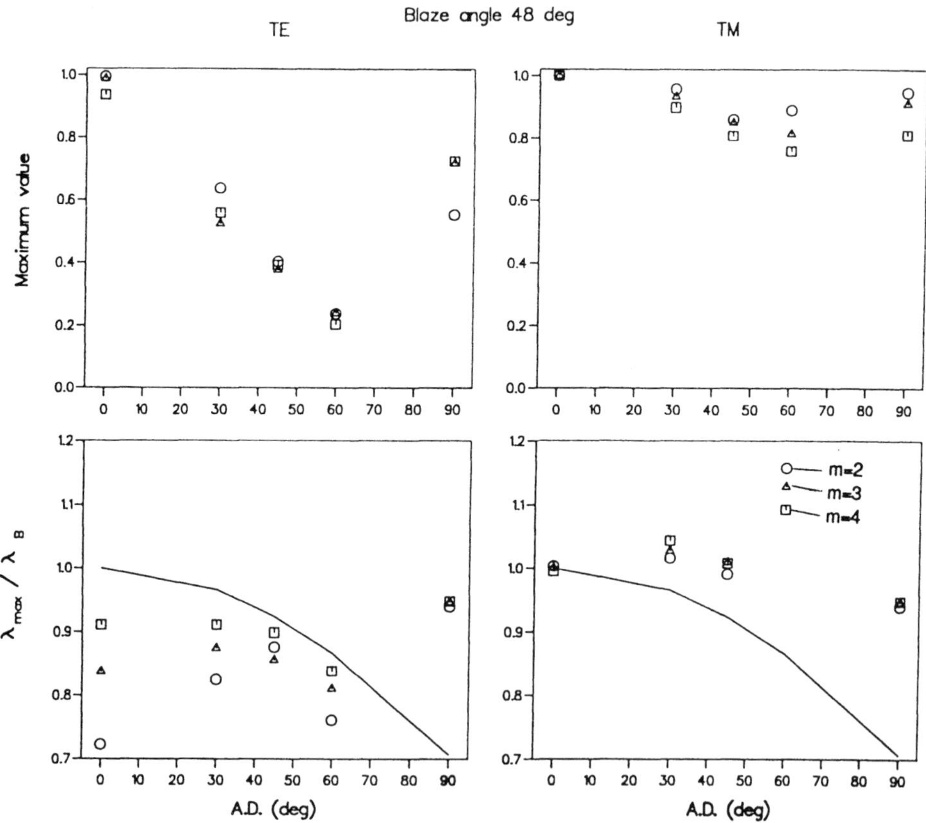

Fig.4.21 Same as Fig.4.16, except the blaze angle is 48°.

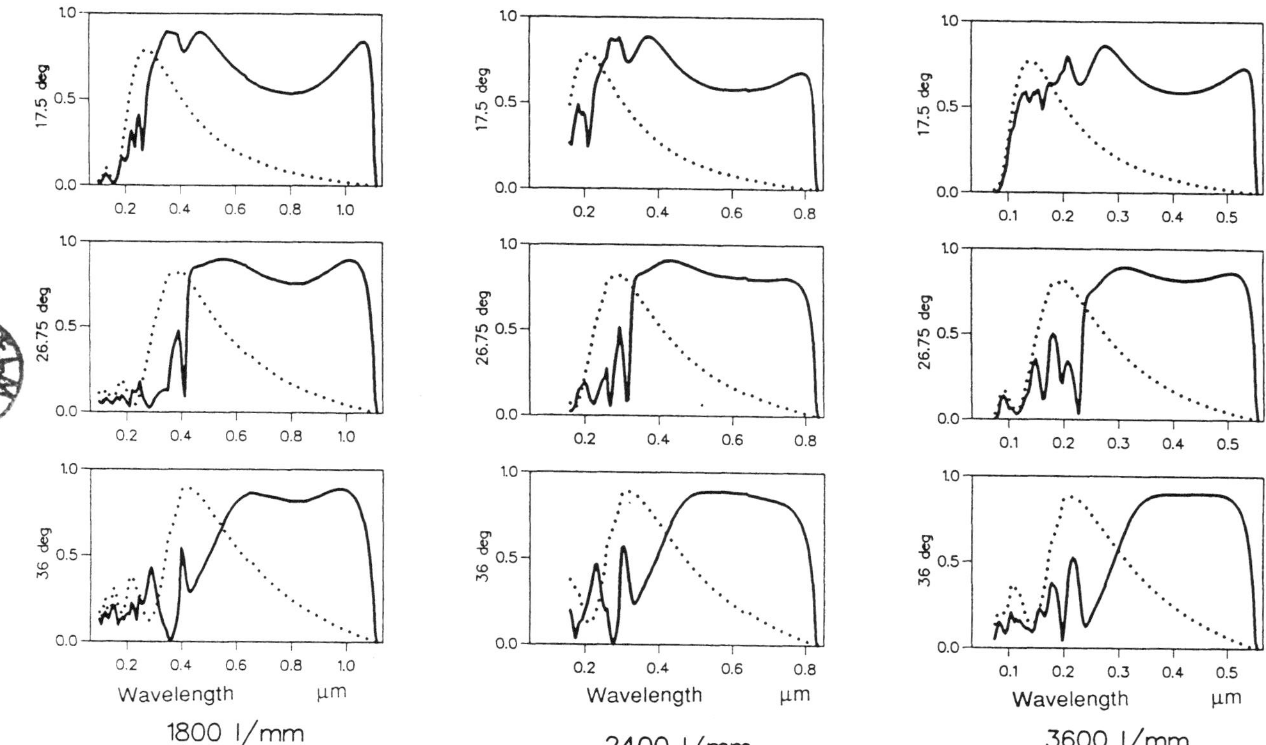

Fig.4.22 Absolute first order efficiency of aluminum echelette grating, A. D. 8°, for 3 blaze angles and three groove frequencies as marked on left. Solid lines TM and dotted lines TE plane values.

applies at 26.75° and 36°, except for the strong anomalies in the region from 0.1 to 0.24 μm.

The effect of switching from aluminum to gold or silver can be noted by comparing Fig.4.23 with Fig.4.5. At 17.5° groove angle the effect of gold is to reduce quite significantly the TE efficiency peaks and the TM values below 0.6 μm, which is to be expected from the reflectance behavior of gold. The effect

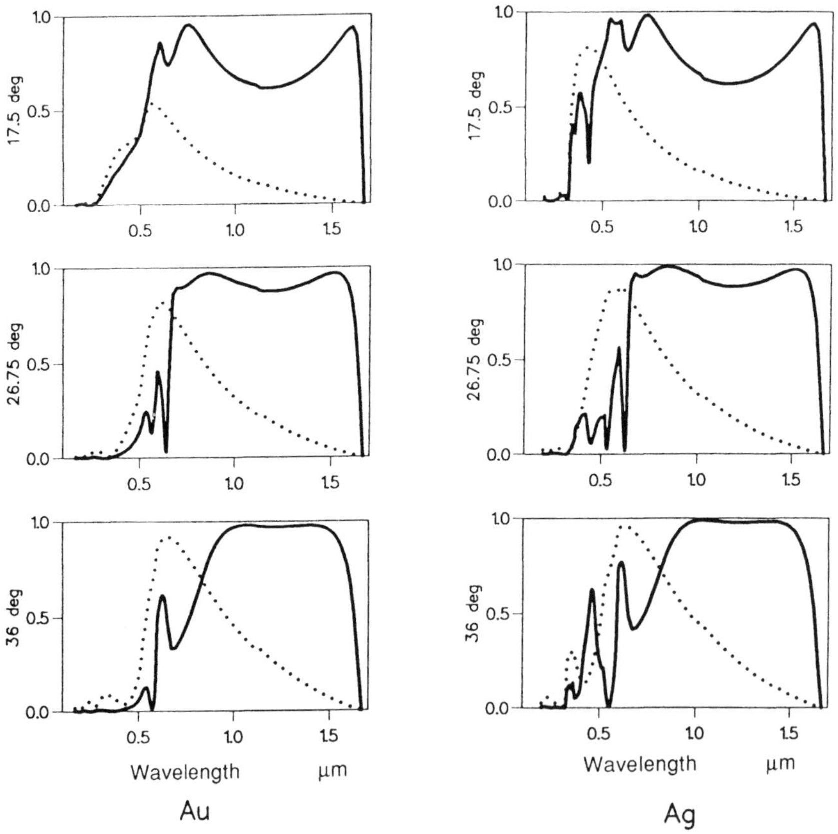

Fig.4.23 Absolute first order efficiency of 1200 gr/mm echelette with blaze angles as marked on left. TM and TE as indicated before. Left-hand curves have gold coatings and right-hand curves silver. A.D. 8°.

of silver is to enhance the TM efficiency at $\lambda > 0.5\mu m$.

At 26.75° the effect of gold and silver is small in the TE plane, while in TM gold avoids the aluminum reflectance dip at 0.8μm, boosting the efficiency peak to 95%. Silvers' maximum reflectance enhances this effect and leads to 99% efficiency near 0.8 μm. Not visible in the small figures is that gold and silver produce small shifts in the location of the 0.65 μm anomaly dip that have

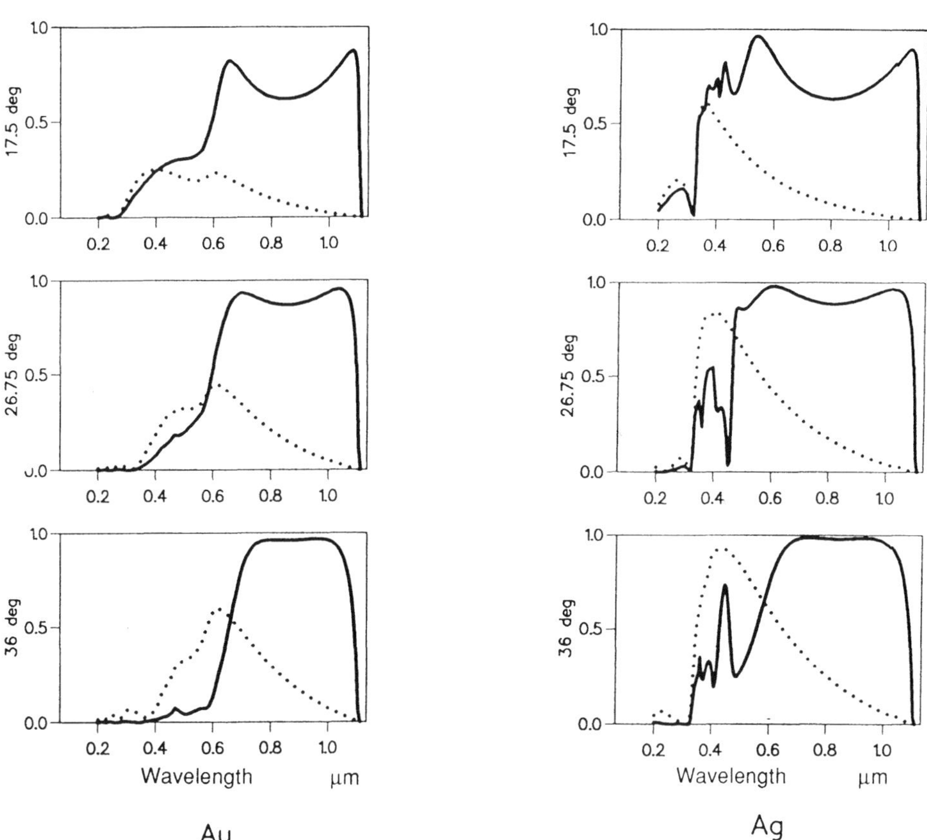

Fig.4.24 Same as Fig.4.23, except the groove frequency is 1800 gr/mm.

been precisely matched in experiments, a particularly effective indication of the accuracy of the theoretical calculations [4.7].

For 36° gratings gold and silver provide small but visible boosts in TE plane peaks, but in TM the changes are well delineated throughout the anomaly region and the high efficiency plateau from 1.0 to 1.5 μm.

Figs.4.24 and 4.22 allow for easy comparison between similar gratings when the groove frequency increases from 1200 to 1800 gr/mm. At 17.5° blaze

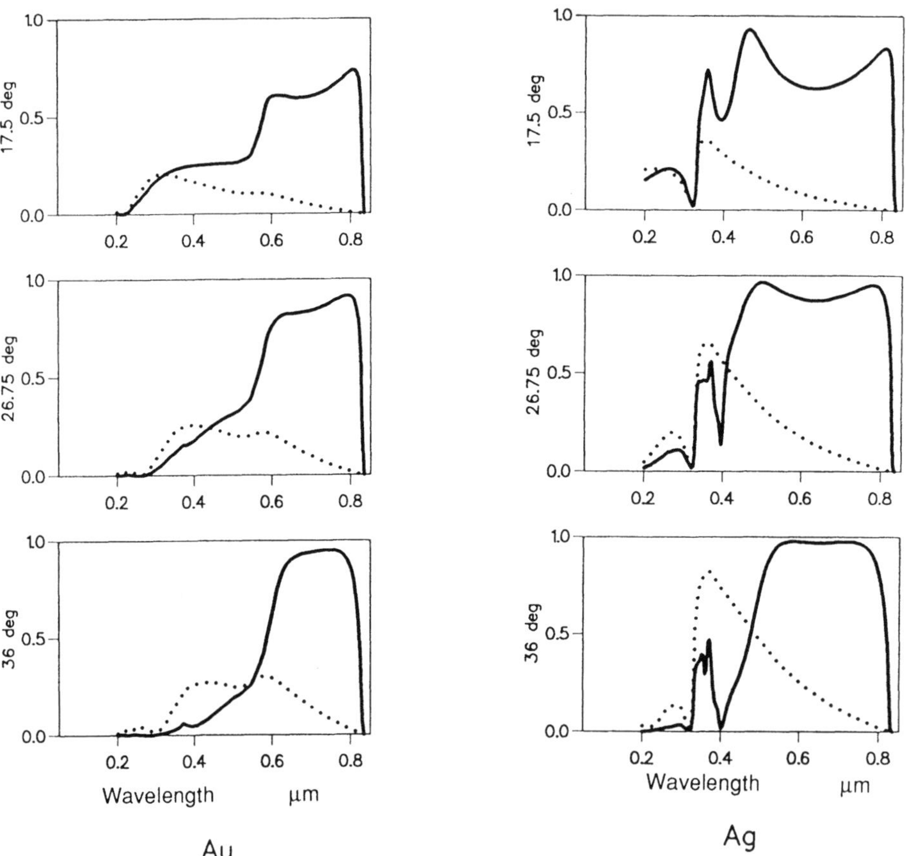

Fig.4.25 Same as Fig.4.23, except the groove frequency is 2400 gr/mm.

neither gold nor silver offer any advantage, because the reflectance of silver drops rapidly at wavelengths below 0.45 μm, and gold even more. A 26.75° grating is used at longer wavelengths, which explains the small difference in going to silver, and gold shows a sharp drop below 0.65 μm. Unusual is that this drop greatly exceeds the reflectance drop of a gold mirror (Fig.4.1b), an unexpected result that has been observed experimentally. Similar observations apply to the 36° grating.

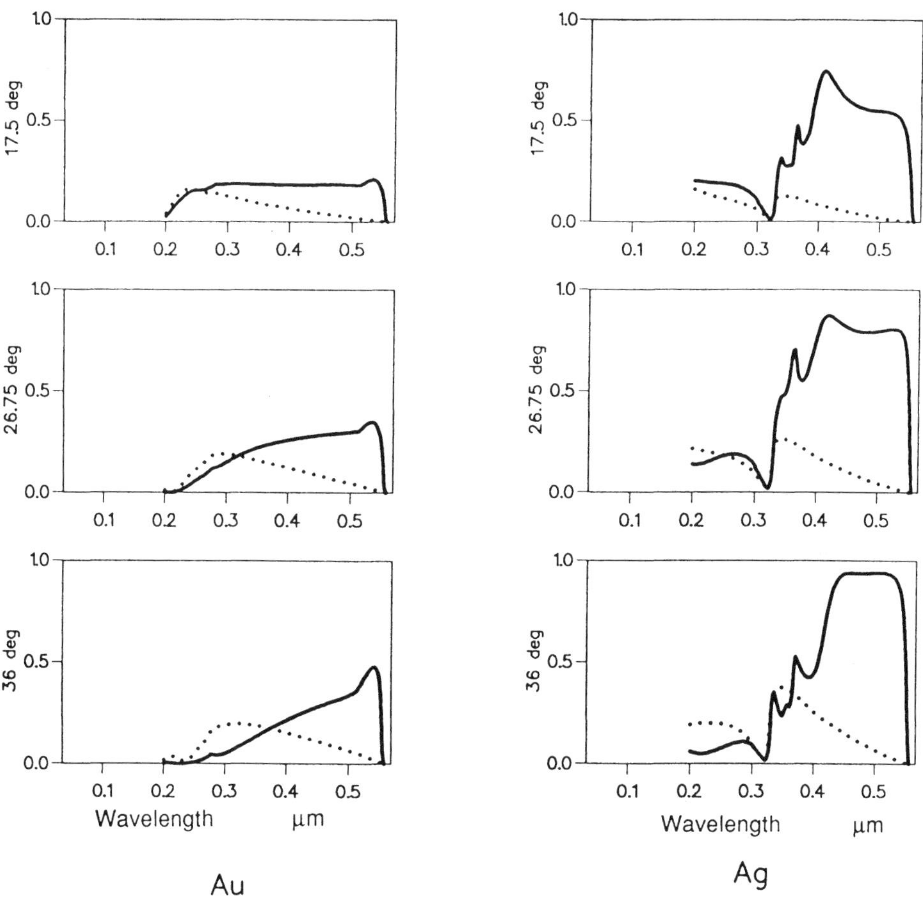

Fig.4.26 Same as Fig.4.22, except the groove frequency is 3600 gr/mm.

In looking at the 2400 gr/mm gratings of Fig.4.25, the same general remarks hold as for Fig.4.24, except that all the effects show up more strongly, because, compared to the 1200 gr/mm grating, the wavelength band has dropped in half instead of by one third.

The 3600 gr/mm curves of Fig.4.26 represent a 3:1 drop in wavelength compared to 1200 or a 2:1 drop compared to 1800 gr/mm curves of Fig.4.24, and confirm that at these wavelengths neither gold nor silver should be used.

4.5 Effect of Groove Apex Angle on Echelette Efficiency

Various opinions can be found in the grating literature concerning the role that departures of the 90° groove apex angle might play, the standard for echelettes. To explore this aspect for otherwise perfect grooves of 1200 gr/mm aluminum echelettes, at 8° A.D., the effect was studied at two blaze angles, 17.5° and 26.75°, Fig.4.27. The apex angle was varied from 85° to 120°.

The first conclusion is that there is remarkably little difference over a range of 85° to 100°, except that the efficiency saddle between the two TM plane peaks sags more and more as the apex angle increases. This is expressed more at the 17.5° angle than 26.75°.

As the apex angle increases further this effect becomes even more pronounced, the first TM efficiency peak is progressively lowered and the TE peak is visibly reduced. There seems to be no detectable advantage to increasing the apex angle beyond the nominal 90°. Reducing it is even less indicated, since it can cause replication problems. These remarks do not necessarily apply directly to the set–up of a ruling engine, because here the operator must make allowances for the plastic flow of the metal surface. To accomplish that he might well find that tools with other than 90° included angles sometimes give superior results.

4.6 Plane Sinusoidal Reflection Grating Behavior

Sinusoidal groove shape gratings have been manufactured commercially since about 1970, ever since successful processes were developed to produce them in photoresist layers by interference (holographic) methods. Rigorous efficiency calculations, as with echelettes, are based on their use in collimated light.

Although the majority of these gratings are in the form of concaves there is sufficient use of plane gratings to justify covering their efficiency behavior here. In addition, the solutions provide an approximate insight into the behavior of concave gratings, even though in that case diffraction conditions are far more

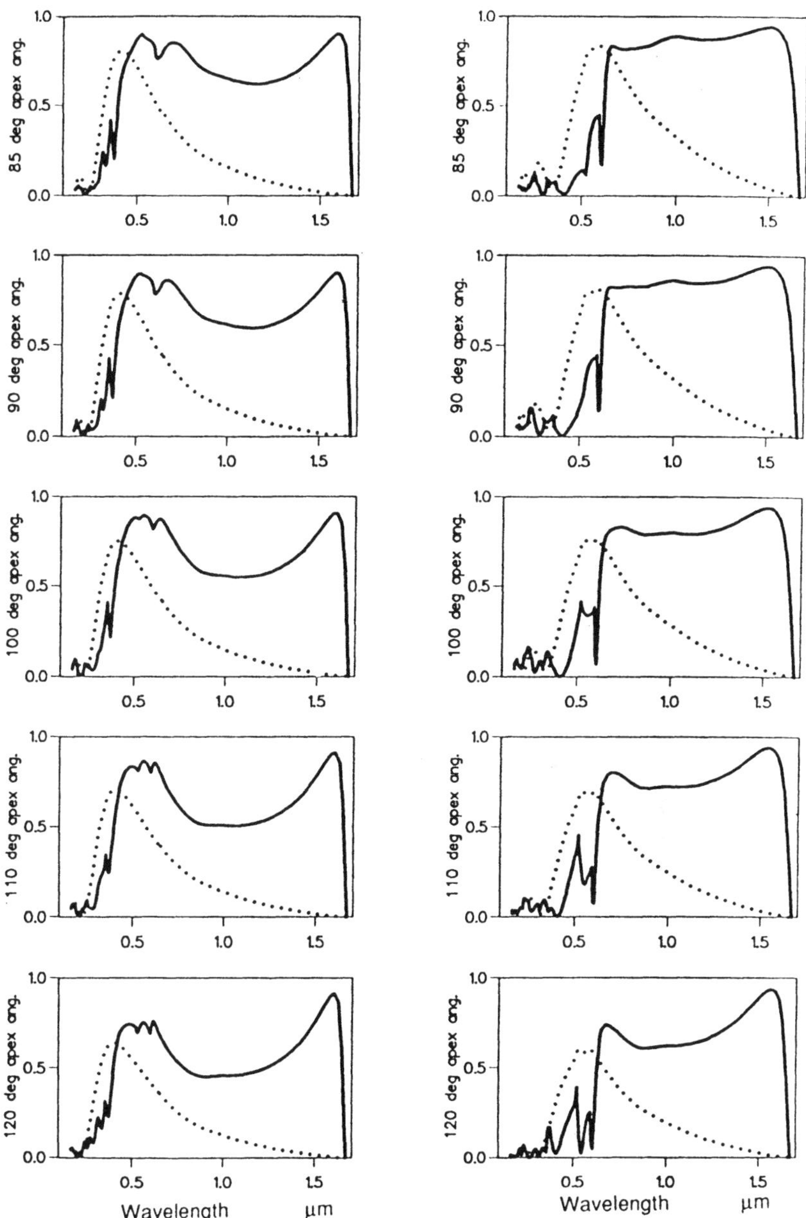

Fig.4.27 Absolute first order efficiency curves for 1200 gr/mm aluminum echelette. 17.5° blaze on left and 26.75° on right, for five apex angles as marked. TE and TM as indicated before.

complex, not only because their method of manufacture tends to lead to non-uniform depth modulation, but especially because incident rays diverge in two planes, instead of being collimated.

As with blazed echelette gratings efficiency behavior is largely a function of the λ/d ratio and the groove depth modulation (h/d). If the groove frequency is specified (1/d), the results can be plotted as a function of wavelength, just as was done with echelettes. The depth modulation h/d is used directly to define these gratings, while with echelettes the groove definition is in terms of the groove angle φ_B, from which the depth is derived from $h = d\sin\varphi_B \cdot \cos\varphi_B$.

An important observation is that the efficiency behavior of deep modulation sinusoids is very similar to that of equivalent echelettes, as discussed in section 4.2.3. As the modulation decreases sharp differences emerge, especially significantly reduced efficiency. However, one should not conclude that these gratings become useless.

Again as with echelettes, sinusoidal grating properties can be analyzed as a scalar problem when the modulation is low enough (< 0.05), and become progressively more and more dominated by electromagnetic properties as the modulation and diffraction angles increase. The results, again as with echelettes, are presented in families that share basic properties, as the modulation increases from 0.05 to 0.50. Once more, most of the results are for an aluminum surface and for a groove frequency of 1200 gr/mm. Also covered are the effects of going to higher groove frequencies, and exchanging aluminum for silver and gold. A special section describes the properties in orders two to four.

Unlike echelettes that are routinely supplied with groove frequencies from 20 to 3600 gr/mm, the majority of sinusoidal groove gratings have values 1200 and higher, and seldom go below 100.

4.6.1 Absolute Efficiency of Plane 1200 gr/mm Aluminum Sinusoidal Gratings

The absolute efficiencies of 1200 gr/mm sinusoidal gratings are shown as a function of wavelength, from about 0.15 to 1.67 μm, in Figures 4.28 to 4.35. Each figure contains 10 sets of curves, each with TE and TM data, with modulations from 0.05 to 0.50 in steps of 0.05. All are for an aluminum surface, except Fig.4.28, where infinite conductivity is assumed, for reference purposes. The successive figures differ only in their progressive increase in A.D. from 0 (Littrow) to 90°, just as was done for echelettes.

The first impression gained from Fig.4.28 is that compared to echelettes the efficiency at low modulations (< 0.20) is remarkably low, although in return the unpolarized efficiency for medium modulation remains fairly constant over a wide range of wavelengths. The effect of electromagnetic behavior becomes noticeable at $h/d \geq 0.15$, and comparison with echelettes of Fig.4.3 shows anomaly spikes to be much sharper, an effect confirmed experimentally. In making such comparisons we should note that the 0.15 modulation corresponds to the depth of a 9° blaze and that 0.45 corresponds to a depth of a 26.75° blaze.

Especially noteworthy is the low efficiency peak at very low modulations (0.05). Based on scalar analysis it can be shown that its theoretical maximum is 0.338 (see eq.(10.5)), and is always located at a wavelength of 3.412 h. An increase to 0.1 modulation, which corresponds to a 6° blaze, is sufficient to raise the efficiency peak slightly, to 0.38, and to show EM behavior beyond 0.6 μm, in the form of anomaly spikes and polarization effects.

As the modulation increases, so does efficiency. Of special practical interest is the 0.35 modulation, because it offers the highest efficiency over a reasonably wide wavelength band, especially in TM plane. In a confirmation of the equivalence rule, its efficiency is seen to be remarkably similar to that of a 36° echelette, Fig.4.3, although the latter has a maximum grove depth 36% deeper. The 0.4 and 0.45 modulations are quite similar, and are a close match to a 48° echelette. The 0.50 modulation is about the highest that has practical use in a reflection grating, although some exceptions are discussed in sec.4.8.

The role played by angular deviation can be followed by comparing Figs.4.29 to 4.35. Like echelettes there is a range of modulations, from 0.3 to 0.4, where increasing A.D. between 15° and 30° reduces anomaly spikes, but at medium modulations (0.2 to 0.25) one can observe just the opposite effect.

When A.D. increases still more it can be seen to depress efficiencies more than is the case with echelettes (compare Fig.4.33 with 4.8). At A.D. 90° (compare Fig.4.35 with 4.10) there is great similarity with echelettes at steeper groove angles, but no sinusoidal grating can be found to match the 9° echelette. Unpleasant feature of sinusoidal groove gratings is the oscillation in TM efficiency that can be observed with all modulations above 0.15 in the wavelength region below 0.5 μm ($\lambda/d = 0.6$).

Comparing Fig.4.28 with 4.30 allows us to judge the effect that Al coating has, since all other conditions remain the same. At low modulations the effect is merely to depress values in accordance with the loss of reflectance, but at 0.30 and 0.35 modulations there is a sharp efficiency reduction when $\lambda > 0.7$ μm. It exceeds the well known reflectance drop of aluminum at 0.8 μm. At the highest modulations the effect is once again reduced to the reflectance loss. It is a good demonstration of how precarious it can be to extrapolate from existing information.

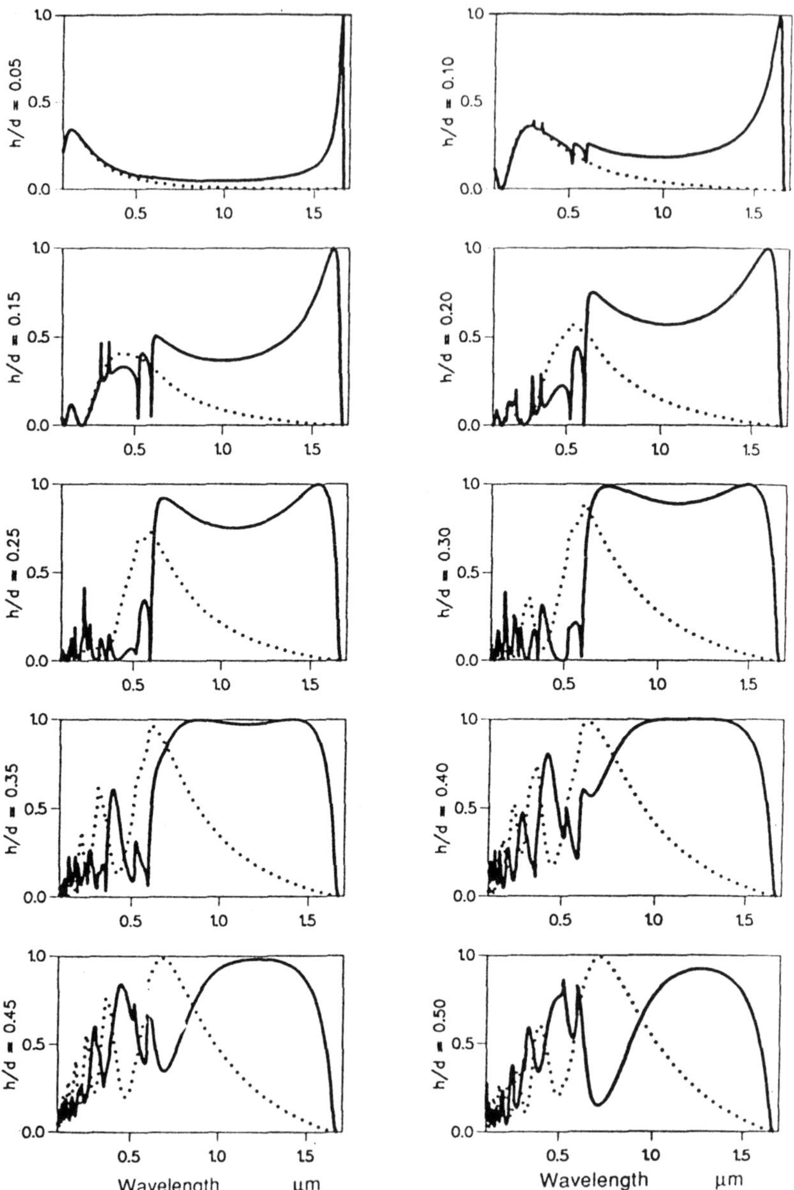

Fig.4.28 Absolute first order efficiency of 1200 gr/mm sinusoidal gratings as a function of wavelength in μm. Values for perfect conductivity, and A.D. 8°. Solid lines TM and dotted lines TE polarization. Ten modulations depths h/d, as marked.

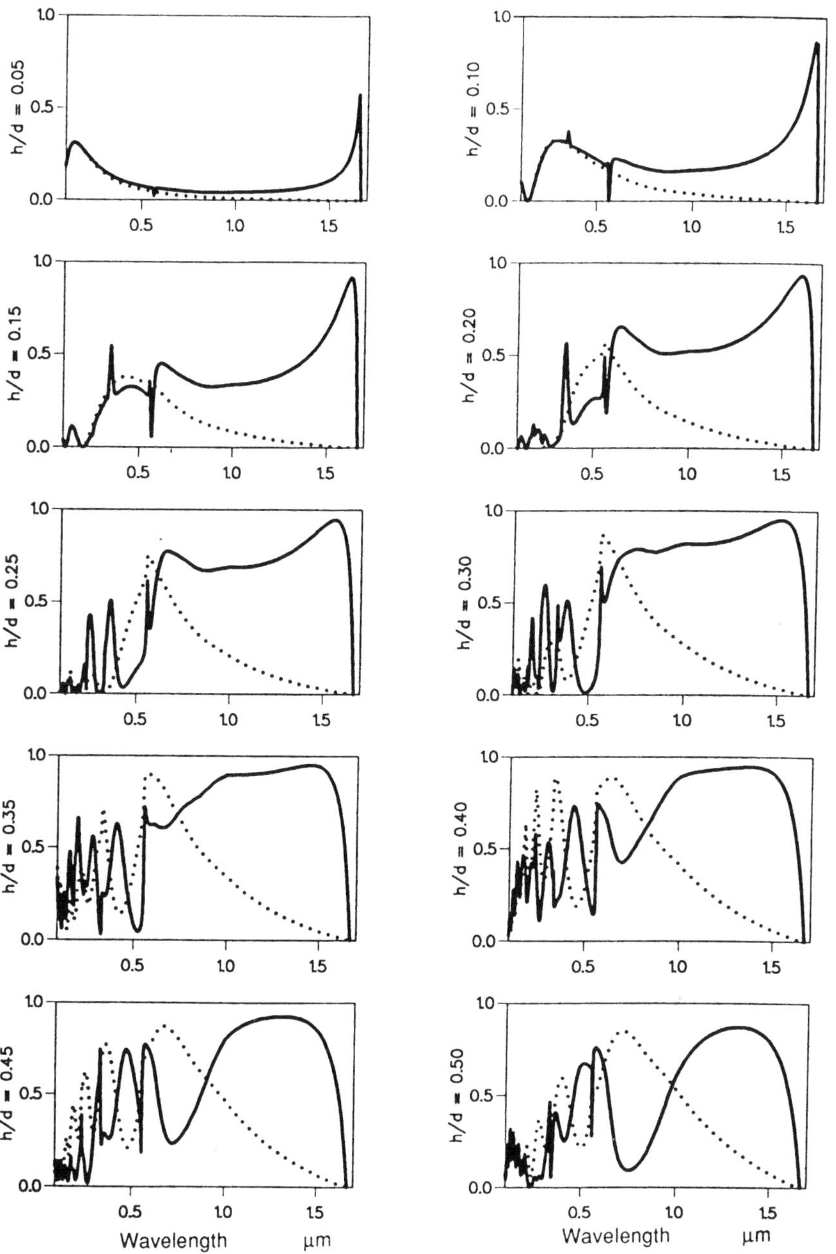

Fig.4.29 Same as Fig.4.28, except aluminum surface and A.D. is 0°.

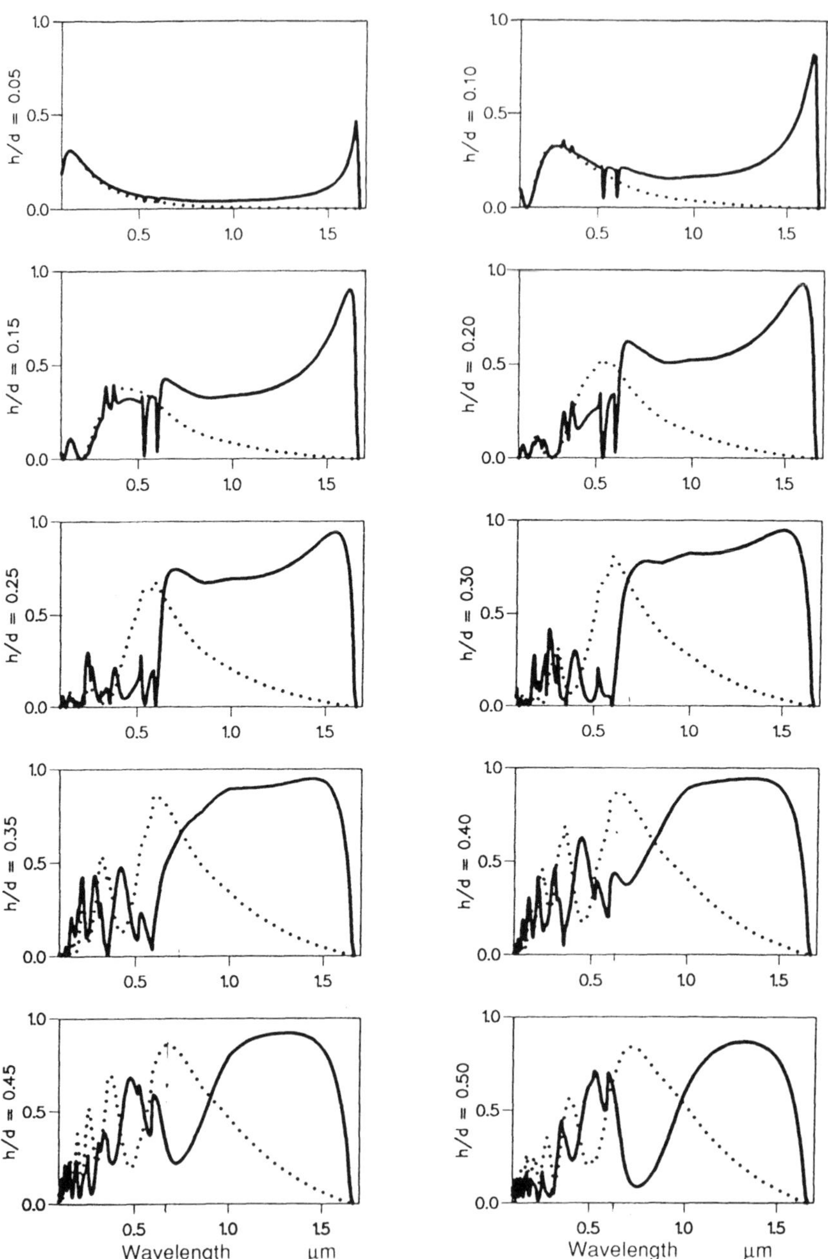

Fig.4.30 Same as Fig.4.29, except that A.D. is 8°.

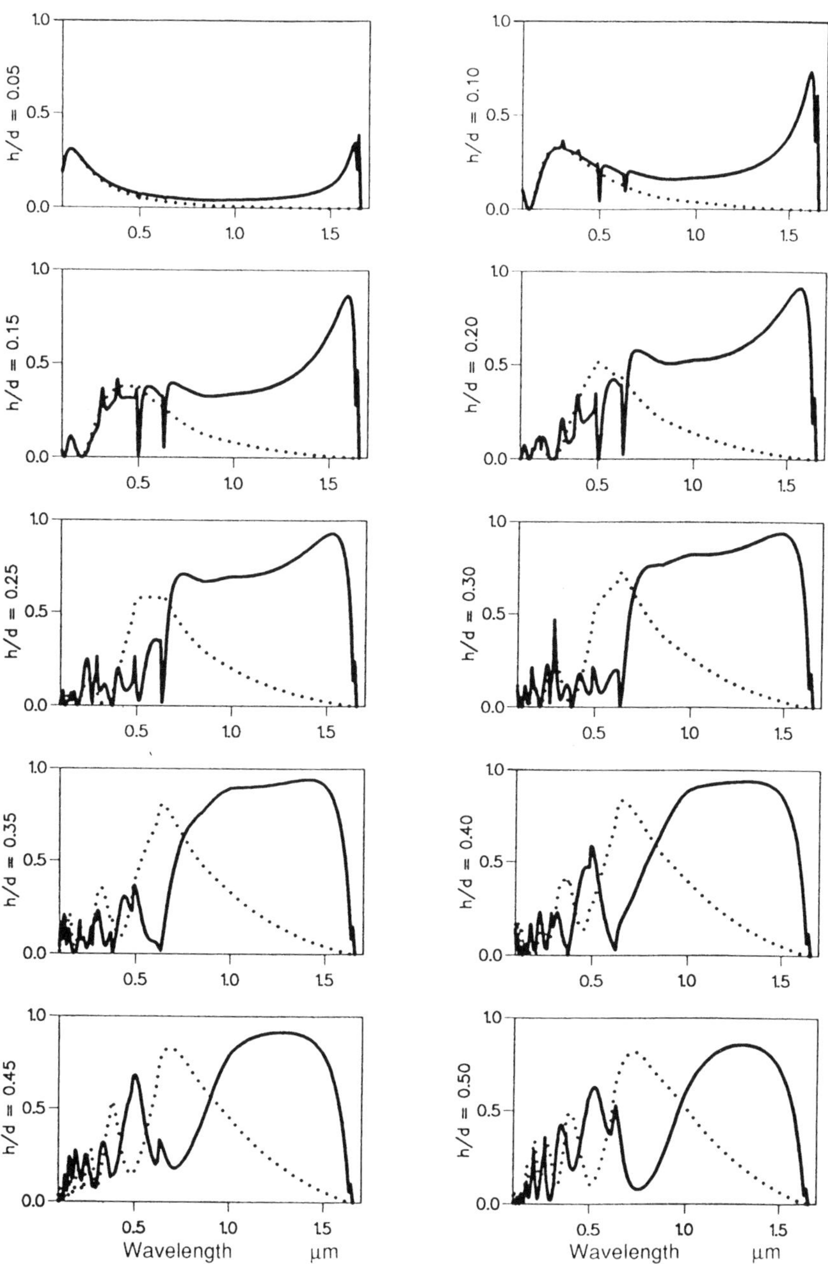

Fig.4.31 Same as Fig.4.29, except that A.D. is 15°.

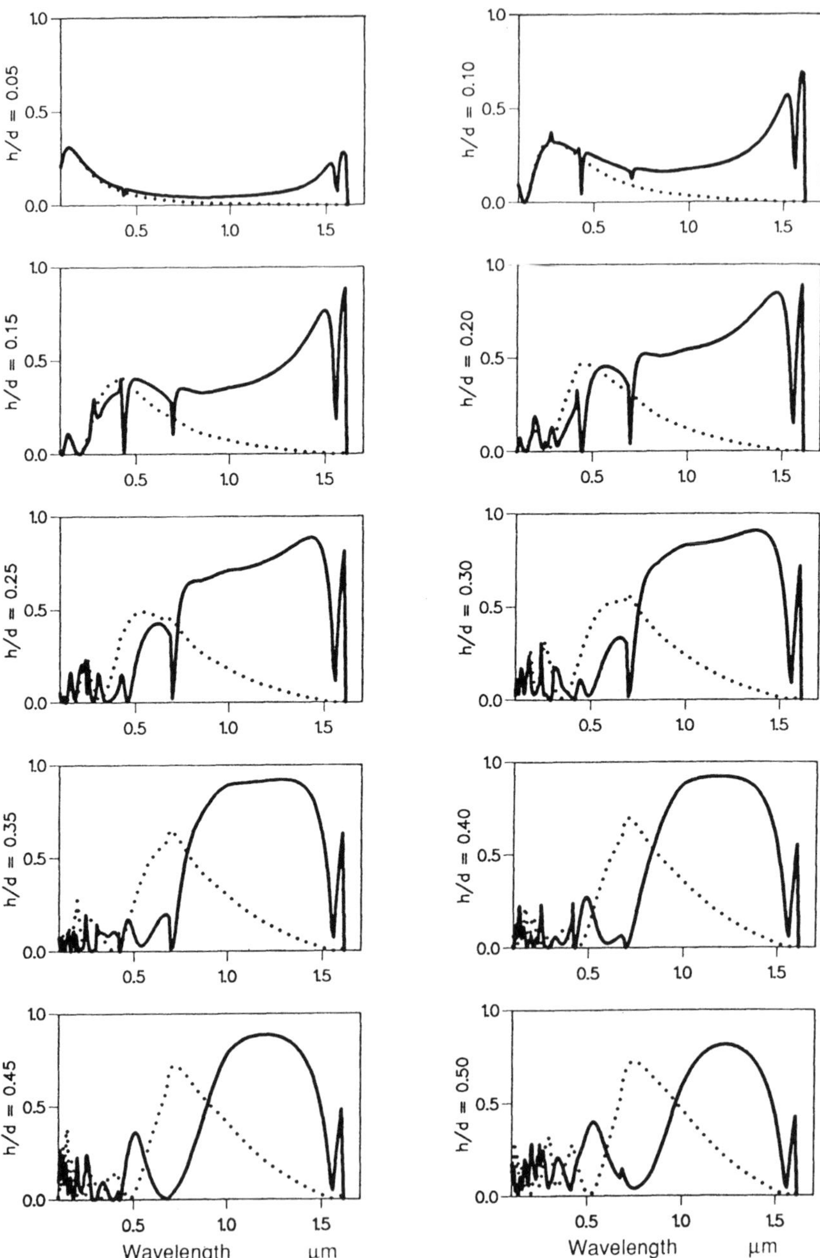

Fig.4.32 Same as Fig.4.29, except that A.D. is 30°.

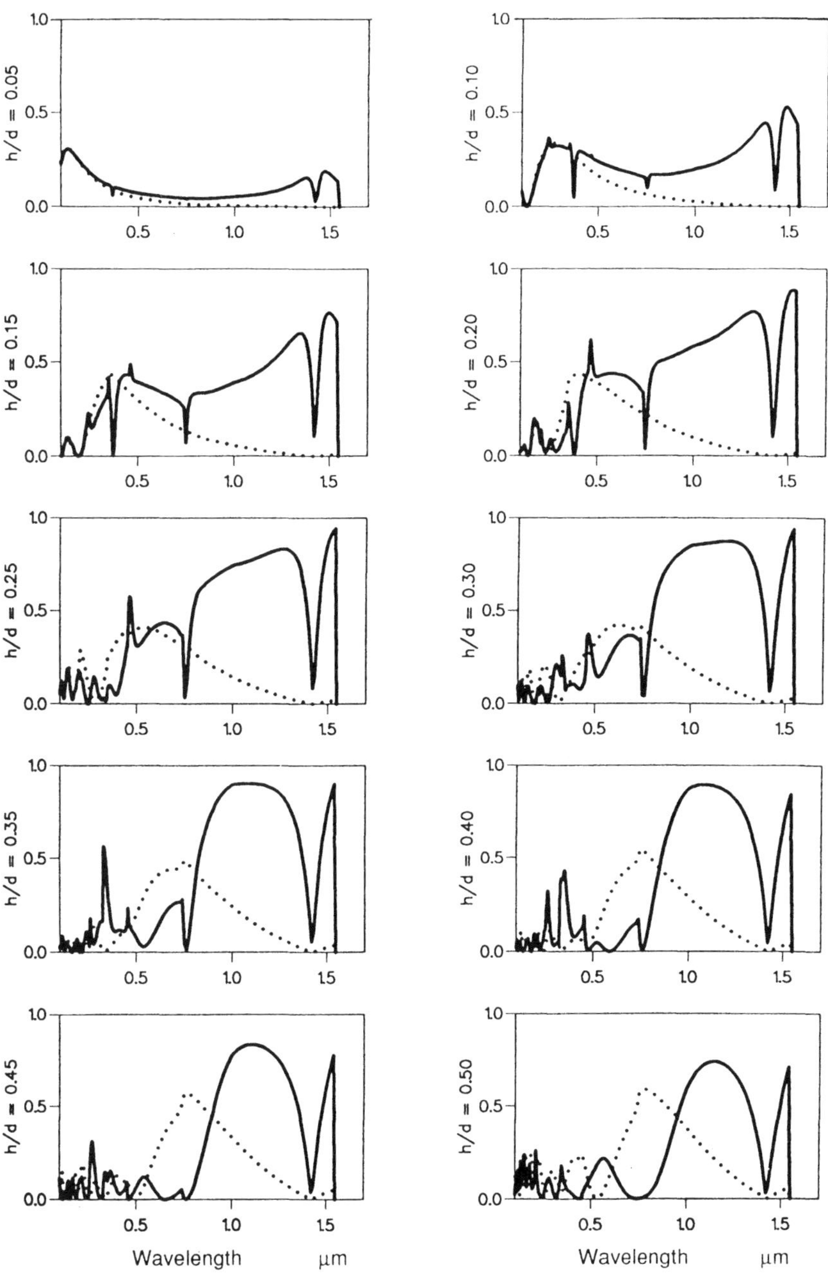

Fig.4.33 Same as Fig.4.29, except that A.D. is 45°.

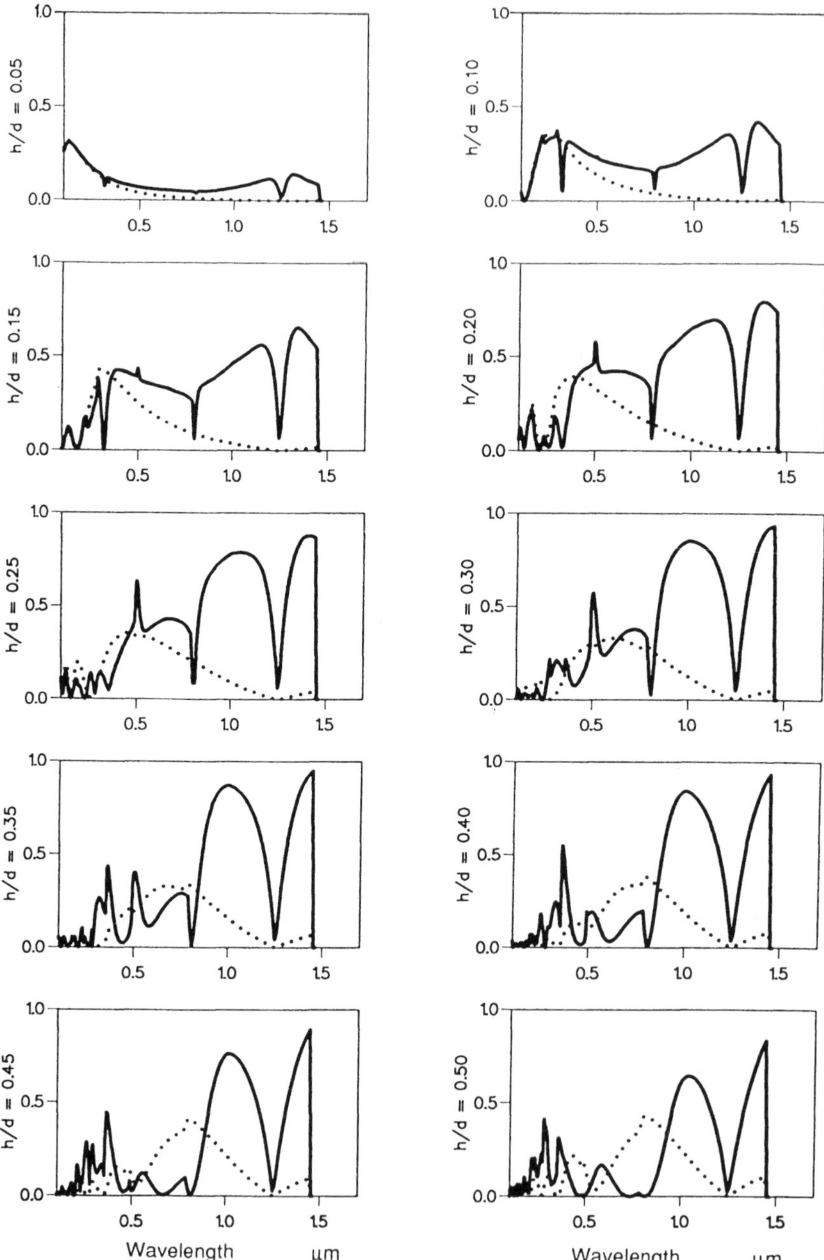

Fig.4.34 Same as Fig.4.29, except that A.D. is 60°.

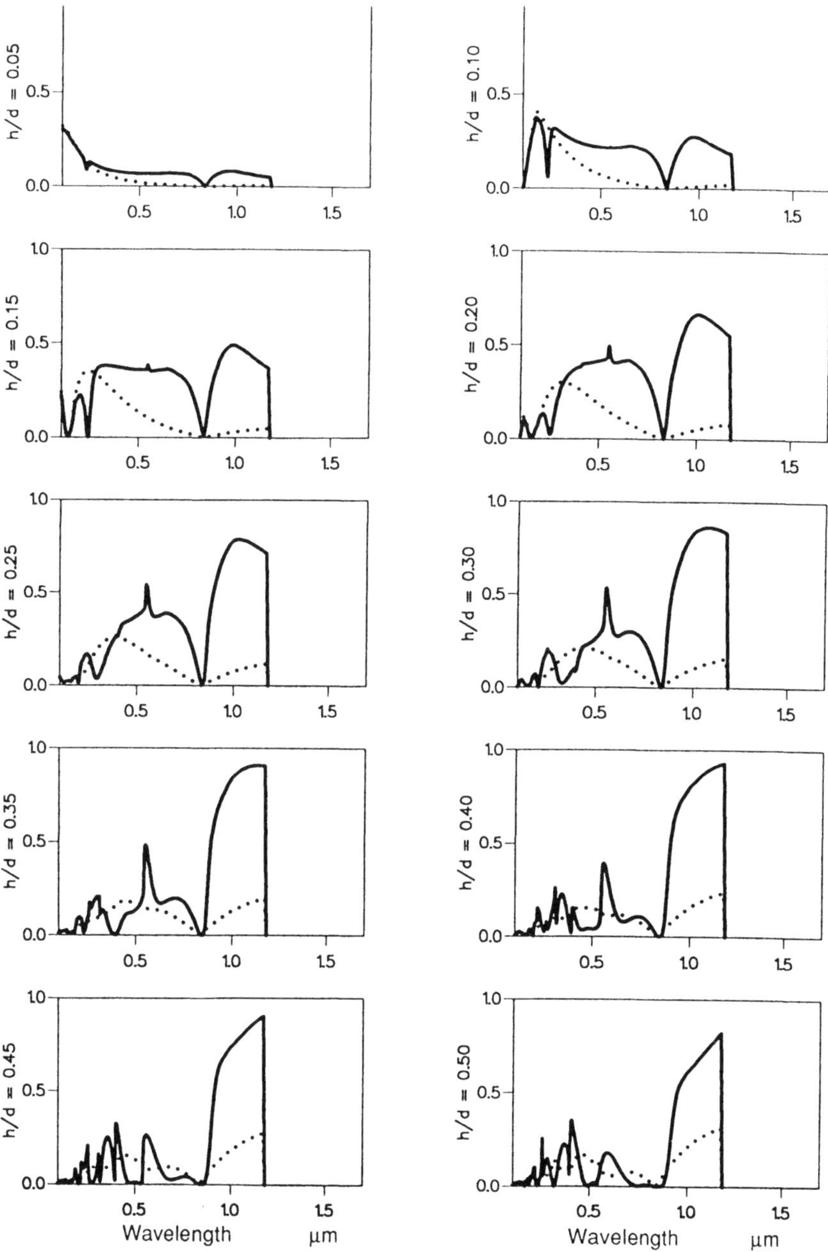

Fig.4.35 Same as Fig.4.29, except that A.D. is 90°

4.6.2 *Absolute Efficiency of 1200 gr/mm Sinusoidal Reflection Gratings in Orders 2 to 4*

Reflection efficiencies in orders 2 to 4 are presented in Figures 4.36 to 4.38, one for each order, since, unlike the case of echelettes, the curves are too complex to combine orders in one figure. For a better comparison of the shapes of the curves, they are again plotted against mλ/d and perfect conductivity is assumed to avoid interactions with changes in the complex index. Not included are the effect of departing from the 8° value for A.D. chosen for comparison. The three sets of curves should be compared to Fig.4.28, which displays equivalent first-order data. Plotting against mλ/d provides 1:1 comparison of shapes.

Most interesting at low modulations is comparing the peak efficiencies and their relative location. For 0.05 modulation the maximum efficiencies steadily decrease from 0.338 in first order to 0.17 in the fourth. This stands in marked contrast with echelettes where there is no such reduction, Fig.4.11. Corresponding peak locations (in terms of the dimensionless ratio mλ/d) increase from 0.17 in first order to 0.24 in the fourth, again in contrast to the echelette behavior, which never shows such shifts. These values may be hard to detect from the small figures, but are real and agree with experiment.

The picture changes slightly when increasing the modulation from 0.05 to 0.15. The relative shift in peak locations remains about the same, but the drop in peak efficiency is somewhat less, 38% instead of 48% in going from first to fourth order. In all cases intermediate orders and modulations fall logically between these numbers.

The important conclusion is that the higher order behavior of sinusoidal gratings is again seen to be entirely different from the simple one of echelettes, where successive order peaks bear a simple and direct relationship to the order. Both TE and TM properties show this difference, and this explains why sinusoidal gratings are not often used in orders higher than two.

4.6.3 *Absolute Efficiency of Aluminum Sinusoidal Gratings at Higher Groove Frequencies (1800, 2400, 3600 gr/mm)*

The effect of going to higher groove frequencies is shown in Figs.4.39 to 4.41, for 1800, 2400, and 3600 gr/mm respectively. Their shapes can be directly intercompared because their wavelength scales are modified in exact proportion to their frequencies, which also allows adequate comparison with the corresponding first order data of Fig.4.30. Looking at data for the often used 1800 gr/mm frequency, Fig.4.39, we note that at modulations below 0.25 there is little difference from 1200 gr/mm. At higher modulations the TE plane peaks higher and the TM curve is better behaved.

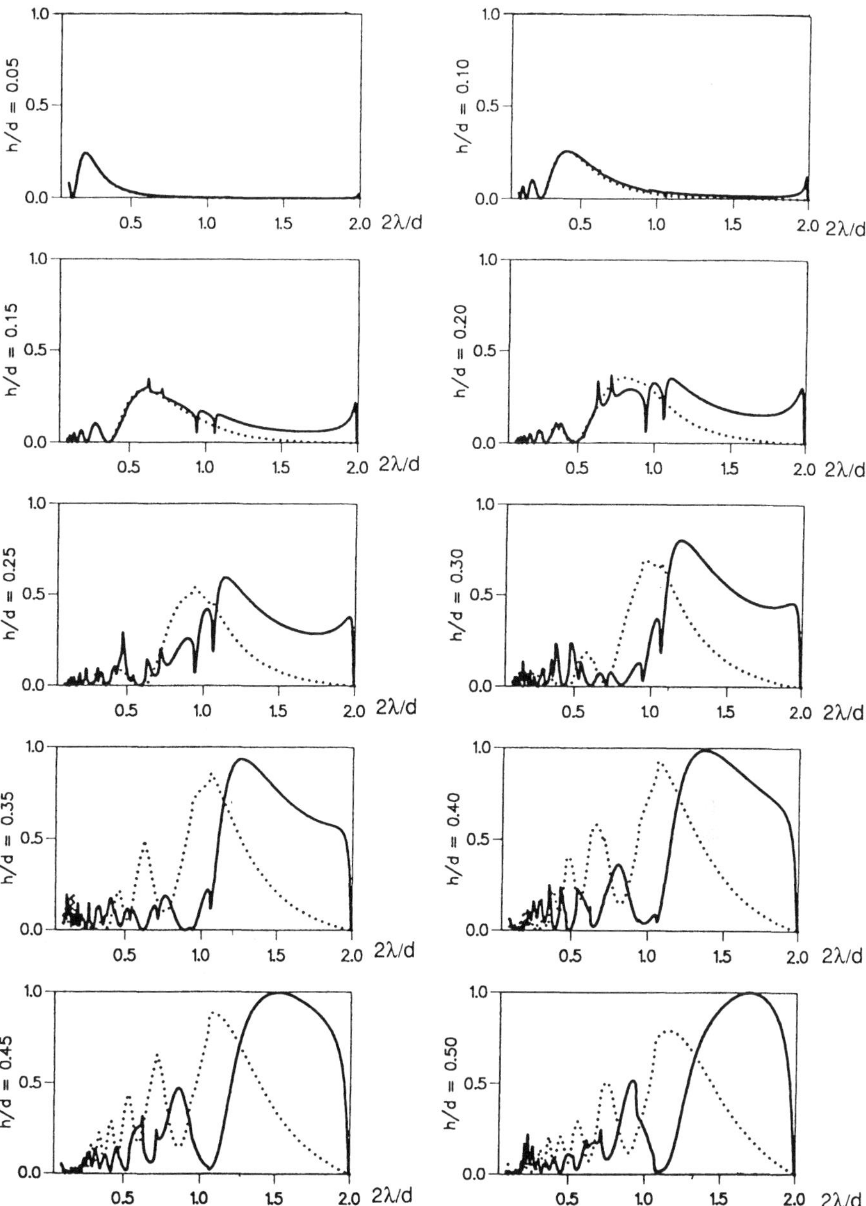

Fig.4.36 Absolute efficiency of 1200 gr/mm sinusoidal gratings in second order, as function of mλ/d. Perfect conductivity and A.D. = 8°. TE and TM polarizations as designated before. Ten levels of modulation h/d as marked.

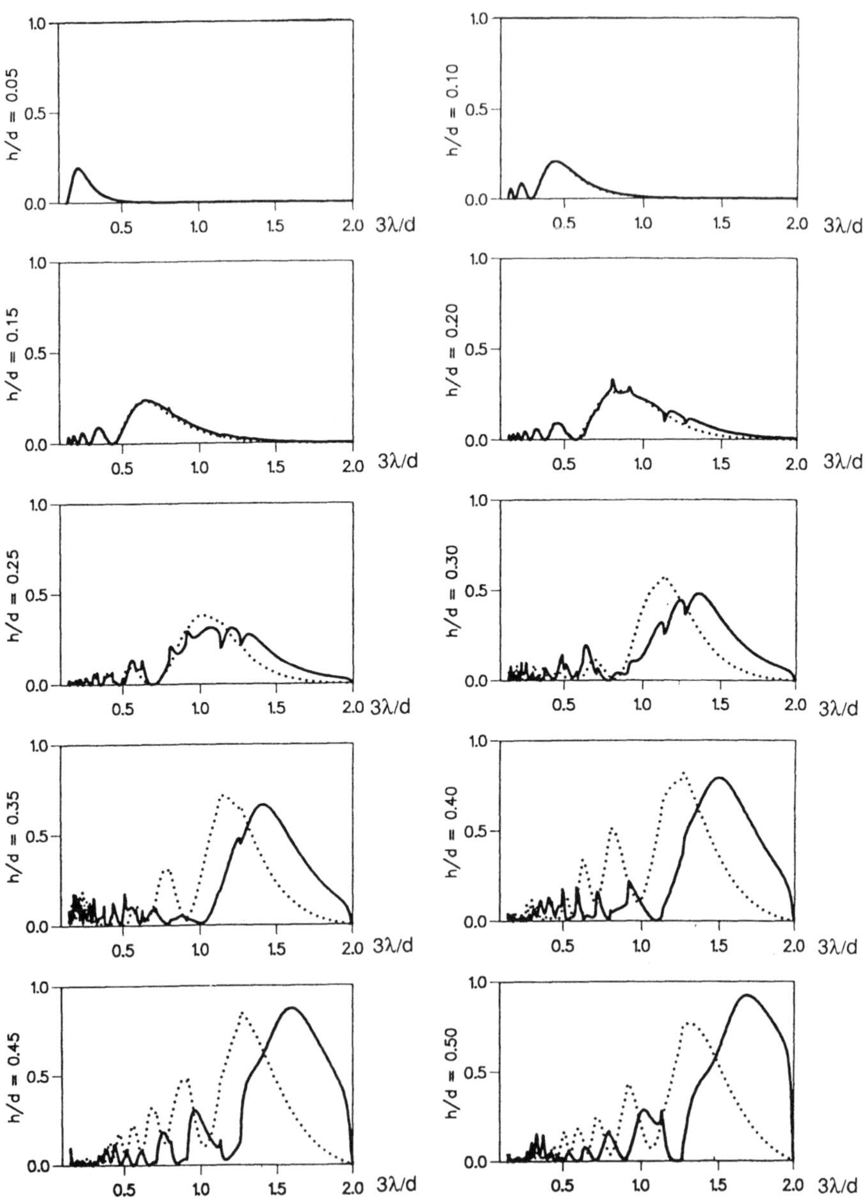

Fig.4.37 Same as Fig.4.36, except in order three.

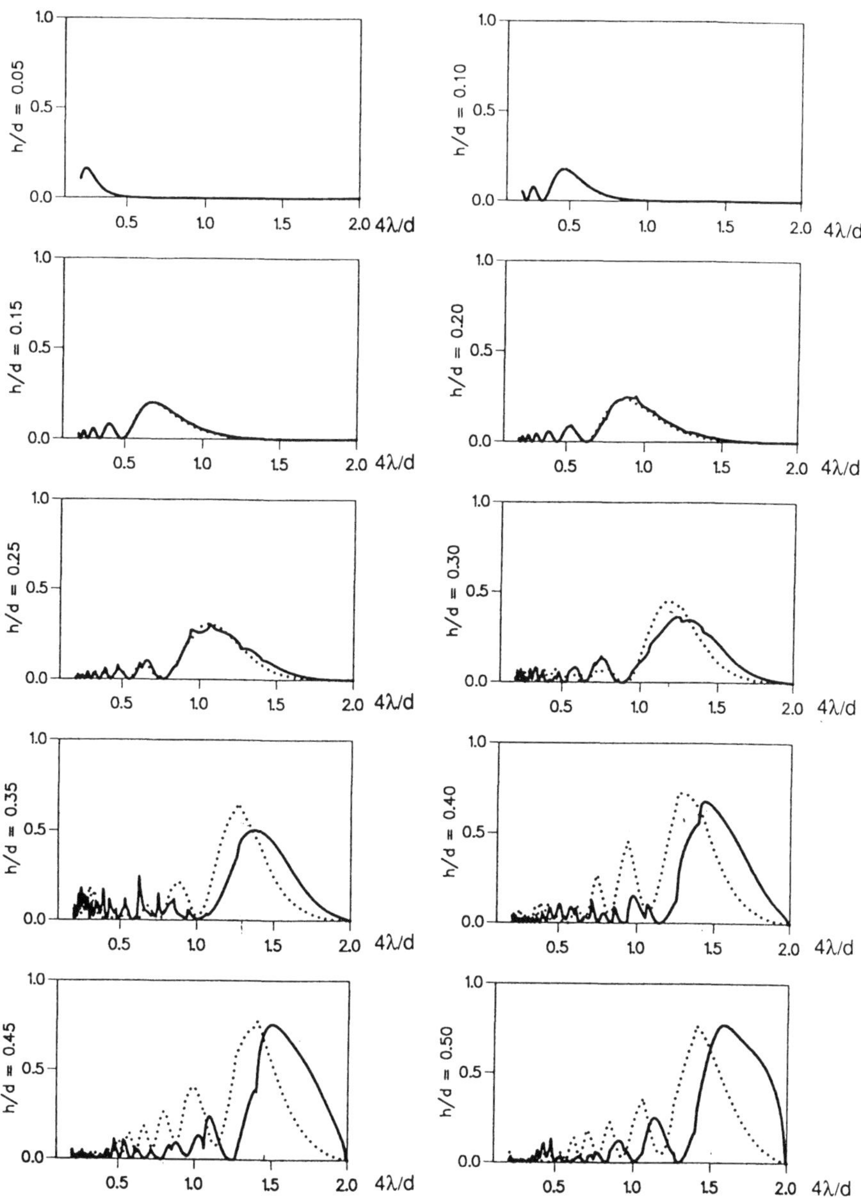

Fig.4.38 Same as Fig,4.36, except in order four.

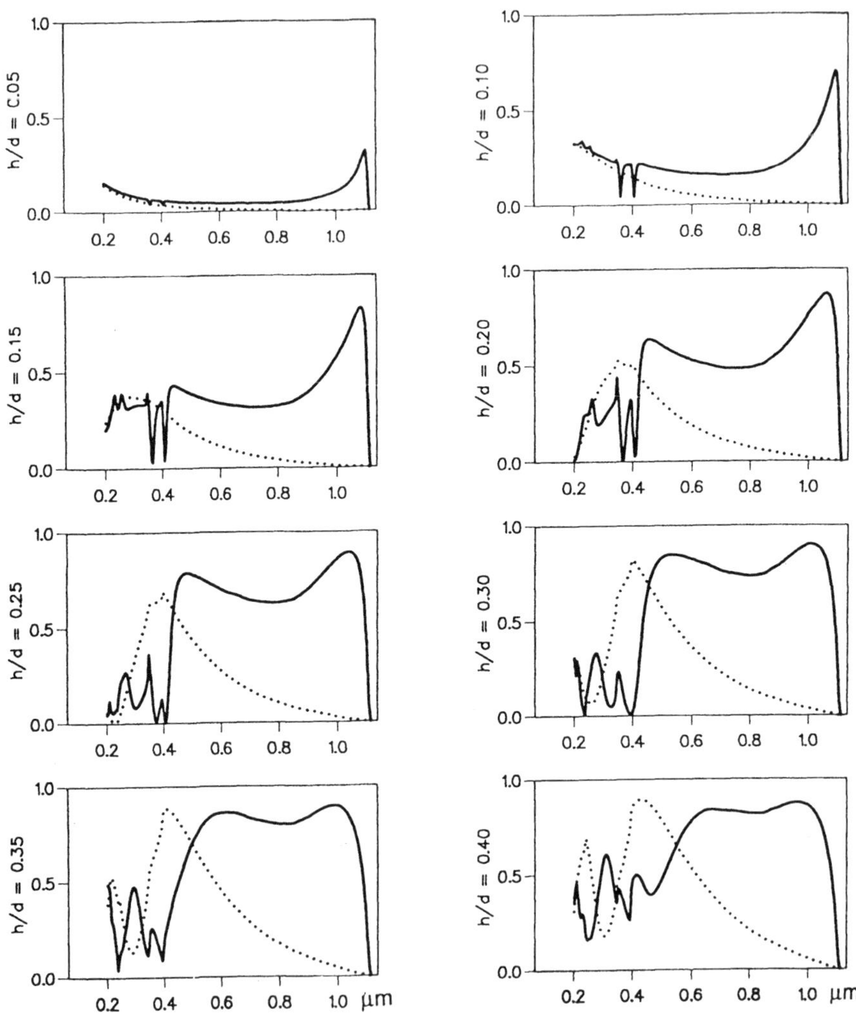

Fig.4.39 First order efficiency for 1800 gr/mm sinusoidal grating as a function of wavelength in μm. Aluminum surface, A.D. 8°. Eight levels of modulations h/d, as marked.

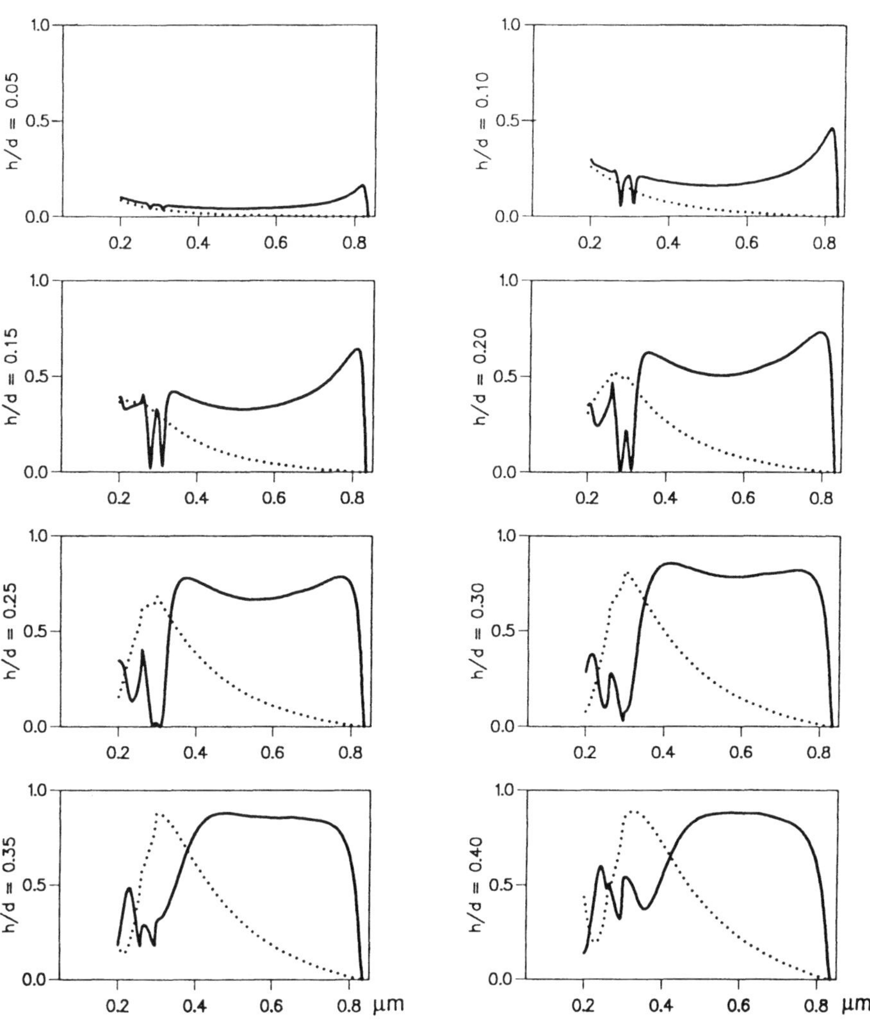

Fig.4.40 Same as Fig.4.39, except groove frequency is 2400 gr/mm.

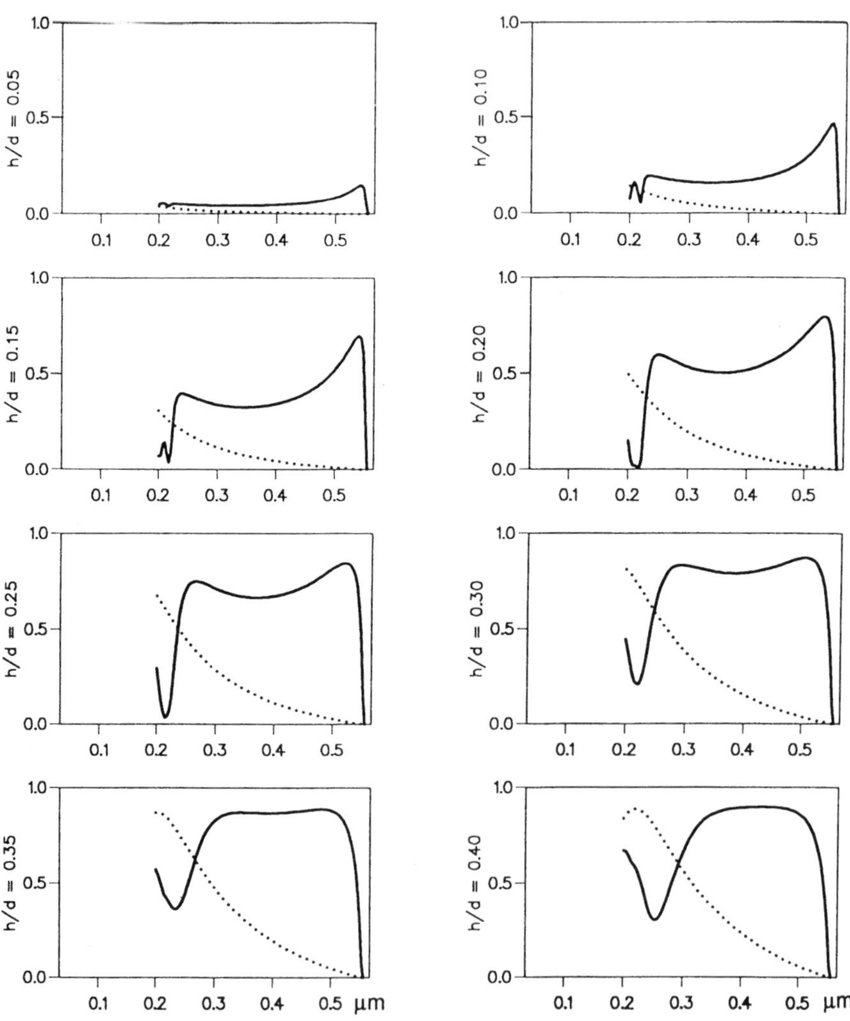

Fig.4.41 Same as Fig.4.39, except groove frequency is 3600 gr/mm.

The 2400 gr/mm curves are almost identical to the 1800 ones, despite the shorter wavelengths. The only visible difference at 3600 gr/mm is the improved flatness of the TM plane values at wavelengths above 0.3 μm (0.6 μm for 1800).

4.6.4 Absolute Efficiency of Higher Groove Frequency Sinusoidal Gratings with Silver Overcoating

To investigate the role that silver overcoatings might play in raising efficiency a set of calculations are presented in Figs.4.42 to 4.44, which may be compared to their aluminum equivalents in Figs.4.39 to 4.41.

At lower modulations there is no basic difference at 1800 gr/mm, except for a clear shift of the TM anomalies towards longer wavelengths. At the common modulations of 0.3 and 0.35 the changes in TM behavior show up positively, boosting the efficiency in the 0.4 to 0.5 μm region, somewhat more than the increase in reflectance, while in the 0.6 to 1 μm region the difference merely reflects the reflectance increase. At 2400 gr/mm the curves for silver look almost identical to the 1800, except for a sharp drop below 0.45 μm, where the low reflectance of silver would have predicted it, Fig.4.43. At 3600 gr/mm, Fig.4.44, it becomes clear that such a grating should never be silver coated, confirmed by comparison with aluminum, Fig.4.41.

4.6.5 Absolute Efficiency of Higher Groove Frequency Sinusoidal Gratings with Gold Overcoating

The effect of giving fine pitch sinusoidal gratings a gold overcoat is shown in Figs.4.45 to 4.47, for 1800, 2400, 3600 gr/mm, respectively. The simple conclusion is that such gratings are superior to aluminum gratings only at wavelengths longer than 0.7 μm, as one would expect from the relative reflectance data of gold vs. aluminum. Because of the wavelength ranges of the three gratings, this shows a small value at 1800 gr/mm, a minor one at 2400 gr/mm and none at all at 3600 gr/mm, because such a grating cannot diffract when $\lambda > 0.55$ μm.

4.7 The Efficiency Surface

The efficiency curves presented in this chapter are the most widely used, simply because a majority of gratings are used in monochromators where A.D. is held constant, and wavelength is the obvious variable. In spectrographs it is the angle of incidence that is fixed, so that coherent presentations would plot efficiency as a function of wavelength, as above, but in families of progressively increasing angle of incidence. Unless angles of diffraction are

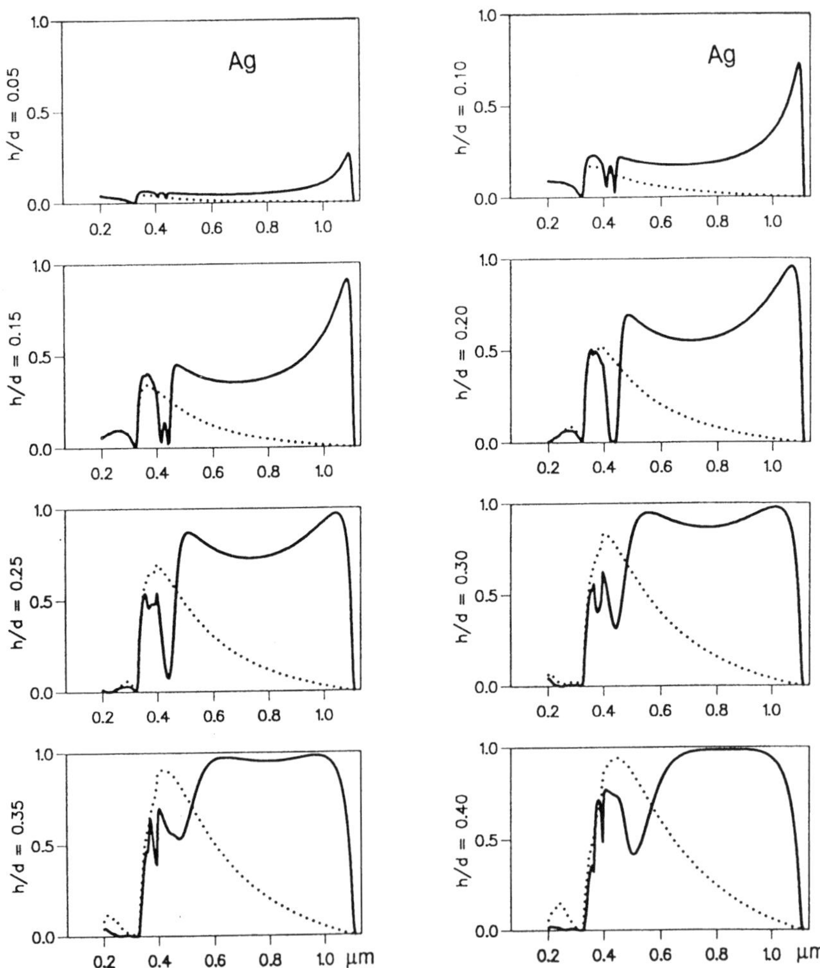

Fig.4.42 Same as Fig.4.39 (1800 gr/mm), except surface is silver.

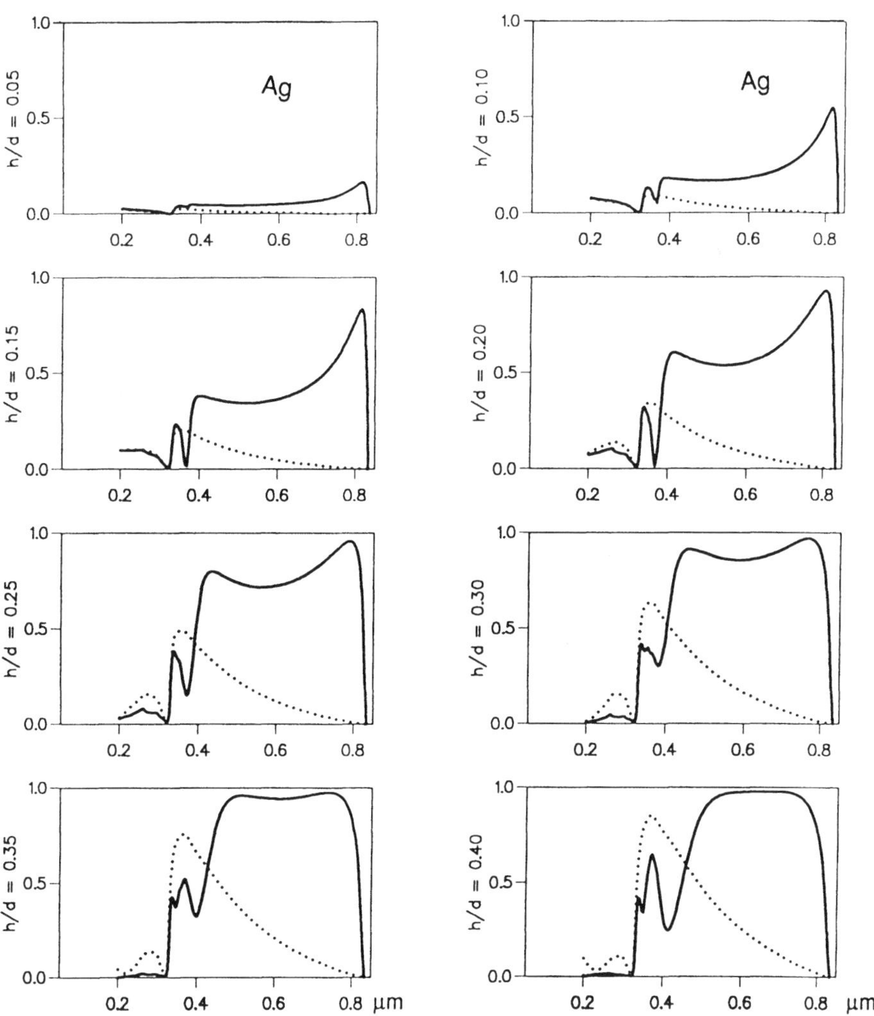

Fig.4.43 Same as Fig.4.40 (2400 gr/mm), except surface is silver.

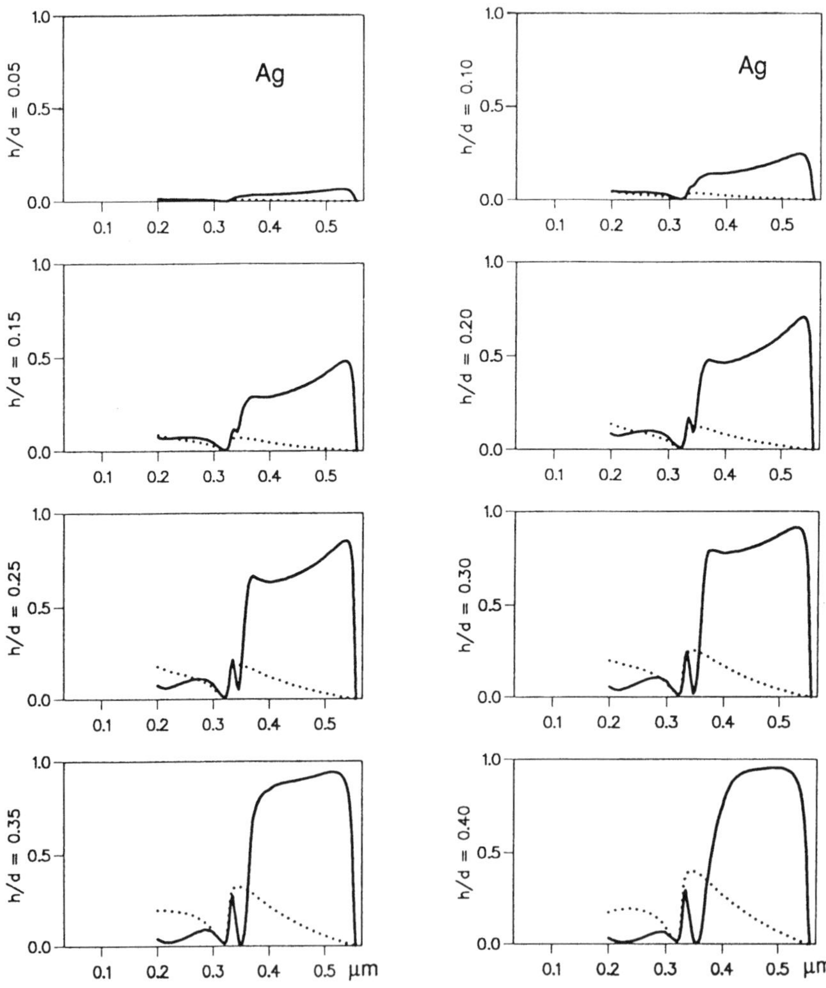

Fig.4.44 Same as Fig.4.41 (3600 gr/mm), except surface is silver.

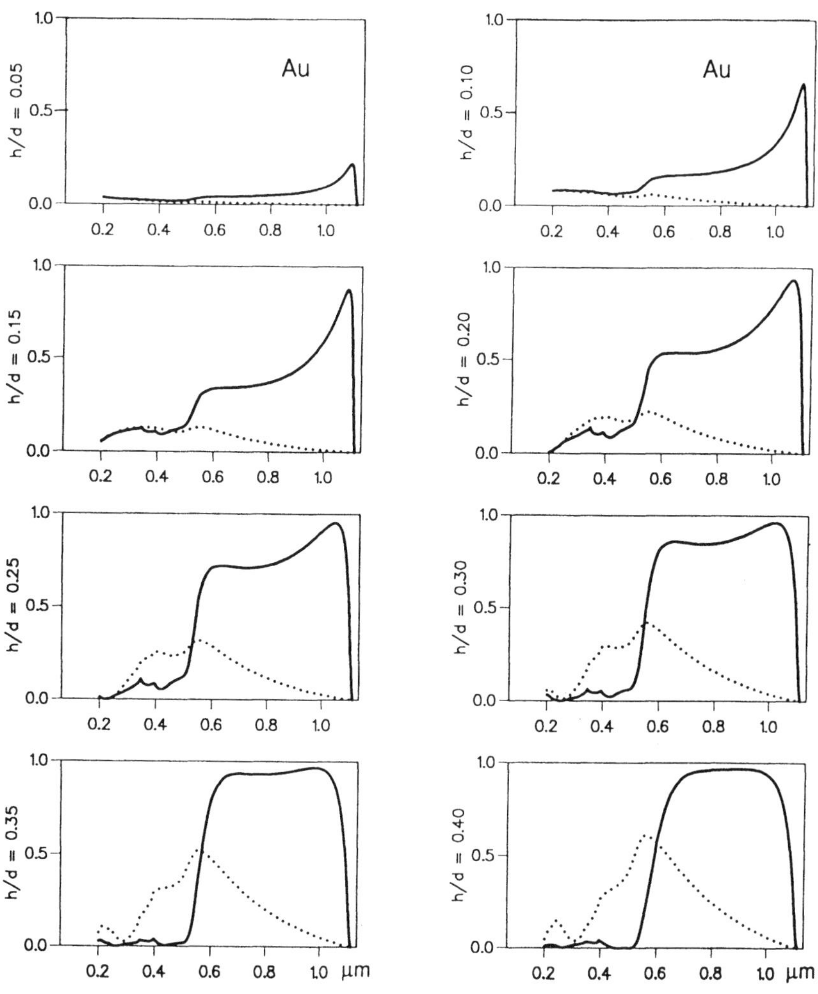

Fig.4.45 Same as Fig.4.39 (1800 gr/mm), except surface is gold.

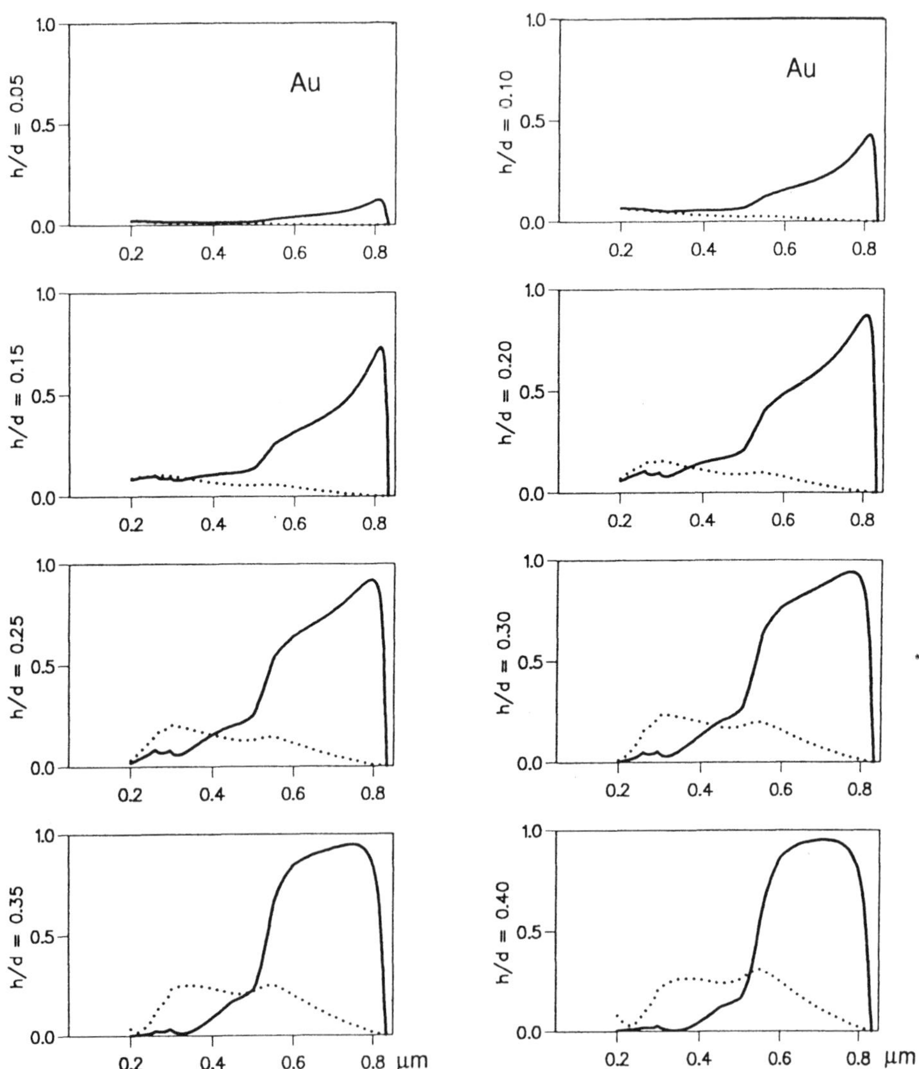

Fig.4.46 Same as Fig.4.40 (2400 gr/mm), except surface is gold.

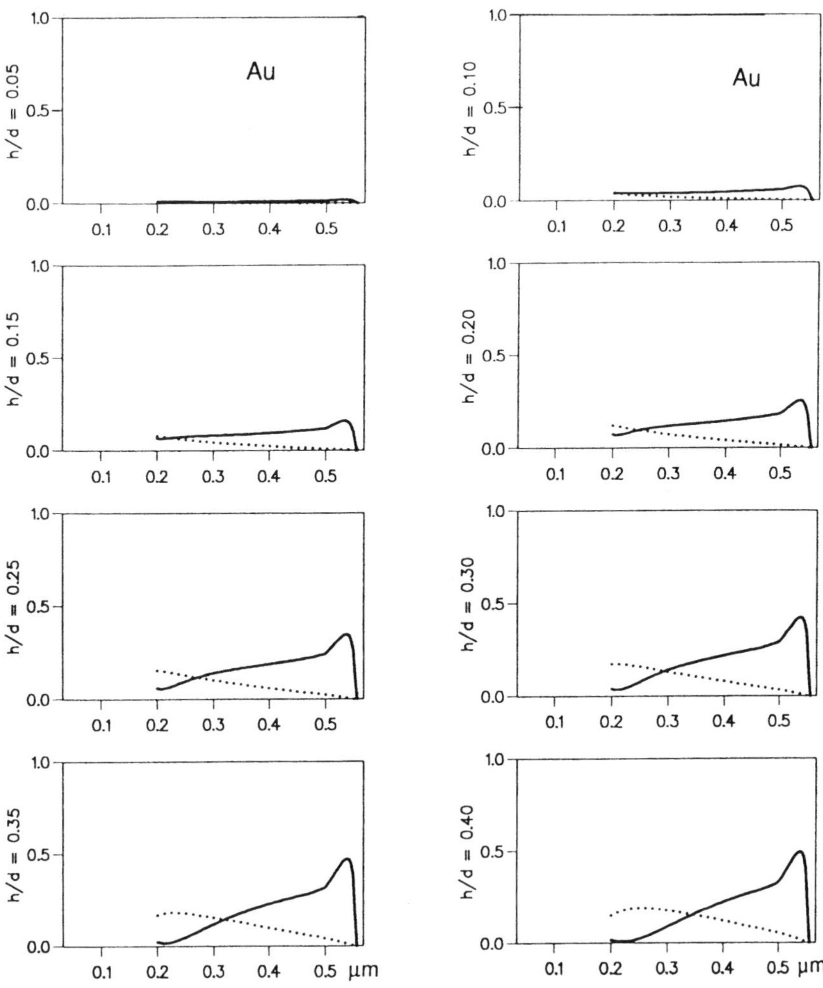

Fig.4.47 Same as Fig.4.41 (3600 gr/mm), except surface is gold.

large, the differences will not be too great.

There is a third mode of presentation useful for experimenters, although not so much for instrument designers. Here the wavelength is held constant and the angle of incidence varied from –90° to +90°. What makes this approach interesting is that it lends itself to exploring grating efficiency properties with lasers [4.8].

For a given grating all three methods must have a common base, which is a 3–dimensional efficiency surface, from which any of the others can be derived by taking an appropriate section. To collect data, or make the corresponding theoretical calculations, is not especially difficult, but becomes rather time consuming with the large number of data points necessary, especially in the TM plane. Hutley and Bird conducted such an experiment with an 830 gr/mm sinusoidal grating of medium modulation [4.9]. Ten detailed wavelength scans were conducted with as many lasers and the efficiencies plotted. The graphs were glued to cardboard and cut along the efficiency curves. They were then glued to a base board along which the X–axis represents the angles of incidence and the Y axis the respective wavelengths, Fig.4.48. The TE curves are so well behaved that the surface is easily visualized with just six of the wavelengths, but all ten are needed to picture the TM surface. The nature of the efficiency variations, especially the sharp anomalies, are very well displayed.

To picture efficiency for a constant angle of incidence requires passing a plane normal to the base at the desired angle and parallel to the Y–axis. To picture efficiency for constant angular deviation skewed planes are required, as indicated by the vertical lines drawn on the TM set.

Modern computer graphics tools would simplify the task of converting such data to a 3-D display.

4.8 Efficiency Behavior of Very Deep Gratings

Commercial gratings seldom exceed 50% modulation depth, because while not impossible to make they present problems in controlling the depth and groove profile. The difficulty lies in the fact that the casting replication process does not work for grooves with very steep slopes – the replica ends up being glued to the master. However, the study of deep groove gratings is of more than academic interest because lithographic gratings can be optically replicated (see Ch.16) at reasonable cost, and there are applications where the use of holographic masters may be justified.

Working in *transmission* under *Bragg conditions* such deep gratings can have very high efficiency (see Chs.5.10 and 5.11). Reflection gratings can achieve similar performance with relatively moderate groove depths (h/d close

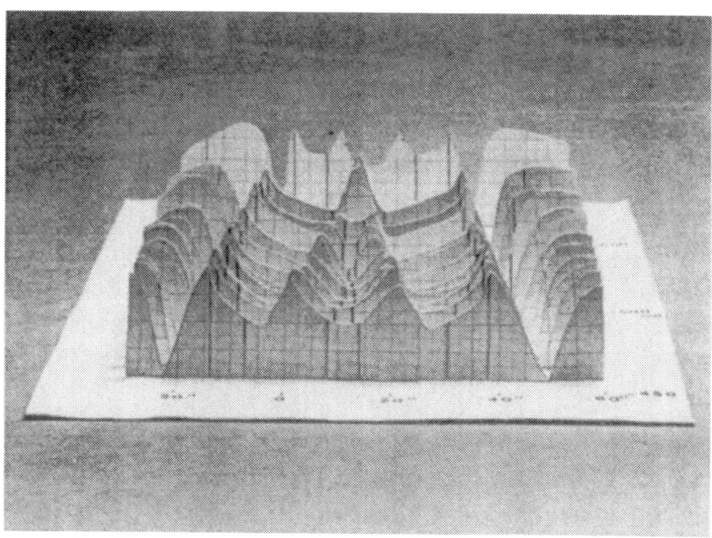

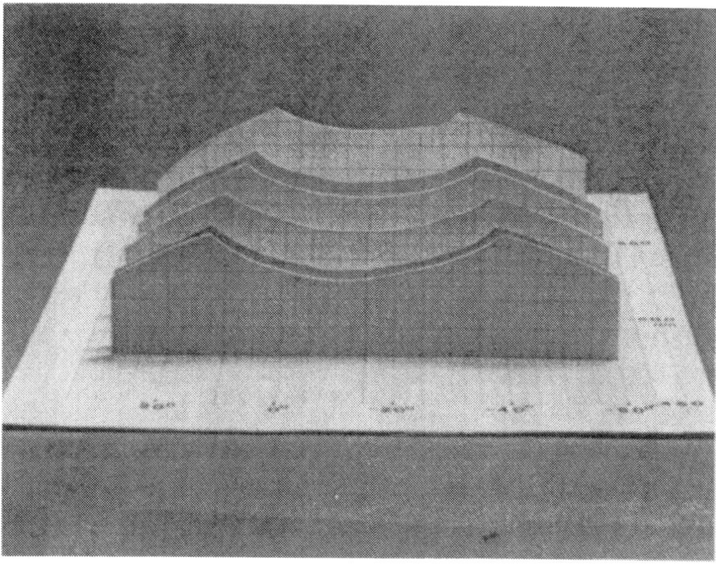

Fig.4.48 Experimental presentation of the efficiency surface of an 830 gr/mm sinusoidal diffraction grating. TM efficiencies at the top and TE below (after [4.9]).

to 0.35), but have asymmetrical response in TE and TM planes (e.g., Fig.4.28), so that non–polarized (NP) incident light will be partially polarized after diffraction, the degree strongly depending on the wavelength. Moreover, the spectral region of high NP efficiency coincides with cut–off and resonance anomalies.

The groove depth dependence of efficiency for sinusoidal gratings (Fig.4.49) reveals two regions of high NP efficiency: when h/d is close to 0.35 and to 1. The quasi periodical behavior in TE and TM plane is due to formation of curls of Poynting vector inside the grooves, as shown in Chapter 8. When considering the spectral dependence of efficiencies (Fig.4.50), the maximum in TE plane for h/d ~ 1 is moved to longer wavelengths as does the region of maximum efficiency in NP light. Thus the region of anomalies can be avoided

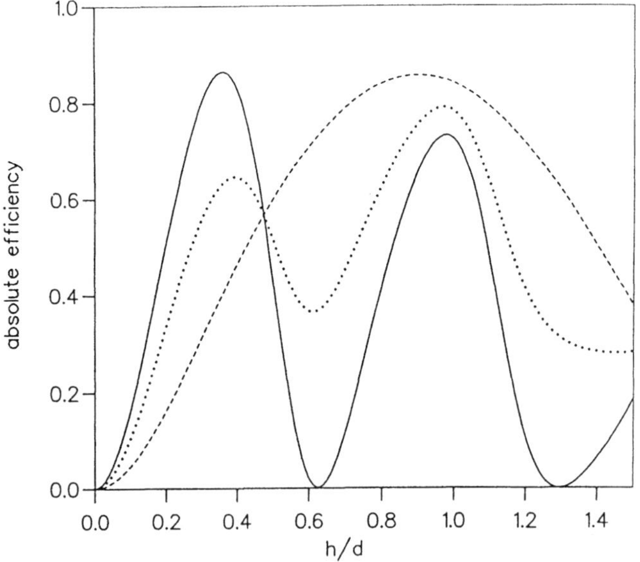

Fig.4.49 Efficiency of a sinusoidal aluminum grating with 1800 gr/mm at λ = 0.6328 μm as a function of modulation depth. Solid line TM plane, dashed line TE, dots nonpolarized light.

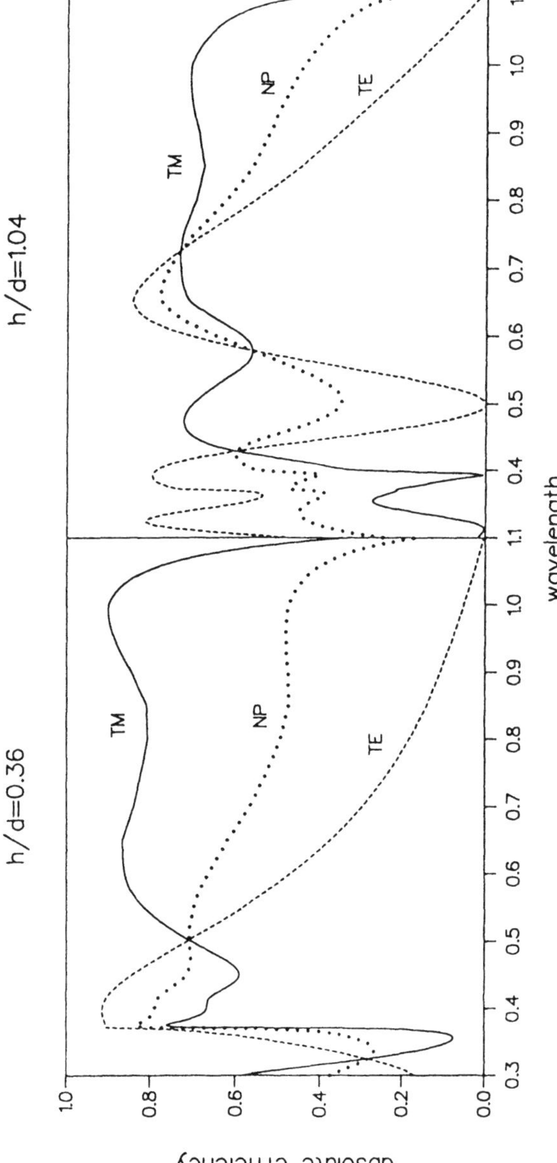

Fig.4.50 Spectral dependence of efficiency of a sinusoidal aluminum grating at two modulation depths, indicated on top of the figure. Solid line – TM plane, dashed line – TE plane, dots nonpolarized light (1800 gr/mm).

and the degree of polarization kept below 10% with a NP efficiency over 70%, while for the grating with moderate modulation depth the corresponding degree of polarization exceeds 25%.

Gratings with other profiles may have similar performance, but the equivalence rule cannot be safely used for such large groove depths. Of special interest are the properties of very deep lamellar gratings. Numerical investigations [4.10] show that they can have simultaneous blazing in TE and TM planes when the lamellar width is 5 – 10 times less than the period. Combined with the requirement to have only a single dispersive order, this puts severe limitations on manufacturing such gratings for the visible region.

4.9 Efficiency Behavior in Grazing Incidence

Reflection gratings are widely used in grazing incidence as the tuning element in dye lasers. It constitutes the classical mount for giving reasonable efficiencies in the x–ray domain, as discussed in the next section. Grazing incidence increases dispersion and the grating also acts as a beam expander/compressor. Laser applications require polarized light so that it is convenient that this mount automatically excludes the TE plane, due to its very low efficiency.

When the angle of incidence is large, most of the incident light is reflected in the specular order, just as with a plane mirror. Numerical methods again can serve to optimize grating parameters. The groove depth dependence of −1st order efficiency of sinusoidal and echelette gratings is given in Fig.4.51 [4.11] for progressively increasing groove frequency. Maximum performance is obtained at h/d = 0.2 for the holographic grating and a groove angle $\varphi_B = 12°$ for the echelette grating (the latter also corresponding to a modulation of 20%, following the equivalence rule). The numbered curves in Fig.4.51c correspond to a gradual increase of material extinction coefficient, i.e., absorption losses, and the dashed lines represent the influence of a 5 nm thick oxide layer.

Comparison of the spectral response of sinusoidal and blazed gratings (Fig.4.52) shows that blazed gratings have higher efficiency in the spectral region where there are several propagating orders ($\lambda/d<1$) while sinusoidal gratings have better performance at longer wavelengths.

When going to higher orders their maxima appear at higher modulation depths (Fig.4.53). Their use evidently increases the dispersion with an efficiency comparable to first order, but with much weaker influence of the oxide layer. However, the usable wavelength region becomes progressively narrower in higher orders.

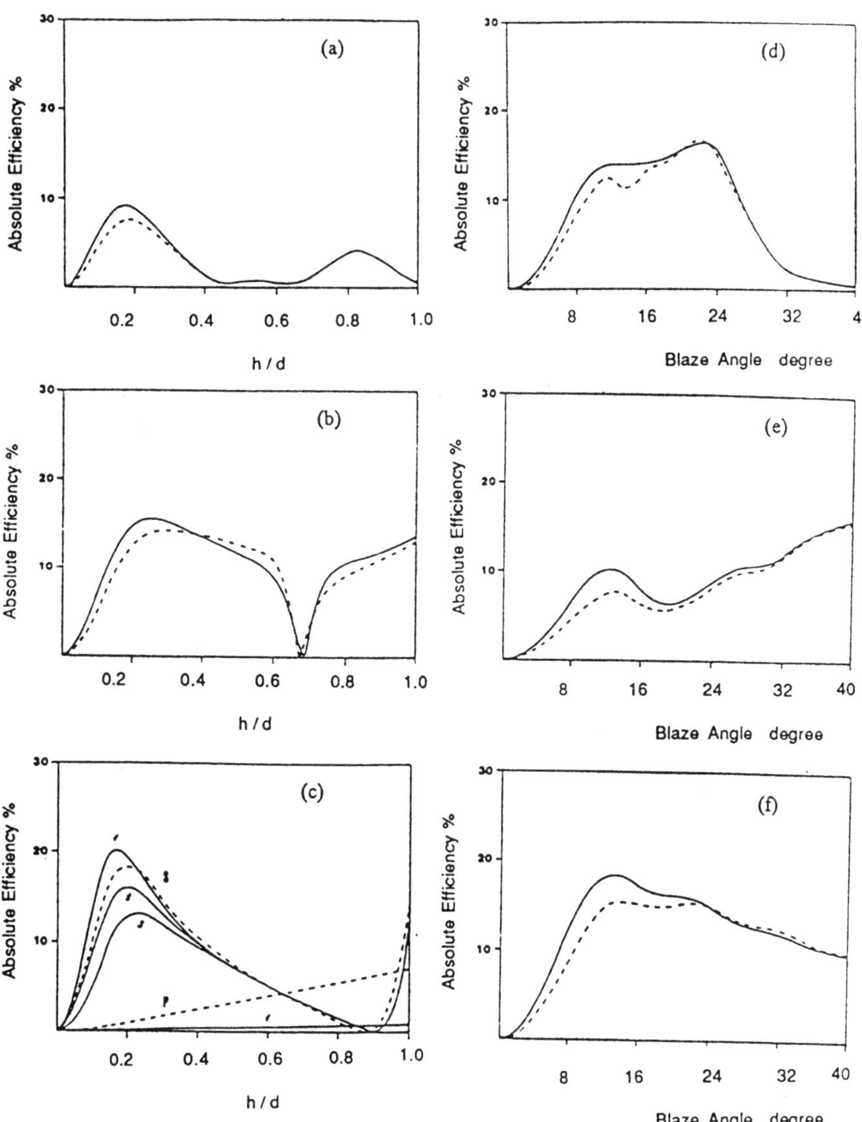

Fig.4.51 Efficiency of aluminum grating at 89° angle of incidence in TM polarization at $\lambda = 0.6328$ μm: (a) – (c) sinusoidal gratings, (d) – (f) blazed grating with 90° apex angle. (a) and (d) 1000 gr/mm, (b) and (e) 2000 gr/mm, and (c) and (f) 3000 gr/mm. Dashed curves represent the influence of a 5 nm thick oxide layer (after [4.11]).

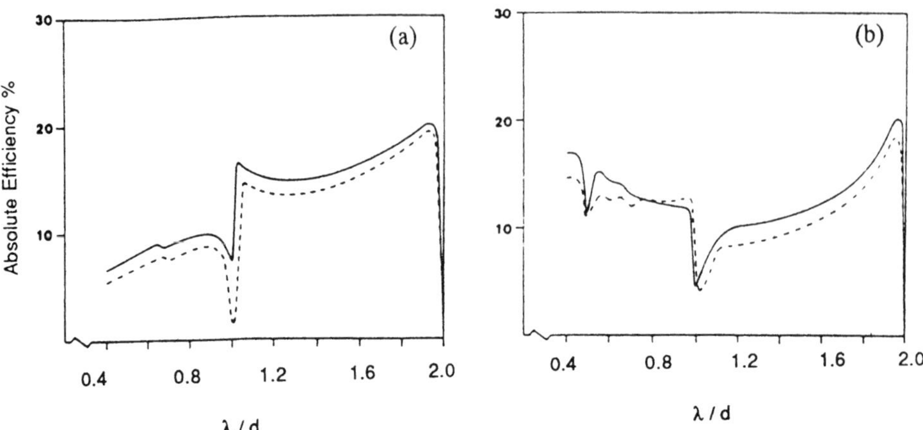

Fig.4.52 Spectral behavior of first order efficiency of aluminum grating in TM polarization at 89° angle of incidence: (a) sinusoidal profile with h/d = 0.20, (b) blazed grating with φ_B = 12° and apex angle 90° (after [4.11]).

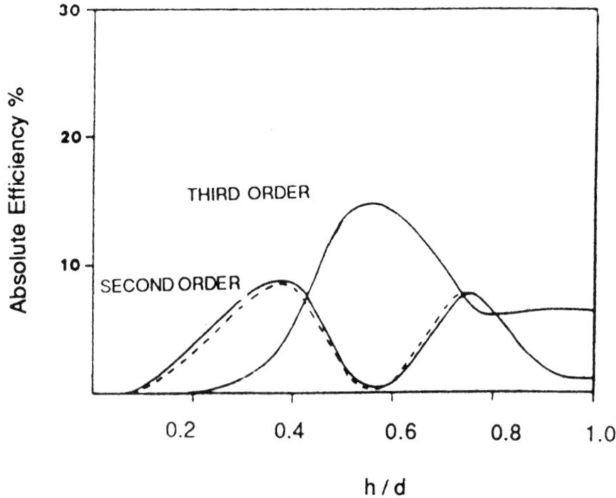

Fig.4.53 Higher order efficiencies (in %) for sinusoidal aluminum grating with 1000 gr/mm, as a function of modulation depth. TM polarization, λ = 0.6328 μm, 89° incidence angle (after [4.11]).

4.10 X–Ray Gratings

Although quite limited in their applications, gratings for the x–ray domain are a true challenge in both manufacturing and theory. Maximum performance, as in most grating applications, requires high dispersion and high efficiency, which can hardly be expected at these low wavelength to period ratios (typically 10^{-4} to 10^{-3}) and low material reflectivities. It is not even clear *a priori* whether the optical macro characteristics (refractive indices) can successfully describe x–ray scattering. Fortunately, comparison between theory and experiment [4.12] shows that Maxwell's equations with homogeneous refractive indices can be safely used down to 1Å wavelength.

In order to have reasonable reflectance it is customary to go to grazing incidence, which puts severe demands on the theory. Although appearing to be in the scalar domain grazing incidence requires electromagnetic theories, as in the case of echelles (Ch.6). Deviation from scalar theory predictions [4.13] can exceed 100% in some instances. The great number of diffraction orders, possible blazing in high orders, and usage of multilayer coatings requires special methods to avoid very large dense matrices and overflows due to growing exponential terms [4.14].

Unfortunately, there is no systematic study of grating properties in the x–ray domain, which is easily explained: wavelengths vary over 3 orders of magnitude, the grating period – at least 10 times, and material reflectivity can also change several orders of magnitude. Only recently has numerical optimization become possible [4.14], so there is no extensive comparative study.

Historically the first x–ray gratings made with were grooves scratched into glass, later followed by gold with blaze or lamellar profile (Fig.4.54a). Highest efficiency values, of 0.1 to 5% (depending on the wavelength), are obtained in low orders under grazing incidence, which calls for very low groove angles. Such shallow grooved gratings can be as difficult to rule as deep ones, due to profile form and depth control problems, so that they are sometimes derived from moderately deep gratings by material ablation using ion beams.

There is a special mounting (GMS) in which the incident light is almost parallel to the grooves (Fig.4.54b). In this mount the grating can diffract 30 to 60% of the light into dispersive orders [4.15], but the angular dispersion under these conditions is much lower than under grazing incidence perpendicular to the grooves. In the latter case the angular separation Δ_θ between the specular and the –1st order is inversely proportional to the cosine of the angle of incidence:

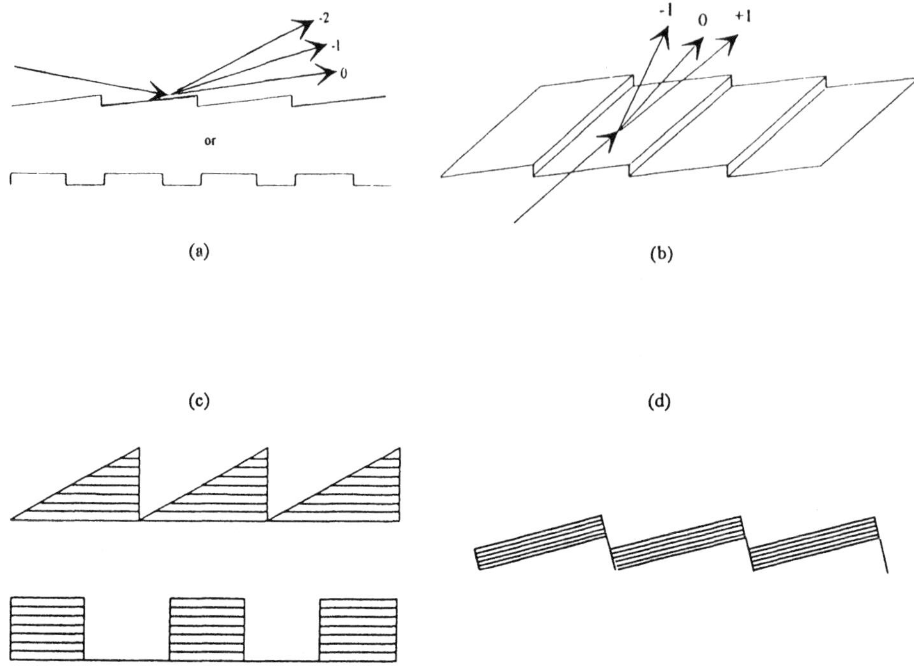

Fig.4.54 Schematic representation of several x–ray gratings: (a) gold blazed or lamellar grating in grazing incidence; (b) bare grating in GMS mount; (c) multilayer etched gratings; (d) multilayer coating deposited on a blazed grating.

$$\sin\Delta_\theta = \frac{\lambda}{d\cos\theta_i} \quad , \tag{4.3}$$

contrary to GMS mount:

$$\sin\Delta_\theta = \frac{\lambda}{d} \quad . \tag{4.4}$$

At large θ_i the difference can become significant, for example 11.5 times for 85° incidence.

Further increase of dispersion can be obtained by shortening the period or by going to higher diffraction orders. The former is quite limited and periods less than 0.1 – 0.2 μm are rarely reported. Soft x-ray achromatic holographic lithography can reach periods small as 50 nm, but the grating is limited to small areas such as 90 x 20 μm [4.16].

Blazing in higher orders requires larger groove angles and this lowers reflectivity due to the effectively smaller angle at which the incident light 'hits' the large facet. It is possible to increase reflectivity with an overcoat of alternating layers of lower and higher refractive indices (typically tungsten and carbon, with a total number of bilayers of at least 50), [4.17]. The two basic concepts are presented in Figs.4.54c and d. In the first case the grating is etched into a plane multilayer coating, while in the second approach the coating is deposited onto a blazed grating. The two types can have similar performance in lower orders (20 to 30%) [4.14, 18], but the second type can be blazed in very high orders (50 to 100) with much larger dispersion. However, it faces much stronger technical limitations. Profile deformations can lower grating performance significantly, readily observed in higher orders. Moreover, it is well known that lithographic methods (see Ch.16) lead to "stitching" errors, which can degrade the performance of the grating and multilayer coatings at these short wavelengths, so that it is advisable to use classically ruled gratings for the system presented in Fig.4.54d, because of their much smoother groove facets.

Fortunately, due to the very low λ/d ratio and the low refractive index, resonance guided wave excitation plays no role here, in contrast to what happens in the visible spectrum, Ch.8.

4.11 Single Wavelength Efficiency Peak in Unpolarized Light

A frequent concern in applications, such as wavelength multiplexing, is to find a grating that has high efficiency at a single wavelength, in both planes of polarization and at an angle of diffraction that gives the necessary dispersion. The problem is sometimes termed "perfect blazing". It is easy to obtain in the scalar domain, where the angle of diffraction is $< 5°$, but in this instance we need ~ 30° (or $\lambda/d \sim 1$). Typically the efficiency desired will be > 85%, perhaps even 90%. It is routinely obtainable in the TM plane, so the problem is to try to combine it with the TE. One advantage is that the systems in question will usually operate under Littrow conditions (A.D. 0°).

The efficiency curves in this chapter cover both triangular and sinusoidal groove shapes. The first one of interest is Fig.4.3, which shows that for perfectly conducting surface the required efficiency is obtainable near λ/d = 0.9, even though this figure is for A.D. 8°, which means that for A.D. = 0 we can expect a small increase in efficiency. For gold coatings in the infrared (1.3 or 1.5 μm is typical) reflectance is near 98%. Thus we can expect such gratings to deliver as much as 90% efficiency, provided the wavelength band width is restricted to ±2%. Standard ruled gratings have been shown to provide such performance.

Given the great interest in gratings made by holographic procedures it is important to look for sinusoidal gratings that can accomplish the same results. From Fig.4.28 we can conclude that at 0.35 modulation the efficiency behavior is almost identical to that of Fig.4.3. However, a close look at the adjacent modulations indicates that there is little leeway in the groove depth, unlike for the ruled grating. The situation can be improved by adopting a modified groove geometry. It was shown by Iida *et al.* that if one makes a normal exposure of a resist coated blank one can take advantage of the non-linear properties of the resist to obtain a more tolerant profile [4.19]. They were able to obtain 95% efficiency at 1.3 μm. It is interesting to note that a similar horn-like profile was proposed much earlier in 1980 by A. Roger [4.20] as a result of a numerical solution of the inverse problem. The groove profile is determined to correspond to maximum efficiency at a given wavelength. Theoretical profiles which ensure 100% relative efficiency are found for gratings supporting two and four diffraction orders. The main technological problems are the tight tolerances and the repeatability of the results.

Section 4.8 presented a different choice: very deep grooves with $h/d \sim 1$, which moves the peak efficiency in unpolarized light towards longer wavelengths.

4.12 Conclusions

The purpose of this chapter has been to produce families of efficiency curves that cover in steps sufficiently small the effective behavior of almost any plane grating likely to be used in spectrometric instrument design. Differences between triangular grooves (blazed echelettes) and sinusoidal (holographic) are readily perceived. The role played by varying angular deviation is well illustrated.

Most of the curves are for 1200 gr/mm gratings, but the effect of going to groove frequencies up to 3600 is made evident. Not included is the effect of going to lower groove frequencies, since any changes in curve shape will be minor as corresponding wavelengths increase. This means their behavior progressively approaches that of perfect reflectors. The wavelength scale can be ratioed in direct proportion to the groove spacing.

Higher order behavior, up to fourth, is also displayed and shows distinctive differences between triangular and sinusoidal groove shapes.

The possible advantages of going to silver or gold overcoatings are easily appreciated for $\lambda > 0.5\mu m$, but show up even more at longer wavelengths not presented here.

Some special cases are described, such as gratings with very deep grooves, as well as the concerns that arise with gratings for the x–ray region of the spectrum.

References

4.1 E. G. Loewen, M. Nevière, and D. Maystre: "Grating efficiency theory as it applies to blazed and holographic gratings," Appl. Opt. **16,** 2711-2721 (1977).

4.2 D. Maystre, M. Nevière, and R. Petit: "Experimental verification and applications of the theory," in *Electromagetic theory of gratings*, R. Petit, ed., Vol.22, Topics in Current Physics, Springer, 1980, ch.6.

4.3 E. Popov, "Light diffraction by relief gratings: a microscopic and macroscopic view", in *Progress in Optics*, ed. E. Wolf (Elsevier, Amsterdam, 1993) v. **XXXI**, pp.139–187.

4.4 *American Institute of Physics Handbook*, 2nd edition (McGraw-Hill, New York, 1963), pp.6-107.

G. Hass and J. E. Waylonis: "Optical constants and reflectence and transmittance of evaporated aluminum in the visible and ultraviolet," J. Opt. Soc. Am. **51**, 719-722 (1961).

R. P. Madden, L. R. Canfield, and G. Hass: "On the vacuum-ultraviolet reflectance of evaporated aluminum before and during oxidation," J. Opt. Soc. Am. **53**, 620-626 (1963).

G. Jacobus, R. Maddden, and L. Canfield: "Reflecting films of Pt for the VUV," J. Opt. Soc. Am. **53**, 1084-1088 (1963).

L. Canfield, G. Hass, and W. Hunter: "The Optical properties of evaporated gold in the VUV from 30 to 200 nm," Jl. de Physique, **25**, 124-129 (1964).

L. R. Canfiled and G. Hass: "Reflectance and optical constants of evaporated copper and silver in the vacuum ultraviolet from 1000 to 2000 Å," J. Opt. Soc. Am. **55**, 61-64 (1965).

J. Samson, J. Padur, and A. Sharma: "Reflectance and relative transmittance of laser deoporited Ir in the VUV," J. Opt. Soc. Am., **57**, 966-967 (1967).

J. Cox, G. Hass, and J. Waylonis: "Further studies on LiF overcoated Al mirrors with highest reflectance in the VUV," Appl. Opt., **7**, 1535-1539 (1968).

J. Osantowski, W. Hunter, and G. Hass: "Reflectance of Al overcoated with MgF2 and LiF in the wavelength region of 160 to 30 nm at various angles of incidence," Appl. Opt., **10**, 540-544 (1971).

D. Burge, H. Bennett, and E. Ashley: "Effect of atmospheric exposure on the infrared reflectance of silvered mirrors with and without protective coatings," Appl. Opt., **12**, 42 (1973).

J. Cox, G. Hass, J. Ramsey, and W. Hunter: "Optical constants of evaporated Os in

VUV," J. Opt. Soc. Am., **63**, 435-438 (1973).

M. Blanc and A. Malherbe: "Applications spatiales de traitements réfléchissants dans la bande 30 - 200 nm," J. Optics (Paris), **8**, 195-199 (1977).

W. Hunter, D. Angel, and G. Hass: "Optical properties of evaporated Pt films in the VUV from 220 to 15 nm," J. Opt. Soc. Am., **69**, 1695-1699 (1979).

4.5 A. Marechal and G. W. Stroke: "Sur l'origine des effets de polarisation et de diffraction dan les réseaux optiques," C. R. Ac. SC **249**, 2042 – 2044 (1980).

4.6 M. Breidne and D. Maystre: "Equivalence of ruled, holographic, and lamellar gratings in constant deviation mountings", Appl. Opt., **19**, 1812–1817 (1980).

4.7 E. G. Loewen, D. Maystre, R. C. McPhedran, and I. Wilson, "Correlation between efficiency of diffraction gratings and theoretical calculations over a wide spectral range," Proc. ICO Conf. On Optical Methods, Japan Jl. Appl. Physics **14**, Suppl.1, 143–152 (1975).

4.8 R. C. McPhedran and D. Maystre, "A detailed theoretical study of anomalies of a sinusoidal diffraction grating," Optica Acta **24**, 413 – 421 (1974).

4.9 M. C. Hutley: *Diffraction Gratings*, (Academic Press, London, 1982), ch.5.

4.10 J. L. Roumiguieres, D. Maystre, and R. Petit: "On the efficiency of rectangular–groove gratings," J. Opt. Soc. Am. **66**, 772-775 (1976).

4.11 L. B. Mashev, E. K. Popov, and E. G. Loewen: "Optimization of the grating efficiency in grazing incidence," Appl. Opt. **26**, 4738-4741 (1987).

4.12 V. Martynov, B. Vidal, P. Vincent, M. Brunel, D. V. Roschupkin, Yu Agafonov, A. Erko, and A. Yakshin: "Comparison of modal and differential methods for multilayer gratings," Nucl. Instr. Meth. Phys. Res. A **339**, 617-625 (1994).

4.13 A. J. F. DenBoggende, P. A. J. DeKopte, P. H. Videler, A. C. Brinkman, S. M. Kahn, W. W. Craig, C. J. Hailey, and M. Nevière: "Efficiency of X–Ray reflection gratings," in *X–Ray Instruments, multilayers and sources*, ed. J. F. Marshall, SPIE, **982**, 283-298 (1988).

4.14 M. Nevière: "Multilayer coated gratings for x-ray diffraction: differential theory," J. Opt. Soc. Am. A **8**, 1468-1473 (1991).

4.15 P. Vincent, M. Nevière, and D. Maystre, "X–ray gratings: the GMS Mount," Appl. Opt., **18**, 1780–1783 (1979).

4.16 M. Wei, E. H. Anderson, and D. T. Attwood: "Fabrication of ultrahigh resolution gratings for x–ray spectroscopy," Proc. OSA Workshop on Diffractive Optics, Rochester, June 1994, 91–94 (1994).

4.17 E. Spiller: "Evaporated multilayer dispersion elements for soft x-rays," AIP Conf. Proc. **75**, *Low Energy X-ray Diagnostics* (Monterey), D. T. Attwood and B. L. Henke, eds., pp. 124-130, AIP, New York (1981).

4.18 A. I. Erko, B. Vidal, P. Vincent, Yu. A. Agafanov, V. V. Martynov, and D. V.

Roshupkin: "Multilayer gratings efficiency: numerical and physical experiments", Nucl. Instr. Meth. Phys. Res. A **333**, 599–606 (1993).

4.19 M. Iida, H. Hagiwara, and H. Asakura: "Holographic Fourier diffraction gratings with a high diffraction efficiency optimized for optical communications systems," Appl. Opt., **31**, 3015–3019 (1992).

4.20 A. Roger: "Grating profile optimization by inverse scattering methods," Opt. Commun. **32**, 11-13 (1980).

Additional Reading

M. Breidne: "Influence of the groove profile on the efficiency of diffraction gratings," Thesis, The Royal Institute of Technology, Stockholm (1981).

M. Breidne and D. Maystre: "A systematic numerical study of Fourier gratings", J. Optics (Paris) **13**, 71-79 (1982).

R. Boyd, J. Britten, D. Decker, B. Shore, B. Stuart, M. Perry, and L. Li: "High-efficiency metallic diffraction gratings for laser applications," Appl. Opt. **34**, 1697-1706 (1995).

A. Hessel, J. Schmoys, and D. Y. Tseng: "Bragg-angle blazing of diffraction gratings," J. Opt. Soc. Am. **65**, 380-384 (1975).

M. C. Hutley and V. M. Bird: "A detailed experimental study of the anomalies of a sinusoidal diffraction grating," Opt. Acta **20**, 771-782 (1973).

M. C. Hutley and D. Maystre: "Total absorption of light by a diffraction grating," Opt. Commun. **19**, 431-436 (1976).

E. V. Jull, J. W. Heath, and G. R. Ebbeson: "Gratings that diffract all incident enegry," J. Opt. Soc. Am. **67**, 557-560 (1977).

H. A. Kalhor and A. R. Neureuther: "Effect of conductivity, groove shape and physical phenomena on design of diffraction gratings," J. Opt. Soc. Am. **63**, 1412-1418 (1973).

R. C. McPhedran and M. D. Waterworth: "Blaze optimization for triangular profile gratings," Opt. Acta 20, 3, 177-191 (1973).

D. Maystre and R. Petit: "Brewster incidence for metallic gratings," Opt. Commun. **17**, 196-200 (1976).

D. Maystre, M. Cadilhac, and J. Chandezon: "Gratings: a phenomenological approach and its applications, perfect blazing in a non-zero deviation mounting," Opt. Acta **28**, 457-476 (1981).

M. A. Ordal, L. L. Long, R. J. Bell, S. E. Bell, R. R. Bell, R. W. Alexander, Jr., and C. A. Ward: "Optical properties of the metals Al, Ca, Cu, Au, Fe, Pb, Ni, Pd, Pt, Ag, Ti, and W in the infrared and far infrared," Appl. Opt. **22**, 1099-1119 (1983).

A. Roger: "Grating profile optimization by inverse scattering method," Opt. Commun. **32**, 11-13 (1980).

A. Wirgin and R. Deleuil: "Theoretical and experimental investigation of a new type blazed grating," J. Opt. Soc. Am. **59**, 1348-1357 (1969).

R. W. Wood: "Anomalous diffraction gratings," Phys. Rev. **48**, 928-936 (1935).

R. W. Wood: "On a remarkable case of uneven distribution of light in a diffraction grating spectrum," Phil. Mag. **4**, 396-402 (1902).

X-Ray Gratings

T. W. Barbee, Jr.: "Combined microstructure x-ray optics," Rev. Sci. Instrum. **60**, 1588-1595 (1989).

M. Berland, P. Dhez, M. Nevier, and J. Flamand: "X-ray ultraviolet grating measurements at LURE: comparison with electromagnetic theory predictions," SPIE *Reflecting Optics for Synchrotron Radiation*, v. **315**, 155-159 (1981).

W. C. Cash, Jr.: "X-ray optics. 2: A technique for high resolution spectroscopy," Appl. Opt. **30**, 1749-1759 (1991).

P. G. Harper and S. K. Ramchurn: "Multilayer theory of x-ray reflection," Appl. Opt. **26**, 713-718 (1987).

W. R. Hunter: "Diffraction gratings and mountings for vacuum ultraviolet spectral region," ch.2, pp. 63-176, *Spectrometric Techniques*, v. **IV**, G. Vanasse, ed. (Academic, London, 1985).

W. Jark and M. Neviere: "Diffraction efficiencies for the higher orders of a reflection grating in the soft x-ray region: comparison between theory and experiment," Appl. Opt. **26**, 943-948 (1987).

U. Kleineberg, K. Osterried, H.-J. Stock, D. Menke, B. Schmiedeskamp, D. Fuchs, P. Müller, F. Scholze, K F. Heidemann, B. Nelles, and U. Heinzmann: "Mo/Si multilayer-coated ruled blazed gratings for the soft-x-ray region," Appl. Opt. **34**, 6506-6512 (1995).

M. P. Kowalski, T. W. Barbee, Jr., R. G. Cruddace, J. F. Seely, J. C. Rife, and W. R. Hunter: "Efficiency and long-term stability of a multilayer-coated, ion-etched blazed holographic grating in the 125-133-Å wavelength region," Appl. Opt. **34**, 7338-7346 (1995).

J. M. Lerner, J. Flamand, A. Thevenon, and M. Neviere: "Discussion of the relative efficiency in the vacuum ultraviolet of diffraction gratings with laminar, sinusoidal and triangular grooves," Opt. Engin. **22**, 220-223 (1983).

E. G. Loewen and M. Neviere: "Simple selection rules for VUV and XUV diffraction gratings," Appl. opt. **17**, 1087-1092 (1978).

S. Mrowka, Ch. Martin, St. Bowyer, and R. Malina: "Evaluation of gratings for the Extreme Ultraviolet Explorer," SPIE, v. **689**, 23-23 (1986).

M. Neviere and J. Flamand: "Electromagnetic theory as it applies to X-ray and XUV gratings," Nucl. Instr. Meth. **172**, 273-279 (1980).

M. Neviere, J. Flamand, and J. M. Lerner: "Optimization of gratings for soft X-ray monochromator," Nucl. Instr. Meth. **195**, 183-189 (1982).

M. Neviere, P. Vincent, and D. Maystre: "X-ray efficiencies of gratings," Appl. Opt. **17**, 843-845 (1978).

J. F. Seely, R. G. Cruddace, M. P. Kowalski, W. R. Hunter, T. W. Barbee, Jr., J. C. Rife, R. Eby, and K G. Stolt: "Polarization and efliciency of a concave multilayer grating in the 135-250-Å region and in normal-incidence and Seya-Namioka mounts," Appl. Opt. **34**, 7347-7354 (1995).

J. F. Seely, M. P. Kowalski, W. R. Hunter, T. W. Barbee, Jr., R. G. Cruddace, and J. C. Rife: "Normal-incidence efliciencies in the 115-340-Å wavelength region of replicas of the Skylab 3600-line/mm grating with multilayer and gold coatings," Appl. Opt. **34**, 6453-6458 (1995).

E. E. Scime, E. H. Anderson, D. J. McComas, and M. L. Schattenburg: "Extreme-ultraviolet polarization and filtering with gold transmission gratings," Appl. Opt. **34**, 648-654 (1995).

B. Vidal and P. Vincent: "Metallic multilayers for x rays using classical thin-film theory," Appl. Opt. **23**, 1794-1801 (1984).

P. Vincent, M. Neviere, and D. Maystre: "Computation of the efficiencies and polarization effects of XUV gratings used in classical and conical mountings," Nucl. Instr. Meth. **152**, 123-126 (1978).

Chapter 5

Transmission Gratings

5.1 Introduction

Although the majority of spectrometric instruments are designed around reflection gratings there are a number of situations where transmission gratings are preferred. One of them arises from the fact that any camera or telescope can be converted into a spectrograph by interposing a transmission grating in front of the objective. Typical applications arise when the source presents itself as a luminous point or line, like falling meteors, lightning, or solar eclipses. Normally these gratings are formed on plane blanks, but they can also be generated on the face of a prism. As grating prisms, or GRISMs, they are especially convenient for telescope prime focus spectrographs, in association with array detectors, where a great advantage is to have the central wavelength with no deviation.

Another important application of transmission gratings has no connection with spectrometry. This encompasses their use as a director of monochromatic beams of light, which may involve just a single beam, but more commonly the interest is in two or three beam systems, acting as narrow angle beam splitters. An opposite case requires generating a large number of beams of equal intensity, so-called fan-out gratings, which are used in optical computing, lens testing and other applications.

Transmission gratings are usually made as plastic film replicas on a glass substrate. However, sometimes they are formed by deposition of regular patterns of dielectric bars, with the aid of suitable masking. An inverse method is to etch groove patterns into the glass. Another approach is to photograph a stationary interference fringe field in a high resolution material such as photoresist or photopolymer. While most transmission gratings have a use limited to the visible spectrum, it is possible to extend their performance into the near UV (250 nm) as well the near IR (2.5 μm), with choice of appropriate materials. Groove frequencies for standard gratings seldom exceed 600 gr/mm, because over the typical wavelength domains involved, their blaze angles reach or even exceed the total reflection limit. In other words efficiency drops off rapidly at higher groove frequencies. With GRISMs the limit increases to 1200 gr/mm. Transmission gratings are thus excluded from high dispersion applications, except in the special cases of echelles and Bragg diffraction. In the latter configuration angles of incidence and diffraction are equal, groove frequency and depth both high, and exceptionally high efficiency is obtainable but limited to a single wavelength.

Transmission gratings are usually made with triangular, sinusoidal or rectangular groove shapes. Triangular, or blazed grooves are designed to direct as much light as possible of a specified wavelength band into one of the low orders, usually the first, or to deliver a particular ratio of first to zero order at a given wavelength. At normal incidence the symmetry of rectangular grooves leads to equal energy in both plus and minus first orders, while the fraction devoted to the zero order at one wavelength can vary from near 0 to over 90%. Sinusoidal gratings share the property of symmetry, but not quite the wide control over zero order. All these types are generally designated as phase gratings, because their behavior is controlled by the physical phase retardation between light originating from successive grooves. Other groove forms may also be used, such as Vs, or a special geometry for multiple order generation.

Amplitude gratings, or Ronchi rulings, have a line pattern that is alternately opaque and transmitting. Their main application is in dimensional metrology, where low diffraction efficiency plays a minor role, and where the ability to replicate by lithographic printing methods is a great advantage. They are also useful in certain optical testing.

5.2 Transmission Grating Physics

Incident light is usually perpendicular to either the back or front surface of transmission gratings (i.e., $\theta_i = 0$), in which case the grating dispersion equation simplifies to

$$m\,\lambda = d\,\sin\theta_d \quad , \tag{5.1}$$

where m is the grating order, λ the wavelength, d the groove spacing, and θ_d the diffraction angle with respect to the grating normal. This equation does not depend in any way on the shape of the groove, nor on whether the grating is of the phase or amplitude type. The ray path is shown in Fig.5.1.

Diffraction efficiency behavior of course depends on groove geometry, with the physics of transmission gratings somewhat simpler than for reflection gratings because there is no metal surface. In addition they operate largely in the scalar domain. One consequence is the near absence of polarization effects, often a useful attribute.

The peak efficiency of a blazed (triangular) groove transmission grating in the scalar region occurs when the refraction of the incident beam through the mini-prism that constitutes a groove is in the same direction as that given by the diffraction equation (5.1).

Unlike reflection gratings, where the blaze angle and groove angle are approximately the same (at least in the Littrow mount) the groove angle here is much larger than the blaze angle. Blaze angle is defined as the diffraction angle of the wavelength whose efficiency is at a maximum.

A simple approach to transmission grating behavior is to apply Snell's law to the interface between the groove facet and air:

$$n_R \sin\varphi = \sin(\varphi + \theta_B) = \sin\varphi \cos\theta_B + \cos\varphi \sin\theta_B \quad , \tag{5.2}$$

where n_R is the refractive index of the grating resin at the desired wavelength λ_B and θ_B is the corresponding diffraction angle, and φ is the groove angle (Fig.5.1). Dividing by $\cos\varphi$ we get

$$n_R \operatorname{tg}\varphi = \operatorname{tg}\varphi \cos\theta_B = \sin\theta_B \quad , \tag{5.3}$$

so that

$$\operatorname{tg}\varphi = \frac{\sin\theta_B}{n_R - \cos\theta_B} \quad , \tag{5.4}$$

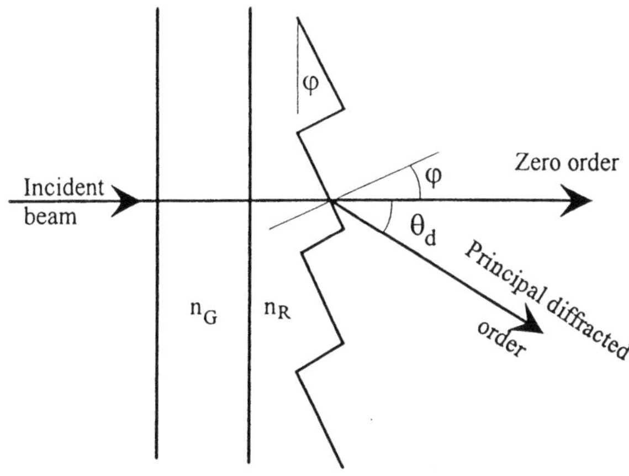

Fig.5.1 Ray path of transmission grating as normally used, but modulated surface can also be on the front.

where θ_B is derived from eq.(5.1) with $\lambda = \lambda_B$ [5.1]. One can also combine the equations and solve for the first order blaze wavelength λ_B:

$$\lambda_B = d\,tg\varphi \frac{n_R - \sqrt{1+\left(1-n_R^2\right)tg^2\varphi}}{1+tg^2\varphi} \quad . \tag{5.5}$$

For most practical purposes it is not necessary to use this transcendental equation. A simple approximation is available that relates the blaze wavelength of a reflection grating with that of a corresponding transmission grating, i.e., the same grating with the metal layer removed. It is based on the fact that blazing (maximum efficiency) occurs at a wavelength for which the phase retardation between successive grooves is λ or an integral multiple. For reflection gratings this means the groove depth is $\lambda/2$ for the Littrow blaze condition in first order. For a transmission grating this blaze wavelength is reduced by the ratio $(n_R - n_A)/2$ compared to a reflection grating, because the optical path difference instead of being doubled by reflection is generated by the difference in the two index values in single pass. Here n_R is the refractive index of the grating surface, usually the replica resin, and n_A is that of the surrounding medium. If n_A is taken as 1, as nearly always the case, and n_R taken as 1.58, a transmission grating will be blazed shorter than the corresponding reflection grating by a factor of 3.7. This turns out to be a useful rule of thumb, rarely in error by more than 10%, provided the groove angle is less than 27°. It is equivalent to replacing the parenthesis of eq.(5.5) by (0.57sinφ). This makes it simple to select possible transmission gratings from a reflection grating catalog.

The choice of groove angle φ has an implied upper limit φ_{max} given by total internal reflection effects

$$\sin\varphi_{max} = \frac{n_A}{n_R} \quad . \tag{5.6}$$

This points to 40° as the upper limit, though in practice electromagnetic wave behavior that becomes noticeable when groove dimensions approach the wavelength, and together with residual roughness, combine to soften the normally sharp cut-off behavior of total internal reflection that is familiar on a macroscale.

For angles of incidence other than normal, calculations are complicated by additional refraction effects at the interfaces. However, the angles involved are usually small enough so as not to significantly alter either choice or behavior.

5.3 Scalar Transmission Efficiency Behavior

As with any grating the object of transmission gratings is to control the division of incident light into the various orders. The number of possible orders tends to be large, compared to most reflection gratings, because groove frequencies are typically 4 times less. In addition, a certain amount of light will be diffracted backwards towards incidence, which almost doubles the number of possible orders. In most spectrometric applications only one order is used, typically the first. In other applications orders may be used in pairs or triplets. In that case the remaining orders are regarded as parasitic, except when multiple beams of equal intensity are desired. Even then there are a certain number of parasitic orders.

A somewhat modified form of scalar transmission gratings formulation gives excellent match with experiments [5.2]. It is based on a single integral:

$$\eta_m = \left| \frac{1}{d} \int_0^d t(x)\, e^{i\left(k_{0_y}^R - k_{m_y}^A\right) f(x) - im\frac{2\pi}{d}x}\, dx \right|^2 , \tag{5.7}$$

where η_m is the efficiency in the m-th transmitted order, t(x) is the local transmission Frensel coefficient the surface (if in each point the grating profile is replaced by a flat surface tangential to the profile), and f(x) is the profile function. $k_{0_y}^R$ and $k_{m_y}^A$ are the corresponding vertical components of the incident and m-th order transmission wave vector, and d the groove spacing:

$$k_{0_y}^R = \frac{2\pi}{\lambda} n_R \cos\theta_i \quad , \tag{5.8}$$

$$k_{m_y}^A = \frac{2\pi}{\lambda} n_A \cos\theta_m \, , \tag{5.9}$$

and θ_m the diffraction angle.

The above formulation can be safely applied when the grating period is large compared to the wavelength, although the limit is not as sharp as that of reflection gratings, due to the much smaller influence of resonance phenomena. In the following sections the scalar approach is applied to study the efficiency behavior of blazed gratings and multiple transmission orders gratings, whereas steep groove angle gratings with only few orders require more sophisticated electromagnetic treatment.

5.4 Efficiency Behavior of Blazed Transmission Gratings

The scalar approach presented in the previous section can be successfully used to calculate transmission grating behavior over the spectral regions of greatest interest. All orders can be accounted for, both positive and negative, forwards as well as back-reflected. It is possible also to calculate the effect of making light incident on the grating surface rather than the back of the blank. Spectrometric efficiency behavior would logically be plotted as a function of wavelength, but that would lead to numerous families of groove frequencies convoluted with different blaze angles (as in Chapter 4 for reflection gratings). Fortunately, as long as groove angles are < 22°, and provided 3% accuracy is acceptable, it is possible to utilize just a single set of curves. Efficiency is simply plotted as a function of the dimensionless parameter $\lambda/[d(n_R-1)\sin\varphi]$, where φ is the groove angle, and n_R the index of the grating surface. Note that $d\sin\varphi$ equals the maximum groove depth h. In order to simplify translating this parameter into specific values for λ, for a given

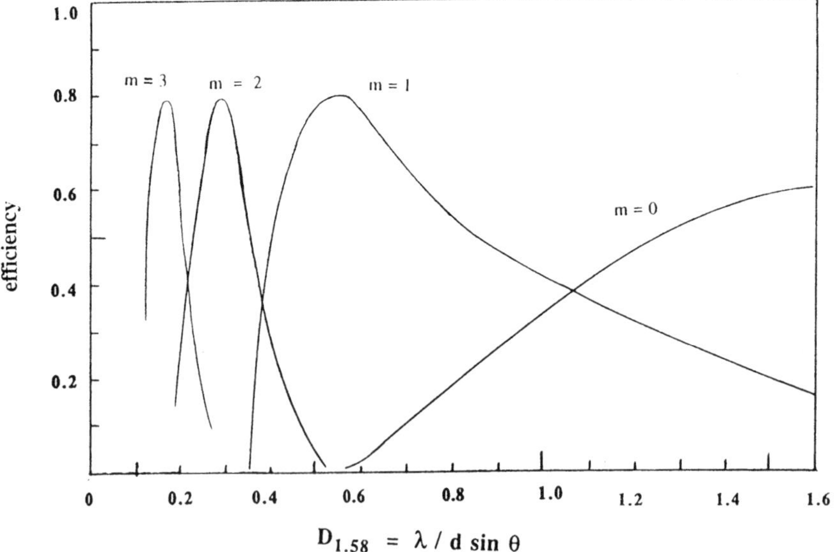

Fig.5.2 Theoretical absolute efficiency of blazed transmission gratings, in orders 0, 1, 2, 3. Zero absorption assumed.

value of d and φ, or with the variables reversed, it is necessary to pick a fixed value for n_R, 1.58 in the case of Fig.5.2. Should a more accurate value be necessary it can be taken from Fig.5.8, and the axis scale shifted by a factor $(n_R-1)/(1.58-1)$ which seldom exceeds 5%. This leads to a dimensionless parameter D_n, defined by

$$D_n = \lambda / (d \sin\varphi) \ . \tag{5.10}$$

One of the results that at first seems surprising in Fig.5.2 is that theoretical efficiencies never exceed 80%, in contrast to metallic diffraction gratings, where under scalar conditions 100% relative efficiency is expected. This reduced ceiling has long been observed experimentally. The explanation derives from the fact that Fresnel reflection at the grating-air interface leads to a complete set of backward in addition to the forward diffracted orders, whose effect becomes especially noticeable as groove angles increase (see later Figs.5.8 and 5.9a) [5.3]. The effect cannot be eliminated with the AR coatings familiar on unmodulated reflectors, because it introduces numerous resonance anomalies, as discussed in Chapter 8. A more rigorous explanation why transmission gratings cannot provide the 96% limit predicted by the scalar theory is given in Chapter 2.7.4 based on some electromagnetic considerations.

The curves of Fig.5.2 represent real gratings quite accurately in zero and first orders, except for a gradual reduction when $\varphi > 22°$. In second order the match is not quite as accurate, in that actual peaks do not exceed 70%, and third order is largely academic, because in the visible they call for blaze angles so steep and groove frequencies so high, as to depart from the scalar domain. Recent diffractive optics applications and large Fresnel lenses and zone plates, however, require some empirical rules to be used when the wavelength-to-period ratio is not negligible, but small enough to allow for multiple order propagation, which increases significantly the computation time for rigorous electromagnetic theories to be applied for profile optimization. Such an approach which leads to spectacularly good results even for λ/d ratios as large as 0.2, is discussed in section 5.6 in connection with Fresnel lens efficiency.

It is common practice to illuminate transmission gratings from the back of the blank, which is normally given an AR coating. It is equally possible to illuminate from the front, with little change in efficiency behavior for groove angles < 12°. However, at larger groove angles there is a distinct advantage to front illumination, because the total backscattered light is significantly reduced. For example, at a 22° groove angle total backscatter reduces from 11 to 5%, and at 30° it reduces from 20% to 3%, by making this simple switch from back illumination.

5.5 Transmission Grating Prisms

For certain applications, such as direct vision spectroscopes or compact astronomical spectrographs, it is useful to have a dispersing system that provides in-line viewing at one central wavelength. This can be achieved by replicating a transmission grating onto the hypotenuse face of a suitably chosen right angle prism, Fig.5.3. The light diffracted by the grating is bent back in line by the refracting effect of the prism, or vice versa. The device is sometimes known as a GRISM or Carpenter prism, and constitutes an elegant way to convert a camera into a long slit spectrograph. Although generally used in the visible spectrum they have also found use in the IR, by generating them into high index materials like Si or Ge [5.4].

The derivation of the required prism angle combines the diffraction equation with Snell's law. In the simplest case we can assume that $n_G = n_R$. Then

$$m\lambda / d = n_G \sin \theta_i - n_A \sin \theta_d \ , \tag{5.11}$$

where n_G is the index of refraction of glass, n_A the index of air, θ_i the angle of incidence, θ_d the angle of diffraction (negative here, because it is on the opposite side of the grating normal with respect to θ_i). Since we can take n_A = 1, and since for the central wavelength we set $\theta_i = \theta_d = \varphi = \alpha$, where α is the

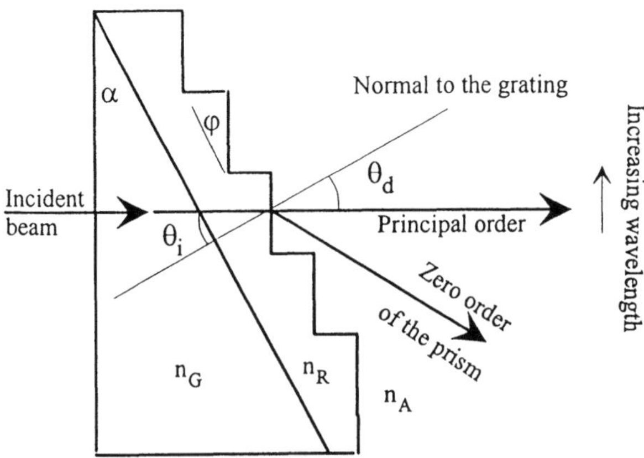

Fig.5.3 Ray path of grating-prism (GRISM or Carpenter prism).

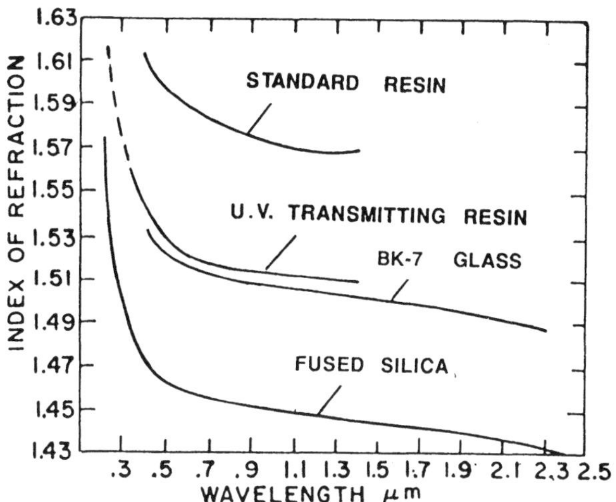

Fig.5.4 Index of refraction vs. wavelength for two types of replica resins and two typical substrates (courtesy David Richardson Grating Lab of Spectronic Instruments Co.).

prism angle, eq. (5.11) reduces to

$$\sin \varphi = \sin \alpha = m \lambda / d (n_G - 1) \quad . \tag{5.12}$$

It is evident that the dispersion of a grating prism cannot be linear, due to the superposition of prism refraction and grating diffraction. The following steps are used in the initial design of a grating prism:

1. Select the prism material desired and obtain the index for the straight-through wavelength from Fig.5.4 or other references.

2. Select the grating spacing, or groove frequency, for the approximate dispersion desired.

3. Determine the prism angle α or φ from eq.(5.12).

4. For maximum efficiency in the straight-through direction, select from a catalog the grating that has the chosen groove frequency and whose groove angle most closely approximates the angle α.

If angles α and φ do not match exactly the only effect is to slightly modify the energy distribution. The straight-through wavelength itself remains the same, but may not quite coincide with the peak efficiency.

5.6 Fresnel Lenses and Zone Plates

Increasing throughput of imaging optics in extereme applications, such as high-contrast night vision cameras, normally requires enlarging the entrance aperture and thicker lenses with larger diameter. This not only increases the requirements for optical purity, and thus costs, but leads to much heavier instruments, suffering more from vibration and shock. An alternative is to project the lens onto a plane piecewise, the result termed a Fresnel lens, Fig.5.5. Its focusing properties are more or less identical with the original lens, provided the optical path difference between successive beams is an integer multiple of $2\pi\lambda$ [5.5]. Whether this is done by refraction or diffraction has given birth to extensive speculation, somehow smoky as all refraction processes result from interference of diffracted waves.

The extreme case involves a geometry when only the 2π phase difference matters, each segment consisting of a single-level binary grating with

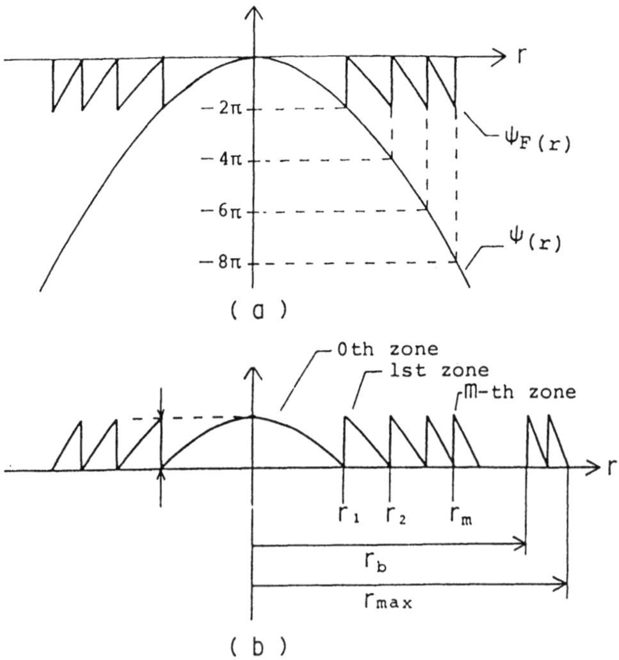

Fig.5.5 A schematic presentation of a Fresnel lens: a) phase shift function; b) thickness distribution (after [5.5]).

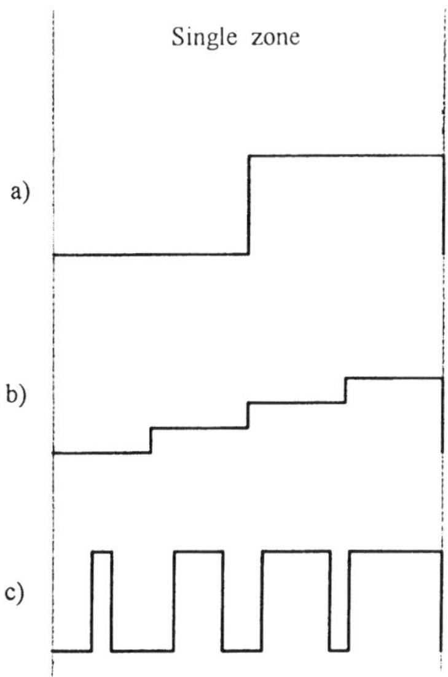

Fig.5.6 Three alternatives of zone profiles for Fresnel zone plates: a) lamellar profile with a 50% duty cylce; b) binary profile; c) multiple duty cycle within a single zone.

a 50% duty cycle, Fig.5.6a, called Fresnel zone plate. Intermediate solutions cover both multi-level binary grooves or multiple duty-cycle profiles [5.6], Fig.5.6b and c.

5.6.1 Geometrical Properties of Plane Lenses

With a plane incident wave focusing is obtained at a distance f from the lens center, Fig.5.5 where a circular groove geometry is assumed, although cylindrical lenses can be successfully made. If two consecutive zones differ by N times 2π phase shift we obtain the relationship:

$$\frac{2\pi}{\lambda}\left(\sqrt{r_M^2 + f^2} - f\right) = 2\pi MN \quad , \tag{5.13}$$

where M is the number of the zone. Typically N = 1, but recent applications have involved higher diffraction orders. From here the exact value of the M-th zone radius r_M can be evaluated rigorously, independent of the numerical aperture (N.A.):

$$r_M = \sqrt{2MN\lambda f + (MN\lambda)^2} \quad (5.14)$$

or, alternatively, the focus length as a function of the zone radius is obtained from:

$$f = \frac{r_M^2 - (MN\lambda)^2}{2MN\lambda} \quad . \quad (5.15)$$

The narrow lenses approximation ($r_M << f$) yields:

$$r_M \approx \sqrt{2MN\lambda f} \quad (5.14')$$

and

$$f = \frac{r_M^2}{2MN\lambda} \quad , \quad (5.15')$$

which shows that Fresnel lenses and zone plates are condemded to much larger chromatic aberrations, as demonstrated later.

In order to preserve the optimal phase shift given by eq.(5.13), all lamellar grooves in Fig.5.6a must have a different depth H_M, given by

$$H_M = \frac{N\lambda}{n_G - \cos\theta_M} \quad , \quad (5.16)$$

determined by the zone number M, the index of the grating material n_G and the diffraction angle, determined from the geometry:

$$tg\theta_M = \frac{r_M}{f} \quad . \quad (5.17)$$

A blazed triangular groove must have a groove angle φ determined from eq.(5.4) with

$$\sin\theta_B = M\,N\,\lambda \quad , \quad (5.18)$$

and its groove depth

$$H_M = d_M \, tg\,\varphi \quad , \quad (5.19)$$

is again given by eq.(5.16), d_M representing the width of the M-th zone:

$$d_M = r_{M+1} - r_M \quad . \tag{5.20}$$

For lenses with low N.A. the cosine of the diffracted angle differs only slightly from unity so that the groove depth is almost constant over the surface, a fact important for photolithographic manufacturing procesess. It can be easily shown that d_M depends on the zone radius and the focal length:

$$d_M = N\lambda \frac{r_M}{\sqrt{r_M^2 + f^2}} \quad . \tag{5.21}$$

The width of the outermost zone for small aperture lenses depends only on the wavelength and the numerical aperture:

$$d_{min} \approx \frac{N\lambda}{N.A.} \quad . \tag{5.22}$$

Thus the increase of the numerical aperture can be done either by a trivial increase of zone width or by going to higher diffraction orders, which requires deeper grooves for reasonable efficiency.

5.6.2 Imaging Properties

The diffraction limited spot width W at 1/e level is estimated to be proportional to the F-number of the lens (equal to the focus length divided by the lens diameter) [5.5]:

$$W = 1.64\lambda F \quad . \tag{5.23}$$

This relation reveals that use of lower F-numbers, to reduce the device dimensions, also leads to smaller focus spots. The limit in this direction is given by the width of the outermost element d_{min}:

$$F_{min} = \frac{1}{2}\sqrt{\left(\frac{N\lambda}{d_{min}}\right)^2 - 1} \quad . \tag{5.24}$$

For large F-numbers spherical aberrations are negligible, but eq.(5.15') reveals that the focal length is inversely proportional to the wavelength, thus the chromatic aberration is quite large, an order of magnitude greater than for a classical glass lens, Fig.5.7.

Smaller F-numbers require use of eq.(5.15) instead of (5.15'), and

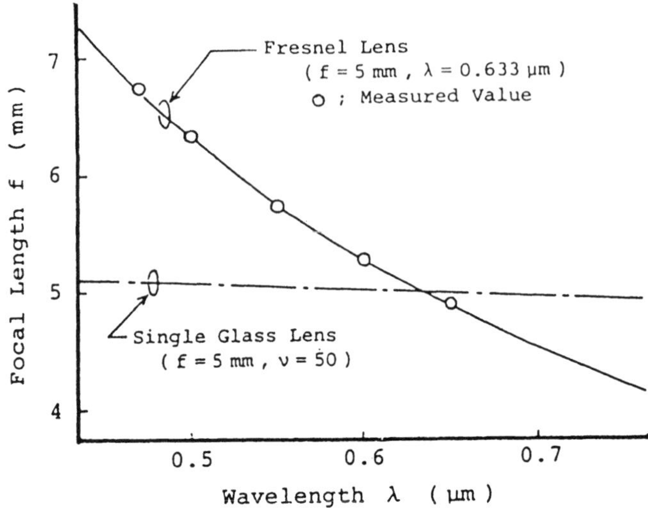

Fig.5.7 Dependence of the focal length on the wavelength for a classical lens (border curve) and of a Fresnel lens (after [5.5]).

spherical aberration can become significant at wavelengths that differ from the designed value.

5.6.3 Diffraction Efficiency

Diffraction efficiency is usually the most important question in the design of Fresnel lenses and zone plates, in particular, and diffractive optic elements in general. The problem is due to the relatively large characteristic periods involved. It is possible to speak in terms of diffraction gratings, because typicallly the devices consist of small features with characteristic length changing slowly along the surface so that a quasiperiodicity can be applied. This is important, because it enables the use of well developed grating theories. Unfortunately, the tendency is to decrease the periods so that the intuitive geometrical optics approach becomes vulnerable. Experience has shown that even when diffraction orders number 40 or more, and when the groove angle (depth) is large, geometrical optics is unable to correctly predict efficiencies. Neither is it practical to always use rigorous methods (if available), because they require long computation times. The situation becomes even more critical when applications involve periods varying from 1 μm to several mm.

A 50% duty cycle lamellar grating (Fig.5.6a), when optimized, is capable of reaching 40% efficiency in a multiple-order regime, without taking into account Fresnel reflection losses. If only few orders propagate (as in the outermost zones) efficiency can reach a theoretical value of 50% (see section 5.8), but then dependence on the technology becomes rather critical. This points to blazing the grooves. Not allowing for technological difficulties in manufacturing, the maximum available efficiency is around 96%, but numerical modelling and experiment have shown that when λ/d exceeds 0.05, deviation from geometrical optics expectations becomes noticable, if not critical. One of the consequences was discussed in section 5.4 with regard to Fig.5.2. The maximum value of efficiency is reduced and the spectral position of the maximum is shifted towards shorter wavelengths, Fig.5.8. An alternative to rigorous theoretical calculations can hardly be found when efficiency is critical, and the geometry involves high groove angles and short periods. Fortunately, a simple formula can be extracted using the scalar approach [5.7], valid for normal incidence on groove profiles presented in Fig.5.5. The starting point is eq.(5.7) with a transmission $t(x) \equiv T_M$ determined from the Fresnel transmission coefficient and geometrical considerations:

$$\eta_M = T_M \left| \frac{1}{d} \int_0^d e^{i\frac{2\pi}{\lambda}\left[(n_G - \cos\theta_M)\mathrm{tg}\varphi - iM\frac{\lambda}{d}\right]x} dx \right|^2 , \qquad (5.25)$$

with

$$T_M^{TE} = \frac{\left[\cos(\theta_M + \varphi) + \xi\right]}{(n_G \cos\varphi + \xi)^2} \frac{n_G}{\cos\theta_M} , \qquad (5.26a)$$

and

$$T_M^{TM} = \frac{\left[\cos(\theta_M + \varphi) + \xi\right]}{(\cos\varphi + n_G \xi)^2} \frac{n_G}{\cos\theta_M} , \qquad (5.26b)$$

with $\xi = \sqrt{1 - (n_G \sin\varphi)^2}$, equal to $\cos(\varphi + \theta_M)$ at the blaze wavelength.

These formulas are not truly scalar as they take into account polarization of the incident light. The difference between the two fundamental polarizations can exceed 10 to 20% for higher groove angles and refractive indices, where Brewster's effect plays a greater role. Although quite approximate, they depict the efficiency behavior fairly well. If only the phase factor in eq.(5.25) is considered, the blaze wavelength is given by eq.(5.5) through simply zeroing

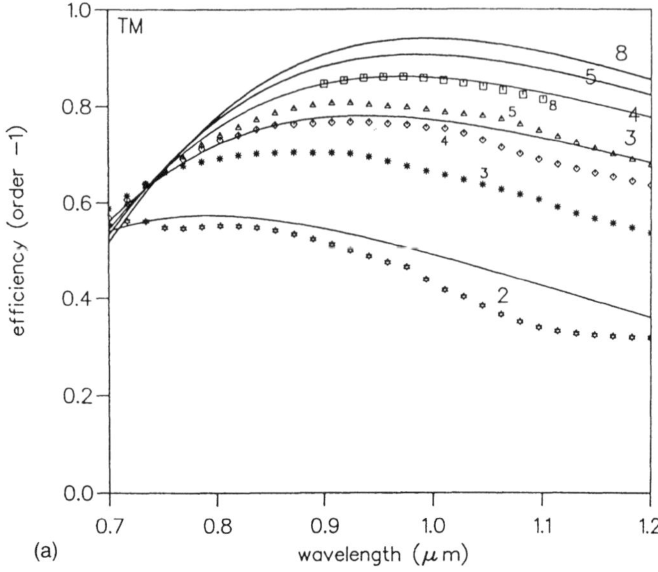

Fig.5.8 Spectral dependence of diffraction efficiency of a transmission grating with a profile given in Fig.5.5a. Normal incidence from the substrate with a refractive index $n_G = 1.46$. Groove angle varies with the period and diffraction order number as given by eqs.(5.4) and (5.18) with a geometrical blaze wavelength of 1 μm. a) Order -1 with groove periods indicated on the curves in micrometers; b) fixed groove period of 15 μm and different diffraction orders as numbered. Scatter - rigorous electromagnetic data, solid line - eqs.(5.26a and b).

the phase, eq. (5.2), and is equal to 1 μm for all the curves in Fig.5.8a and b. The transmission factors (5.26a, b) not only lead to reduction of the maximum value, Fig.5.9a, sometimes by a significant amount, but they cause also a slight shift of the position of the maximum, as far as transmission is higher for shorter wavelengths, where the angle of diffraction is smaller.

When the angle of diffraction is close to the angle of refraction at the working facet (i.e., when close to the blaze wavelength), a slightly different expression for the transmission coefficients (5.26a, b) follow from a more intuitive geometrical ray interpretation [5.7]:

$$T_M^{TE} = 4\frac{\cos^2(\theta_M + \varphi)}{\left[n_G \cos\varphi + \cos(\theta_M + \varphi)\right]^2}\frac{n_G}{\cos\theta_M}, \qquad (5.26'a)$$

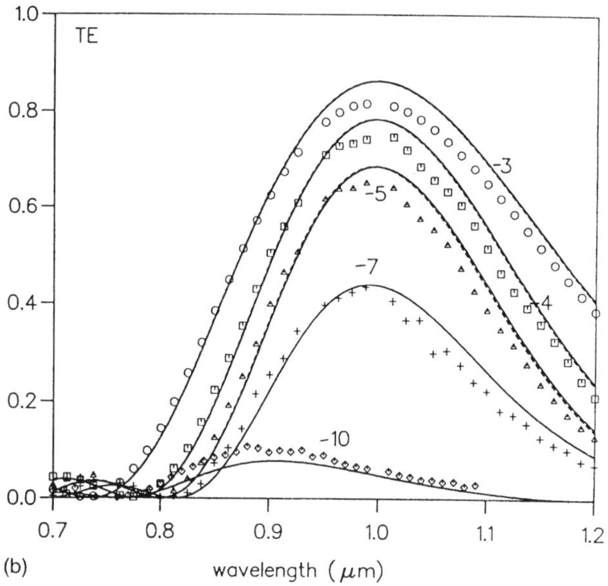

and

$$T_M^{TM} = 4\frac{\cos^2(\theta_M+\varphi)}{\left[\cos\varphi + n_G\cos(\theta_M+\varphi)\right]^2}\frac{n_G}{\cos\theta_M} \ . \quad (5.26'b)$$

They are obtained from (5.26) by substituting ξ with the angle of refraction on the large facet $\cos(\varphi+\theta_M)$.

The difficulties of producing triangular grooves with controlled spacings and depths leads to a desire to approximate them by using binary gratings with step-like profiles. As discussed, a single step cannot yield more than 40% efficiency, so that quasi-triangular grooves are usually made with 3 or sometimes a maximum of 4 levels, as shown in Fig.16.1. The greater the number of steps, the higher the theoretical efficiency but the finer the feature dimensions. In addition efficiency will always be reduced by some factors similar to eqs.(5.26). In practice total efficiency of a planar lens rarely exceeds 60% and the width of the outermost zone is never less than 1 μm.

Single step zones are used in x-ray optics. A multilayered reflection coating is deposited on a flat surface and then etched as in Fig.4.54c [5.8]. The upper theoretical limit of 40% efficiency is hardly an obstacle in this domain and is compensated by the gain in the smoothness and homogenity of the multilayers when compared to deposition over blazed profiles as in Fig.4.54d.

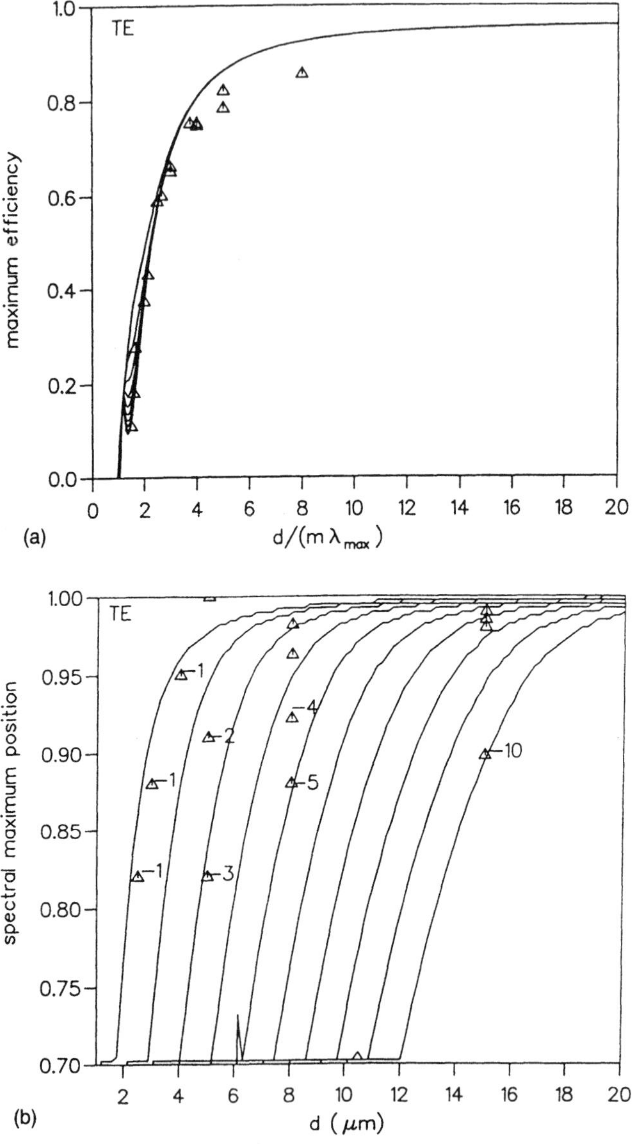

Fig.5.9 Decrease of efficiency maximum (a) and shift (b) of its spectral position with respect to the ideal one (λ_{max} = 1 μm) as a function of the groove period d and diffraction order number m. Groove parameters the same as in Fig.5.8 and the groove angle varies with d and m according to eqs.(5.4) and (5.18). Scatter - rigorous electromagnetic data, solid line - eqs.(5.26).

Theoretical modelling is based on the Born approximation [5.9] as in this spectral region the indices differ only slightly from unity. Comparison with rigorous theoretical results have recently become available [5.10].

5.7 Blazed Transmission Gratings as Beam Dividers

Blazed transmission gratings can serve as efficient and compact beam dividers for monochromatic light, especially when the angle between the beams is to be small and well controlled. At wavelengths longer than the blaze peak of order 1, nearly all transmitted light will be either in the zero (straight-through) or in the first order, see Fig.5.2. Splitting into the three beams is described in section 5.8 and multiple-beam sampling in section 5.9.

Controlling the ratio of zero to first orders is usually the important parameter. It depends on the phase retardation between successive grooves (i.e., the dimensionless ratio D_n). It will be near zero when D_n equals 0.55, increasing rather steeply as D_n increases, as seen in Fig.5.10. Zero and first orders are equal when $D_{1.58}$ is 1.08, and it should be noted how accurately the

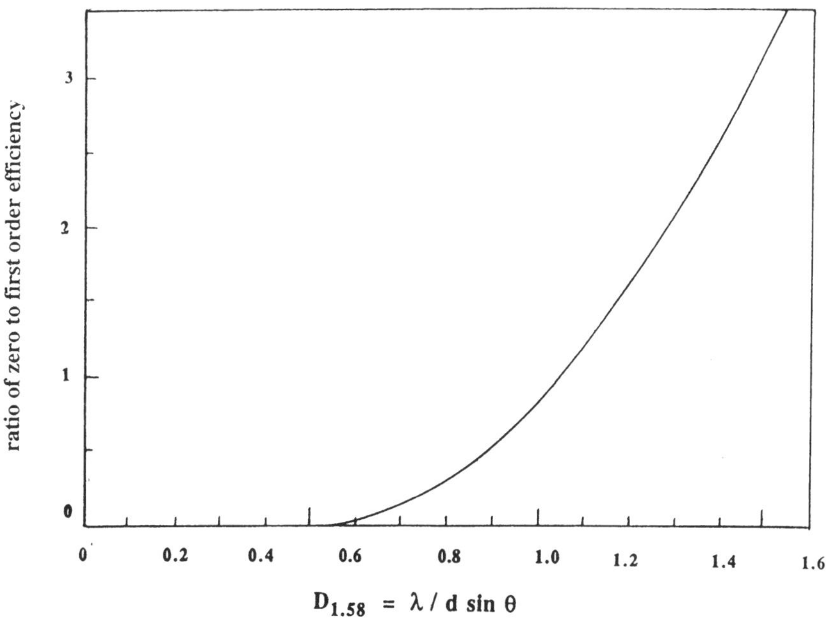

Fig.5.10 Zero to first order beam splitting ratio for blazed transmission gratings, as a function of D (for $n_R = 1.58$).

groove depth needs to be controlled when the order ratio is critical. For example, the groove depth for a transmission grating designed to split first and zero orders evenly at 0.8μm, for n_R of 1.58, is 0.74 μm. If this ratio is to be held within 10%, then the groove depth must be held to a variation of only 2.5%, or about 20 nm. Alternately a change in either wavelength or index n_R of 2.5% has the same effect. Obviously a high level of process control is required to maintain close control of the zero to first order ratio. The smaller this ratio is, the more tolerances can be relaxed.

5.8 Trapezoidal Gratings as Beam Splitters

Symmetrical groove transmission gratings play a minor role in spectrometry, but are used extensively as beam splitters for optical disk readers, where the wavelength of the laser source is constant. As a rule the zero order beam reads the track and the two first order beams read adjacent tracks to keep the head both centered and focused. By controlling groove depth the ratio of zero to first orders transmission can be varied over a factor of 10, and a high degree of symmetry is inherent. It is relatively easy to suppress all even orders from the second, with 50% duty cycle rectangular groove shape, and higher orders always drop off rapidly. As usual, diffraction angles are a function of λ/d only. Quite often the profile differs from rectangular form. Fig.5.11 is a sketch of a generic trapezoid, the actual shape defined by its c/d ratio. Sinusoidal grooves are typically produced interferometrically in photoresist, and are most

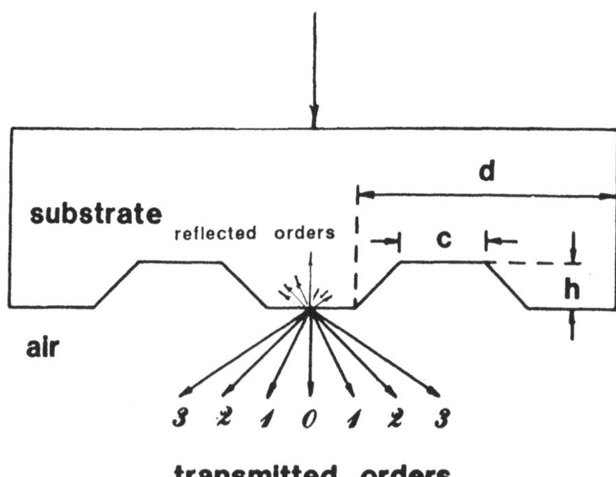

Fig.5.11 Symmetrical transmission grating schematic (after [5.11]).

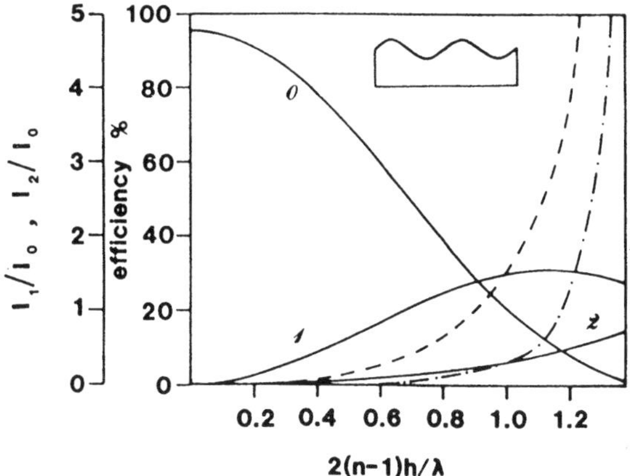

Fig.5.12 Absolute efficiency behavior of sinusoidal groove transmission grating. Solid lines: orders as labeled. Order ratios I_1/I_0 and I_2/I_0 shown (– – –) and (– · – · –) respectively. Calculations based on 200 gr/mm groove frequency (after [5.11]).

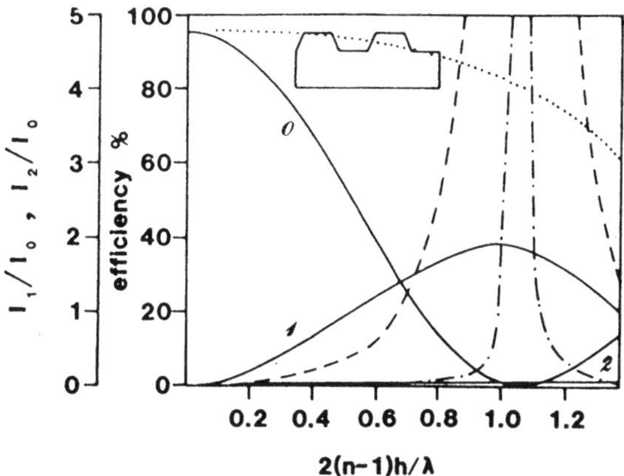

Fig.5.13 Same as Fig.5.12, except for trapezoidal grooves with c/d ratio of 0.4. The dotted curve shows the total theoretical amount of light transmitted (after [5.11]).

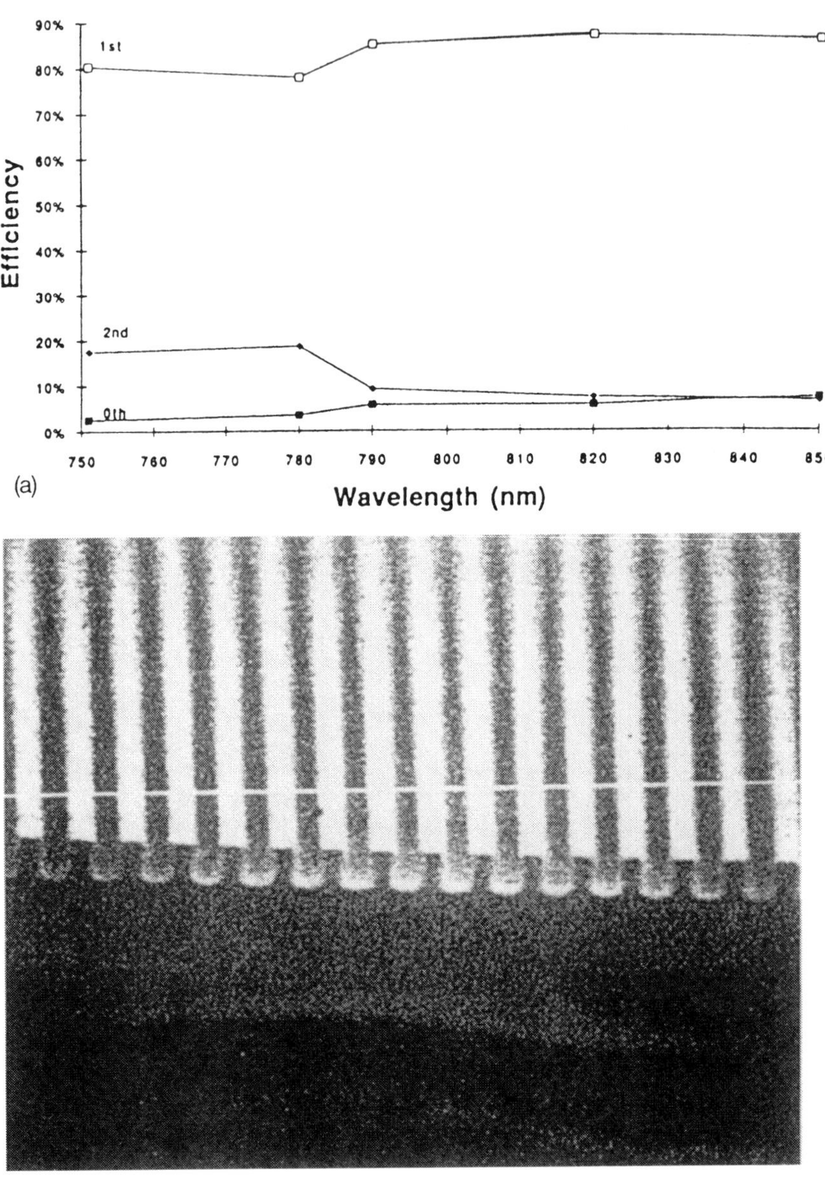

Fig.5.14 Beam splitters with lamellar gratings (after [5.12]): a) experimental results of a transmission-type beam splitter grating with a period d = 2 μm, depth 0.8 μm and duty cycle of 45%; b) scanning electron micrograph of a glass transmission beam splitter grating with a period of 1 μm and 50% duty cycle.

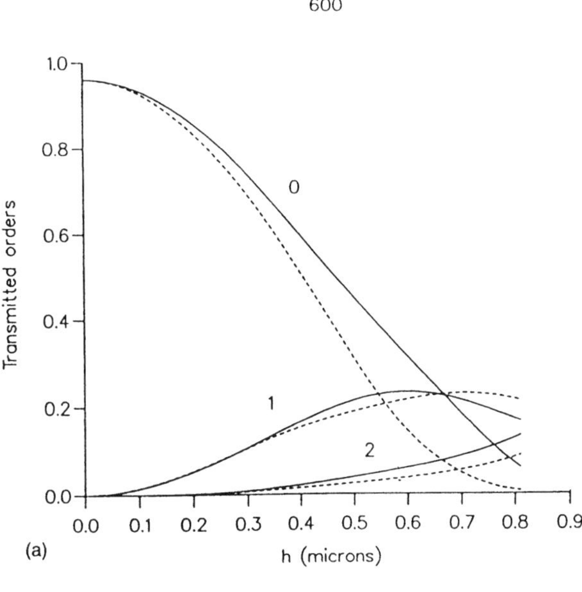

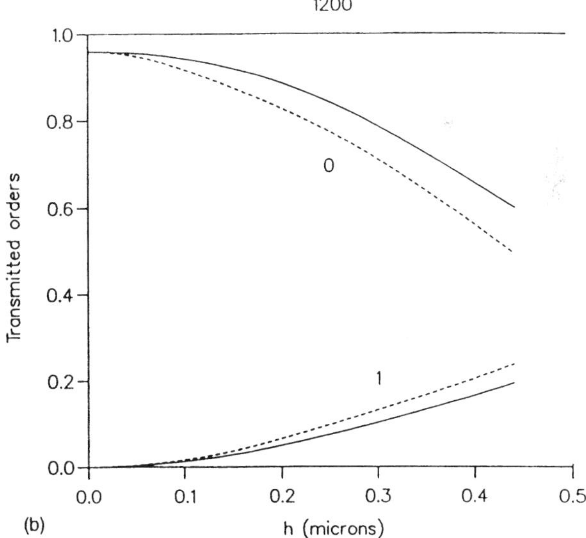

Fig.5.15 Theoretical diffraction efficiency of a transmission sinusoidal grating under normal incidence from the substrate side. Solid line - TM case, dashed line - TE case, λ = 632.8 nm. The number of orders as indicated. (a) 600 gr/mm; (b) 1200 gr/mm.

useful at relatively fine pitches (> 200 gr/mm). Rectangular grooves, the most commonly used, are well approximated by a 0.4 c/d ratio, even though the exact value is 0.5. V-grooves, which are difficult to produce to high symmetry, are closely matched by a trapezoid with c/d = 0.1, and have diffraction properties that barely differ from sinusoidal ones.

To illustrate the diffraction behavior of these gratings, their efficiency for the zero and first two diffracted orders as calculated by theoretical methods, is shown in Figs.5.12 and 5.13 for the two most readily produced shapes of sinusoidal and near rectangular. The calculations were made for a 200 gr/mm frequency and 632 nm wavelength, and a resin index n_R of 1.55 [5.11]. The ratios of first to zero order, and second to zero order are also shown, since the former is usually specified and the latter often required to be minimized. The rectangular groove approximation of Fig.5.13 shows its superiority for obtaining maximum first order transmission, combined with minimum second order. The 38% theoretical maximum in first order is accounted for by relatively large amounts of light back reflected for this fine pitch grating (note the drop in total transmitted light at the larger groove depths). A point of interest is that when zero and first order efficiencies are equal, a frequent requirement, a 6 nm change in groove depth is sufficient to alter the ratio by 10 %. In transcribing to lower groove frequencies it should be noted that with lower diffraction angles the efficiencies will increase a few percent because of reduced backscatter.

Beam splitting angles can be increased up to a point by simply going to finer pitch gratings, although lithographic gratings with trapezoidal (or quasirectangular) profile are rarely available with groove frequency greater than 500 lines/mm because technological problems do not permit production of the optimized profile parameters and the 2-nd order is not negligible (Fig.5.14a, ref.[5.12]). An extreme solution is to further reduce the period to 1 μm (Fig.5.14b) so that only the 0-th and ±1st orders propagate, making it much easier to optimize the groove form. Holographic gratings with quasi-sinusoidal profile can easily be manufactured with smaller periods but they cannot eliminate the "parasitic" zero order, as becomes obvious when comparing Fig.5.12 with 5.15. Small periods have increased polarization effects and their efficiency behavior cannot be fully predicted by simple scalar considerations but requires electromagnetic theories.

5.9 Multiple Order Transmission Gratings (Fan–Out Gratings)

A special type of transmission grating can be used to generate an entire family of orders with a groove shape designed to make their intensities as equal as possible. This gives us multiple beam splitters, which may have 5 or even 20 orders on both sides of zero. A top view of such gratings under working conditions (with a laser input) gives rise to the term of "*fan-out gratings*". Applications are found in scanning reference planes for construction use, optical computing, and others [5.13].

The difficulty in making such gratings lies in achieving a groove shape that leads to a sufficient degree of efficiency uniformity among orders, especially if they are to function over a finite wavelength range. Two obvious candidates are cylindrical sections or an approximation of this shape in the form of a wide angle V with several segments of different angles. The choice may lie with the availability of the corresponding diamond tools. They have also been produced by holographic methods.

Such gratings tend to have large groove spacings (10 to 100 μm) and low depth modulations. The energy distribution can be derived using the integral representation (5.7), taken in normal incidence so that $\sin\theta_i = 1$ and $\sin\theta_d = m\lambda/d$. Since the grooves are rather shallow and smooth, without sharp edges and steep slopes, the transmission coefficient in most cases can be taken simply equal to 0.96. In can be easily shown that the argument of the exponent

$$\frac{2\pi}{\lambda}\left[\left(n_R - \cos\theta_d\right)f(x) + x\sin\theta_d\right] \tag{5.27}$$

is equal to the optical path difference between the incident wave and the diffracted order.

A cylindrical groove that covers the entire groove spacing is specified by its initial angle ϕ (Fig.5.16). The radius of the groove R is given by

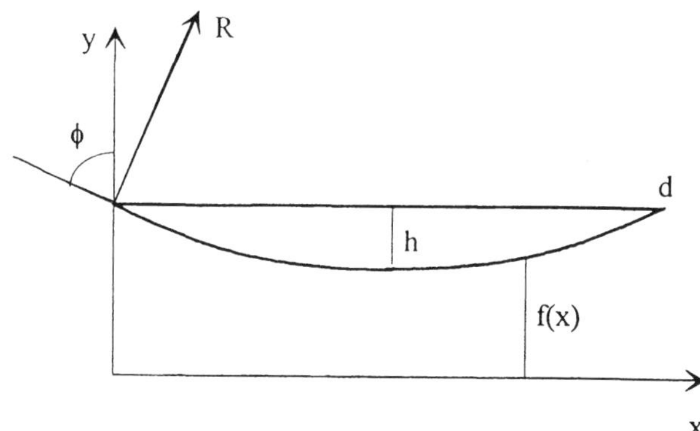

Fig.5.16 Geometry of a circular groove section.

$$R = \frac{d}{2\sin\phi} \tag{5.28}$$

and the maximum groove depth h by

$$h = R(1-\cos\phi) \approx \frac{R\phi^2}{2} \ . \tag{5.29}$$

Theoretical efficiency calculations based on equations (5.27 - 5.29) are shown in Figure 5.17, for orders zero to ±7 and the combined excess in all orders

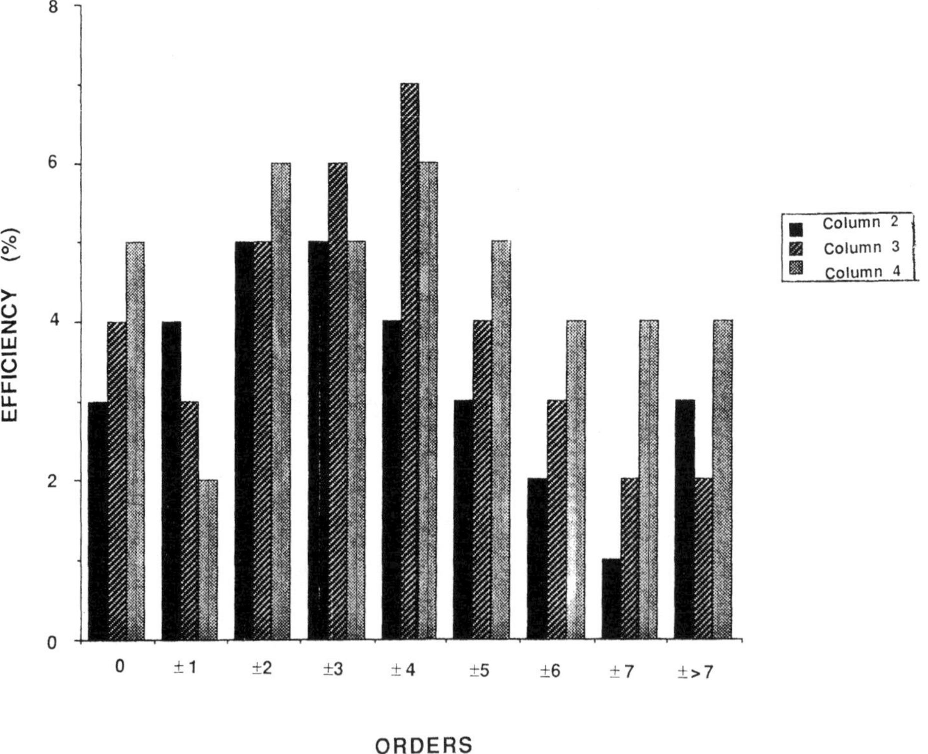

Fig.5.17 Efficiency of multiple order transmission gratings, with circular groove shape and three depth modulations: 5.4%, 3.5%, 2.0%, in columns 2, 3, 4, respectively. Resin index 1.5.

beyond, for three different values of R/d, with their equivalent depth modulations indicated. Note that the results depend on these ratios only, while the angles between orders are a function of λ/d only. The greater the demand for order uniformity the more difficult it becomes to find and achieve an appropriate groove shape.

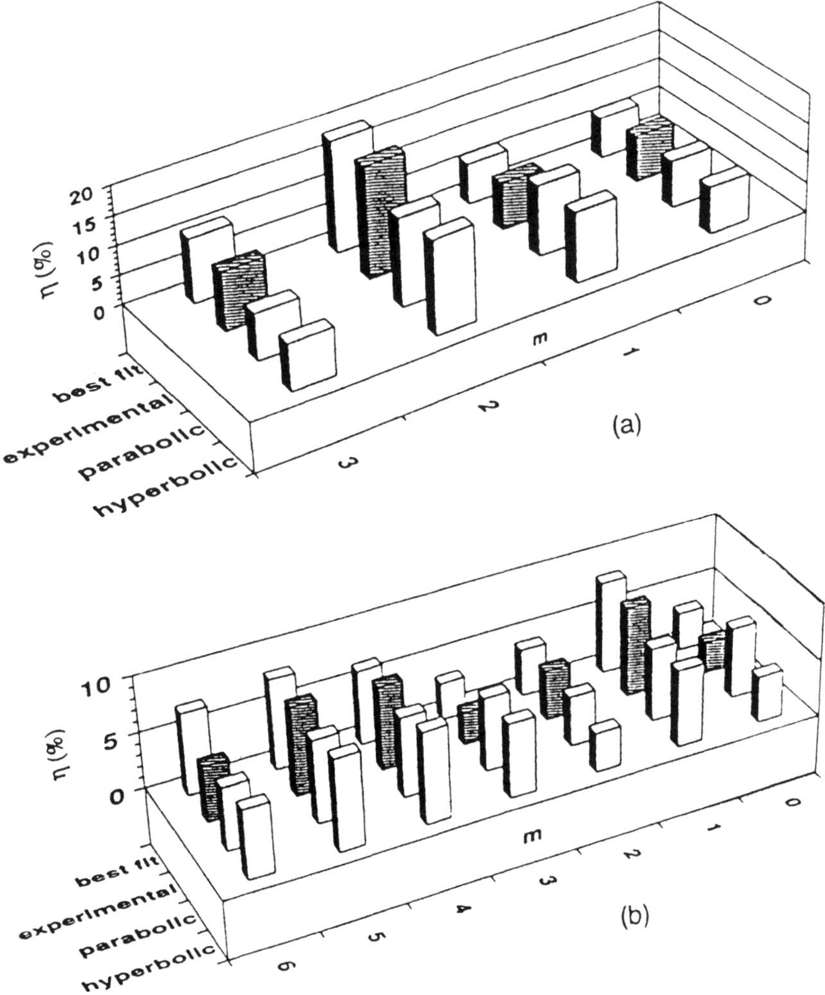

Fig.5.18 Comparison of theoretical efficiency of multiple beam-splitter gratings having hyperbolic and parabolic profiles with the experimental data of the grating shown in Fig.5.19. (a) 7-beam splitter; (b) 13-beam splitter (after [5.14]).

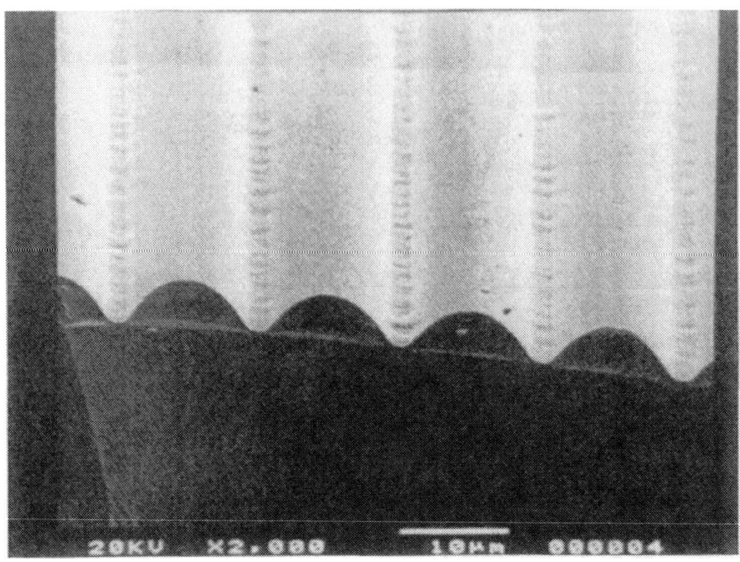

Fig.5.19 SEM photograph of a quasiconic section profile grating recorded in photoresist (after [5.14]).

An alternative design includes holographic recording with special treatment of the photoresist to obtain the profile desired to produce orders with uniform efficiency. Calculations show [5.14, 15] that a parabolic groove form will diffract with the most uniform light distribution between orders (Fig.5.18). The examples presented have profiles given in Fig.5.19 with a period of 16.4 μm. By changing the groove depth most of the incident light can be distributed either among 7 beams (Fig.5.18a with $h = 1.33$ μm) with their sum totaling 77%, or among 13 beams (Fig.5.18b with $h = 2.4$ μm) containing 80% of energy. Unfortunately, due to the natural photoresist behavior, the groove form cannot be easily manipulated so that results differ from sample to sample and the profile may end up closer to hyperbolic rather than parabolic, as one can judge by comparing the energy distribution in Fig.5.18.

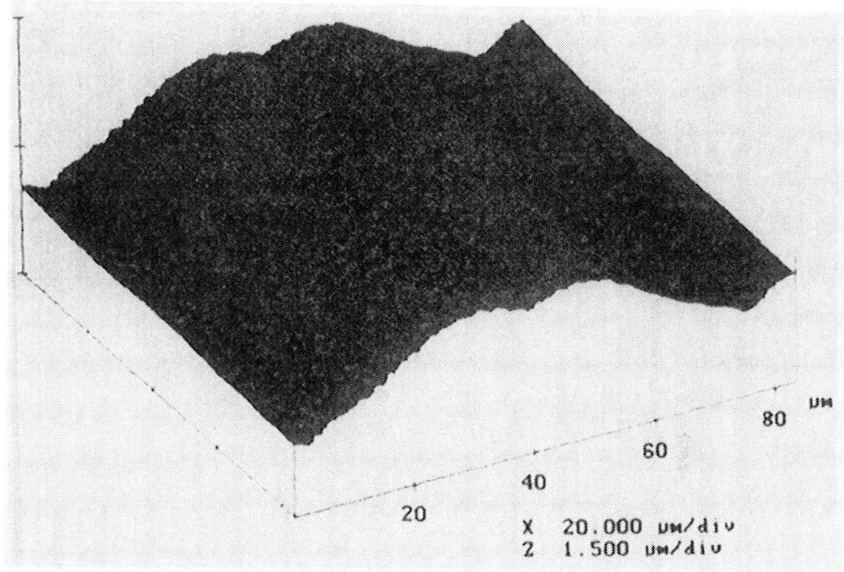

Fig.5.20 AFM image of one period of a continuous surface relief grating produced with laser-beam writing which diffracts 95% of incident energy into 9 beams almost equally (after [5.16]).

Laser beam writing can maintain better profile control, at least for periods large enough so that a 0.5 to 1 µm spot diameter can be considered sufficiently small. Such a grating [5.16] with a period of 72 µm and an exotic profile given in Fig.5.20 can diffract 95% of the incident light into 9 transmitted orders.

By using two such gratings face to face, at right angles to each other, an accurately defined 2–D array of rays is generated, which can be used for such applications as calibrating the image field distortion of a precision lens, or in robotic vision systems, or parallel optical computing. Cross-ruled gratings can also serve in such applications. The naturally large periods necessary to generate a great number of rays permits using techniques well-known in computer holography to optimize the surface relief pattern for maximum homogeneity of the diffracted beams. The simplest configuration involves a single-level binary (Dammann) crossed grating produced by e-beam writing and optically replicated by means of contact lithography. Fig.5.21 [5.16] presents an example of such grating with a basic cell of 50 x 50 µm and small feature dimensions 2 x 2 µm, which can generate an array of 19 x 19 beams of

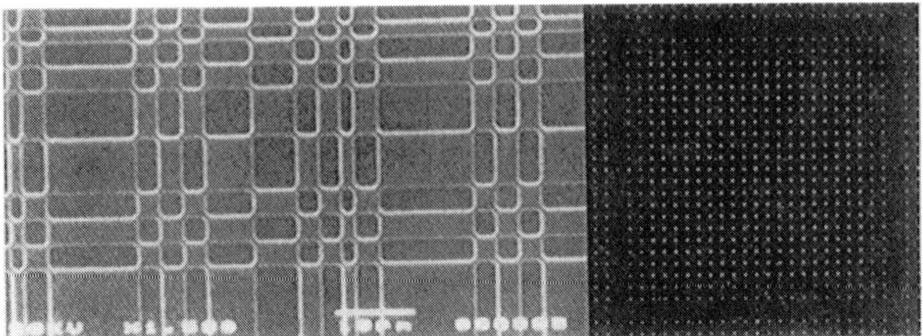

Fig.5.21 SEM picture of a 2-level binary crossed grating designed to produce an array of 19 x 19 beams, as shown at the right (after [5.16]).

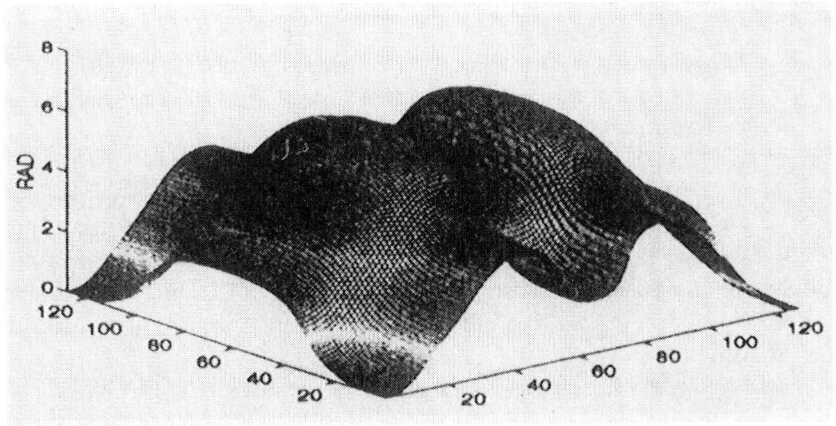

Fig.5.22 One period of a continuous profile crossed grating designed to diffract an array of 3 x 7 beams (after [5.16]).

almost equal intensity. When a smaller number of orders is required, the systems must have more complex continuous rather than binary groove design. These can be generated by laser beam writing (Fig.5.22). A small high efficiency fan-out grating can sometimes be constructed from an assembly of adjacent glass fibers [5.17] (see Fig.3.4).

5.10 Bragg Transmission Gratings

Bragg-condition transmission gratings represent an interesting special case for surface modulated gratings, in that they deliver exceptionally high first order efficiency, in both planes of polarization, and at high diffraction angles. They are characterized by the first order (there will be no higher) being diffracted in a direction symmetrical to the zero order, with respect to the grating normal, Fig.5.23.

While useful in constant wavelength applications, such as laser scanning elements, or pulse compression grating pairs [5.18], the Bragg mount is unfortunately of little value in spectrometric instrument design. This is because high efficiency comes at too high a price in terms of instrumental complexity. Since high efficiency requires maintaining symmetry of input and output beams with respect to the grating, the wavelength tuning requires that two elements must rotate. This may be the two beams (with grating fixed) rotating in exactly opposing directions, or it may be accomplished with the grating rotating, in which case either incident or diffracted beam must rotate at exactly twice the angle. In addition wavelength range is restricted by the high angle of diffraction.

High efficiency, i.e., above 90%, requires groove modulations much deeper than that used for reflection gratings, often ~ 100%. This means they are more difficult to make and almost impossible to replicate if accurate groove placement is to be maintained. This is because when grooves are so deep it becomes difficult to separate master from replica, unless compliant replica tooling is used, which impacts geometric fidelity of replication.

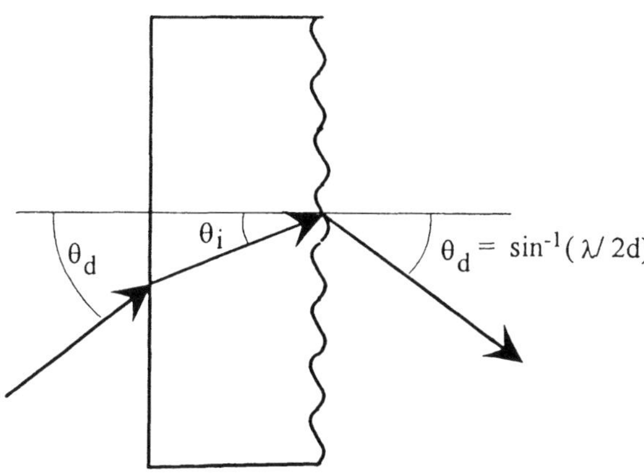

Fig.5.23 Schematic of Bragg diffraction grating.

Bragg gratings are sometimes produced by replacing surface modulation with index modulation in a photopolymer or dichromated gelatin. The latter is not an attractive candidate for high accuracy, because it is difficult to process gelatin in such a way that the modulation pattern is accurately reproduced after a wet–dry processing cycle.

Under Bragg conditions transmission gratings have behavior quite similar to that of metallic reflection gratings. As modulation depth increases, first order efficiency rises to a maximum value, and then decreasing beyond that, Fig.5.24. The main difference is that for dielectric gratings the TE-efficiency is reached for shallower gratings, in comparison to metallic gratings.

Fortunately, the maximum value does not depend significantly on the profile, although it does influence the groove depth value responsible for 'perfect blazing', Fig.5.24. This fact is of great importance when making such gratings, as it is difficult to obtain deep gratings with a carefully specified profile. A detailed study may be found in [5.19], from which basic rules can be summarized:

1. Contrary to blazed metallic gratings, the highest efficiency values are obtained for symmetrical profiles.

2. Spectral dependence of diffraction efficiency in TM polarized light resembles efficiency curves of metallic gratings for TE polarization and vice versa, Fig.5.25.

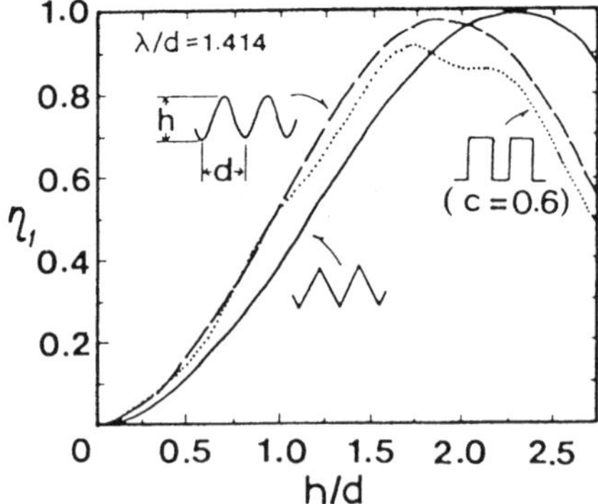

Fig.5.24 Groove depth dependence of 1st order transmission efficiency with normal incidence for different groove profiles indicated; index n = 1.66, $\lambda/d = 1.414$, $\theta = 45°$. For rectangular grooves, aspect ratio c is the ratio of groove width to period. TE polarization (after[5.19]).

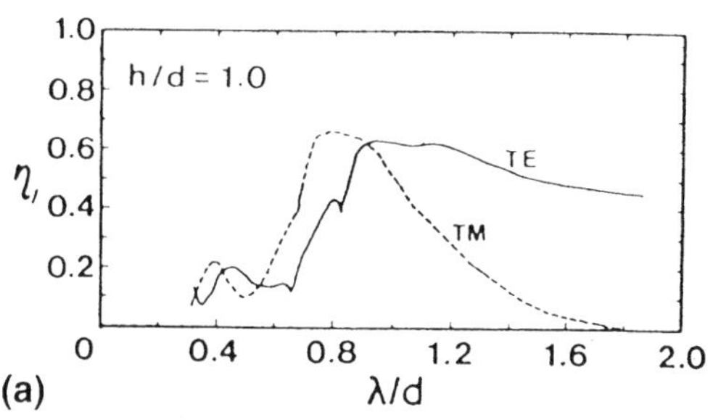

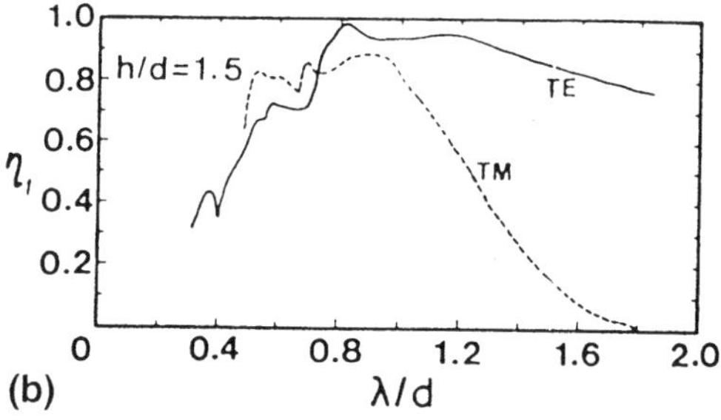

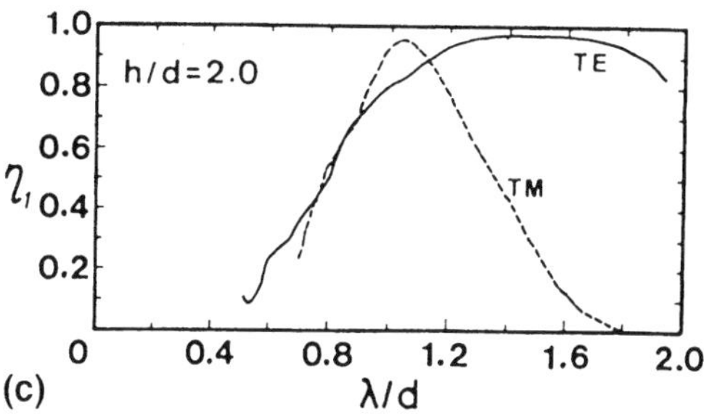

Fig.5.25 Spectral dependence of transmitted first order efficiency for a sinusoidal Bragg grating, n = 1.66. Solid line - TE polarization, dotted line - TM polarization. Figs.(a), (b), (c) for depth modulations h/d of 1.0, 1.5, 2.0, respectively (after [5.19]).

3. Maximum value of diffraction efficiency depends somewhat on the refractive index of the grating, in that lower index leads to greater efficiencies, due to reduced backward diffraction that results from lowered reflectivity. Unfortunately, decreasing the index also calls for increased modulation depth. For example, in the case of sinusoidal grooves, maximum efficiency is attained for h/d values of 1.3, 1.85, 2.3 for index values of 2, 1.66, and 1.5 respectively. The corresponding theoretical peak efficiencies are 95.3, 96, and 99% respectively. Experimental efficiency measurements closely match theory, except that values are about 4 % less [5.19].

5.11 Transmission Gratings Under Total Internal Reflection

In section 5.10 it was shown that dielectric gratings may have efficiencies exceeding those of metallic gratings. However, in some applications, such as laser tuning, maximum efficiency is desired under autocollimating conditions. Keeping in mind that the number of propagating orders needs to be minimized, this can be accomplished by combining the properties of a total reflecting prism with those of a grating [5.20]. Such a prism is shown in Fig.5.26.

Light is incident from the substrate side under an angle greater than the critical one for total internal reflection

$$n_R \sin \theta_i > 1 \quad , \tag{5.30}$$

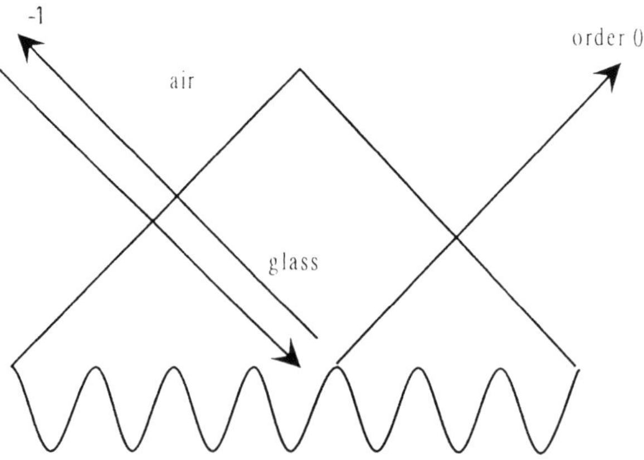

Fig.5.26 Littrow grating prism.

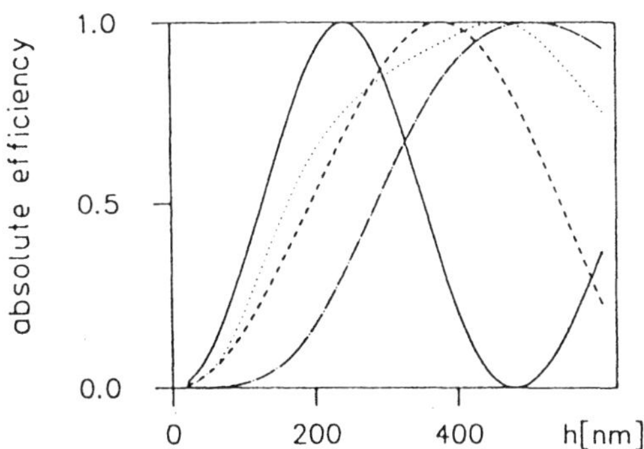

Fig.5.27 Groove depth dependence of –1st order backscatter from sinusoidal grating (n_R=1.5), Littrow mount, light from back side: Solid line TE, dotted TM polarization both at 550 nm. Dashed line TE and dotted–dashed line TM, both at 650 nm (after [5.20]).

where n_R is substrate refractive index and no 0-th transmitted order propagates in air. The grating period is made small enough to ensure that only the 0-th and 1-st orders can propagate in glass. In that case only evanescent orders are allowed in air. In Littrow mount the necessary condition is expressed as

$$3\,\lambda/d > 2\,n_R \quad . \tag{5.31}$$

It is most suitable to use such a grating under 45° incidence, in which case the grating can be manufactured onto a 45° glass prism face by replication.

As can be expected, groove depth dependency of diffraction efficiency is similar to that of metallic gratings supporting two diffraction orders, Fig.5.27, except that

1. Maximum *absolute* efficiency is 100%, while for metallic gratings one can obtain no more that 100% *relative* efficiency.

2. TE polarization efficiency is attainable at shallower 'groove depths than for TM polarization.

3. Optimal groove depths required for 100% absolute efficiency are much larger than for metallic gratings.

Spectral dependencies are shown in Fig.5.28. Provided that groove depth is properly chosen, 100% efficiency is attained over a narrow spectral interval. The sharp short wavelength edge is due to the appearance of diffraction orders propagating in air. Most unusual is that, contrary to metallic gratings, the spectral behavior of the two polarizations is very much alike. This

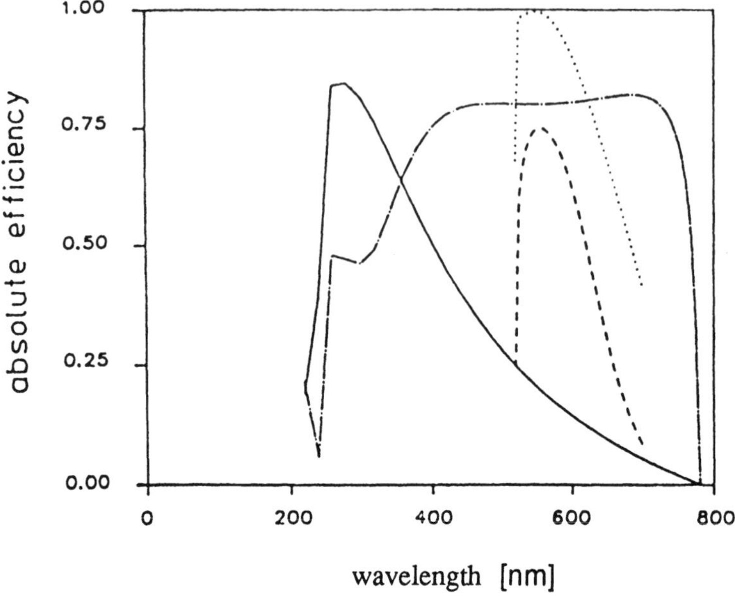

Fig.5.28 Spectral dependence of diffraction efficiencies of dielectric grating (dotted curve TE, dashed curve TM polarization, h = 240 nm), used on substrate side in Littrow prism mount, d = 260 nm, and for an aluminum grating (solid curve TE, border line TM polarization, h = 92.6 nm) (after [5.20]).

peculiarity may be understood by taking account of the fact that the broad plateau of spectral dependence in TM polarization of metallic gratings is due to the existence of 'non-Littrow perfect blazing' that is associated in a peculiar manner with surface wave excitation. Such waves are forbidden along a bare dielectric interface, so that the behavior of the two fundamental polarizations are not sharply differentiated.

5.12 Zero Order Diffraction (ZOD) Microimages

It is often quietly assumed that the zeroth order of a grating is non-selective. That is quite true as far as its propagation direction is concerned, which of course is independent of wavelength. However, if we are concerned with control of transmitted wavelengths, the zero order can play a useful role. In fact the possibility of varying intensity from zero to unity at a specific wavelength enables construction of high contrast optical transmission filters.

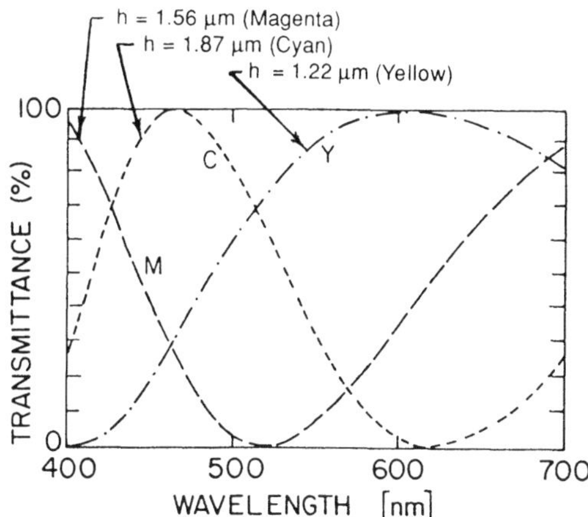

Fig.5.29 Zeroth order transmittance for rectangular grating as a function of wavelength. Groove depth h as indicated (after [5.21]).

Spectral range and absolute value of transmissivity are determined by groove density, profile, and depth. Light that is *not* diffracted in higher orders can be found in the zero order, so that a grating can work like a subtractive color filter. With incident white light the three principal colors that are used in subtractive color systems, cyan (minus red), magenta (minus green), and yellow (minus blue), can be readily obtained by the proper choice of groove depth, form, and period, having the corresponding spectral dependencies, Fig.5.29 [5.21].

Superimposing such gratings by appropriate screening results in full color pictures, without the use of any dyes. It requires first that three different gratings be formed, on three different blanks, using standard techniques (see Chapters 15 and 16). Then over the picture area, regions are formed where one, two, or three gratings are present or absent, using standard lithographic methods: Each blank is covered with a photoresist layer and exposed to light through a screen positive transparency, each one corresponding to a principal color. After processing, the grating structure is destroyed over the illuminated area. Removing the remaining resist, Fig.5.30, gives a master ready for embossing [5.22].

Slides made by plastic embossing are called '*zero order micro-images*'. They demonstrate their color only when projected in a slide projector, whose projection lens has an f-number such that it will not accept any of the first diffracted orders, Fig.5.31.

A total luminous range of 50:1 is obtained experimentally, and images

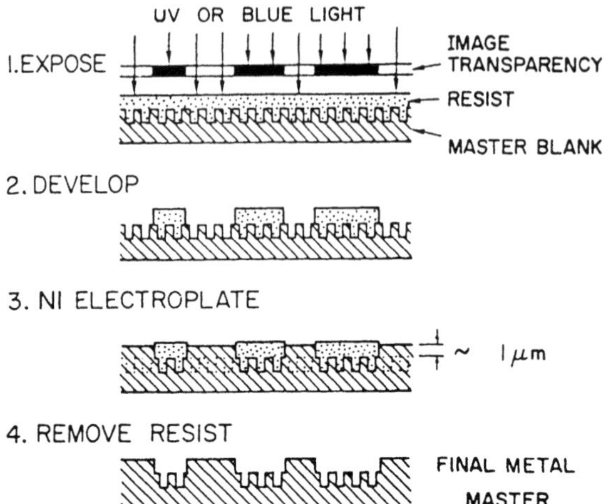

Fig.5.30 Manufacturing steps of master for ZOD micro-image (after [5.22]).

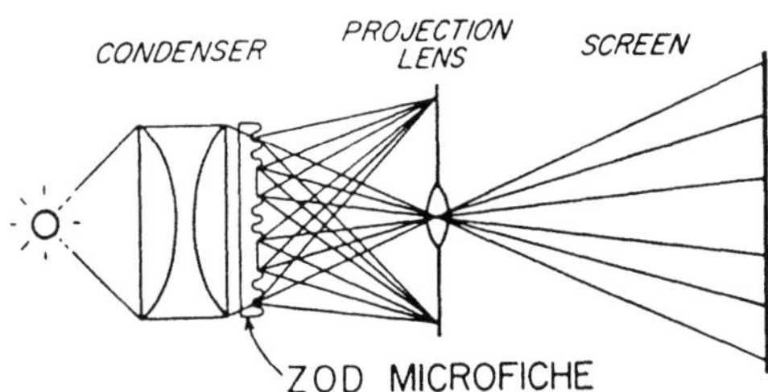

Fig.5.31 Ray path for ZOD micro-image projector (after [5.21]).

can be displayed with near 100% transmission (i.e., as bright as is possible with the projector). Since no dyes are involved there is no bleaching as a function of time or light. Also since little light is absorbed there will be no thermal damage even under intense illumination.

5.13 Ronchi Rulings

Ronchi rulings are amplitude gratings, in which opaque and transmitting areas alternate. The opaque areas may be developed photographic grains, which limits resolution somewhat, and which can only be used in transmission. More commonly they are made in chrome patterns which have much more sharply defined edges, as made by photo-lithographic methods, and can be used both in transmission and low efficiency reflection.

Diffraction directions are governed by the same grating equation as all gratings, and of major interest here is the efficiency behavior, which is a function mainly of the ratio of opening width a to the line spacing d. Efficiency η is given by the equation below:

$$\eta_m = \frac{a}{d} \frac{\dfrac{\sin^2(\pi am/d)}{(\pi am/d)^2}}{\sum\limits_m \dfrac{\sin^2(\pi am/d)}{(\pi am/d)^2}} , \tag{5.32}$$

which is a direct consequence of equation (2.25). The results are plotted in Fig.5.32. The use of this simple scalar formula is possible because Ronchi gratings are used with large periods [5.23].

The most interesting observation from Fig.5.32 is that the maximum first

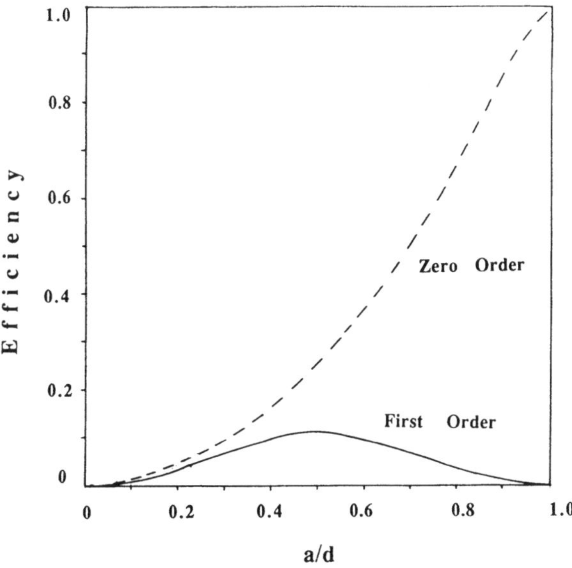

Fig.5.32 Transmission efficiency in zero and first orders of a Ronchi ruling, as function of the ratio a/d.

order efficiency is 11.1%, and naturally occurs when the opening ratio a/d is 50%. Under this condition the sum of zero and two first orders is 47.6%, which implies that the sum of all remaining orders is only 2.4%. This is because even orders vanish under ideal conditions and odd orders decrease rapidly with m^2, although when illuminated with a small diameter laser beam there will be a slight departure from the simplifying assumption that an infinite number of lines are being illuminated and that the lines are infinitely thin. As a result faint even orders are usually detectable under laser light, but may be hard to measure. The strong dependence of duty cycle on even order diffraction has been used as a sensitive tool for determining the width of fine lines [5.24, 25].

References

5.1 M. C. Hutley: *Diffraction Gratings*, (Academic Press, London, 1982) p.39.

5.2 E. K. Popov, L. V. Tsonev, and E. G. Loewen: "Scalar theory of transmission relief gratings," Optics Comm., **80**, 307–311 (1991).

5.3 M. Nevière: "Electromagnetic study of transmission gratings," App. Opt., **30**, 4540–4547 (1991).

5.4 H. U. Käufl: "N–Band long slit grism spectroscopy with TIMMI at the 3.6 m telescope," The ESO Messenger, **78**, 4–7 (Dec.1994).

5.5 H. Nishihara and T. Suhara: "Micro Fresnal lenses," in *Progress in Optics*, ed. E. Wolf, v. **XXIV**, ch.I, pp.1-37 (Elsevier, North-Holland, Amsterdam, 1987).

5.6 S. A. Weiss: "Single etch step produces efficient grating," Photonics Spectra **29**, 30-32 (1995).

5.7 M. Rossi, G. Blough, D. Raguin, E. Popov, and D. Maystre: "Diffraction efficiency of high N.A. continuous-relief diffractive lenses," O.S.A.Techn. Digest Series **6**, paper DTuD3 (1996).

5.8 V. Aristov, A. Erko, and V. Martinov: "Principles of Bragg-Fresnel multilayer optics," Rev. Phys. Appl. **23**, 1623-1630 (1988).

5.9 A. Sammar and J.-M. André: "Diffraction of multilayer gratings and zone plates in the x-ray region using the Born approximation," J. Opt. Soc. Am. A **10**, 600-613 (1993).

5.10 F. Montiel and M. Neviere: "Electromagnetic theory of Bragg-Fresnel linear zone plates," J. Opt. Soc. Am. A **12**, 2672-2678 (1995).

5.11 E. G. Loewen, L. B. Mashev, and E. K. Popov: "Transmission gratings as three–way beam splitters," Trans. SPIE, **815**, 66–72 (1981).

5.12 S. Walker, J. Jahns, L. Li, W. Mansfield, P. Mulgrew, D. Tennant, C. Roberts, L. West, and N. Ailawadi: "Design and fabrication of high-efficiency beam splitters and beam deflectors for integrated planar micro-optic systems," Appl. Opt. **32**, 2494-2501 (1993).

5.13 L. P. Boivin: "Multiple imaging using various types of simple phase gratings," Appl. Opt., **11**, 1782–1792 (1972).

5.14 P. Langlois and R. Beaulieu: "Phase relief gratings with conic section profile in the production of multiple beams," Appl. Opt. **29**, 3434-3439 (1990).

5.15 D. Shin and R. Magnusson: "Diffraction of surface relief gratings with conic cross–sectional gratings shapes," J. Opt. Soc. Am. A **6**, 1249–1253 (1989).

5.16 NOI Bulletin, v.**5**, no.2, July 1994, Québec, Canada.

5.17 H. Machida, J. Nitta, A. Seko, H. Kobayashi: " High efficiency fiber grating for producing multiple beams of uniform intensity," Appl. Opt., **32**, 330–332 (1984).

5.18 J. Agostinelli, G. Harvey, T. Stone, and C. Gabel: "Optical pulse shaping with a grating pair," Appl. Opt., **18**, 2500–2504 (1979).

5.19 K. Yokomori: "Dielectric surface relief gratings with high diffraction efficiency," Appl. Opt., **23**, 2303–2310 (1984).

5.20 E. Popov, L. Mashev, and D. Maystre: "Backside diffraction by relief gratings," Opt. Commun. **65**, 97-100 (1988).

5.21 K. Knop: "Diffraction gratings for color filtering in the zero diffracted order," Appl. Opt., **17**, 3598–3603 (1978).

5.22 M. Gale and K. Knop: "Surface relief images for color reproduction," in *Progress Reports in Imaging Science 2*, (Focal Press, London, 1980).

5.23 C. Meyer: *The Diffraction of Light, X-rays and Material Particles* (J. W. Edwards Co, Ann Arbor, MI, 1949).

5.24 G. Mendes, L. Cescato, and J. Frejlich: "Gratings for metrology and process control–I, a simple parameter optimization problem," Appl. Opt., **23**, 571–583 (1984).

5.25 W. Bösenberg and H. Kleinknecht: " Linewidth measurement on IC masks by diffraction grating test patterns," Solid State Technology., **25**, 10, 110–115 (1982).

Additional Reading

A. Baranne: "Sur l'emploi des réseaux par transmission en optique astronomique," C. R. Acad Sc. Paris B **291**, 205-207 (1980).

J. Bengtsson, N. Eriksson, and A. Larsson: "Small-feature-size fan-out kinoform etched in GaAs," Appl. Opt. **35**, 801-806 (1996).

G. Bouwhuis and J. Braat: "Video disk player optics," Appl. Opt. **17**, 1993-2006 (1978).

H. Dammann: "Spectral characteristic of stepped-phase gratings," Optik **53**, 400-417 (1979).

H. Dammann and K. Gorler: "High-efficiency in-line multiple imaging by means of multiple phase holograms," Opt. Commun. **3**, 312-315 (1971).

R. C. Enger and S. K. Case: "High-frequency holographic transmission gratings in photoresist," J. Opt. Soc. Am. **73**, 1113-1118 (1983).

M. T. Gale and K. Knop: "Reliefbilder im Mikroformat," Newe Zurcher Zeitung, Forschung und Technik, **186** (1976).

M. T. Gale, J. Kane, and K. Knop: "ZOD images: Embossable surface-relief structures for color and black-and-white reproduction," J. Appl. Photogr. Eng. **4**, 41-47 (1978).

M. Gale, M. Rossi, H. Schütz, P. Ehbets, H. Herzig, and D. Prongué: "Continuous-relief diffractive optical elements for two-dimensional array generation," Appl. Opt. **32**, 2526-2532 (1993).

M. Gupta and S. Peng: "Diffraction characteristics of surface relief gratings," Appl. Opt. **32**, 2911-2918 (1993).

M. Jocse and D. Kendall: "Rectangular-profile diffraction gratings from single crystal silicon," Appl. Opt. **19**, 72-76 (1980).

K. Knop: "Rigorous diffraction theory for transmission phase gratings with deep rectangular grooves," J. Opt. Soc. Am. **68**, 1206-1210 (1978).

M. G. Moharam, T. K. Gaylord, G. T. Sincerbox, H. Werlich, and B. Yung: "Diffraction characteristics of photoresist surface-relief gratings," Appl. Opt. **23**, 3214-3220 (1984).

M. Neviere, D. Maystre, and J. P. Laude: "Perfect blazing for transmission gratings," J. Opt. Soc. Am. A **7**, 1736-1739 (1990).

M. Neviere: "Echelle grism: an old challenge to the electromagnetic theory of gratings now resolved," Appl. Opt. **31**, 427-429 (1992).

E. Noponen, J. Turunen, F. Wyrowski: "Synthesis of paraxial-domain diffractive elements by rigorous electromagnetic theory," J. Opt. Soc. Am. A **12**, 1128-1133 (1995).

J. Saarinen, E. Noponen, J. Turunen, T. Suhara, and H. Nishihara: "Asymmetric beam deflection by doubly grooved binary gratings," Appl. Opt. **33**, 2401-2405 (1995).

W. J. Tomlinson and H. P. Weber: "Scattering efficiency of high-periodicity dielectric gratings: Experiment," J. Opt. Soc. Am. **63**, 685-688 (1973).

Z. Zhou and T. J. Drabik: "Optimized binary, phase-only, diffractive optical element with subwavelength features for 1.55 μm," J. Opt. Soc. Am. A **12**, 1104-1112 (1995).

Chapter 6

Echelle Gratings

6.1 Introduction

Echelle gratings, or simply echelles, are defined as coarse, but precisely ruled gratings used only at high angles of diffraction and in high spectral orders. Typical groove frequencies are 316 gr/mm or less, with 20 gr/mm a rough lower limit, and angles of use that vary from 63° to 78°, but occasionally go as low as 40°. While normally used in reflection there are special applications where they can be used in transmission. Rarely are spectral orders used below 10, but the upper limit may reach 600, although 100 is more common.

Echelles are considered to be among the most difficult gratings to rule, because not only do high diffraction angles demand exceptional ruling accuracy, but this has to be achieved under the high tool loads that accompany coarse groove spacings. In addition, the use at high orders require blaze faces to be flat to nanometer tolerances if the peak diffracted energy is to be concentrated in one blaze order.

Echelles have two special properties that define their applications. Most obvious is the high dispersion that leads to compact optical systems with a high throughput as well as high resolution for a given sized grating. Unique to echelles is the fact that because they are never used far from the blaze direction, efficiency remains relatively high over a large spectral range. Finally, in higher orders at least, they are nearly free of polarization effects. With so many advantages it is almost a forgone conclusion that some penalty must be accepted, which is that multiple orders will overlap. Therefore, some type of order separation is necessary, most commonly with cross-dispersion. This leads to a compact two-dimensional display, well matched to photographic recording but especially to array detectors.

In astronomical spectrometry the capabilities of echelles have been responsible for the virtual demise of the large coudé spectrographs formerly considered the only instrument capable of achieving maximum dispersion. *Compact Cassegrain echelle* spectrographs have taken their place. Similar thoughts apply to modern *inductively coupled plasma* (ICP) spectrographs.

Another application of echelles is in precision laser wavelength tuning where the high dispersion and damage resistance of echelles has proved useful.

This happens to be the only application in which a grating operates under exact Littrow conditions [6.1].

6.1.1 History

The first publication that described the use of a high order grating combined with cross–dispersion for convenient display of spectral data was by R. W. Wood [6.2], following a suggestion by his friend Edward Shane of the Lick Observatory. However, its development as a practical tool, with coarse groove spacings but high accuracy, is largely due to Harrison [6.3], who conceived it as a highly useful intermediate device between a Michelson echelon and an ordinary grating, often termed echelette. Echelons have virtually disappeared, because they are extremely difficult to make and have an inconveniently short free spectral range [6.4]. For maximum resolution Fabry-Perot etalons are still in wide use, but have severe limitations in spectral transmission, a variable dispersion, and at high intensities their dielectric coatings are subject to damage. Echelle resolutions exceeding 10^6 are readily attained, which is more than adequate for virtually all atomic and stellar spectrometry. In some respects echelles can be thought of as ruled interferometers.

The first applications of echelles were in astronomy, both satellite and ground-based, with specially designed instruments. Later came commercial atomic spectrographs, particularly for ICP applications. Harrison's work with ruling engines (see Chapter 14) was largely motivated by his desire to rule perfect echelles, an often frustrating task because every success was quickly followed by demands for still greater performance. His work occupied some

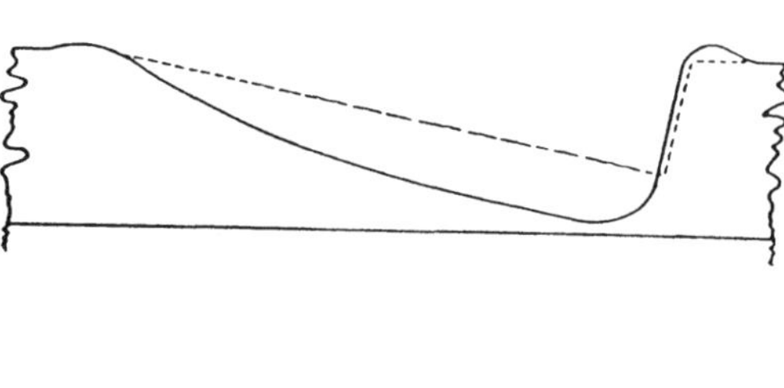

Fig.6.1 Schematic of typical echelle groove showing metal flow at groove edges.

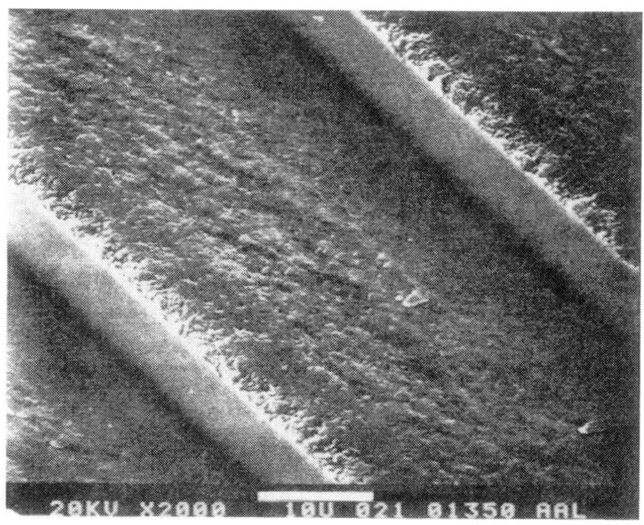

Fig.6.2 SEM photograph of 31.6 gr/mm echelle, 76° blaze angle (courtesy Spectronics Instruments Co.).

three decades and culminated in the ruling of a record 400 x 600 mm, 79 gr/mm echelle, 125 mm thick and weighing 100 kg [6.3], which was ruled on his C-engine. The early work was done with his A-engine, originally built by Michelson, modified by Harrison, but since dismantled. The B-engine, which has ruled most of the high accuracy echelles in the world, is still in full use. Its features are also described in Chapter 14.

6.2 Production of Echelles

In principle echelles are ruled just like standard echelette gratings. However, since it is the steep face that now must be optimized, the quality of the flat face must be sacrificed and the ruling has to proceed backwards (i.e., from right to left in Fig.6.1), opposite to the choice for echelettes. Another key difference is that the grooves are much deeper, which in turn requires metallic coatings abnormally thick compared to those in any other branch of optics. Reference has already been made to the increased ruling accuracy required.

Deep grooves require a large amount of plastic deformation around the tool as it is dragged under heavy load across the metal film. Aluminum is the almost universal choice because it combines good deformability with good adherence to the substrate and has relatively low internal stress, which means that it can to be deposited in thick layers. The major difficulty arises from the need to combine thicknesses of 10 to 30 μm with flatness tolerances around 50 to 100 nm. Fig.6.1 shows the kind of groove profile to be expected under these conditions, with the ideal shape shown dotted. An SEM photograph of the edge profile of a 31.6 gr/mm echelle, 76° blaze angle, is shown in Fig.6.2.

6.3 Physics of Echelles

6.3.1 The Grating Equation

As with standard gratings the angular relationship between input and output beams for echelles is given by the grating equation. However, echelles are more likely to be used under conditions of conical diffraction, which allows

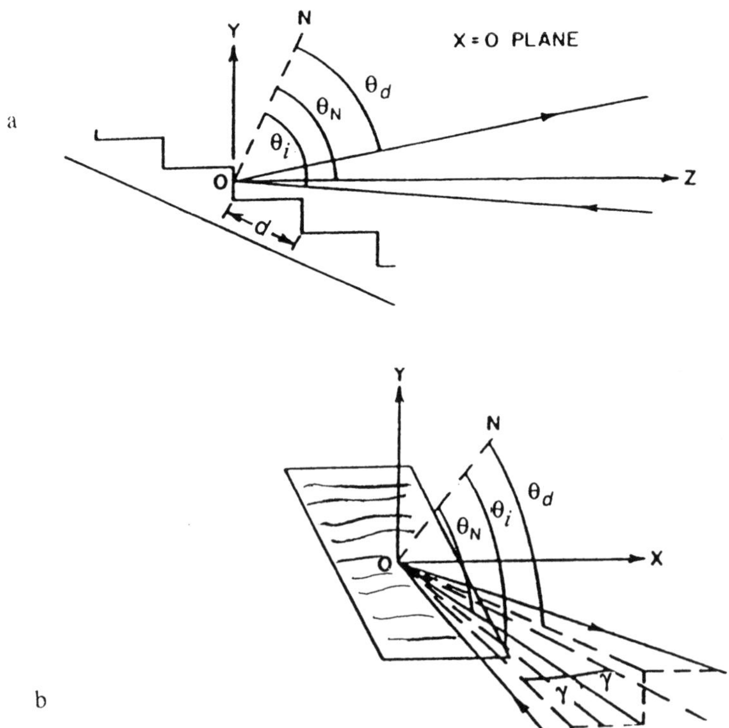

Fig.6.3 a) Coordinate system for echelle incident and diffracted rays, where the z-axis is perpendicular to the echelle facet. ON is the echelle normal; b) out-of plane ray system (after [6. 5]).

beams to be separated while maintaining near Littrow diffraction conditions, Fig.6.3. The grating equation is then written as

$$m\lambda / d = \cos\gamma (\sin \theta_i + \sin \theta_d) \quad , \tag{6.1}$$

where m is the order of diffraction, λ the wavelength, d the groove spacing, θ_i and θ_d the angles of incidence and diffraction respectively, as measured in the x-z plane, and γ the angle between the incident ray and the x-z plane. With echelles θ_i and θ_d are always on the same side of the grating normal from which they are measured, so there are no sign problems [6.5].

6.3.2 Angular Dispersion

From the grating equation, the angular dispersion for constant values of θ_i and γ, is derived by differentiation

$$\frac{d\theta_d}{d\lambda} = \frac{m}{d \cos\gamma \cos\theta_d} \quad . \tag{6.2}$$

Since in most cases $\cos \gamma \sim 1$, and since it is more useful to express the equation in angular terms, simple substitution leads to

$$\frac{d\theta_d}{d\lambda} = \frac{\sin\theta_i + \sin\theta_d}{\lambda \cos\theta_d} \quad . \tag{6.3}$$

The significance of this expression is more readily appreciated when simplified for the Littrow conditions ($\theta_i = \theta_d$), which are always closely approached in the case of echelles, in which case (as previously described in Chapter 2, eq.(2.14)):

$$\frac{d\theta_d}{d\lambda} = \frac{2 \operatorname{tg}\theta_d}{\lambda} \quad . \tag{6.4}$$

It should be noted that when γ is finite the image of an entrance slit will be rotated by an angle χ, given by

$$\operatorname{tg} \chi = 2 \operatorname{tg} \varphi \sin\gamma \quad , \tag{6.5}$$

where φ is the groove angle.

It is evident from eq. (6.4) why high angles of diffraction hold the key to high dispersion, and why echelles are often described by their r-values, where r = tgφ. With an r-4 echelle, dispersion is at least 10 times that of a typical first order echelette. Note that order number is not the determining factor. Another feature of echelles also derives from eq.(6.4), namely that due to the high values of θ_d the relationship between wavelength and dispersion becomes noticeably non–linear. When $\theta_i \neq \theta_d$ (i.e., operating off Littrow), dispersion will increase according to eq.(6.3) by about 2% per degree of angular deviation for an r–2 echelle and twice that for an r–4.

6.3.3 Free Spectral Range

The concept of free spectral range is particularly important in the case of echelles, because they work in high orders. Scanning through a spectrum with an echelle involves scanning through a succession of orders whose span of wavelength is one free spectral range (FSR).

Free spectral range is defined as the interval between two wavelengths λ_m and λ_{m+1}, which diffract in the same direction, but in successive orders, i. e., $m\lambda_m = (m+1)\ \lambda_{m+1}$. The difference $\Delta\lambda_m = \lambda_{1+m} - \lambda_m$ is the free spectral range, which, as previously given in eq. (2.16) depends on the order number:

$$\Delta\lambda_m = \lambda_{m+1} / m \quad . \tag{6.7}$$

Under the simplifying Littrow conditions, and substituting from the grating equation, leads to the alternate expression

$$\Delta\lambda = \lambda^2 / 2d \sin \theta_d \quad . \tag{6.7'}$$

It is obvious that in wavelength units the FSR increases rapidly with wavelength. In some cases it is convenient to specify the free spectral range in terms of wavenumbers ξ or (1/λ), i. e. FSR_ξ:

$$FSR_\xi = \Delta\lambda / \lambda^2 = 1 / m\lambda = 1 / 2d \sin \theta_d \quad . \tag{6.8}$$

In terms of wavenumbers the FSR for a given echelle will be constant. In practice it will be found to lie between 150 and 1800 wavenumbers (cm^{-1}). Since the value of sin θ_d varies little for high angles, FSR_ξ is seen to be almost directly related to the groove frequency 1/d.

When echelles are used for laser tuning the spectral range covered is typically less than one FSR, so that the normally confusing order overlap is of no concern. In all other instances orders must be separated, either by filtering out all but one FSR or by cross dispersion.

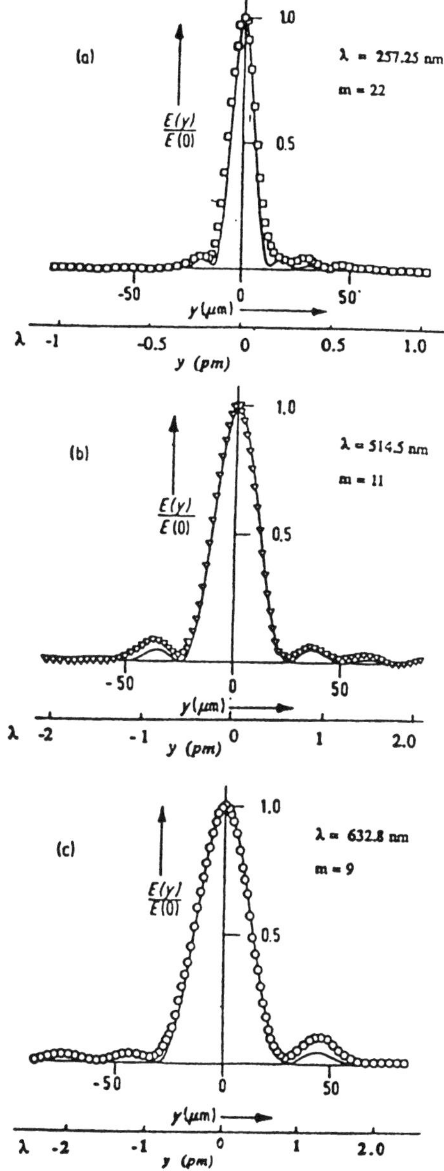

Fig.6.4 Normalized instrumental function of 5m vacuum spectrograph, with 254 mm echelle, 316 gr/mm, 63.5° blaze. Solid lines are theory, and (a), (b), and (c) represent 257.25, 514.5, and 632.8 nm from frequency stabilized He-Ne and Ar^+ lasers. Exit slit motion is shown in μm motion and the equivalent in pm (after [6.7]).

6.3.4 Resolution

The resolution of gratings tends to be utilized much closer to the theoretical limit in the case of echelles, because applications are centered around atomic spectrometry. Angles of diffraction are steeper, and many are utilized in large ruled widths, particularly in astronomical spectrographs. This becomes clear from recalling the defining relationship for R, the resolution in dimensionless terms (see Chapter 2):

$$R = 2W \sin \theta_d / \lambda \quad , \tag{6.9}$$

where W is the ruled width and θ_d the angle of diffraction under Littrow conditions. Physically it is equal to the number of fringes ($\lambda/2$) accommodated in the projected width of the grating. The criterion is the classical one of Rayleigh, when two monochromatic line images of equal intensity can just be distinguished [6.6].

Resolution actually attained is so close to theoretical that it is difficult to measure departures with accuracy. What can be seen will always be convoluted with instrumental deficiencies (i.e., the instrumental function). The most careful work in this direction appears to be that of Nubbemeyer and Wende, who tested a 254 mm wide echelle with 316 gr/mm and 63.5° blaze angle in a 5 meter focal length vacuum spectrograph with entrance and exit slits adjusted to a width of only 2.5 μm [6.7, 8]. The only suitable sources of sufficiently monochromatic light are frequency stabilized lasers, frequency doubled in one case. Results are shown in Fig.6.4, a, b, c. The diffraction side lobes from single slit diffraction are rarely seen in standard instruments, although well known from diffraction theory. Measurements showed that at the three wavelengths, in sequence 632.8, 514.5, and 257.25 nm, the resolutions attained were 710, 000; 860, 000; 1, 420, 000. Such an increase is to be expected from eq.(6.9). However, it is interesting to note that in terms of fractions of theoretical resolution (as determined from the image width at 40% of peak) they were 98%, 97%, and 80% respectively. This too is readily understood, since the shorter the wavelength the harder it is to obtain theoretical performance, as residual mechanical errors become a larger fraction of the wavelength. In addition, instrumental aberrations start to play an increased role.

6.3.5 Immersion of Echelles

In order to reduce the size of instruments and improve throughput it has been suggested that echelle dispersion be increased to still higher levels by immersing the ruled surface in a fluid or material of higher index [6.9]. The idea is that within such a medium the effective wavelength is divided by n_f, the

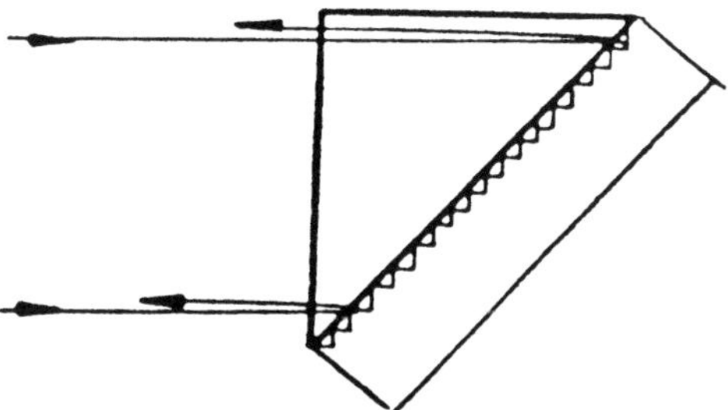

Fig.6.5 Schematic of immersion echelle grating (after [6.9]).

index, so that for a given groove geometry the order and hence dispersion and resolution increase in proportion. For standard type echelette gratings there is little advantage to applying this concept, because the effect is merely to change the apparent blaze angle [6.10]. However, in the case of echelle gratings there is an advantage in that one can now operate in higher orders of diffraction.

For such an approach to work with echelles a prism is necessary to couple the light beams to the ruled surface, Fig.6.5. A plain glass plate, such as one would use to cover the immersion fluid on an echelette grating, is of no use here, because at high angles of incidence the light would be totally reflected. The prism must be of excellent optical quality if high resolution is to be obtained, especially since the beams traverse in double pass. The justification derives from greater dispersion, in particular being able to obtain high performance in a reduced volume.

In such an arrangement the dispersion and resolution formulas become

$$\frac{d\theta_d}{d\lambda} = \frac{2n_f \operatorname{tg}\theta_d}{\lambda} \tag{6.4'}$$

and

$$R = 2W\, n_f \sin\theta_d / \lambda \tag{6.9'}$$

for Littrow conditions, with n_f being the refractive index of the replica film.

Anamorphic Immersion System

Given the desire to increase still further the resolution of an echelle grating for a given beam diameter, use is sometimes made of an alternate immersion system, using a prism as shown in Fig.6.6. The idea is to widen the entrance face of the prism by a factor w, given by

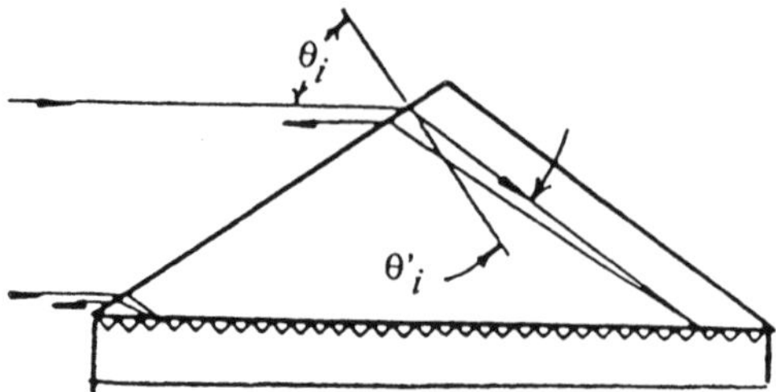

Fig.6.6 Schematic of anamorphic immersion grating (after [6.9]).

$$w = \cos \theta_i' / \cos \theta_i,$$

where θ_i' and θ_i are the angles of incidence and refraction at the entrance face respectively. This makes the theoretical resolution

$$R = 2\, w\, n_f\, W \sin \theta_d / \lambda \quad , \qquad (6.9'')$$

which in a typical system can lead to a boost by a factor of about 2.3. It becomes important to minimize Fresnel reflection from the prism-replica resin interface, which means choosing materials that have closely matching refractive indices.

6.4 Efficiency Behavior of Echelles

Just as with other gratings the efficiency behavior of echelles is of great importance to instrument and system designers. Unlike echelettes (Chapter 4), the efficiency varies in a cyclic fashion, Fig.6.7. The simple model is to take each groove face as a single slit source, which generates a blaze envelope function proportional to the function $\text{sinc}^2 p = \sin^2 p / p^2$, where p is the phase difference between the midpoint and the edges of the slit. Like any grating this is convoluted with the phase summation of light coming from all the 'slits', i.e. proportional to

$$\sin^2 Np' / \sin^2 p' \quad ,$$

where p' is 1/2 the difference between the phase of the centers of adjacent grooves, and N is the total number of grooves (see Ch.2).

Since echelles operate in a domain where $\lambda/d < 0.15$ it has long been considered safe to assume that they function solely in the scalar region of

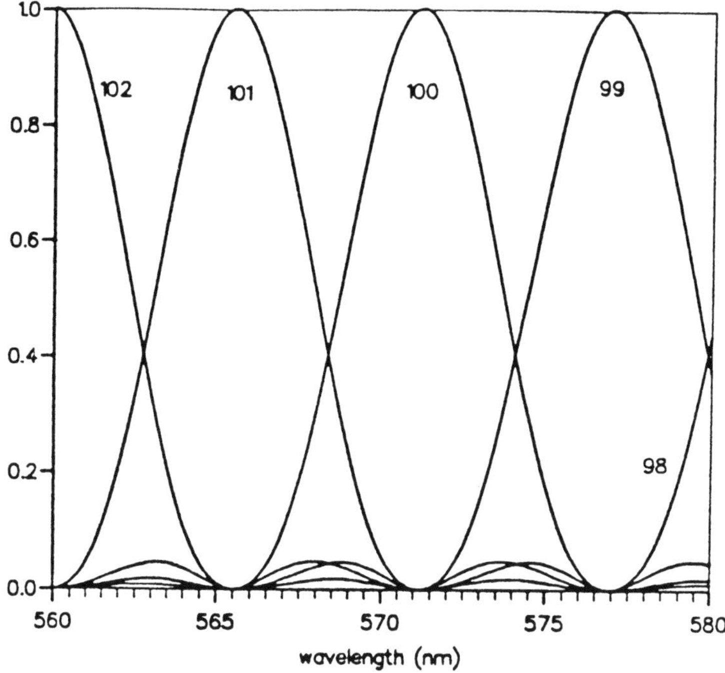

Fig.6.7 Cyclic variation of echelle efficiency as a function of wavelength for 64° echelle. Numerical data for 31.6 gr/mm echelle, fixed incidence at 64°. Order numbers as shown (after [6.11]).

diffraction. This attitude has been encouraged by experimental observations that seemed to confirm it in most instances, as well as by the apparent difficulties of extending electromagnetic theory to cover a grating that supports so many orders. Any measurements to the contrary were simply ignored.

As in other instances, in the world of physics this convenient picture was called into question by a detailed set of experiments with a series of echelles that covered a wide range of groove spacings and blaze angles. A set of lasers provided collimated, monochromatic and highly polarized light to make efficiency measurements in fine intervals of incident angle [6.11]. The most important observation was a degree of polarization (difference between TE and TM efficiency divided by the sum) that could not be ignored. Although often of negligible amount, there were cases where 10% polarization was seen in orders as high as 50. In orders below 10 it could be much greater. In addition there were unexpected discontinuities that looked like anomalies. Fig.6.8 is a

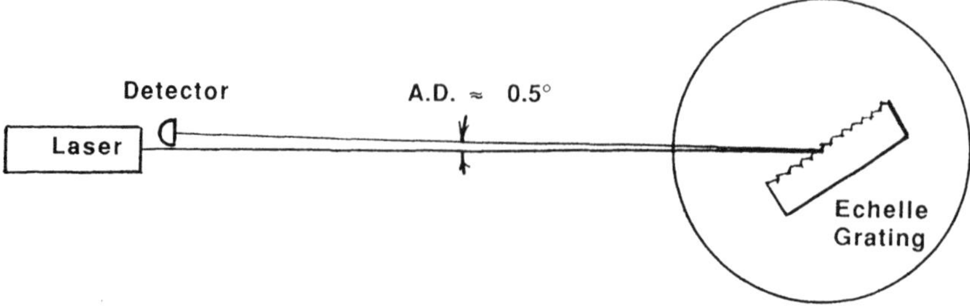

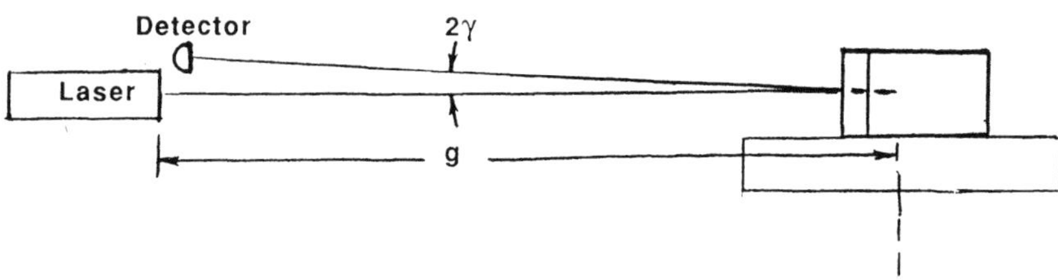

Fig.6.8 Schematic of echelle efficiency tester, in near Littrow mount. Detector is immediately in front of the laser. Distance g is large enough so that incident and diffracted ray remain within 1° of the normal (meridional) plane.

schematic of the optics used to make the experimental measurements.

Finally, it proved possible to extend the integral theory for rigorous treatment of efficiency to cover the case of echelles, and it confirmed the detailed experimental observations, including anomalies [6.11]. This made it possible, and therefore useful, to perform analytical studies of echelle behavior under all sorts of conditions, as treated in the following sections.

6.4.1 Scalar Model for Efficiency

The blaze behavior of gratings is a convolution of the interference function (IF) and the blaze function (BF). As described in many optics textbooks (see also Ch.2) the diffracted intensity $I(\theta_d)$ as a function of the diffraction angle θ_d, is given by [6.6]:

$$I(\theta_d) = \mathrm{BF.IF} = \frac{\sin^2 p}{p^2} \frac{\sin^2 Np'}{\sin^2 p'} \quad , \tag{6.10}$$

where p is the phase difference betwen one groove and the next, given by

$$p = \pi\, b/\lambda\, (\sin\theta_i + \sin\theta_d) \quad . \tag{6.11}$$

Here b is the effective width of the groove; p' is 1/2 the difference between the phase of the centers of adjacent grooves and is equal to

$$p' = \pi\, d/\lambda\, (\sin\theta_i + \sin\theta_d) \quad . \tag{6.12}$$

The intensity function (IF) is at a maximum when $p' = m\pi$, where m is an integer (the spectral order), and as such represents the grating equation.

The blaze function (BF) is the intensity of the diffraction patterns from a single slit of width b. It is a maximum when $p = 0$, where $\theta_i = -\theta_d$, which corresponds to the zero order ($m = 0$). Its first minimum occurs when $p = \pi$.

In an echelle it is typical for $\theta_i > \varphi$, where φ is the groove angle, so that part of the groove facet is not illuminated. Thus its effective width b becomes

$$b = d \cos\theta_i / \cos(\varphi - \theta_i) \quad . \tag{6.13}$$

With a reduced width of groove there will be an increase in the angular width of the blaze function, while the angular dispersion decreases, so that the free spectral range acquires a reduced angular span. Thus the fraction of the BF that corresponds to one FSR will be smaller than in the Littrow mode by the factor $\cos\theta_i/\cos\theta_d$. The effect is to reduce efficiency *variation* across the blaze function when departing from Littrow.

With peak efficiency reduced by the same factor when departing from Littrow more light will necessarily be diffracted into neighboring orders, leading to the broader but slightly reduced efficiency profile.

There is a simple explanation given here for the saddle-shaped efficiency curves seen in Figs.6.13 and 6.16. They arise when the incident beam is not normal to the groove facet, differing by a small amount Δ_θ. Reflection conditions require that the angle of diffraction with respect to the facet normal be equal to Δ_θ, i.e.,

$$\theta_d = \varphi - \Delta_\theta \tag{6.14}$$

when

$$\theta_i = \varphi + \Delta_\theta \quad . \tag{6.15}$$

From the grating equation (6.1) we can deduce that for each Δ_θ a maximum can be expected at a wavelength λ_B such that

$$\cos \Delta_\theta = \lambda_\theta / \lambda_B \quad , \tag{6.16}$$

where λ_B is the blaze wavelength.

For wavelengths less than λ_B, two lateral maxima are observed, as in the two figures referenced above, whose separation $2\Delta_\theta$ is represented by eq.(6.16). When the wavelength is greater than λ_B there will be just a single peak. This circumstance can sometimes serve as an exceptionally accurate means for determining the blaze angle, because the angles can be easily measured to high accuracy and λ_θ is known from the light source used.

6.4.2 Rigorous Electromagetic Efficiency Theory

The great advantage of an accurate, rigorous efficiency theory is that the influence of nearly all the variables can be studied, singly or in combination, without having to resort to difficult and expensive experimentation[1]. The normal range of echelle groove frequencies (20 to 360 gr/mm), groove angles (40 to 79°) are readily included, and orders may go up to at least 600. Of great interest has always been the relationship between groove shape deformations and efficiency, where it is well known that the influence is felt most strongly in the TM plane of polarization. To accurately control or modify geometry is always difficult, and may not even be worthwhile because there are no tools capable of adequately determining the actual microgroove geometry achieved. Useful analytical exploration is now possible, although unfortunately there seems little hope of solving the inverse problem of determining groove shape from efficiency data or to use this knowledge to improve ruling technology.

One of the long standing puzzles has been the exact relationship between the mechanically defined steep facet angle φ and the angle at which efficiency is maximum. In scalar theory they are the same, but it seems clear that in practice there are small differences, which might not matter if they were not also a function of wavelength. This spectral dependence is an important issue in some critical spectrometers.

For practical reasons, the experimental efficiency studies referred to above were conducted in a constant wavelength mode, with the angle of incidence θ_i as independent variable. While this is not the mode of use in instruments, which are usually constant angular deviation (monochromators) or constant θ_i (spectrographs), it serves well to illustrate echelle behavior, and is eminently useful for demonstrating conformance between theory and experiment.

---

[1] At least ignoring the cost of the theory, its numerical implementation, and computation time

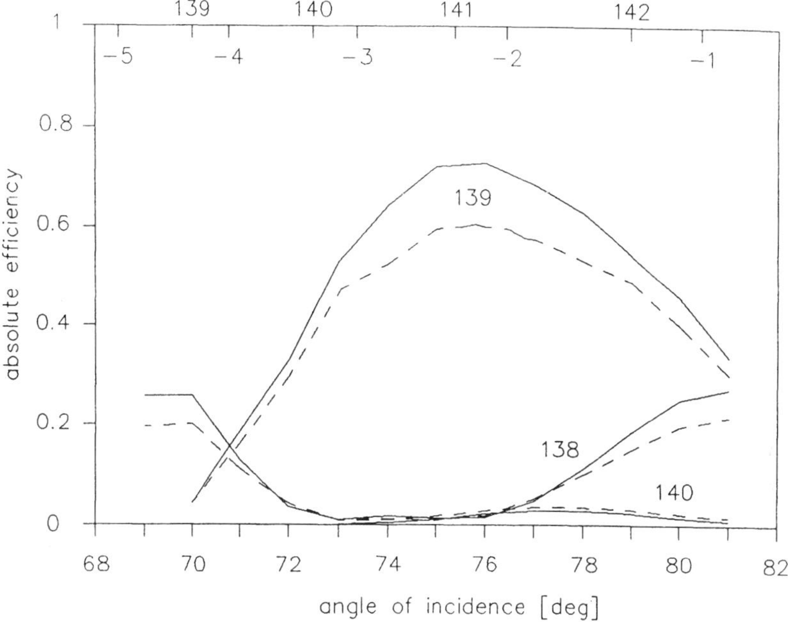

Fig.6.9 Measured absolute efficiency for 31.6 gr/mm r-4 echelle at 441.6 nm, as a function of angle of incidence. TE and TM planes of polarization shown solid, dashed respectively (after [6.11]).

6.4.3 Efficiency Behavior in High Orders

In order to examine high order behavior, use was made of 31.6 gr/mm echelles, at two different and frequently used groove angles 63.5° (r-2) and 76° (r-4), over a wavelength range from 441.6 nm to 676.4 nm, which corresponded to orders from 84 to 139. In every case data was taken in small angular steps, so as not to miss any features, and repeated in both planes of polarization. TE and TM plane data are shown plotted in solid and dashed lines respectively.

Starting with a 76°, echelle data is shown at four different wavelengths, 441.6, 632.8, 496.5, and 676.4 nm respectively in Figs.6.9 to 6.12. Figures 6.9 and 6.10 are similar in that their wavelengths happen to lie close to the peak efficiency angle, easily recognized from the fact that there is little diffracted energy in the next higher and lower orders. The grating equation leads to the conclusion that the apparent blaze peak occurs at an angle of 76.2°. Yet from the lack of perfect symmetry we can conclude that this not quite the correct choice, it seems actually nearer 76° or even slightly less. It is difficult to define

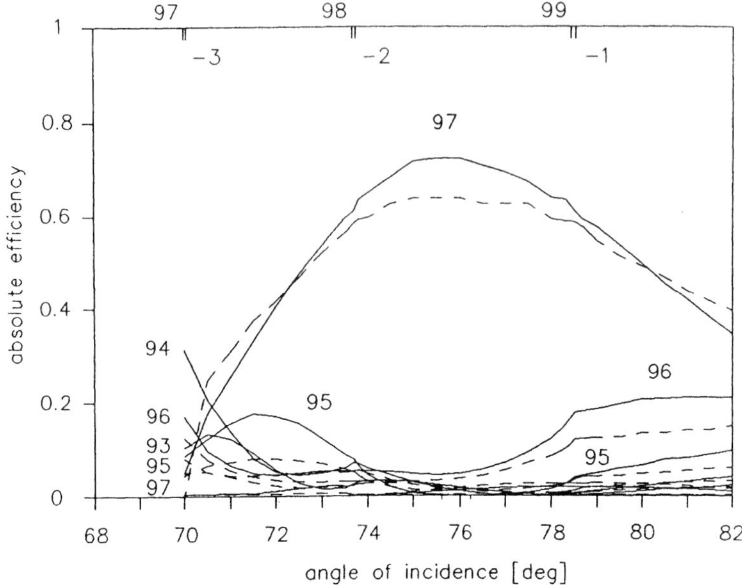

Fig.6.10 Same as Fig.6.9 except wavelength is 632.8 nm (after [6.11]).

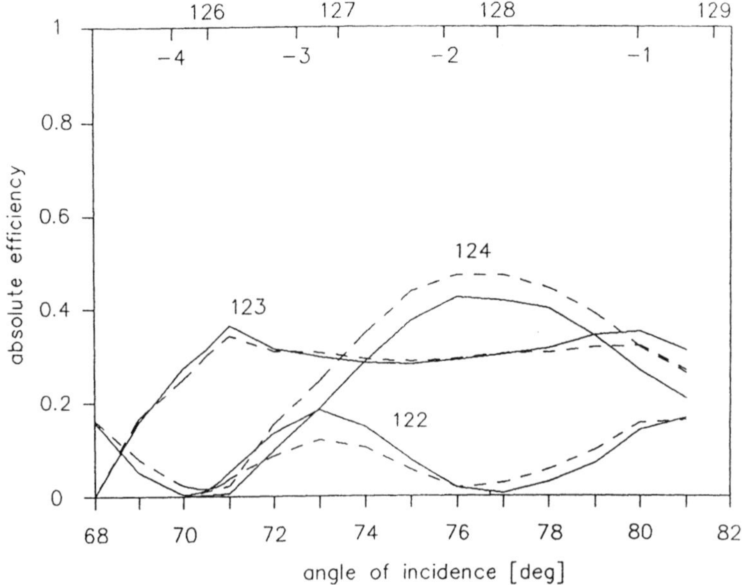

Fig.6.11 Same as Fig.6.9, except wavelength except is 496.5 nm. (after [6.11]).

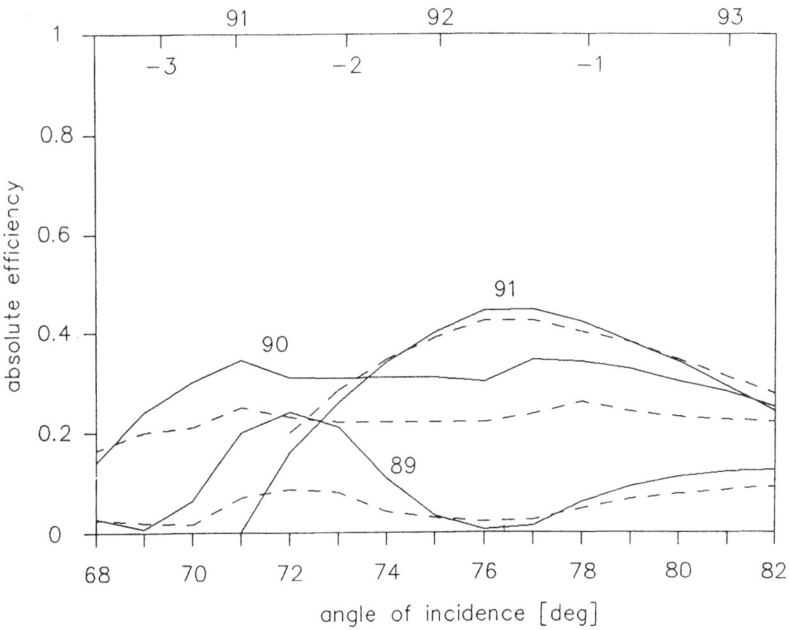

Fig.6.12 Same as Fig.6.9 wavelength is 676.4 nm (after [6.11]).

it closer. At 632.8 nm, Fig.6.10, there is a profusion of lower orders, unlike Fig.6.9, but no higher ones because the next higher order (98) cannot diffract when $\theta_i < 78.5°$. The (+) markers indicate the angles below which the specific higher orders cannot diffract, while the (–) markers give the limits above which the specific negative orders cannot diffract. Both Figures 6.9 and 6.10 show polarization levels of 6% which seems unexpectedly high for orders near 100.

In Fig.6.11 (496.5 nm) we find ourselves near the 1/2 order position between 123 and 124, estimated at 123.6. A rather similar situation can be seen in Fig.6.12, where λ = 676.4 nm, except that here the polarization roles are reversed, in that it is the higher order (91) that is virtually free of it, rather than the lower one.

In Figs.6.13 to 6.16, the same wavelengths are used to study a similar echelle, except with a groove angle of 64.4° (r-2). The 441.6 nm wavelength appears to fall almost exactly in the center of the 129th order (Fig.6.13), confirmed by the nearly perfect symmetry of all four orders detected. Of special interest is the saddle in the center of order 129, corresponding exactly to the efficiency peak in order 130. In Fig.6.15a we have the same condition, in that the 632.8 nm wavelength happens to correspond closely with the center

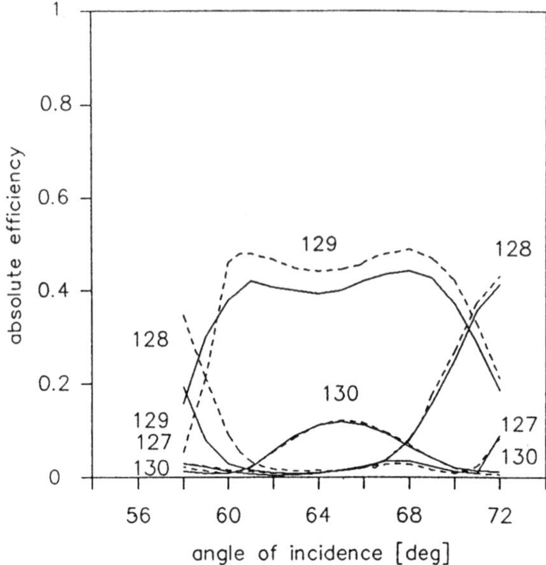

Fig.6.13 Measured absolute efficiency for 31.6 gr/mm r-2 echelle at 441.6 nm, as a function of angle of incidence. TE and TM planes of polarization shown solid, dashed respectively (after [6.11]).

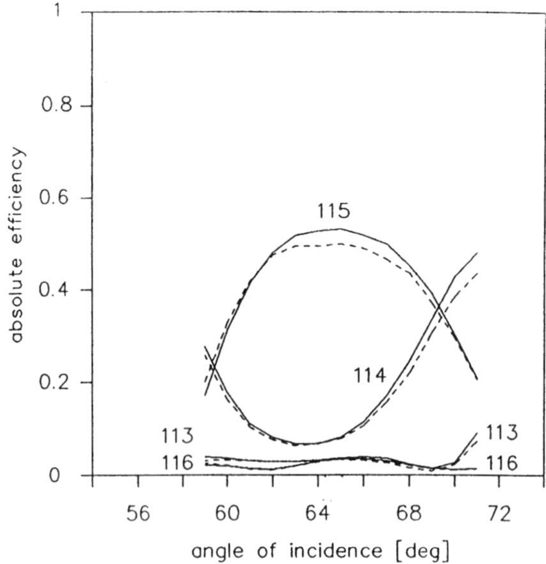

Fig.6.14 Same as Fig.6.13 except wavelength is 496.5 nm (after [6.11]).

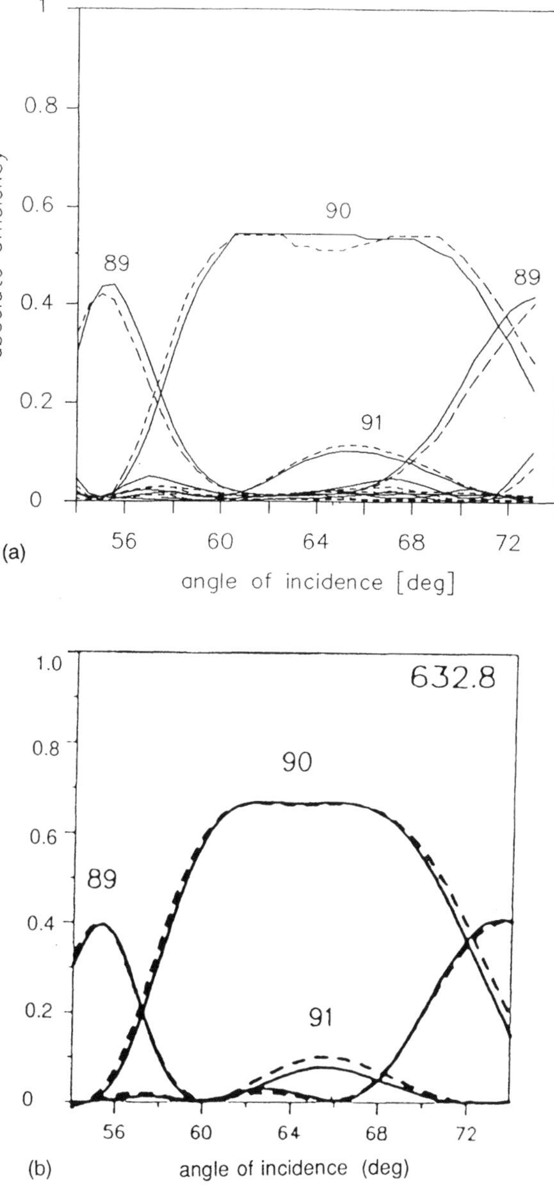

Fig.6.15 a) Same as Fig.6.13, except wavelength is 632.8 nm; b) theoretical absolute efficiency for ideal 63.4° groove geometry. Rigorous theory for wavelength of 632.8 nm (after [6.11]).

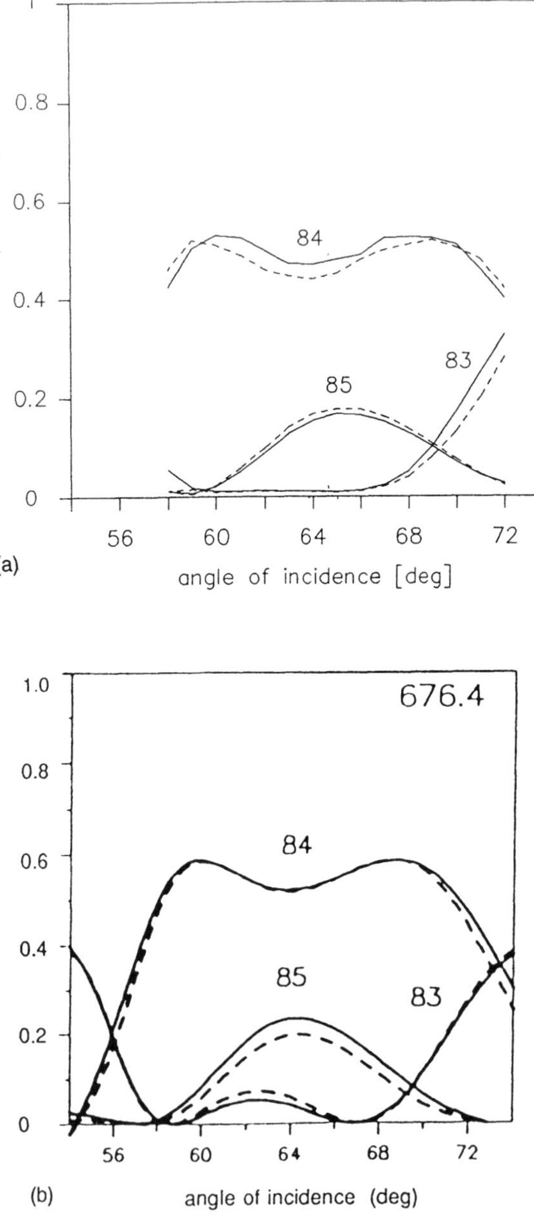

Fig.6.16 a) Same as Fig.6.13 except wavelength is 676.4 nm; b) same as 6.15b, except wavelength is 676.4 nm (after [6.11]).

of order 90, with order 91 not quite symmetrical, but robbing enough light to depress the central peak. The 89th order is also not perfectly symmetric, from which we conclude that there is a slight departure from the assumed 64.4° groove angle. At 496.5 nm, Fig.6.14, the order that corresponds to 64.4° is 114.8, which explains the prominence of 114 with respect to 115. Finally, at 676.4 nm, Fig.6.16a, the nominal order (from the grating equation) is 84.3, which explains the energy in order 85. For comparison this wavelength is evaluated by rigorous theory, including the complex index for aluminum, but assuming a perfect triangular groove, with results shown in Fig.6.16b. The match between theory and experiment is nearly perfect in orders 83 and 85, while in the dominant order 84 the measured values look similar, but are reduced by a rather uniform 17%. Polarization properties are perfectly reflected. A similar comparison with theory can be noted in Fig.6.15b, for 632.8 nm. Again there is good conformance and the same 17% difference in the efficiency values of the principal order seems reasonable in view of the necessary assumptions.

6.4.4 Efficiency Behavior in Medium Orders

This region is represented by the family of 79 gr/mm echelles, probably the most widely used groove frequency of all echelles. The visible portion of the spectrum is covered conveniently by orders 30 to 60, although such echelles also perform well in the UV, using orders from 100 to 180.

Starting with the high angle example (r-4), which in this case happens to have a groove angle near 74°, and observing behavior at the same four wavelengths as in section 6.4.3, gives results shown in Figs.6.17 to 6.20. At 441.6 nm, Fig.6.17, we can see an unusual and sharply defined dip in the efficiency right in the center of the blaze peak. Its edges are defined by markers at the top of the figure which identify the angular limits to the left of which the (m+1) order cannot diffract and to the right of which the −1 order cannot diffract. Between them there is a central ‘window’ where both can diffract, the effect of which is to rob light from the main order. The sharpness with which such a region is bounded is well known in echelette theory, and corresponds to Wood’s anomalies, or specifically the Rayleigh pass-off effect. For the groove angle of this particular echelle, 72.7°, it turns out that just one of the laser wavelengths (514.5 nm) falls close to a blaze peak (order 47), Fig.6.18. It shows the expected symmetry in all three detected orders, and stands out by also showing the lowest degree of polarization of any of the wavelengths tested. The dip at the peak has the same explanation. At 632.8 nm, Fig 6.19, the corresponding order is 38.2, and is accompanied by the highest degree of polarization observed (23%). In this instance the pass-off limits for −1 and 39 orders happen to coincide at the blaze peak, leading to an unusually high

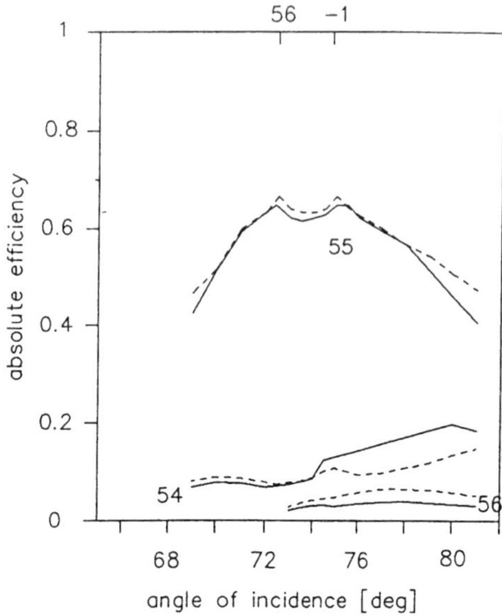

Fig.6.17 Measured absolute efficiency for 79 gr/mm r-4 echelle at 441.6 nm, as a function of angle of incidence. TE and TM planes of polarization shown solid, dashed respectively (after [6.11]).

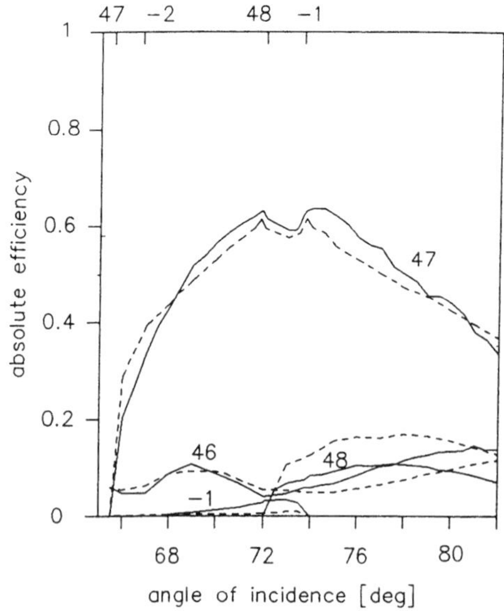

Fig.6.18 Same as Fig.6.17 except wavelength is 514.5 nm (after [6.11]).

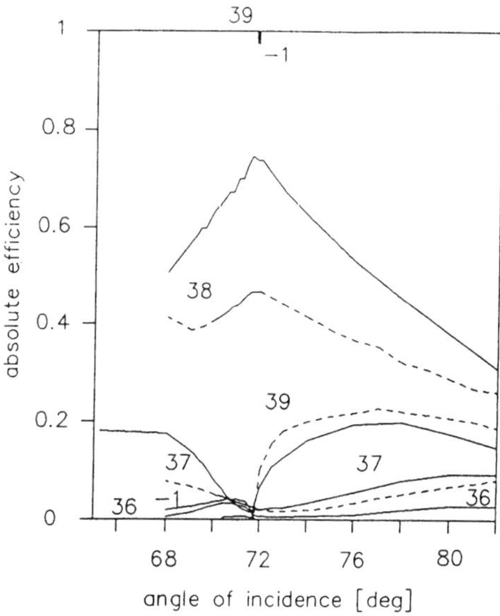

Fig.6.19 Same as Fig.6.17, except wavelength is 632.8 nm (after [6.11]).

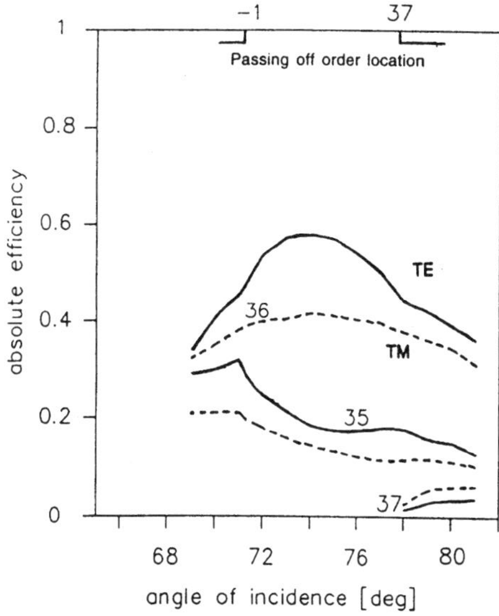

Fig.6.20 Same as Fig.617 except wavelength is 676.4 nm (after [6.11]).

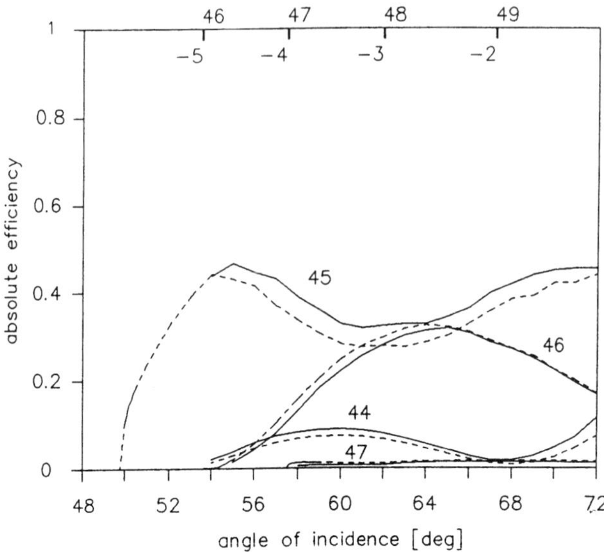

Fig.6.21 Measured absolute efficiency for 79 gr/mm r-2 echelle at 441.6 nm, as a function of angle of incidence. TE and TM planes of polarization shown solid, dashed respectively (after [6.11]).

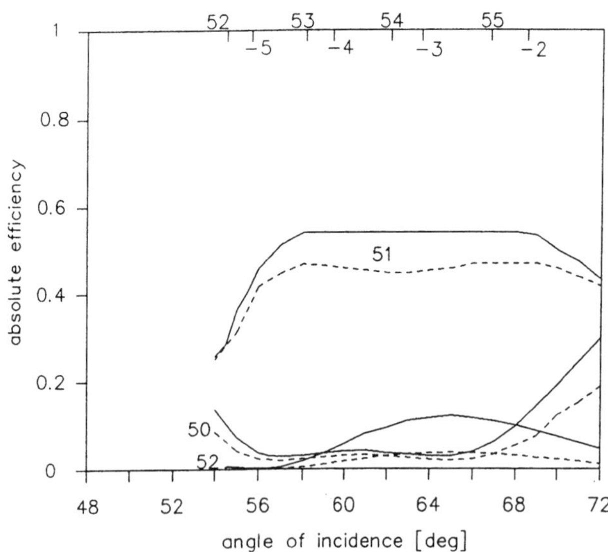

Fig.6.22 Same as Fig.6.21 except wavelength is 496.5 nm (after [6.11]).

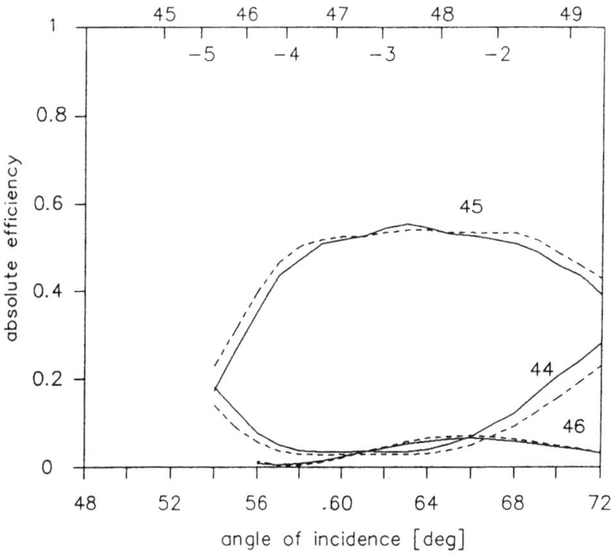

Fig.6.23 Same as Fig.6.21, except wavelength is 501.7 nm (after [6.11]).

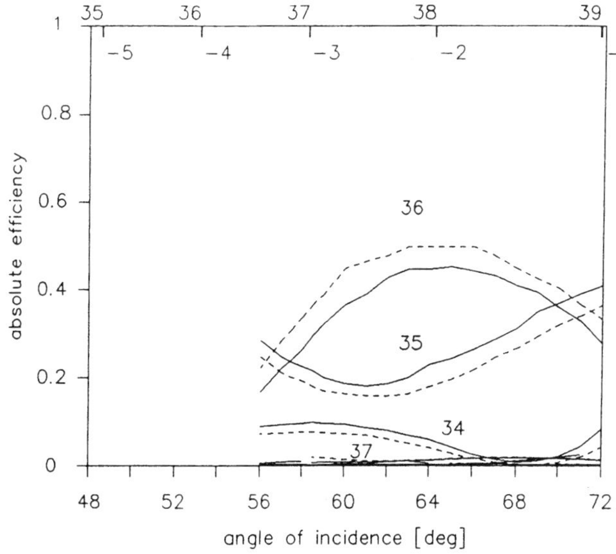

Fig.6.24 Same as Fig.6.21 except wavelength is 632.8 nm (after [6.11]).

TE efficiency peak of 75%. At the longest of the available test wavelengths, 676.4 nm, Fig.6.20, the –1 and (m+1 = 37) pass-off limits now lie on opposite sides of the peak. This means that between θ_i = 70.5° and 78°, neither of these orders can diffract. One would expect this to give a boost to the efficiency in the 36th order, and would do so if it were not for the circumstance that it corresponds to a fractional order (35.7). This results in a significant amount of light going into the 35th order. If the two orders were combined their total in the TE plane reaches a rather unusual efficiency level of 90%, while the equivalent value in the TM plane is only 60%, corresponding to 30%

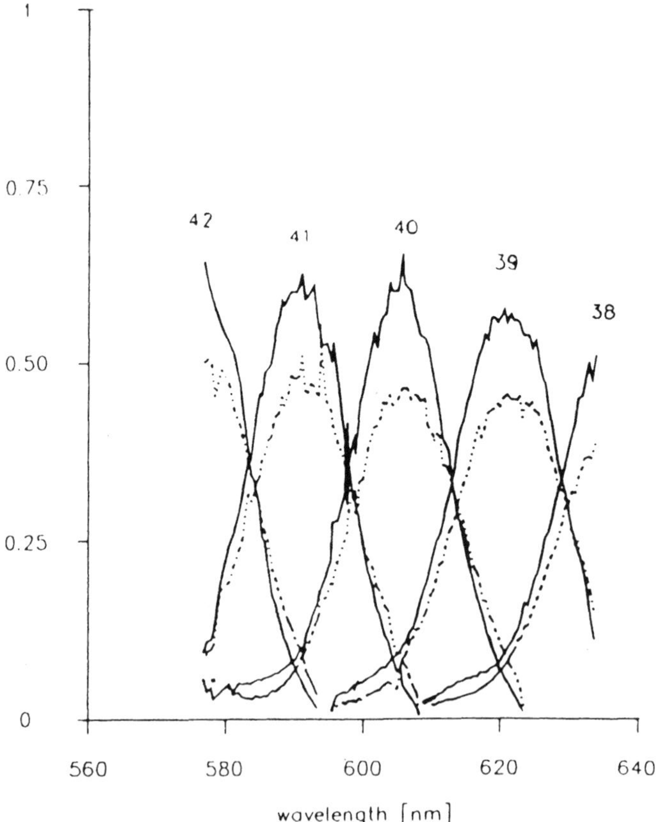

Fig.6.25 Wavelength scan of absolute efficiency with incident angle held constant at the groove angle, over range of 575 to 635 nm (after [6.11]).

polarization. However, it seems evident that if the grating were tuned to a slightly different λ/d ratio it could put an integral order number at the center, in which case the pass-off free window would lead to exceptionally high efficiency, at least over a narrow wavelength band (see section 6.4.9).

When switching to a lower groove facet angle near 63° (r-2) it is evident that pass-off orders no longer play a significant role (Figs.6.21-6.24). This comes from having so many of them close together that no single one has a dominating effect.

One of the wavelengths (501.7 nm) falls very close to the center of an integral order (45), Fig.6.23, as evidenced by low values of both adjacent orders plus a barely detectable level of polarization. At 441.6 nm, Fig.6.21, the nominal order number is 51.1, which explains why order 52 is somewhat stronger. At 496.5 nm, Fig.6.22, we are almost exactly at the half order position, as evidenced by the fact that at the groove angle (63.1°) orders 45 and 46 intersect, their combined value adding up to 60%, typical for such a grating. At 632.8 nm, Fig.6.24, we are at the 35.7 order, so that the 36th order dominates; adding it to order 35 leads to a combined peak value close to 64%.

With this particular grating a second set of measurements was undertaken under conditions of constant input angle and at the nominal groove angle of 63.1°. The input wavelength derived from a dye laser was varied from 555 to 615 nm, and orders 39 to 41 completely traced, Fig.6.25. Each peak corresponded to the same angle, was near 60% efficiency, and polarization was reduced.

6.4.5 Efficiency Behavior in Low Orders

The only echelle grating tested in the low order domain was a 316 gr/mm echelle with 63.4° nominal groove angle, which allowed testing in orders 8 to 13. In this domain polarization can be so strong, up to 40%, that the scalar model clearly cannot apply. The effect of passing-off orders (–1) and (m+1) can become quite strong, which enables a critical comparison with rigorous theory. In general the agreement was good in the TE plane, but in the TM plane magnitudes differed significantly in some instances, although the basic behavior is well described.

Starting with the shortest wavelength of 441.6 nm, Fig.6.26a, the nominal order number is 12.8, so we can observe nearly equal energy flow into orders 12 and 13. Since between θ_i = 54.5° and 59.5° both the –1 and 13th orders can diffract their presence accounts for the depression in the 12th order peak. Corresponding theoretical calculations, Fig.6.26b, match well for both orders in the TE plane, although for the 12th order theory predicts 10% higher vales. In the TM plane the picture reverses, in that it is in order 13 that the theory predicts 10% higher values.

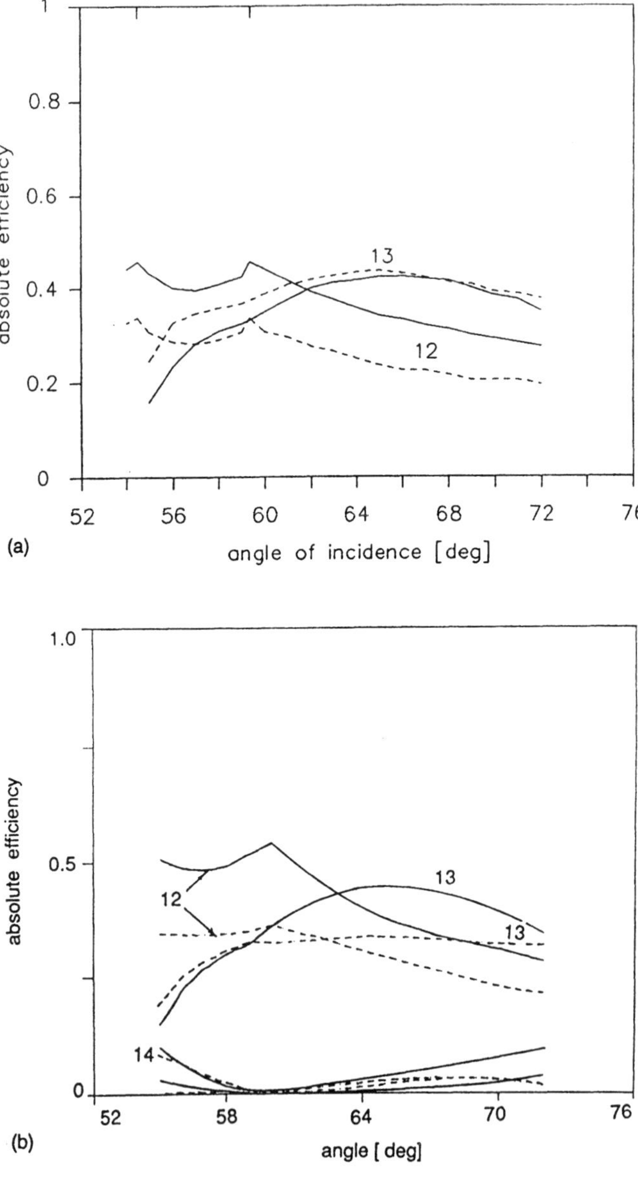

Fig.6.26 Absolute efficiency for 316 gr/mm r-2 echelle at 441. nm, as a function of angle of incidence. TE and TM planes of polarization shown solid, dashed respectively: a) experiment; b) E.M. theory and ideal geometry (after [6.11]).

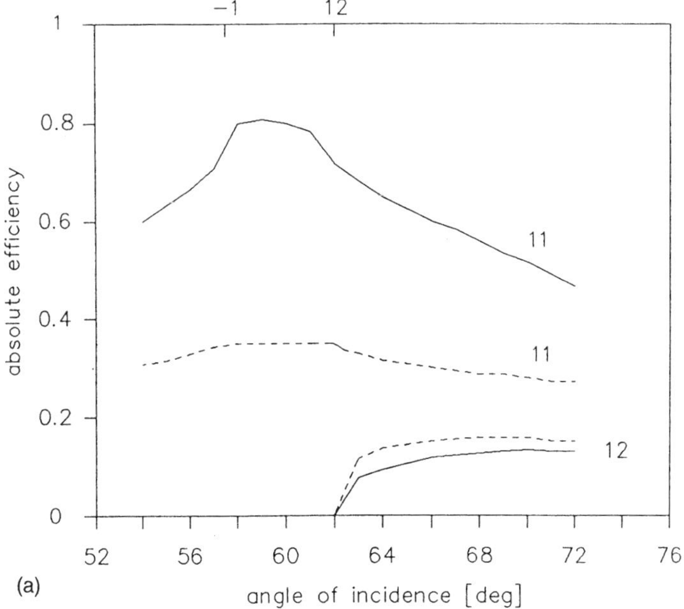

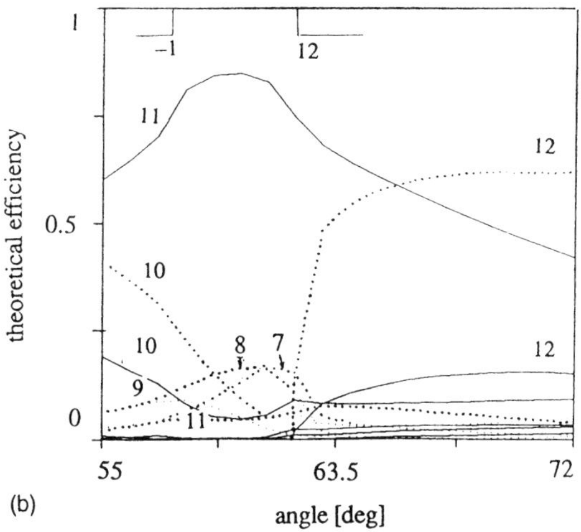

Fig.6.27 Same as Fig.6.26, except wavelength is 496.5 nm; a) experiment; b) theory (after [6.11]).

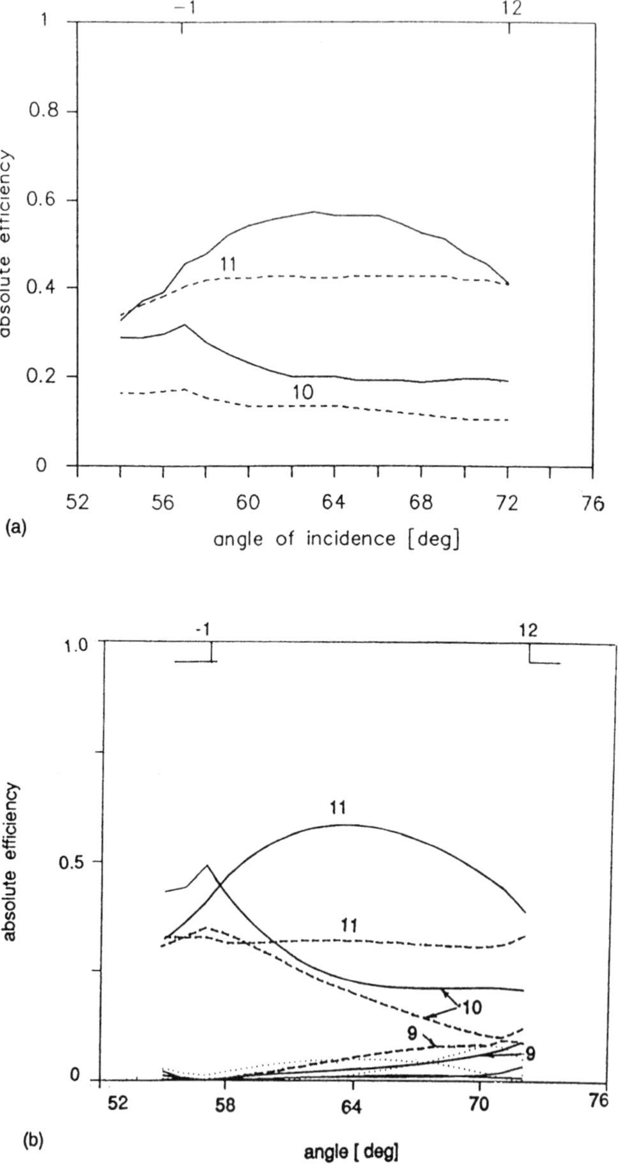

Fig.6.28 Absolute efficiency for 316 gr/mm r-2 echelle at 514.5 nm, as a function of angle of incidence. TE and TM planes of polarization shown solid, dashed respectively. a) Experiment; b) rigorous E. M. theory and ideal geometry (after [6.11]).

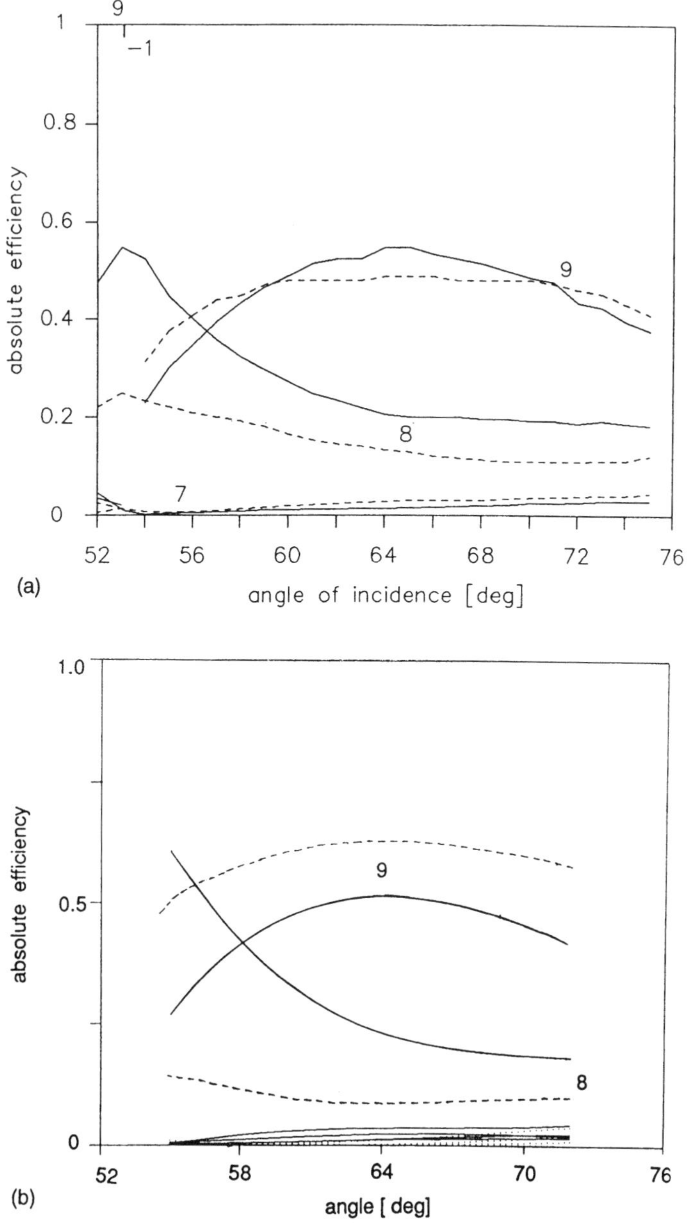

Fig.6.29 Same as Fig.6.28, except wavelength is 632.8 nm. a) Experiment; b) theory (after [6.11]).

At 496.5 nm, we have the 11.4 order (near double order) position, but do not see symmetry because the 12th order cannot diffract below $\theta_i = 62°$, Fig.6.27a. Polarization was at a record 40% in order 11. Compared to the previous case, the location of the pass-off orders was reversed, in that the −1 order cannot diffract above $\theta_i = 57.5°$. The result was a boost for the 11th order from the window between 57.5 and 62°, accurately reflected in the TE plane of Fig.6.27b, which shows the corresponding theoretical values. In the TM plane, theory predicts very low values for order 11, less than 1/3 of those actually measured. Whether this can be ascribed to departure from ideal groove geometry and the theoretical bulk aluminum refractive index is not clear. However, in order 12 the situation was reversed, in that a good TE match was accompanied by much higher predicted values in TM, and a remarkable degree of polarization (60%) that was never observed experimentally.

At 514.5 nm, Fig.6.28a, we have the first case of a good integral order match (at 11.0) for this blaze angle, which may account for the low degree of polarization. At the same time there was a good match to theory, Fig.6.28b, except that in TM plane the observed efficiency was 13% higher than theory. In the 10th order, by contrast, theory predicts enhanced values in both planes of polarization. It is interesting how clearly evidence of the effect of the −1 order passing off can be seen in both theory and experiment.

How quickly the picture can change is seen in Fig.6.29a, at 632.8 nm, which comes close to being an integral order at 9. The −1 and 9th orders happen to pass off together at $\theta_i = 53°$, the effect also visible in order 8. In order 9 the TE curve again matched almost perfectly with theory, Fig.6.29b, while for the TM plane theory predicts 20% higher values. In order 8 there was good match in both planes of polarization.

6.4.6 Confirmation of Theory

Accurately matching theory to experiment is rather difficult. To begin with there is the question of how well we know the actual groove geometry. It is tempting to assume a simple 90° triangular groove shape, as was done so successfully in Chapter 4 with echelettes. Unfortunately, when grooves are so much deeper, there is always severe plastic flow involved in generating them. As the theory itself shows, such effects have considerable influence on diffraction efficiency. This makes it clear that confirmation requires reasonably accurate determination of groove shape. Techniques, based on low temperature cracking of replica films that can be observed in a scanning electron microscope (SEM), give fairly good results, as evident from Fig.6.30.

The second concern is more subtle. Calculations in EM theory necessarily depend on using the complex index of the metallic surface in question, usually aluminum. The n and k values are calculated from reflection

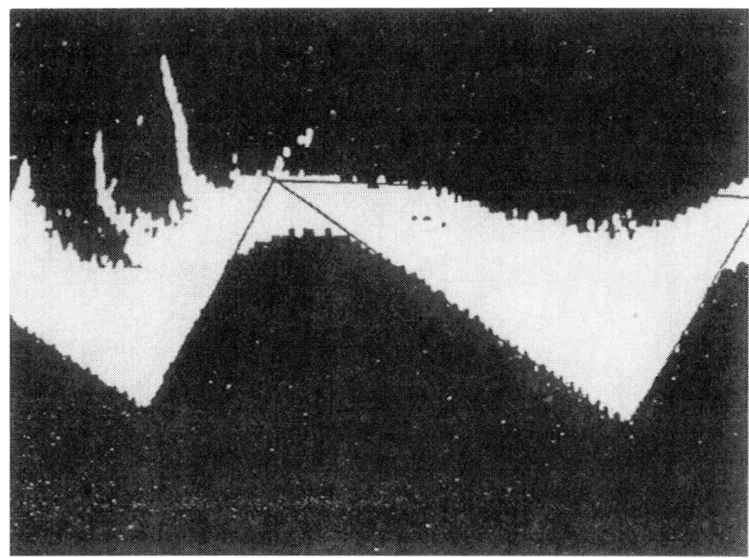

Fig.6.30 SEM of a replica on edge, 79 gr/mm echelle, 63.5° nominal blaze angle (courtesy Spectronic Instruments Co.).

coefficients and absorption measurements of thin films vacuum deposited on plane glass substrates whose surface is smooth. As such they are found in handbooks. To what extent these values apply to the conditions found on the grating surface is an intriguing question, because in the first case the surface is flat and smooth, while in the other is partially rough and severely modulated. In general such considerations are always more prominent in the TM plane of polarization than in the TE.

In order to test the theoretical approach, calculations were made for the echelle shown in Fig.6.30, at the observed groove angle of 62.4°, with wavelength the independent variable. The actual groove shape of an odd generation replica, taken from the figure, is shown as (2) in Fig.6.31, together with the nominal groove shape, labeled (1). The efficiency results are shown in Fig.6.32 for orders 49 to 51, over the wavelength interval from 430 to 450 nm. Note the cyclic changes, the low degree of polarization, but especially the considerable effect that the departure from nominal shape has on the peak efficiency in the center of the order. If we take unpolarized light (i.e., the average of the TE and TM curves), the peak drops from 78% to 52%. Also of interest is the behavior where the orders intersect (i.e., where one crosses over from one order to the next). Here the change in efficiency between the curves

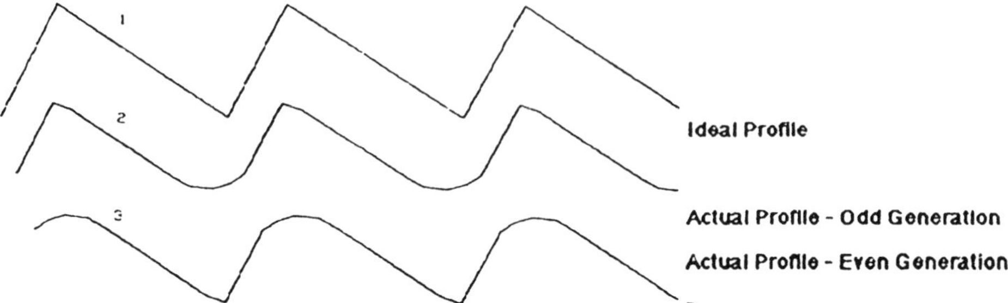

Fig.6.31 Groove shape profile taken from Fig.6.30. **1** The ideal groove shape. **2** Actual groove shape. **3** Inverted groove shape (after [6.11]).

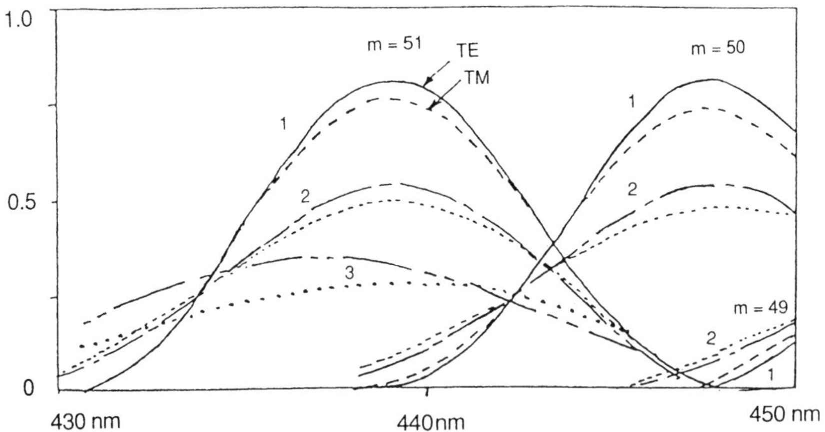

Fig.6.32 Theoretical efficiency curves in orders 49 to 51 of echelle of Fig.6.31 (after [6.11]).

is much less, from 37 to 32% respectively. This means that while for an ideal groove the crossover efficiency drops to 48% of peak, it drops to only 62% of peak for the actual groove in this instance. If uniformity of efficiency across orders is important the latter may actually be preferred. Energy that fails to show up in the principal order is invariably distributed among other orders.

While most of it tends to show up in the two adjacent orders, that does not appear to be quite the case here. The reduction in efficiency that accompanies groove inversion is quite clear (curve 3), which is interesting in that the peak locations are shifted as well, although in opposite directions for the two polarizations.

It has long been a source of annoyance that when efficiency peaks are experimentally determined as a function of wavelength over a wide range, and the grating equation is used to calculate corresponding diffraction angles, these angles do not remain constant. Instead the apparent blaze angle decreases with order number (i.e., decreasing wavelength [6.12]).

The rigorous theory of echelle efficiency provides an explanation, which is based on the different behavior in the two planes of polarization. Results of calculations are shown in Fig.6.33, for two 79 gr/mm echelles, r–2 and r–4, and show the apparent shift in peak wavelength as a function of order number N. A visible effect of the anomalous behavior can be readily observed in Fig.6.33, and also noted in Figs.6.17 and 6.26. The conclusion is that the greater the blaze angle (r–number) and the lower the order, the larger will be the shifts. There is a well defined power dependency that can be extracted from the figure

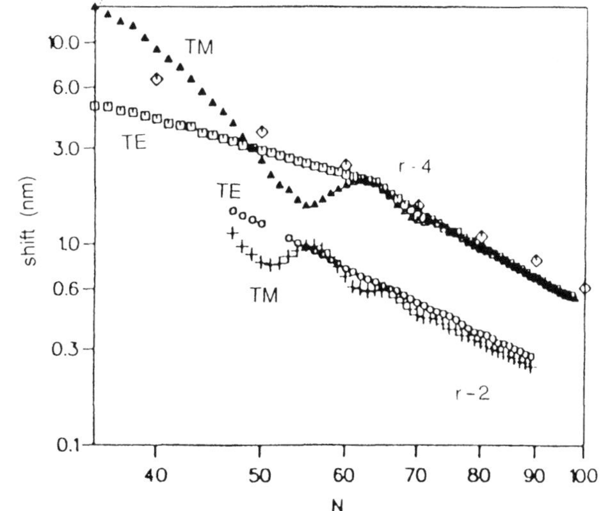

Fig.6.33 Spectral shift Δ_{max} in nm of the position of maximum efficiency from the scalar value, plotted on double logarithm scale. Square and triangles are for 79 gr/mm r–4 aluminum echelle, TE and TM respectively. Circle and x signs are for r–2 echelle, TE and TM respectively (after [6.11]).

$$\Delta_{max}(N) \propto \frac{1}{N^{2+\delta}} \quad , \tag{6.17}$$

where $\delta = 0.6$ for r–2 echelles and for r–4 as well, provided $N > 65$. For $N < 65$ the value of δ becomes zero for the r–4 echelle.

The physical significance is that when the λ/d ratio decreases the difference between the TE and TM efficiency curves reduces faster than the separation between consecutive orders (which is proportional to N^{-2}). We can see that as λ goes to 0 the positions of TE and TM maxima merge into each other, as expected from scalar theory. Unexpected is the abrupt change in behavior for the r–4 echelle below N = 65. This is due to cutoff effects that occur when the blaze wavelength is such that the next higher order cannot diffract. This behavior cannot occur at lower r numbers.

6.4.7 Efficiency Behavior in Spectrometer Modes

Most echelle spectrometers are configured as spectrographs in which the angle of incidence is fixed and the dispersed spectrum recorded by a stationary detector array, which in the early days was photographic film. If the exit beam is directed to a single fixed slit in front of a detector, we have a monochromator mount, and wavelength scanning requires high precision rotation of the echelle. It is rarely practiced because of the problem of isolating single orders.

Fig.6.34 was prepared to show how efficiency of a typical echelle (79 gr/mm, r–2 or 63.5° blaze angle) varies under different conditions. If beams are configured for Littrow (A. D. = 0) at one order (44), we obviously expect maximum efficiency in that order, and observe it in both TE and TM planes of polarization. In practice this requires tilting the incident beam slightly out of the meridional plane (e.g., make γ of Fig.6.3 ≈ 2°, which has only a small effect on efficiency). If A. D. is 10°, a reasonable value to adopt if one wants to remain in plane, it is clear from the figure that efficiencies will drop. Also evident is that efficiency peaks shift somewhat to shorter wavelengths in TM plane, but still more in TE. If incident ray angles are kept equal to the blaze angle, that amounts to adjusting angles to the order and leads to adjacent orders maintaining identical peak values.

Fig.6.35 presents similar calculations for the equivalent r–4 echelle (i.e., 76° blaze angle), with 49 the order at which angles are adjusted. The first impression is how much efficiency behavior departs from the "regular" Gaussian shape of the previous example. These anomalies are explained by higher and the –1 orders passing off, as discussed in section 6.4.4. Polarization effects (i.e., the difference between TM and TE values), are also much more pronounced. The effect of departing by 10° from Littrow is much stronger, as expected for a higher groove angle.

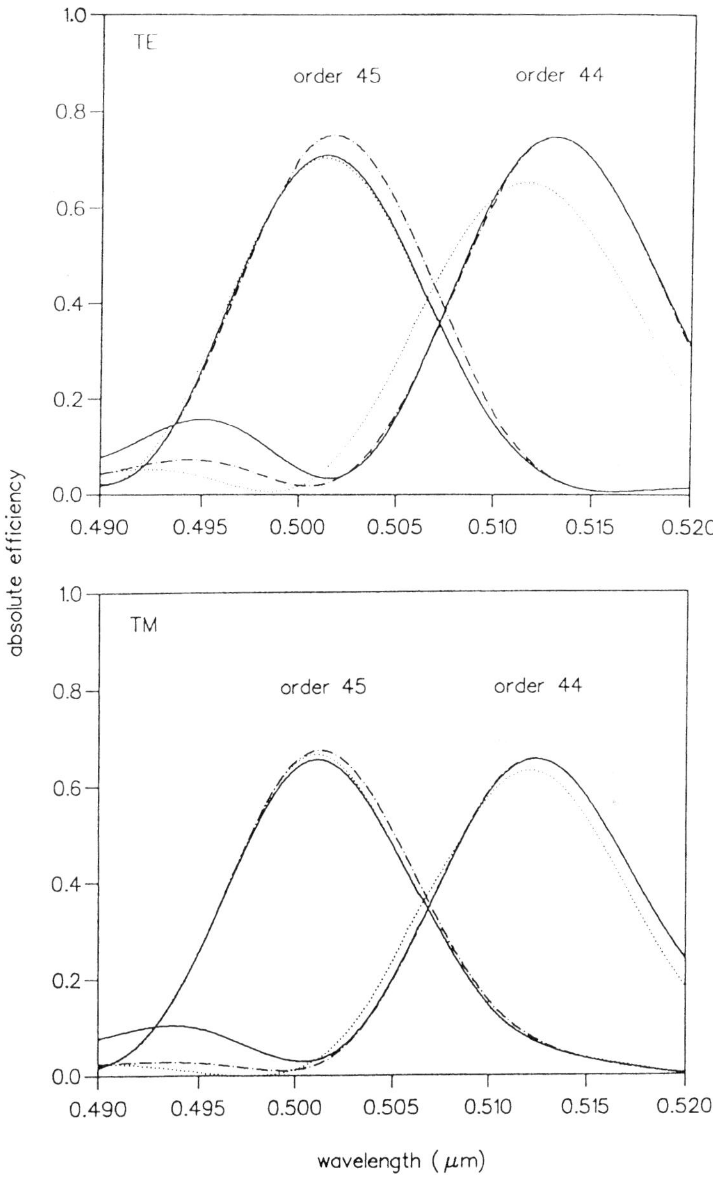

Fig.6.34 Theoretical absolute efficiency of 79 gr/mm r–2 echelle as a function of wavelength in μm. Solid curves: Littrow (A. D.= 0) adjusted for order 44. Dotted curves for A. D. = 10°. Dot–dash curves for $\theta_i = \theta_{Blaze}$.

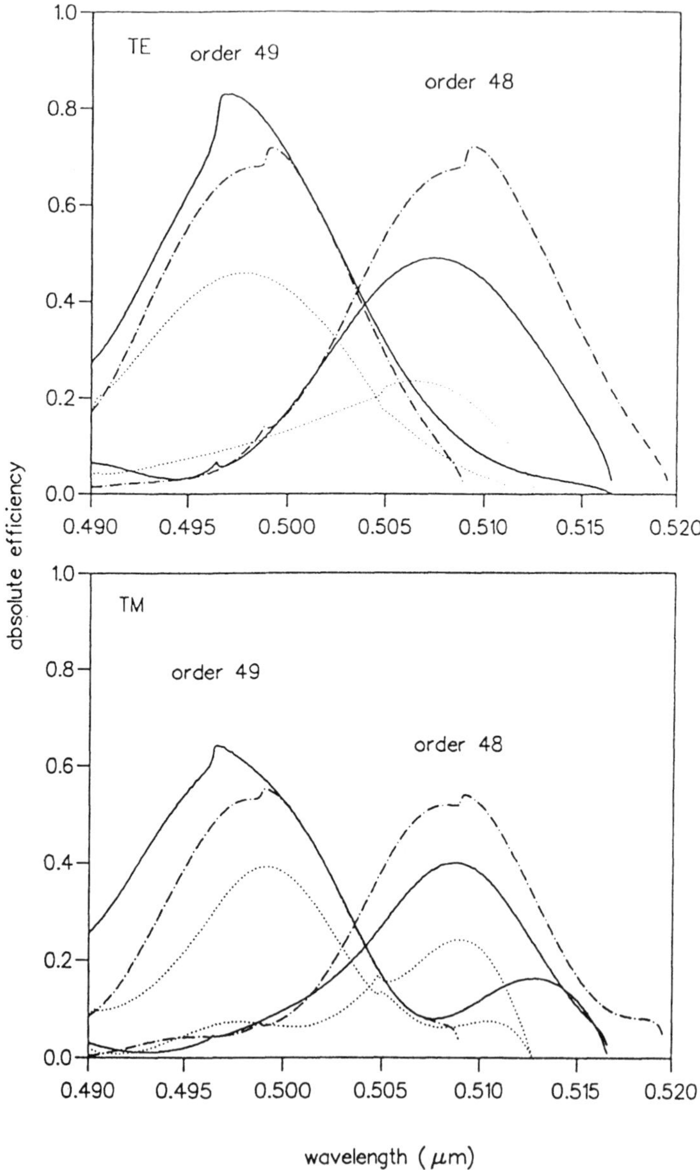

Fig.6.35 Same as Fig.6.34, except that echelle is r–4, 76° blaze angle, with adjustment made for order 49.

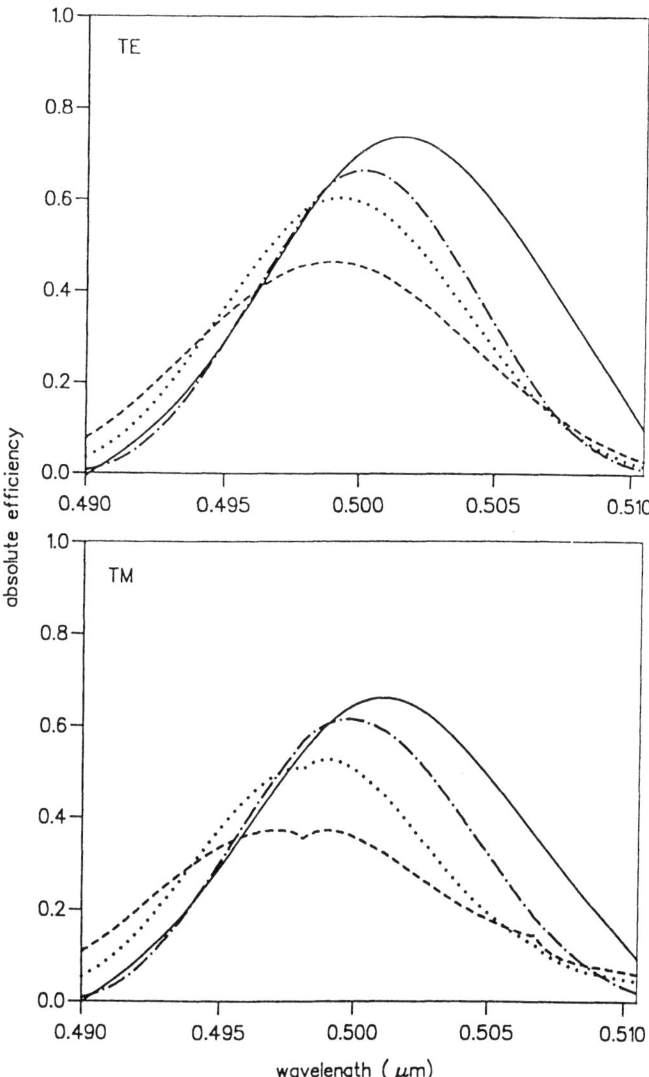

Fig.6.36 Efficiency of 79 gr/mm r–2 echelle in 45th order, as a function of wavelength. Solid curve: standard shape; dot–dash curve: ruled 80% depth. Dotted curve: standard shape with 10% of top flattened; dash curve: with 20% flattened.

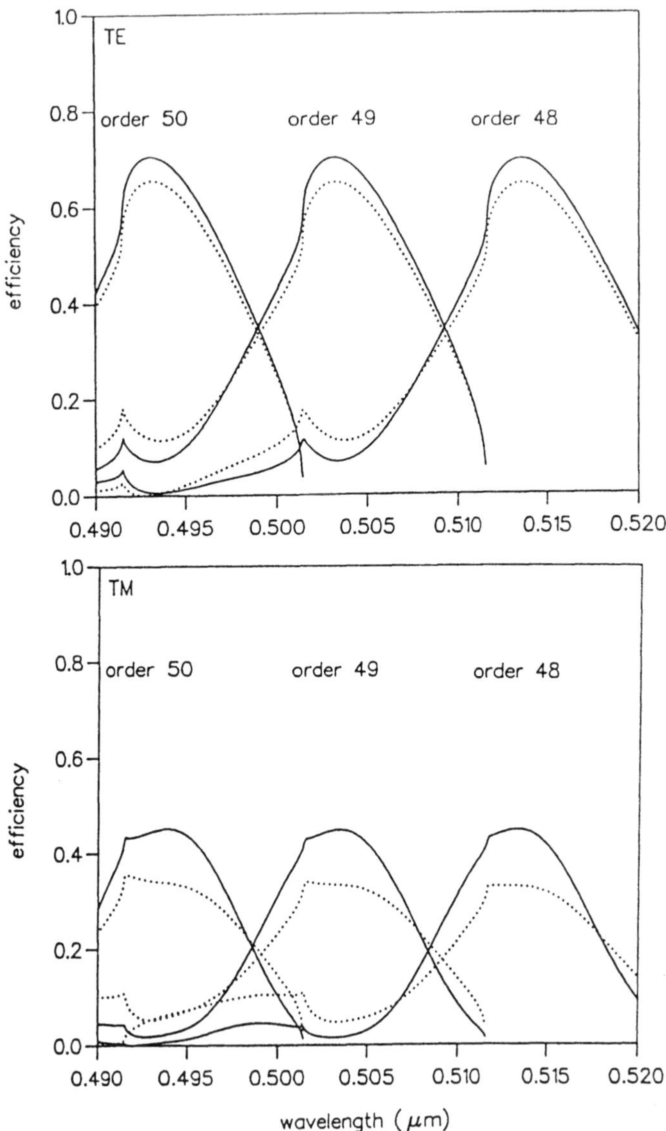

Fig.6.37 Absolute efficiency behavior of 79 gr/mm r–5 (79° blaze) echelle in orders 48 to 50. Solid curves: full groove shape; dotted curve: 20% of bottom not cut (i.e., ruled 80% depth).

6.4.8 Effects of Severe Groove Shape Disturbance

One of the useful attributes of a rigorous efficiency theory is the ability to study analytically the role played by departures from perfect groove geometry (see Figs.6.1, 6.30-6.32). The previous figures point to the role played by the sharpness of the groove tips. To examine this in more detail, we look at the family of curves shown in Fig.6.36. Using again the 79 gr/mm frequency, 63.5° blaze angle echelle of section 6.4.7, the figure shows what happens when the echelle is ruled to only 80% of full depth. We can also observe the sharp reductions in efficiency that occur when either 10 or 20% of the top is removed. As usual the TM plane is more sensitive to departures from ideal groove shape. There is not much difference in behavior between an assumed flat top and the more typical radius observed in practice.

There are useful applications for very high blaze angle echelles, such as r–5, or 79°. A set of corresponding efficiency curves for 79 gr/mm is shown in Fig.6.37. Pass–off anomalies can be seen to deform the curves significantly and again there is a large difference between TM and TE plane behavior. A second feature is that there is only a small efficiency penalty involved in such an echelle should the bottom 20% be left unruled.

6.4.9 A Useful Role for Anomalies

The classical role of efficiency anomalies, seen only in the TM plane, are discontinuities that appear when other orders can no longer diffract (i.e. when their angles of diffraction reach the pass-off angle of 90°). With echelles this has usually been disregarded because of the simultaneous presence of so many orders. However, as shown in sections 6.4.3 to 6.4.5, anomalies do exist, and are visible when one looks closely enough. They can be responsible for lack of symmetric efficiency behavior with respect to the blaze angle. Under some conditions it is possible to take positive advantage of them by designing echelles with groove angles and spacing chosen so that at one specific wavelength (or wavelength band) both the –1 and next higher order are precluded, which leads to a special efficiency peak. The following simple procedure describes such a design.

The first decision is to pick a blaze angle θ_B. Next is the choice of the angle by which the –1 order is to be offset from the Littrow diffraction angle θ_B. A reasonable figure is 2°. This immediately defines the necessary groove spacing d, from the relation

$$\lambda_c / d = 1 - \sin(\theta_B - 2°) \quad ,$$

where λ_c is the central wavelength chosen. Next we need to locate the spectral order to be used, m_c. This is derived from the grating equation

$$m = \frac{2 \sin \theta_B}{\lambda_c / d} \quad . \tag{6.18}$$

Finally the value of m is reduced to to the nearest integral value m_c, and d recalculated

$$d = m_c \lambda / 2 \sin \theta_B \quad . \tag{6.19}$$

To confirm the angular distance by which the (m_c+1) order diffracts (compared to θ_B), its angle of incidence is calculated from

$$\theta_{m_c+1} = \sin^{-1} [(m_{c+1}) \lambda/d - 1] \quad . \tag{6.20}$$

Comparing Fig.6.26 with 6.27 demonstrates the potential benefit to be derived from designing a high angle echelle to work at an optimum efficiency when there is concern for just one wavelength [6.11].

6.5 The Role of Overcoatings

With few exceptions echelles are used in the form of aluminum coated replicas. This is only natural since the wide spectral range covered by echelles matches the special ability of aluminum to reflect well over a wide band of wavelengths. However, there are limitations. The thin layer of Al_2O_3 that normally protects aluminum from environmental effects becomes progressively more and more opaque when $\lambda < 175$ nm. The solution, borrowed from mirror technology, is to overcoat the echelle with a fast-fired layer of aluminum (100 nm) which is followed immediately with an overcoat of MgF_2 just sufficient to prevent oxidization of the aluminum surface. Typically 25 nm thick, it serves at wavelengths down to 120 nm. If the lower wavelength limit is 160 nm it is advisable to increase the coating thickness to 35 nm. These thicknesses are designed to get some $\lambda/4$ enhancement effect on reflectance, although that plays a minor role here. The coating process will be more complex than with a mirror, because of the need to evenly coat the steeply inclined groove faces.

Since aluminum mirrors routinely have reflectance enhanced by multilayer dielectric overcoatings, it is perhaps natural to assume that the same can be done with echelles. However, even if one accepts the limited wavelength intervals for which such coatings are designed, experience has shown that for gratings the only effect is to decrease efficiency. This may be due to the difficulty of properly coating the steep face, or more likely due to electromagnetic effects (see Chapter 8).

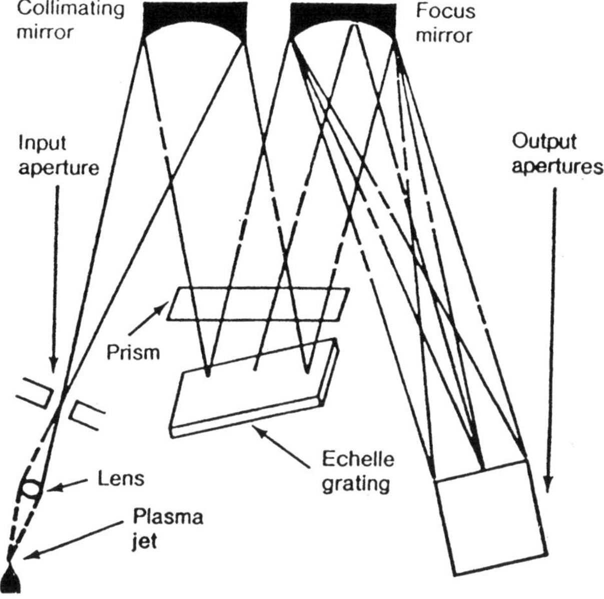

Fig.6.38 Conventional echelle spectrograph design with prism cross-disperser. (after [6.13]).

6.6 Instrument Design Concepts

In the classical design of echelle spectrometers the grating is illuminated in collimated light derived in most cases from a mirror, naturally achromatic. It may be in the form of an off-axis paraboloid or a sphere if the f number is high enough. The diffracted beam is focussed onto the image plane by a similar mirror, termed camera mirror, like the Czerny–Turner mount described in Chapter 12. In order to separate the overlapping orders a cross-dispersing element has to be interposed in one or both beams. This may be a prism or a low dispersion grating, Fig.6.38. For scanning systems the cross-disperser can be replaced by a fore-monochromator.

6.6.1 Choice of Echelle

The first decision is to select the free spectral range (see section 6.3.4), usually based on the availability of catalog echelles. The choice is often contingent on the overall size of the detectors as well as their pixel size combined with the resolution desired. This also involves deciding the collimator focal length, since the linear dispersion is the product of the angular dispersion and the focal length. The single most common groove frequency is 79 gr/mm, and it serves well from the UV to the near IR. To increase the number of orders (i.e., to shorten the FSR), it is common practice to switch to 31.6 gr/mm frequency. For special applications intermediate values are used, and for shorter wavelengths 158 and 316 gr/mm are effective. These odd frequencies derive from the ruling history at MIT, where at one time it was considered safer to make the spacings integral multiples of the He-Ne laser wavelength used to control the engine. The blaze angle in widest use is 63.5°, a value chosen early on, but based on no more than the convenience that its tangent is equal to 2. For increased dispersion, which allows reducing the f–number or increasing the entrance slit width, it is common to switch to r-4, or 76°, with r-5 a normal upper limit. However, intermediate values are possible if design conditions make that sufficiently important.

Size of the grating, and with it the associated optics, depend on the desired throughput and especially the resolution required. Typical values vary from 50 to 400 mm ruled width for single gratings, although mosaic assemblies can increase this (see Chapter 17).

6.6.2 Cross Dispersion: Prisms vs. Gratings

By far the most common choice for cross-dispersion is a prism, sometimes made of glass, but more often of fused silica because of its transparency in the important UV region. Prisms have a number of advantages in this application. Throughput losses are likely to be no more than the Fresnel reflections at the two faces. Ultrapure silica will transmit down to the oxygen cut off near 185 nm. Stray light is low provided high grade material is used. Ideally the dispersion should be linear, so that order separation is uniform, which is especially desirable with CCD detector arrays. Focal lengths must be such that there is enough pixel separation between orders where they are closest together (i.e., at the long wavelength end). This will mean excessive separation at shorter wavelengths, wasting some of the CCD area. Fig.6.39 shows the appearance of the image field, given the usual non-linear dispersion.

For some applications in astronomy, especially at red wavelengths, prisms either do not give enough dispersion or become so large (for big telescopes) as to be impractical. Even though tandem prism systems have been used to get around this limitation there are still cases where standard blazed echelette diffraction gratings are more effective. Here the dispersion is also non-linear, in fact more so, but now at the shorter wavelengths the orders are crowded together. It is also easier to maintain an interchangeable set of gratings in order to optimize their choice for a specific spectral region. There may be

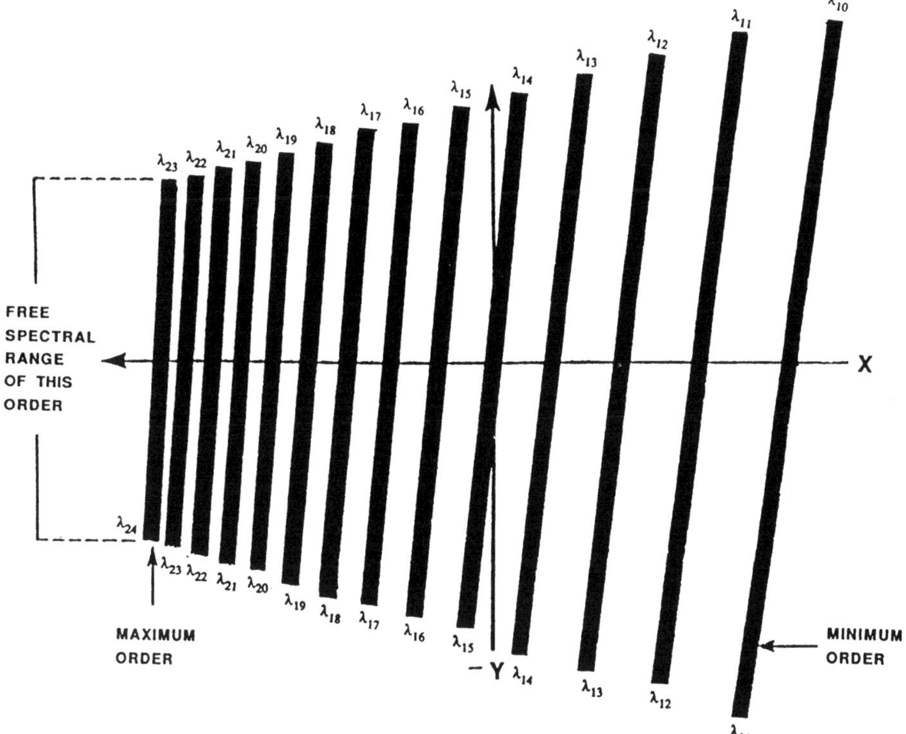

Fig.6.39 Typical image presentation of an echelle spectrograph with prism cross–dispersion. (after [6.13]).

instances where band-pass filters need to be added to avoid interference effects from higher orders.

The lack of linearity of cross-dispersers becomes serious when a very large spectral range must be covered simultaneously. It has been suggested that this could be done by combining the opposite effects of diffraction and refraction, with a GRISM containing a transmission grating replicated on one face of a prism [6.14, 15]. The lower wavelength limit will be determined by the transmission of the replica resin (i.e., about 250 nm).

6.6.3 Examples of Echelle Instruments

Over the years a large number of echelle spectrographs have been built. Most of them were specially designed for astronomical applications, of which

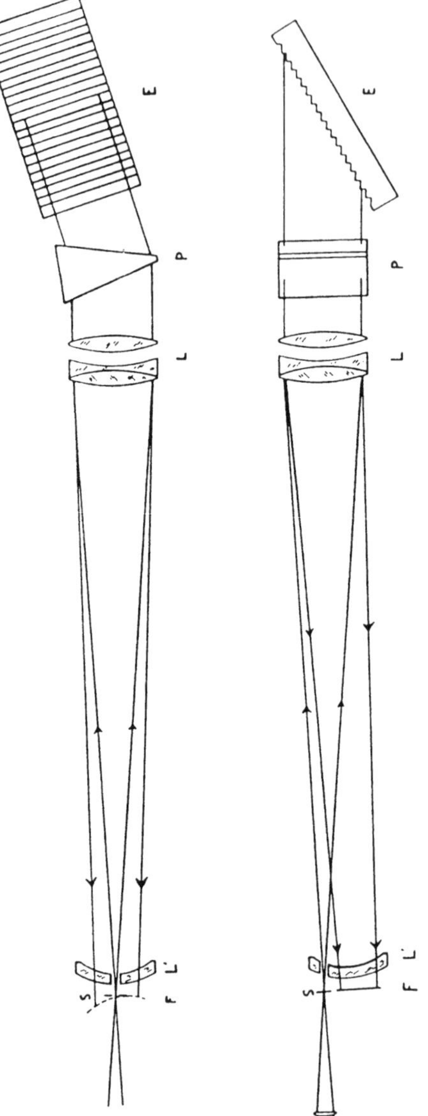

Fig. 6.40 Optical system of echelle spectrograph for near ultraviolet solar spectroscopy from a rocket. Collimating lens L is an achromat of fused silica and LiF, prism P is of fluorite, field flattener L' is of quartz. Echelle E is a replica on fused silica (after [6.17]).

some representative examples have been chosen below, both space and ground based. Most commercial instruments are intended for atomic spectroscopy, where the inherent high dispersion and resolution is used to full advantage. Typical are inductively coupled plasma (ICP) applications or atomic absorption (AA).

UV Rocket Spectrograph

A particularly elegant application was a 3 ft instrument flown by Tousey in an Aerobee-Hi rocket, in one of the first attempts to observe with high resolution the near UV spectrum of the sun [6.16, 17]. A schematic of the instrument is shown in Fig.6.40 and a spectrum returned from an Aug. 1961

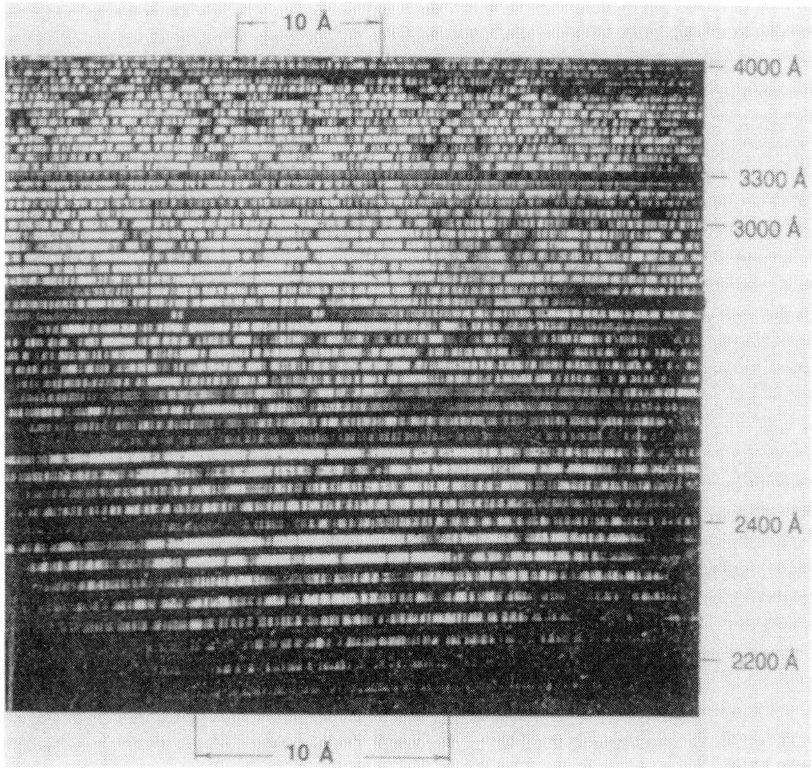

Fig.6.41 Echelle spectrum of sun from 210 to 400 nm. Dark lines are Fraunhofer absorption and bright lines solar emission. Orders are from 60, near 4000 Å, to 120 at 2000 Å (after [6.17]).

flight is given in Fig.6.41. It was recorded on high resolution film just 25 mm square, but in effect recorded a 1.2m long spectrum. An enormous amount of information was thus compressed into a small area, and displayed several thousand solar lines that had never before been seen, at a resolution > 60, 000.

HIRES: High Resolution Echelle Spectrometer

HIRES is the acronym for the high resolution instrument designed for the Keck telescope. As necessary for large telescopes, this is a huge spectrograph, with an entrance beam width of 300 mm. The dispersing element is a specially ruled r–2.6 (69°) echelle with 46.5 gr/mm. It required a ruled width of 835 mm, which is beyond a reasonable capacity to rule. The solution was to assemble three 280 mm echelles onto a single ZeroDur™ substrate in such a way as to maintain coplanarity to about 0.5 μm [6.18]. In addition the grooves must remain parallel to each other within 2 arc seconds and the normals to the ruled face must remain parallel to each other within 1 arc sec. Once cemented into position, the tolerances must be maintained indefinitely. Fortunately it is not necessary to match the phase of the three gratings, because the resolution of a single grating is more than adequate, the remaining two serving only to increase throughput, but of course they must maintain image quality (i.e., diffract into exactly the same spot as the first). A sketch of the system is shown in Fig.6.42.

Incident light is collimated by an off-axis parabolic mirror, and input will be either from an entrance slit or from a family of optical fibers that allows simultaneous observation from multiple sources. Angular deviation is maintained at 10°, in the meridional plane. The cross dispersers are much too large to consider the use of prisms, in fact are twice the size of the largest available gratings (300x400 mm). They too will be in the form of a mosaic assembly, this time of two gratings, simulating 400 mm ruled width with 600 mm length of groove. Low dispersion serves for standard use, while high dispersion gratings give the high order separation needed for the multi-object mode. The camera that focuses the spectral images onto the CCD detectors is a large telescope in its own right, with a 760 mm aperture, operating at f/1.0 [6.19].

Compact High Resolution Spectrograph

A versatile spectrograph designed by Baranne has been applied both to stellar motion studies and ICP analysis [6.20]. A schematic diagram is shown in Fig.6.43. Input is typically via a high efficiency fiber **F**, directed to the first collimating mirror $\mathbf{C}_1$, from which collimated light is directed to a high blaze angle (r–4) echelle **G**. Dispersed light is focused by camera mirror $\mathbf{C}_2$ to folding mirror **M**, which directs it to the final camera mirror $\mathbf{C}_3$. From here the

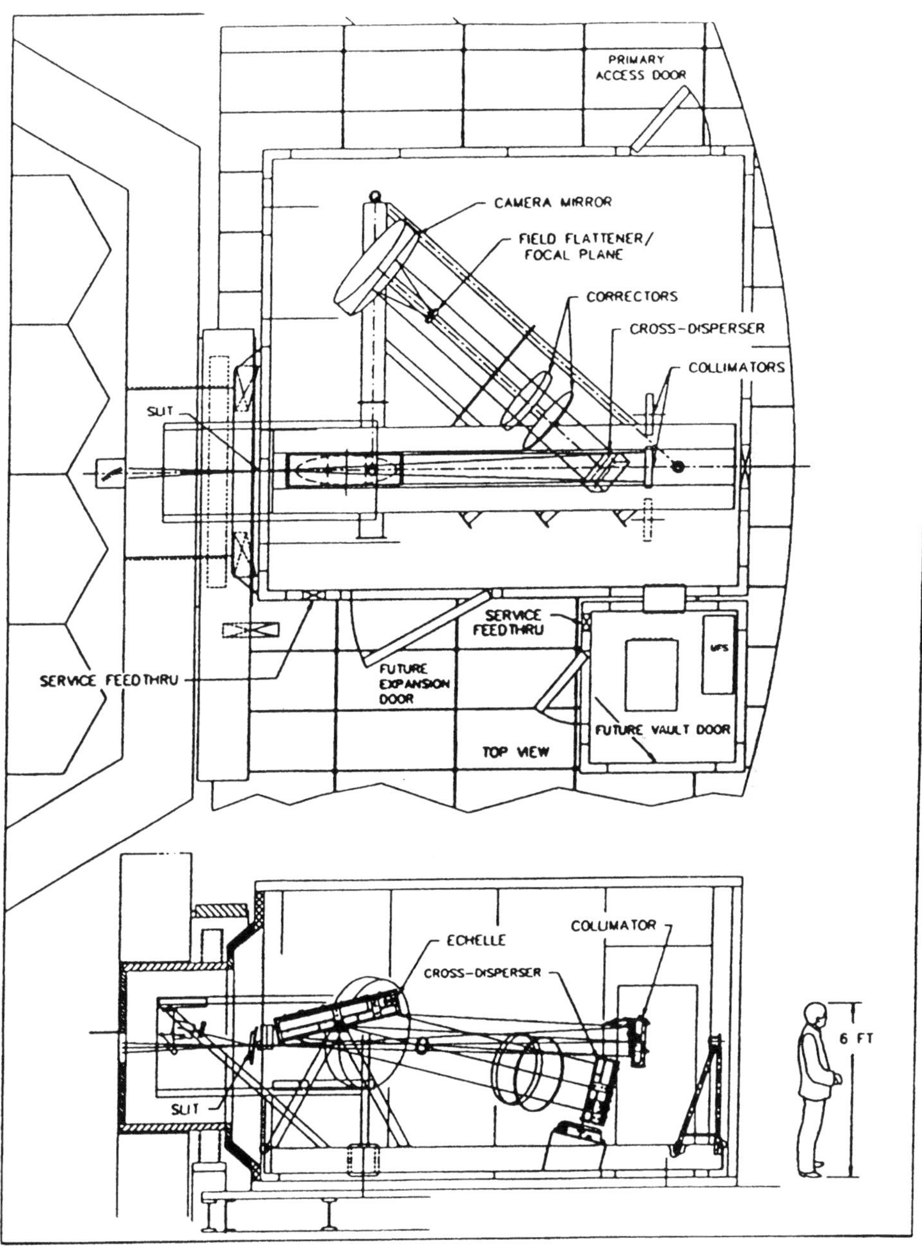

Fig.6.42 Plan and section view of optical layout of HIRES (after [6.18]).

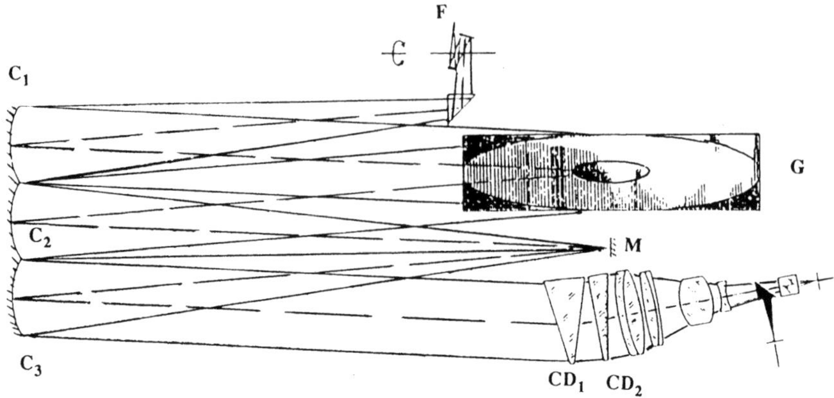

Fig.6.43 Schematic of high resolution echelle spectrograph with double prism cross dispersion (after [6.20]).

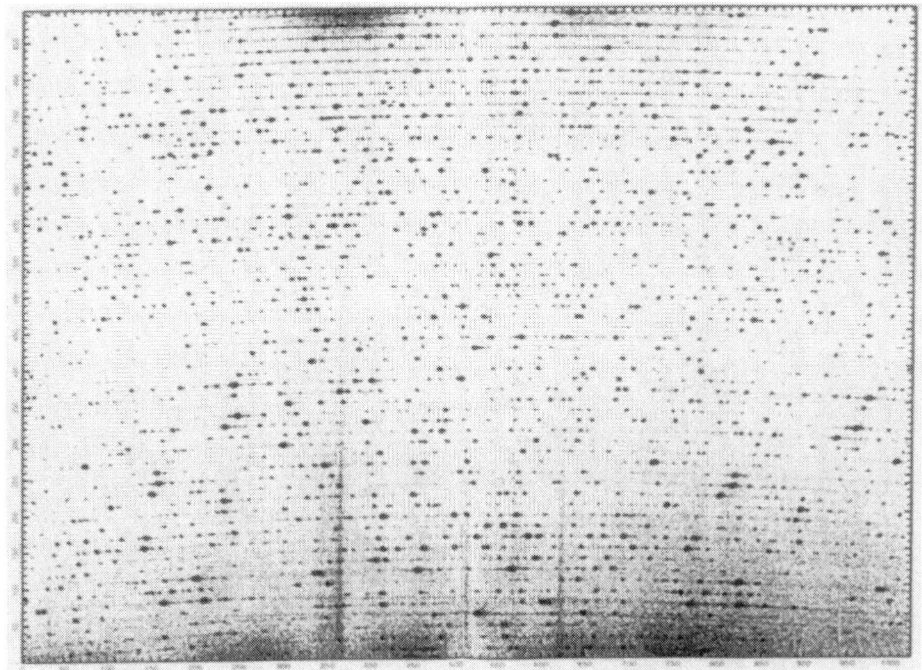

Fig.6.44 Complete display of CCD readout of spectrograph of Fig.6.43. Field covers 350 to 900 nm and is that of a sun–like star (after [6.20]).

collimated light passes through the cross-dispersing combination of a fused silica prism $\mathbf{CD_1}$ and a glass GRISM $\mathbf{CD_2}$. A complex system of lenses focusses the entire spectrum (300 to 900 nm) onto a high resolution 20x20 mm CCD. Depending on source brightness, the CCD integration time will vary from a few seconds to an hour, after which the data is transferred to a personal computer. The appearance of the computer screen is shown in Fig.6.44, and displays about 60 orders with even spacing. Wavelengths can be determined to 0.01Å, but to maintain calibration to that level requires careful thermal shielding and temperature constancy to 0.01°C.

Ultra-Short Wavelength Satellite Spectrograph

A special approach was necessary for a successful high-resolution satellite spectrograph, designed to operate down to 95 nm. The absence of good reflecting materials made it essential to truly minimize the number of reflecting surfaces. A standard echelle given a special overcoating served effectively as the dispersing element, while the 1.8 m focal length parabolic camera mirror was made to serve simultaneously as the cross-disperser by ruling onto its surface an 8-partite 1800 gr/mm grating. The input collimator was made in the form of a multiple grid, thus avoiding a third reflection, Fig.6.45. A special

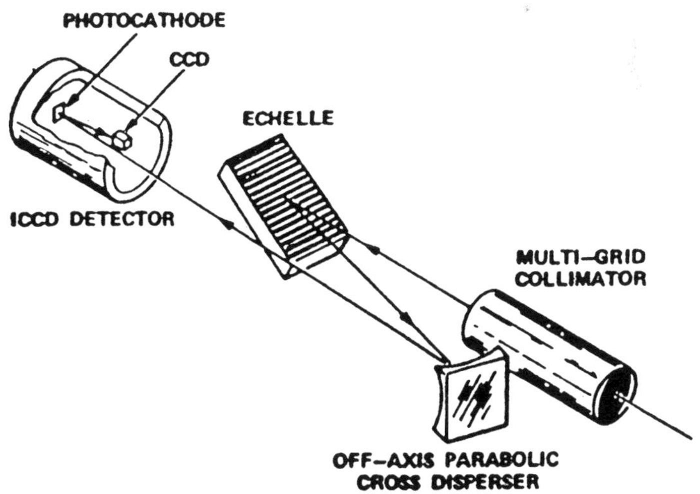

Fig.6.45 High dispersion, high resolution stellar spectrograph for IMAPS satellite (after [6.21]).

CCD detector was used, but covered only 1/4 of the free spectral range. To scan the entire spectrum, the mirror was tilted in 4 steps by means of a motor driven cam [6.21].

6.7 Maximum Resolution Systems

To observe maximum resolution over a wide range is always a difficult undertaking. For example, in order to test the performance of a large r-2 (63.5°)

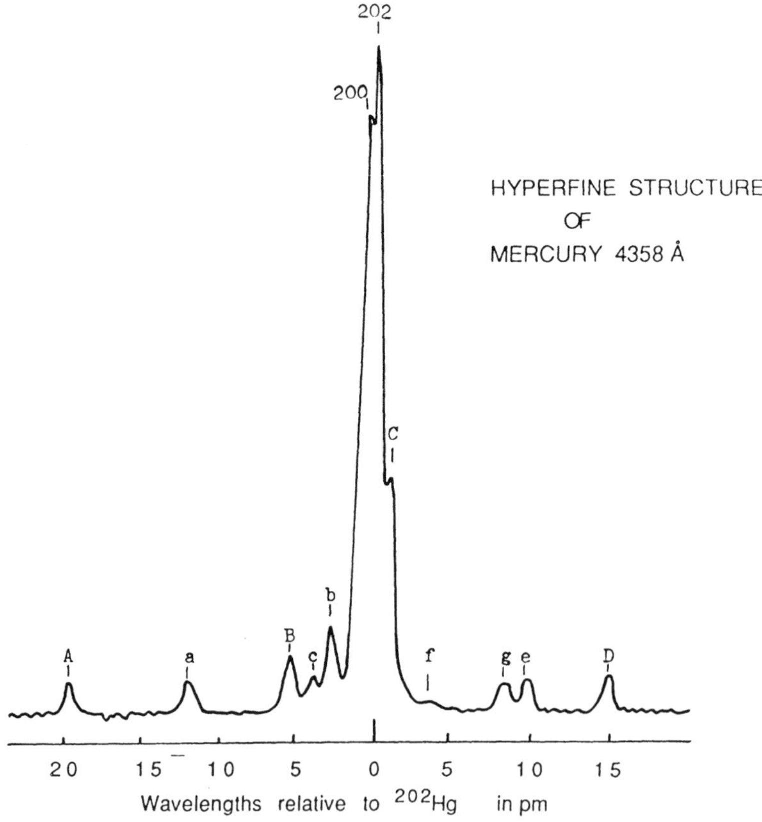

Fig.6.46 Hyperfine scan of 4358Å line from a low pressure air cooled mercury lamp on 10 m Czerny-Turner test bench equipped with 408 mm wide 79 gr/mm r–2 (63.5°) echelle. The (b) and (c) lines are separated 0.11Å, the 200 and 202 lines by 0.0057Å (courtesy Spectronic Instruments Co.).

79 gr/mm echelle with 408 mm ruled width, it is mounted on a 10m Czerny-Turner test bench. Operating at f/23, its mirrors are 430 mm diameter with near perfect 10m radius spheres. Linear scanning in the image plane is 0.088 mm/Å at 4358 Å. The theoretical resolution of 1.8 x 10^6 corresponds to an image displacement of only 0.2 μm, or a wavelength shift of 0.0024Å. The stability to observe at this level calls for a high degree of vibration isolation, careful shielding of the air path, as well as overall temperature stability of 0.2°C at least over the 30 min time required for a scan. The output of a photomultiplier tube (PMT) behind the exit slit is recorded as the hyperfine spectrum shown in Fig.6.46. The Hg 200 and 202 lines are separated by only 0.0057Å, or 0.57 pm. Given that the Doppler broadened width at half intensity is 0.38 pm one can estimate that the echelle must have achieved at least 90% of its theoretical resolution.

6.7.1 The MEGA Spectrometer

The goal of Lindblom and associates was to obtain still higher resolution, as well as a more compact instrument. They adopted a multigrating approach termed *multi echelle grating arrangement* (MEGA) [6.22]. The idea was to use two or more echelle gratings in tandem, Fig.6.47. If the dispersions are made additive, the resolution will increase as the sum of the number of gratings. Even with high efficiency gratings there will be significant light loss from so many reflections, so that the practical limit seemed to be four echelles. Two instruments were built. The first, with 1.2 m focal length collimators, contained four 160 mm echelles. Using the two modes of a 125 mm long He-Ne laser, 2x10^6 resolution was easily demonstrated by observing the width at half intensity, 0.0055 Å. The second instrument, twice the size, contained four 320 mm echelles, and aimed for doubling the resolution, unique for a relatively compact instrument. The echelles needed to be well matched and uniform. To rotate them in synchronism for wavelength scanning would require drive systems of unimaginable accuracy and over a significant range. Instead, scanning was performed by placing the entire optical system in a heavy tank designed to be progressively pressurized with nitrogen, up to 40 atm. Even so, the free spectral range must be kept small to fall within the range that this provides, which is why 31.6 gr/mm was the chosen groove frequency. A typical scan is shown in Fig.6.48. An interesting comparison can be made with Fig.6.46, which scanned the same 4358Å mercury line, except that the lamp was air cooled. Doppler broadening still sets the limit of what one can see. Instead of a cross-disperser this instrument uses a fore-monochromator to maintain a single order scan [6.23].

A commercial multi-echelle spectrometer was based on a design by Mazzacurati, who showed that the output of two co-planar echelles can actually

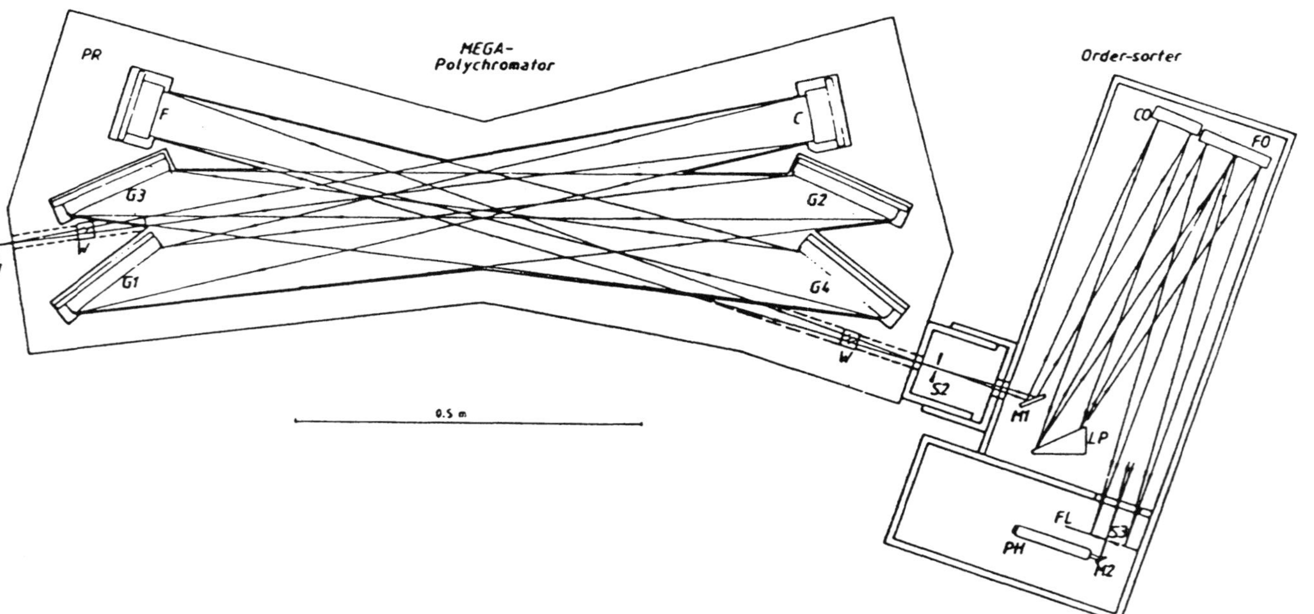

Fig.6.47 Schematic of 1.2 m MEGA spectrometer with 4 echelle gratings. Collimating mirror C, focussing F. Polychromator in pressure vessel.Order sorting fore-monochromator in right (after [6.22]).

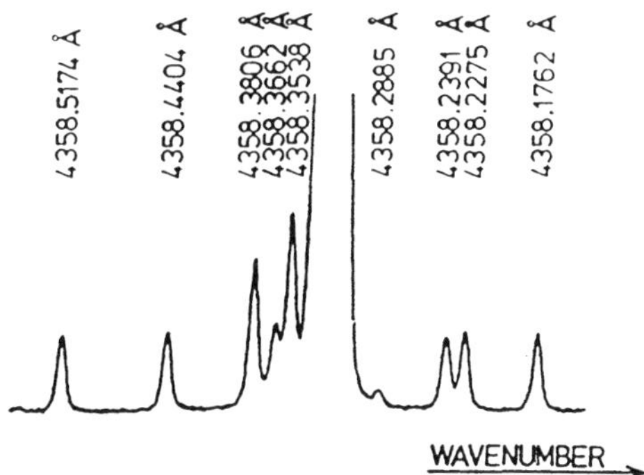

Fig.6.48 Scan of Hg 4358Å line from water cooled electroless discharge lamp operating at 2W (after [6.23]).

double the resolution of a single one, provided their grooves are kept in phase within a linear match of 25 nm. Since it was hardly reasonable to make, let alone maintain a permanent assembly to such a tolerance, this difficult task was accomplished with an active piezo feedback mechanism [6.24].

Another approach to increased resolution is to double pass a single grating. This has been used by Rank to observe the mercury hyperfine structure with a 250 mm echelle [6.25] and by Delbouille [6.26] to develop an atlas for solar absorption spectra. The latter instrument uses a similar grating, but makes use of an intermediate slit to reject virtually all stray light.

6.8 Transmission Echelles

The only way to operate a high dispersion angle echelle in transmission is to make it in the form of a GRISM. This allows input normal the the groove face, just like a Michelson echelon. It is obviously important to minimize

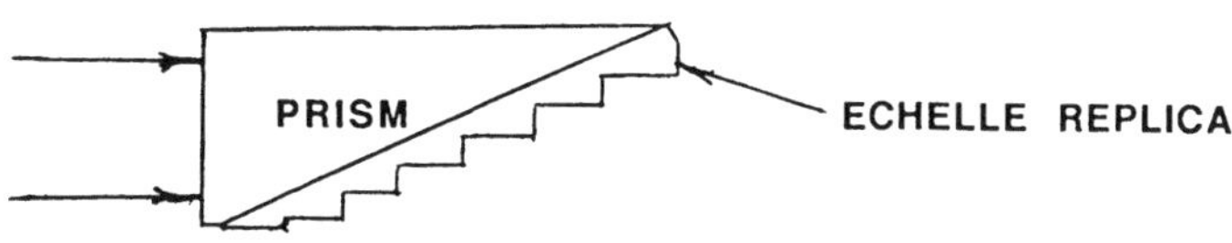

Fig.6.49 Transmission echelle prism.

reflection at the glass–resin interface, which means using a replica resin whose index is a close match to that of the glass. Since dispersion is controlled by the (n-1) phase delay factor in transmission, compared to the doubling that takes place in reflection, dispersion for a given echelle geometry will be 4 times less than in reflection, Fig.6.49. Nonetheless such echelle grating prisms have proven useful for increasing the dispersion of compact transmission spectrographs by a factor of about 2.5 [6.27]. Efficiency remains relatively high, 30% a typical value, and stray light is low. With no metallic surfaces, polarization effects are negligible.

6.9 Comparing Echelles with Holographic Gratings

Since both echelles and fine pitch holographic gratings are capable of high angle diffraction (i.e., high dispersion), there is some natural competition between them, and interesting choices may have to be made. The important attributes, besides dispersion and its uniformity, are signal to noise (i.e., stray light), the diffraction efficiency as a function of wavelength, any associated polarization and anomaly effects. Also to be considered are availability of large sizes, and needs for auxiliary devices. The relative importance of these attributes will depend on the application, even the wavelength range, plus the design of the spectrometer. There is no simple rule of thumb that dictates a preference for one over the other.

For example, holographic gratings have lower inter-order scatter, an issue of special importance in the study of absorption phenomena. However, diffraction angles will vary over a much wider range if a significant part of the spectrum has to be covered, which has a strong effect on design. The peak efficiency of echelles is likely to be greater, and constant over a wider region, although it varies within each order. In addition the need to supply cross dispersion is not only an instrumental complication but also can cost significant amounts of light.

Polarization effects in the case of echelles are usually small, while with holographic gratings they become large at high angles, except in the immediate vicinity of where the TM and TE efficiency curves cross (i.e., near $\lambda/d = 0.87$, or $\theta_{Littrow} = 26°$), which of course is a low angle. For an astronomer it makes a difference whether a target is a star or an extended object. For stellar work the entrance slit is usually larger than the stellar image, in order not to cut off any of the light, which will favor echelles. A comparison based on equal slit limited resolution, with unwanted spectral bands suppressed with narrow band interference filters, gives similar throughput results. Holographic gratings lead to simpler systems since in first order use there is normally no need for cross–dispersion. On the other hand with a cross-dispersed echelle and an array detector there is the capability of recording a great deal of information simultaneously [6.28].

Echelles are more likely to be available in larger sizes, 400 mm being the standard. It is possible to make holographic ones equally large, but this becomes a difficult project at the high groove frequencies typically used (2400 to 3600 gr/mm). In the visible spectrum the highest practical groove frequency is 2400 gr/mm. For shorter wavelengths 3600 gr/mm is available, but if diffraction angles are to be <50° the wavelengths cannot be > 400nm. With echelles the variation of efficiency across each spectral order may complicate photometry somewhat, although polarization effects may do the same for holographic gratings.

References

6.1 U. Sengupta: "Krypton Fluoride excimer laser for advanced microlithography," Opt. Eng. **32**, 2410–2420 (1993).

6.2 R. W. Wood: "The use of echelette grating in high orders," J. Opt. Soc. Am. **37**, 733–737 (1947).

6.3 G. R. Harrison: "The production of diffraction gratings II:The design of echelle gratings and spectrographs," J. Opt. Soc. Am. **39,** 522–528 (1949).

6.4 G. R. Harrison, E. G. Loewen, and R. S. Wiley: "Echelle gratings: Their testing and improvement," Appl. Opt. **15,** 971–976 (1976).

6.5 D. J. Schroeder: "Design considerations for astronomical echelle spectrographs," Astron. Soc. Pacific, **82**, 1253-1275 (1970).

6.6 D. J. Schroeder: *Astronomical Optics* (Academic Press, London, 1987), ch.15.

6.7 H. Nubbemeyer and B. Wende: "Instrumental functions of a 5 m echelle spectrometer with diffraction limited resolving power," Appl. Opt. **16,** 2708-2710 (1977).

6.8 H. Nubbemeyer and B. Wende: "Optical properties of a 5m echelle vacuum spectrometer with resolving power of ~ 10^6,"Appl. Phys, **23**, 254-266 (1980).

6.9 H. Decker: "An immersion grating for an astronomical spectrograph," in *Instrumentation for ground astronomy* in Proc. 1987 Santa Cruz Workshop, pp. 183-188 (Springer, 1988).

6.10 C. G. Wynne: "Immersed gratings and associated phenomena - I," Opt. Commun. **73**, 419-421 (1989), part II: ibid, **75,** 1–3 (1990).

6.11 E. Loewen, D. Maystre, E. Popov, and L. Tsonev: "Echelles: scalar, electromagnetic and real groove properties," Appl. Opt., **34**, 1707-1727 (1995).

6.12 F. Diego: "Blaze angle measurement of 31.6 and 79.01 gr/mm r-2 echelle gratings from B&L," Appl. Opt. **26**, 4714-4716 (1987).

6.13 A. E. Dantzler: "Spectrograph software design," Appl. Opt. **24**, 4504-4508

(1985).

6.14 N. A. Danielsson and K-P. C. Lindblom: "Apparatus and method for the uniform separation of spectral rulers," U. S Patent No.3,922,089 (1975).

6.15 W. Traub: "Constant–dispersion grism spectrometer for channeled spectra," J. Opt. Soc. Am. A **7**, 1779–1791 (1990).

6.16 R. Tousey: "The extreme ultraviolet - past and future," Appl. Opt. **1,** 679-694 (1960).

6.17 R. Tousey, J. Purcell, and D. Garrett: "An echelle spectrograph for middle ultraviolet solar spectroscopy from rockets," Appl. Opt., **6**, 365-372 (1967).

6.18 S. S. Vogt and G. D. Penrod "HIRES: A High resolution echelle spectrometer for the Keck 10 m telescope," in *Instrumentation for Ground Astronomy*, Proc. 1987 Santa Cruz Workshop, pp. 68-103 (Springer, 1988).

6.19 H. W. Epps and S. S. Vogt: "Extremely achromatic fV10 all spherical camera constructed for the high resolution echelle spectrometer of the Keck telescope," Appl. Opt. **32,** 6270-6279 (1993).

6.20 A. Baranne: "Un nouveau montage à pupille blanche (A new white pupil system)," C. R. Acad. Sci. Paris, **t. 312**, Serie II, 1521–1526 (1991).

6.21 E. Jenkins, C. Joseph, D. Long, P. Zucchino, G. Caruthers, M. Bottema, and W. Delamere: "IMAPS (Interstellar-medium absorption-profile spectrograph): A high resolution echelle spectrograph to record far-ultraviolet spectra of stars from sounding rockets," SPIE, **932**, (Ultraviolet Technol. II) 213-229 (1988).

6.22 S. Engman and P. Lindblom: "MEGA spectrometer: A monochromator with supermillion resolution," Appl. Opt. **23,** 3341–3348 (1984).

6.23 O. Gustavsson and P. Lindblom: "Test performed with the improved MEGA spectrometer," Appl. Opt. **27,** 147-151 (1988).

6.24 V. Mazzacurati and G. Rucco: "The super-gratings: how to improve the limiting resolution of grating spectrometers," Opt. Comm. **76,** 185-190 (1990).

6.25 D. H. Rank, G. Skorinko, D. P. Eastman, G. D. Saksena, T. K. McCubbin Jr., and T. A. Wiggins: "Hyperfine structure of some HgI lines," J. Opt. Soc. Am. **50**, 1045–1052 (1960).

6.26 L. Delbouille and G. Roland: "High resolution spectra of the sun," Phil. Trans. Royal Soc. A **264**, 171–182 (1969).

6.27 D. Enard and B. Delabre: "Two design approaches for high efficiency low resolution spectroscopy," Proc SPIE v.**445**, 522-529 (1984).

6.28 D. Dravins: "High-dispersion astronomical spectrographs with holographic and ruled diffraction gratings," Appl. Opt. **17**, 404-414 (1978).

Additional Reading

T. W. Barnard, M. Crockett, J. Walsi, and P. Lundberg: "Design and evaluation of an echelle grating optical system for ICP–OES," Anal. Chem. **65**, 1225-1230 (1993).

M. Bottema: "Echelle efficiencies: theory and experiment; comment," Appl. Opt. **20**, 528-530 (1981).

M. Bottema, G. W. Cushman, A. W. Holmes, and D. Ebbets: "UV–Grating performance in the high resolution spectrograph," SPIE *Instrumentation in Astronomy V*, **445**, 452-460 (1985).

R. A. Brown, R. L. Hillard, and A. L. Phillips: "Actual blaze angle of the Bausch&Lomb R4 echelle grating," Appl. Opt. **21**, 167-168 (1982).

W. M. Burton and N. K. Reay: "Echelle efficiency measurements in the ultraviolet," Appl. Opt. **9**, 1227-1229 (1970).

T. K. McCubbin, R. P. Grosso, and J. D. Mangus: "A high resolution grating prism spectrometer for the IR," Appl. Opt., **1**, 431-436 (1962).

N. A. Danielsson and K-P. C. Lindblom: "Apparatus and method for the uniform separation of spectral rulers," U. S. Patent No 3,922,089 (1975).

A. D. Dantzler: "Echelle spectrograph software design aid," Appl. Opt. **24**, 4504–4508 (1985).

D. Dravins: "Diffraction gratings - holographic and ruled," Proc. 4th Triest Astrophysical Colloq.: High Resolution Spectrometry, pp.1-21 (Trieste, 1978).

S. Engman and P. Lindblom: "Blaze characteristics of echelle gratings," Appl. Opt. **21**, 4356-4362 (1982).

S. Engman and P. Lindblom: "Multiechelle grating mountings with high spectral resolution and dispersion," Appl. Opt. **21**, 4363-4371 (1982).

S. Engman and P. Lindblom: "Blaze angle of the Bausch&Lomb R4 echelle grating," Appl. Opt. **22**, 2512-2513 (1983).

E. F. Erickson, S. Matthews, G. Augeson, J. Houck, M. Harwit, D. M. Rank, and M. R. Haas: "All-aluminum optical system for a large cryogenically cooled far infra-red echelle spectrometer," SPIE **509**, 129-140 (1984).

P. Hansen and J. Strong: "High Resolution Hadamard Transform Spectrometer," Appl. Opt. **11**, 502–506 (1972).

G. R. Harrison, J. Archer and J. Camus: "A fixed focus broad range spectrograph of high speed and resolving power," J. Opt. Soc. Am., **42**, 706–712 (1952).

R. Hoekstra, T. M. Kamperman, C. W. Wells, and W. Werner: "Balloon born ultraviolet echelle spectrograph," Appl. Opt. **17**, 604–613 (1978).

R. Hoekstra: "Basic solutions and new techniques in high resolution astronomical spectrometry," Proc. 4th Intern. Colloqu. on Astrophysics, Trieste, 46-71 (1978).

E. Hultén and H. Neuhaus: "Diffraction gratings in immersion," Arkiv för Fysik **8**, 343–353 (1954).

B. Gentry, L. L. Strow, and C. L. Korb: "High resolution cooled optics infrared grating spectrometer," Appl. Opt. **23**, 2401–2407 (1984).

S. W. McGeorge: "Imaging systems: detectors of the past, present, and future," Spectroscopy **2**, 26–32 (1988).

E. B. Jenkins, C. L. Joseph, D. Long, P. M. Zuchino, G. R. Carruthers, M. Bottema, and W. A. Delamere: "IMAPS: a high resolution echelle spectrograph to record far–ultraviolet spectra of stars from sounding rockets," SPIE **932**, 213–229 (1988).

K. Kawaguchi, Y. Yoshimura, and A. Mizuike: "Some characteristics of a commercial echelle spectrometer," Spectrochimica Acta **41B**, 295–300 (1986).

P. N. Keliher: "Applications of echelle spectrometry to multi-element atomic spectrometry," Res. and Devel. **27**, No. 6, 26-28 (1976).

J. Kielkopf: "Echelle and holographic gratings compared for scattering and spectral resolution," Appl. Opt. **20**, 3327-3331 (1981).

R. C. M. Learner: "Spectrograph design 1918-68," J. Sci Instr. (J. Phys. E) Series 2, v.**1**, 589-594 (1968).

W. Liller: "High Dispersion Stellar spectrograph with echelle grating," Appl. Opt.,**9**, 2332–2336 (1970).

P. Lindblom and F. Stenman: "Resolving power of multigrating spectrometers," Appl. Opt. **28**, 2542-2549 (1989).

D. H. McMahon, W. A. Dyes, R. F. Cooper, W. C. Robinson, and A. Mahapatra: "Echelon grating multiplexers for hierarchically multiplexed fiberoptic communication networks," Appl. Opt. **26**, 2188–2196 (1987).

R. Masters, C. Hsiech, H. L. Pardue: "Advantages of an off–Littrow mounting of an echelle grating," Appl. Opt. **27**, 3895–3897 (1988).

C. F. Meyer: *The Diffraction of Light, X-rays and Material Particles*, (J. W. Edwards Co, Ann Arbor, MI, 1949), Ch.6.

M. Neviere: "Echelle grism: an old challenge to the electromagnetic theory of gratings now resolved," Appl. Opt. **31**, 4, 427-429 (1992).

E. H. Pinnington: "Simple order sorter for use with diffraction gratings blazed for high orders," Appl. Opt. **6**, 1655–1657 (1967).

W. A. Rense: "Techniques for rocket solar UV and for UV spectroscopy," Space Science Rev. **5**, 234-264 (1966).

D. J. Schroeder: "An echelle spectrometer-spectrograph for astronomical use," Appl. Opt., **6**, 1976-1980 (1967).

D. J. Schroeder and R. L. Hillard: "Echelle efficiencies: theory and experiment," Appl. Opt. **19**, 2833-2841 (1980).

D. J. Schroeder: "Echelle efficiencies: theory and experiment; authors reply to comment," Appl. Opt. **20**, 530-531 (1981).

R. K. Skoberboe and I. T. Urasa: "Evaluation of the analytical capabilities of a dc plasma-echelle spectrometer system," Appl. Spectroscopy **32**, 527–532 (1978).

R. G. Tull: "Planetary Spectra with the 107 inch telescope," Sky and Telescope, **38**,156–160 (1969).

R. G. Tull: "A comparison of holographic and echelle gratings in astronomical spectrometry," Proc. 9-th Workshop on Instrumentation of ground-based optical astronomy, Santa Cruz, July 1987, ed. L. B. Robinson (Springer 1988), pp.104-117.

D. L. Wood, A. B. Dargis, and D. L. Nash: "TV direct reading spectrometer," Am. Laboratory **11**, 3, 16–25 (1979).

A. A. Wyler and T. Fay: "URSIES: an ultrahigh resolution single interferometer echelle scanner," Appl. Opt. **11**, 1152–1162 (1972).

F. Zhao: "A diffraction model for echelle gratings," J. Mod. Opt. **38**, 2241-2246 (1991).

Chapter 7

Concave Gratings

7.1 Introduction

Ever since their invention by Rowland in 1883 [7.1], concave diffraction gratings have played an important role in spectroscopy. Compared with plane gratings they offer the important advantage of providing the focusing or imaging properties that otherwise have to be supplied by additional elements. This advantage enabled concave gratings to dominate the field of spectrometry for many years. When advances in photoelectronics made monochromator-based instruments more attractive, there was a natural shift to plane grating designs. Thanks to Czerny-Turner and Ebert-Fastie mounts (see Ch.12), wavelengths can be easily tuned simply by rotating the grating. Step drives solve the problem of nonlinearity (sine law coming from the grating equation - see Ch.2). While such designs call for focusing mirrors, they have the important advantage of near stigmatic imaging so that maximum resolution and high photometric efficiency are achieved. These advantages proved so great that concave gratings became restricted to just two major fields. One was for work at wavelengths so short (< 110nm) that focusing mirrors introduced energy losses too high to be combined with a grating (for example, two mirrors with 20% reflectivity will reduce the energy throughout of the device by a factor of 25).

The other application of concave gratings was direct reading spectrographs where the Rowland circle configurations utilized families of suitably placed photodetectors, each one supplied with a slit, suitable to allow for the astigmatism and exit slit curvature. Limitations in the quality of concave grating imaging were accepted as necessary evils in the former and played only a minor role in the latter. Because of the defocusing that occurs when rotated for wavelength tuning, an aberrationally non-corrected concave grating is not naturally suited to high resolution monochromators. An optimal compromise was found in the Seya-Namioka mount.

New life for concave gratings is introduced by recent holographic (interferometric) recording techniques; these gratings turn out to be highly suitable when designed according to geometrical aberration theory. Two point-like sources are used instead of collimated recording beams, which provides additional degrees of freedom to decrease some of the natural aberrations of concave gratings. The development of solid-state array detectors brings two new factors into the picture. They require flat field imaging, which is not natural for concave gratings. However, given the many aplications where resolution requirements are modest and detector sizes limited to 6 - 25 mm, the combination works well together. Low dispersion concave gratings are the general tool for spectral intervals of 200 to 800 nm per detector array. The large size of the single detector elements makes it clear that resolution is determined by imaging quality rather than diffraction limits.

Classically ruled gratings are not capable of good imaging outside the Rowland circle; thus flat-field imaging can succeed only by changing the position of the focal (imaging) curves. The holographic aberrationally reduced recording is the most common solution, although ruling of specially curved grooves by computer controlled engines, or possibly electron beams, may also serve as an expensive alternative. Additional improvement may be achieved through sophisticated substrate designs. *Spherical* blanks are the simplest, but significant reduction of astigmatism can be achieved by having two different radii of curvature in the vertical and horizontal planes (i.e., *toroidal* substrates [7.2]). More complicated *aspherical* blanks are rarely justified [7.3], because the advantage is usually too small to match their high cost.

Within the limits of concave grating applications, the main concern comes from their comparatively low diffraction efficiency. Low aberrations require small angular deviations of the beams from the grating axis. Combined with a large working spectral interval and the small sizes of the array detectors, this leads to low dispersion values so that the grating grooves may need to be blazed to ensure high efficiency in the working diffraction orders. Such blazing is relatively simple for plane gratings, but becomes difficult for concave gratings. During mechanical ruling one has to take into account both blank and groove curvature and the change of blaze direction due to incident wave direction variation along the grating. This requires varying the aspect angle of the diamond tool which can be achieved only by multipartite grating ruling where the ruling is interrupted, either once or more usually twice, to reset the tool angle. Phase matching between the multiple rulings is beyond the present state of technology, but is not critical since resolution is limited predominantly by defects in imaging.

Holographic recording, while being more flexible for aberration reduction, is less accommodating to blazing. Since the direction of the recording beams is fixed with respect to the substrate, it is not possible to utilize recording with large asymmetry to blaze the grating (see Ch.15). This limits diffraction efficiency of such gratings to less than 30-40% in the common cases when diffraction angles are small. Ion-beam etching appears to be the optimal choice for enhancing efficiency by converting sinusoidal grooves into equivalent triangular ones.

7.2 Aberrations in Concave Gratings [7.4]

7.2.1 Aberration Function of Concave Gratings

Fig.7.1 demonstrates schematically the general mountings used for concave diffraction gratings together with some notations: Cartesian coordinate system 0xyz connected with the grating blank, the source A and the image B coordinates, and the image plane with its own coordinate system. The y-axis, perpendicular to the grooves at the centre of the blank is called *meridional coordinate*, and the z-axis the *sagittal coordinate*. The source coordinates are denoted by a subscript *a*. It is useful to represent them in a polar coordinates (r_a, α, z_a) and in normalized Cartesian coordinates:

$$\left(\frac{x_a}{r_a} \equiv \cos\alpha, \frac{y_a}{r_a} \equiv \sin\alpha, \frac{z_a}{r_a}\right). \tag{7.1}$$

The exit pupil coincides with the grating surface because it is the last (and only) optical element. A point on the grating surface is denoted by P and its coordinates acquire index *p*, but they are usually denoted with (x_p, w, l). The ideal image B_0 of the source A has coordinates denoted by an index *b*. The image plane is perpendicular to the projection of the principal ray OB_0 on the meridional plane. The non-central rays APB form the image spot B on the image plane. An image coordinate system is defined in the image plane with origin in B_0 and axes D_v and D_h, indices v and h standing for vertical and horizontal, respectively. They can be normalized with respect to the ideal image distance r_b from the grating apex.

When the imaging properties of the optical system are ideal the images

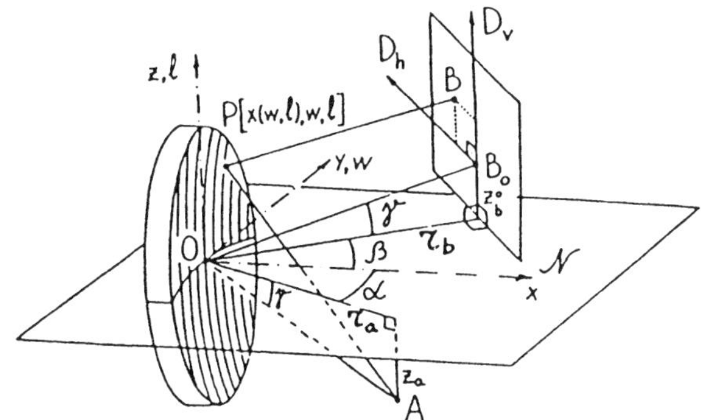

Fig.7.1 Concave grating with a point source A and its image B on the image plane. Some notations and coordinate systems are discussed in the text.

of the point source formed by different parts of the grating coincide (i.e., all the rays APB intersect the image plane in the ideal image B_0). The ideal image is formed when the different waves arrive at the image point in phase, giving rise to the grating equation, which can be expressed in terms of *optical paths* F(P) and F(O) along the rays APB and AOB:

$$F(P) - F(O) = N_{OP} m\lambda \ , \tag{7.2}$$

where m denotes the diffraction order and N_{OP} is the number of grooves that separate the points O and P. Unfortunately, aberration free images are rarely possible and the images of the point source are spread over the image plane forming the real *image spot*. Then the real optical path F(P)=APB differs from the ideal optical path $F_0(P)$ = APB_0. It depends on the position of the pupil point P. The deviation between the optical paths of the central (AOB_0) and non-central (APB) rays

$$\delta F \equiv F(P) - F_0(P) \tag{7.3}$$

is called the *aberration function* and defines grating aberrations. According to Fermat's principal, rays move along the minimum optical path. For points B and B_0 to be as close as possible (i.e., to minimize the aberrations), it is necessary to minimize the aberration function δF. If the source point position is fixed it determines the position of the ideal image. The aberration function depends on the coordinates of the grating point P, on the system parameters, and on the source position. It is useful to divide these dependencies by expanding δF in power series with respect to the pupil coordinates *w* and *l*. Taking into account that the third (x) coordinate of P is given by the grating blank equation $x_P = g(w,l)$, the first several terms are given by:

$$\begin{aligned}
\delta F &= \sum_{i,j} \frac{w^i l^j}{i!j!} \left[F_{ij0} + F_{ij1} + F_{ij2} \right] \\
&= w(F_{100} + F_{102}) + l(F_{011} + F_{012}) \\
&+ \frac{w^2}{2}(F_{200} + F_{202}) + wlF_{111} + \frac{l^2}{2}(F_{020} + F_{022}) \\
&+ \frac{w^3}{6}(F_{300} + F_{302}) + \frac{w^2 l}{2}F_{211} + \frac{wl^2}{2}(F_{120} + F_{122}) + \frac{l^3}{6}F_{031} \\
&+ \frac{w^4}{24}(F_{400} + F_{402}) + \frac{w^3 l}{6}F_{311} + \frac{w^2 l^2}{4}(F_{220} + F_{222}) + \frac{wl^3}{6}F_{431} + \frac{l^4}{24}(F_{040} + F_{042}) \\
&+ \dots \quad .
\end{aligned} \tag{7.4}$$

There are several different conventions for the coefficients in this expansion [7.5-7]. In our notations we use 3-indices coefficients. The first two indices represent the power terms with respect to the pupil coordinates, and the third index is the height z_a of the point source A from the meridional plane, so that F_{ij0} is independent of z_a, F_{ij1} is linearly proportional to z_a, and F_{ij2} depends on z_a^{ν}, where $\nu \geq 2$. The explicit dependence on the source height is determined by its importance as a coordinate of the source point in the entrance slit in real devices.

The dependence of the aberration function on the grating parameters and on the other two source coordinates is implicitly included in the F coefficients and the minimization of aberrations is performed with respect to them. The requirement of the linear terms to be zero

$$\begin{aligned} \frac{\partial}{\partial w}\delta F &= F_{100} + F_{102} = 0 \\ \frac{\partial}{\partial l}\delta F &= F_{011} + F_{012} = 0 \end{aligned} \tag{7.5}$$

determines the direction of propagation of the central ray OB_0, usually defined through the angles β and γ such that $\cos\beta = y_b/r_b$ and $\cos\gamma = z_b/r_b$. Taking into account that both F_{102} and F_{012} depend on z_a, it follows that *β and γ also depend on z_a*, and an important conclusion can be drawn: *a straight entrance slit is usually imaged into a curvilinear exit slit* (Fig.7.2a). This is called *line curvature* and is usually not considered an aberration.

If the point source lies near the meridional plane, the terms F_{102} and F_{012} are very small (F_{102} is proportional to the square of (z_a/r_a) and F_{012} to its cube) and the central diffracted ray is determined from the simple equations $F_{100} = F_{011} = 0$. The first term corresponds to the *standard grating equation* and the second one expresses a simple *mirror-like reflection* relation in sagittal direction:

$$\begin{aligned} \sin\alpha + \sin\beta &= m\frac{\lambda}{d} \\ \frac{z_a}{r_a} &= -\frac{z_b}{r_b} \end{aligned} \tag{7.6}$$

The diffraction spot dimensions are defined through the image coordinates D_h and D_v. Rigorous formulae can be found elsewhere, and their simplified form is:

$$\frac{D_h}{r_b} \approx \frac{1}{\cos\beta}\frac{\partial}{\partial w}\delta F$$
$$\frac{D_v}{r_b} \approx \frac{\partial}{\partial l}\delta F \quad . \tag{7.7}$$

If series expansion (7.4) is substituted into (7.7), after regrouping the terms of equal order (i+j) with respect to the pupil coordinates, the following general representation of the diffraction spot dimensions is obtained:

$$\frac{D_h}{r_b}\cos\beta = A_h^{(1)} + A_h^{(2)} + A_h^{(3)} + \ldots$$
$$\frac{D_v}{r_b} = A_v^{(1)} + A_v^{(2)} + A_v^{(3)} + \ldots \quad , \tag{7.8}$$

where the terms of different orders are represented as:

$$A_h^{(1)} = w(F_{200} + F_{202}) + lF_{111}$$
$$A_h^{(2)} = \frac{w^2}{2}(F_{300} + F_{302}) + wlF_{211} + \frac{l^2}{2}(F_{120} + F_{122}) \tag{7.9}$$
$$A_h^{(3)} = \frac{w^3}{6}(F_{400} + F_{402}) + \frac{w^2 l}{2}F_{311} + \frac{wl^2}{2}(F_{220} + F_{222}) + \frac{l^3}{6}F_{131} \quad ,$$

and

$$A_v^{(1)} = wF_{111} + l(F_{020} + F_{022})$$
$$A_v^{(2)} = \frac{w^2}{2}F_{211} + wl(F_{120} + F_{122}) + \frac{l^2}{2}F_{031} \tag{7.10}$$
$$A_v^{(3)} = \frac{w^3}{6}F_{311} + \frac{w^2 l}{2}(F_{220} + F_{222}) + \frac{wl^2}{2}F_{131} + \frac{l^3}{6}(F_{040} + F_{042}) \quad .$$

The first order terms $A^{(1)}$ correspond to *astigmatism* and contain the aberration coefficients F_{ijk} with i+j=2. The second order terms $A^{(2)}$ describe *coma* and depend on the coefficients with i+j=3. The third order terms $A^{(3)}$ correspond to *spherical aberration* and are characterized by i+j=4.

Fortunately, the terms of different orders differ significantly in magnitude so that the minimization of aberrations can be conducted one by one, starting from the lower orders. All terms having the same order must be considered simultaneously. As always, there are exceptions to the rule (if, for example, the grating dimensions are quite different in horizontal and vertical directions and/or under grazing incidence conditions).

7.2.2 Aberrations of Concave Diffraction Gratings

A basis for naming different grating aberrations can be easily found by comparing them with corresponding classical aberrations of lenses. For convenience, the aberrations are often studied using the so called *testing pictures*: the image spot obtained when only some aberration coefficients are non-zero. This is an idealization, but a useful one. While the testing pictures of different aberrations in lenses and concave gratings are similar, several general differences can be formulated:

lenses	*gratings*
1. On-axis point source is characterized by a single aberration - the spherical one. Other aberrations appear only for off-axis sources, assuming properly aligned systems.	1. Even the image of a point source in the meridional plane contains several aberrations (astigmatism, coma, spherical aberration), characterized by the aberration coefficient F_{ij0}. Only a single aberration (distortion, connected with the spectral line curvature) appears when going off-plane.
2. The lowest term in expansion with respect to the pupil coordinates already contains four different aberrations (spherical, astigmatism, coma, and distortion), which are, in general, comparable in magnitude and have a combined effect on the image deformation.	2. The terms of each order with respect to the pupil (grating) dimensions each contain a single aberration. The strongest is the influence of the astigmatism. Coma is of the second order and spherical aberration is the weakest one.

It is enough to analyze concave grating aberrations with only an in-plane source (Figs.7.2b,c,d), taking into account that going off-plane also introduces image curvature (Fig.7.2a). In that case (z_a=0) all the aberration coefficients F_{ijk} with non-zero third index k disappear, and equations (7.9 and 7.10) are significantly simplified.

The dependence of these aberration coefficients on the grating G_{ij} and mounting M_{ij0} parameters can be represented by different terms:

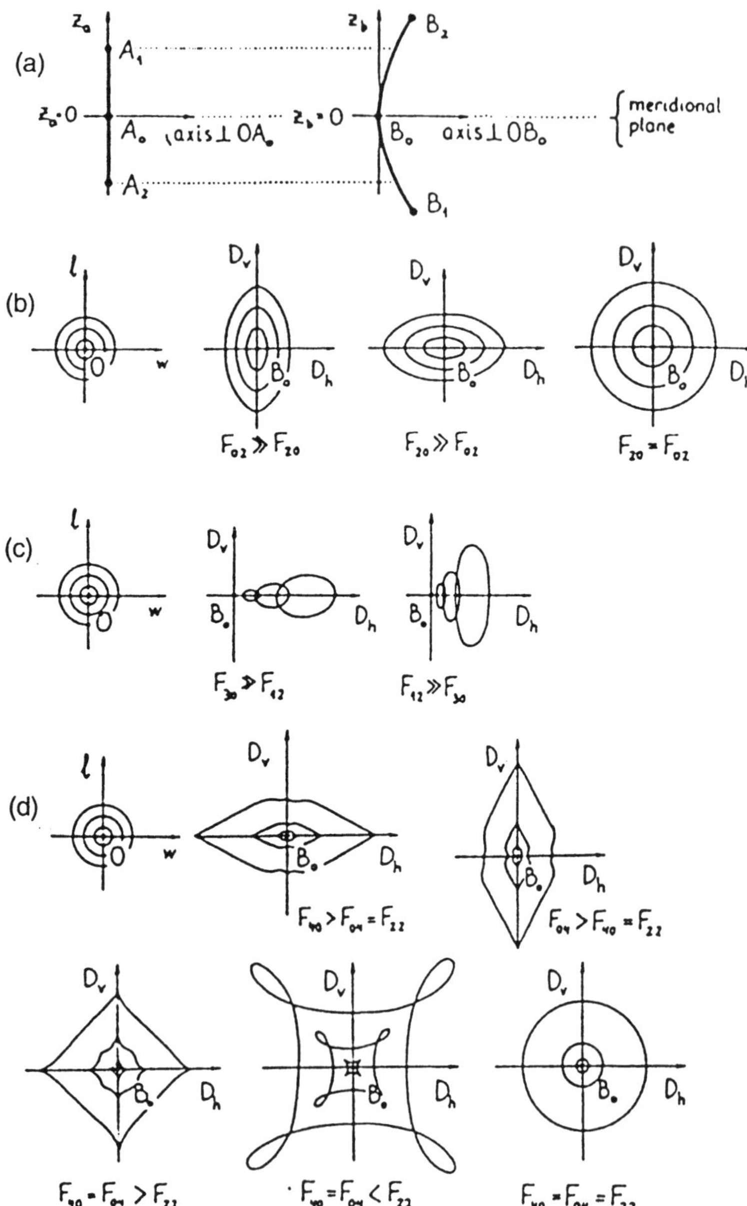

Fig.7.2 Aberrations of a concave grating. Note that the third index k of F_{ijk} is zero and is omitted. a) Distortion (curvature of a slit image); b) astigmatism; c) coma; d) spherical aberration.

$$F_{ij0} = M_{ij0} - m\frac{\lambda}{d}G_{ij} , \tag{7.11}$$

where the form of G_{ij} differs for holographically recorded and for ruled gratings, and the mounting term M_{ij0} is a function of the incident α and diffracted β angles of the principal rays and the source r_a and image r_b distance to the grating apex O. They have simple forms for the most commonly used spherical substrate and are worth writing

$$\begin{aligned} M_{200} &= \left(\frac{\cos^2\alpha}{r_a} - \frac{\cos\alpha}{R}\right) + \left(\frac{\cos^2\beta}{r_b} - \frac{\cos\beta}{R}\right) \\ M_{020} &= \left(\frac{1}{r_a} - \frac{\cos\alpha}{R}\right) + \left(\frac{1}{r_b} - \frac{\cos\beta}{R}\right) \quad , \end{aligned} \tag{7.11a}$$

$$\begin{aligned} M_{300} &= 3\frac{\sin\alpha}{r_a}\left(\frac{\cos^2\alpha}{r_a} - \frac{\cos\alpha}{R}\right) + 3\frac{\sin\beta}{r_b}\left(\frac{\cos^2\beta}{r_b} - \frac{\cos\beta}{R}\right) \\ M_{120} &= \frac{\sin\alpha}{r_a}\left(\frac{1}{r_a} - \frac{\cos\alpha}{R}\right) + \frac{\sin\beta}{r_b}\left(\frac{1}{r_b} - \frac{\cos\beta}{R}\right) \quad . \end{aligned} \tag{7.11b}$$

In case of holographic recording from two point sources, C and D, their position with respect to the blank can be represented by the angles γ and δ between their direction and the grating normal at the blank center, and their distances r_c and r_d to this center. The grating part of the aberration function now has the same form as the mounting part (7.11) with the mounting angles and distances being replaced by the recording ones. There is a confusing convention that recording angles have opposite signs at the two sides of the blank normal, whereas for α and β the opposite convention is used: they both are positive, when the source and the image are on opposite sides of the grating normal.

Astigmatism

The first order term with respect to the pupil coordinates is responsible for the astigmatism. It contains two aberration coefficients:

$$\begin{aligned} \frac{D_h}{r_b}\cos\beta &= wF_{200} \\ \frac{D_v}{r_b} &= l\,F_{020} \quad , \end{aligned} \tag{7.12}$$

and its aberration picture is given in Fig.7.2b. This picture resembles the astigmatism deformations of lens images when F_{200} differs strongly from F_{020}: if $F_{200} >> F_{020}$, the image spot is stretched in the horizontal (meridional) direction, and in the vertical (sagittal) direction, if $F_{200} << F_{020}$. That is why F_{200} is called the *coefficient of meridional astigmatism*, and F_{020} is the *coefficient of sagittal astigmatism*.

If the source position is fixed and the image plane is moved (i.e., varying r_b) it is possible to follow the evolution of the image spot (see Fig.7.3). As far as the mounting part of the aberration coefficients depends on r_b (eq.7.11), but not the grating part, there is a position r_b^M for which the meridional astigmatism becomes zero ($F_{200}=0$). Then the image is extended in the vertical direction and is called the *meridional focal image*. If there are no higher order aberrations, the width of the image should correspond to the width of the entrance slit (magnified by the ratio r_b/r_a). When, for example, the wavelength is varied, r_b^M moves in the meridional plane, forming a curve, called the *meridional focal curve*.

Correspondingly, when $F_{020}=0$, a *sagittal focal image* and a *sagittal focal curve* are defined. The distance between the two focal images is called *astigmatic difference* Δ_{ast}, similar to the case of lenses.

Astigmatism is the main aberration in concave gratings and its influence on the image spot deformations is the most significant. The other aberrations can be noticed only when the astigmatism becomes negligible.

Coma

Coma is described by the equations:

$$\frac{D_h}{r_b}\cos\beta = \frac{w^2}{2} F_{300} + \frac{l^2}{2} F_{120}$$
$$\frac{D_v}{r_b} = wl\, F_{120} \quad , \tag{7.13}$$

and its test pictures are shown in Fig.7.2c. It receives its name from the obvious similarity with coma in lenses. There are two aberration coefficients responsible for coma. When F_{300} prevails, the image is extended fan-like in the meridional direction and F_{300} is called meridional, or the classical coma coefficient. In the opposite case the image is also extended vertically and F_{120} is called the mixed coma coefficient. At the few spectral points with zero astigmatism, coma becomes the main aberration.

Spherical Aberration

Spherical aberration is determined by three coefficients:

$$\frac{D_h}{r_b}\cos\beta = \frac{w^3}{6}F_{400} + \frac{wl^2}{2}F_{220}$$
$$\frac{D_v}{r_b} = \frac{w^2 l}{2}F_{220} + \frac{l^3}{6}F_{040} \tag{7.14}$$

and its test pictures are shown in Fig.7.2d. The similarity with lenses is weaker (only when the three coefficients are equal), but still exists to serve as an argument for calling this term *spherical aberration.* The form of the image deformations names the three coefficients, correspondingly: F_{400} - meridional, F_{040} - sagittal, and F_{220} - mixed coefficient of spherical aberration. In practice, spherical aberration can barely be observed in pure form in concave gratings and leads only to modification of the lower-order aberrations picture.

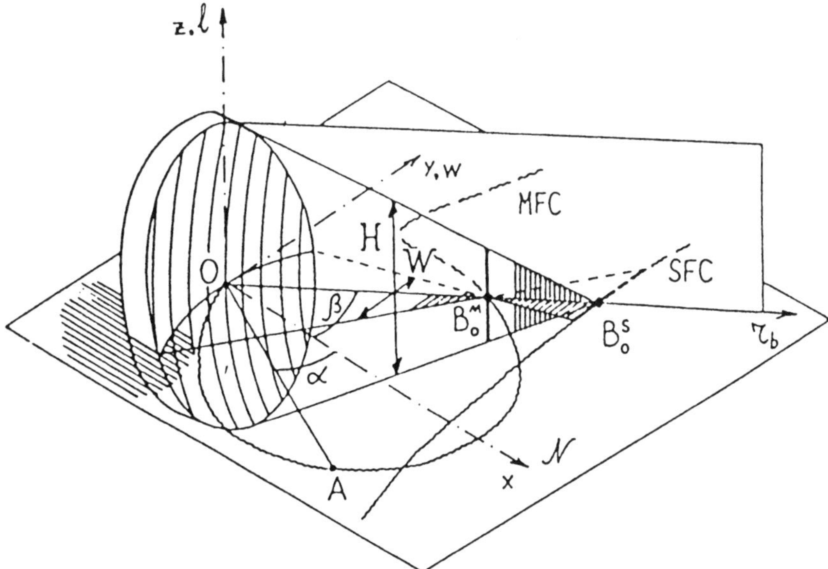

Fig.7.3 Meridional B_0^M and sagittal B_0^S image and the meridional and sagittal focal curves. The meridional image is formed by the meridional beam with width W and the sagittal image - by the sagittal rays (dimension H).

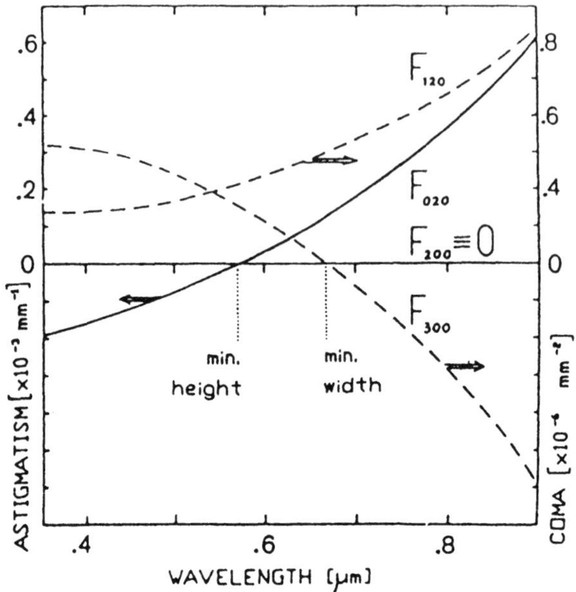

Fig.7.4 a) Spectral behaviour of the lower aberration coefficients of a flat-field holographic concave grating with radius of curvature 210.8 mm, period 1.261 μm and diameter 92 mm (after [7.8]).

7.3 Focal Curves

7.3.1 Definition and Properties

The chief aim of concave diffraction gratings, to separate and focus the wavelengths radiated by the source into different spots, determines the requirements needed. The main direction of the diffracted beam with a fixed wavelength is determined by the grating equation (or, in terms of optical path function, by the equation F_{100}=0). It is desired that the image of the point source be a point or at least a small spot. How small is determined in each particular case. If the evolution of the image is analyzed as a function of the distance (r_b), there is a position (B_0^M) where the horizontal dimensions of the beam are minimal. This position is called the meridional image focus and it is determined by the meridional astigmatism (see previous section). The position B_0^S with minimal image height is called the sagittal image focus and is

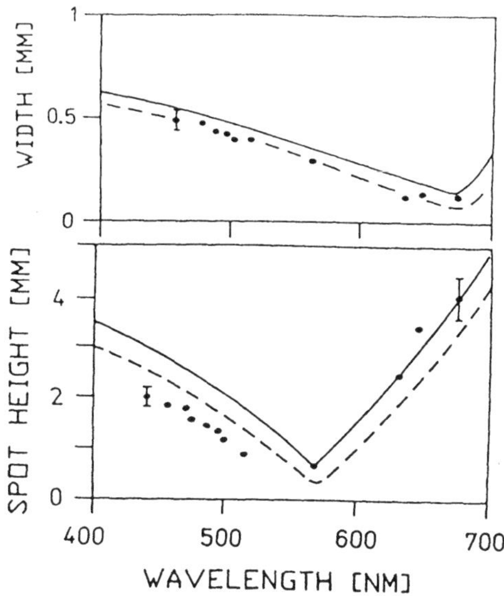

Fig.7.4 b) Spectral dependence of the meridional spot width W and height H of the same grating: points - the measured values, dashed line - ray-tracing results, and solid line - analytical estimation according to eq. 7.16 with values of the aberration coefficients given in Fig. (a) (after [7.8]).

determined by the sagittal astigmatism coefficient (Fig.7.3). With the variation of wavelength, these two image foci form two curves in the meridional plane, called *focal curves*. They play an important role in image analysis and deserve more detailed attention. As far as the spectral selectivity is concerned, it becomes more important to have narrower images at each wavelength (i.e., the detector is usually located at the meridional focus, defined by the equations $F_{200}=F_{100}=0$). The dimensions of the spot in the vertical (W_M) and horizontal (H_M) dimensions can be obtained from the aberration coefficients, using eqs.7.7-11. When the meridional astigmatism is zero on the meridional focal curve, the spot width is dominated by the coma term, while its height H_M is determined by the sagittal astigmatism (Fig.7.4):

$$H_M = \frac{2l_{max}}{r_b^S}\left|\Delta_{ast}\right| \; , \qquad (7.15)$$

where $2l_{max}$ is the height of the grating and Δ_{ast} is the distance between the meridional and the sagittal foci. It is possible to evaluate more general dependencies for a circular blank with diameter Φ:

$$\begin{aligned} W_M &= \frac{r_b^M}{8}\Phi^2\left(\left|F_{300}^M\right| + \left|F_{120}^M\right|\right) \\ H_M &= r_b^M\left(\Phi\left|F_{020}^M\right| + \frac{\Phi^2}{2}\left|F_{120}^M\right|\right) \quad , \end{aligned} \tag{7.16}$$

where the index M reminds that all terms are evaluated in the meridional focus.

A direct consequence of eq.(7.15) (and, also, in a hidden form from the more general eq.(7.16)) is that the distance between the focal curves determines the meridional focal height. Thus most of the designs utilize mountings and gratings that bring the focal curves close to one another. In the classical mounting, when it is necessary to detect only separate wavelength values (e.g., to detect some known spectral light intensities), the form of the meridional focal curve is not so important; it is merely necessary to position the detector slits on this curve, even if it is quite curved. Such is the case with the classical Rowland circle.

Recent concave grating applications include detection by photodiode arrays, which unfortunately are presently offered only mounted on a plane substrate. The choice of the grating is complicated by the need to have a straight part of the meridional focal curve coinciding with the desired spectral region; otherwise image defocusing in the horizontal direction grows linearly with the displacement of the detectors from the meridional focal curve. The knowledge of the grating focal curves becomes important for both grating manufacturers and device designers. It is possible to use computers to minimize the aberrations (or, more generally, the spot dimensions) by ray-tracing, but the problem is so multi-dimensional that a good initial approximation is always necessary, even with a strong sense of intuition and experience.

7.3.2 Types of Focal Curves

The form and position of the focal curves depend on the grating itself: on the nature of the blank shape, groove spacing and spacing variation over the grating, groove curvature, as well as on the mounting (i.e., source location). Fortunately, the dependence on the grating type can be determined using several parameters. If the grating blank is aspherical it can have two different radii of curvature: horizontal R_M and vertical R_S. Grooves can be straight or curved. Curved grooves can be characterized by their radius of curvature in the grating plane. It is sufficient to study this radius just in the middle of the grating

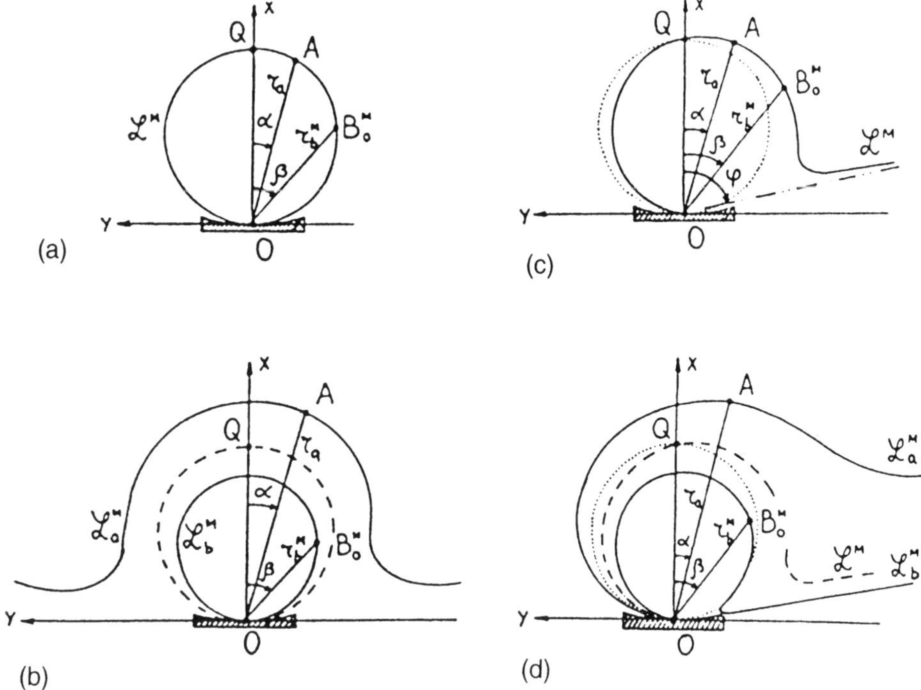

Fig.7.5 Meridional focal curves. $\mathcal{L}^M$ is the self-generating MFC.

ρ_0. The grooves can be equidistant or with a varying period over the grating. A parameter that is used to characterize their aperiodicity is the linear coefficient of aperiodicity μ, equal to 0 for equidistant grooves[1]. Then the form of the *meridional focal curve* is determined from the meridional radius of curvature R_M and the linear coefficient of aperiodicity μ and not from the groove curvature. The *sagittal focal curve* is determined from the sagittal radius of curvature R_S and the groove curvature ρ_0. That is why for some special applications it is possible to use substrates with two radii of curvature in order to "bring" the two focal curves close together. Such a process can be quite expensive.

---

[1] Of course, one can imagine an artificial example, when one part of the grating has one set of parameters and the other(s) - a different set. This can be realized if the grating ruling is changed during the ruling, something quite rare. But then the different parts of the grating can be considered independently.

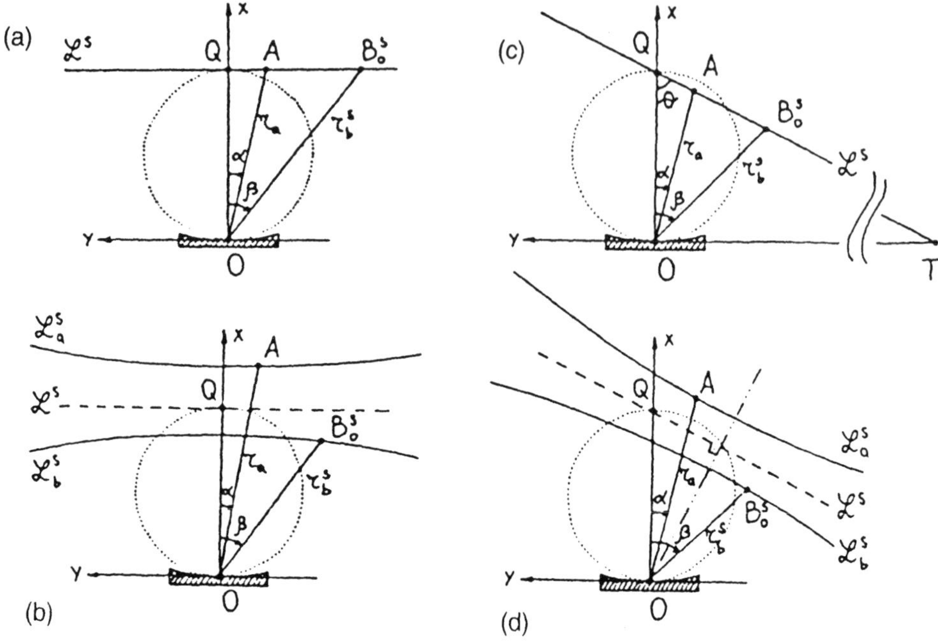

Fig.7.6 Sagittal focal curves. $\mathcal{L}^S$ is the self-generating SFC.

It is difficult to investigate all the combinations of focal curves, but some general rules can be drawn. Equidistant grooves produce symmetrical meridional focal curves. The classical case, called the *Rowland circle* puts the source and the meridional focus on a circle (Fig.7.5a) with a diameter equal to the blank radius R_M, and symmetrical with respect to the grating normal (such focal curves when the source and the image lie on the same curve are called *self-generating*). When the source lies inside the Rowland circle, the focal curve lies outside and has a more complicated form (Fig.7.5b). This also leads to magnification (or reduction) of the image. Variation of the groove spacing over the grating surface deforms the meridional focal curves so that even the self-generating one is asymmetrical (Fig.7.5c). Moving the source closer to the grating can straighten a part of the meridional focal curve (Fig.7.5d), which can be useful in flat-field spectrographs (although magnification of the image can destroy the gain). It must also be taken into account that there are geometrical limitations if the source is something more complicated (slit or fiber) and cannot overlap with the image.

The *sagittal focal curves* are determined from the curvature of the grooves in the grating plane. The self-generating focal curve is a straight line which passes through the center of the curvature (the sagittal one, if R_S is different from R_M) - Fig.7.6a, c. Straight grooves form a sagittal focal curve parallel to the blank (Fig.7.6a). Larger groove curvature inclines the sagittal line with respect to the grating normal (Fig.7.6c). When the source is moved from the self-generating curve, the sagittal focal curve is slightly curved and shifted from the self-generated one (Fig.7.6b, d).

7.4 Grating Image Deformation Estimation and Optimization: Flat-Field Spectrograph and Monochromator

Concave gratings are used predominantly in two mountings: in *flat-field spectrographs* and in *monochromators*, with quite different operating conditions. In the flat-field spectrograph there are no moving parts, except for fine adjustment of the grating and slit positions. Light is incident on the grating from a slit with fixed position and the spectrum is formed (focused) somewhere in space on the surface of a photodiode area. Because the area is flat it is desired that the focal line be as straight as possible. For some applications the vertical deformation of the image is not so important because the area can be extended in height to gather as much diffracted light as possible. The distance between the meridional and the sagittal focal curves is then not so critical. However, in many cases the vertical resolution can be as desirable as the horizontal (spectral) one. Examples are high-speed scanning where it is necessary to distinguish between the spectrum of the different points of the entrance slit, such as in aeroscanning and in crossed-grating spectral recording in astronomy. It now becomes important to reduce the separation of the two focal curves as much as possible. An example is shown in Fig.7.7. Manufacturers rarely provide data for the recording conditions of the gratings and one of these cases is presented in ref. [7.9]. The grating is recorded on a spherical substrate and has curved grooves with varying spacing. The working conditions are moved from the self-generated focal curve ($\mathcal{L}^M$ and $\mathcal{L}^S$) so that the meridional and the sagittal curves are made almost parallel to each other. In the working region (λ between 350 and 900 nm) they intersect in two points λ_1=425 nm and λ_2=710 nm. At these points the astigmatism is zero and image spot deformations are dominated by the mixed coma term (F_{120}). In the lower part of Fig.7.7 the results of the geometrical ray-tracing procedure are given, compared with the estimated upper limits of the spot, determined from eqs.7.16. Outside of the stigmatic points, the image deformations are determined mainly by the sagittal astigmatism (F_{020}), which is kept very small

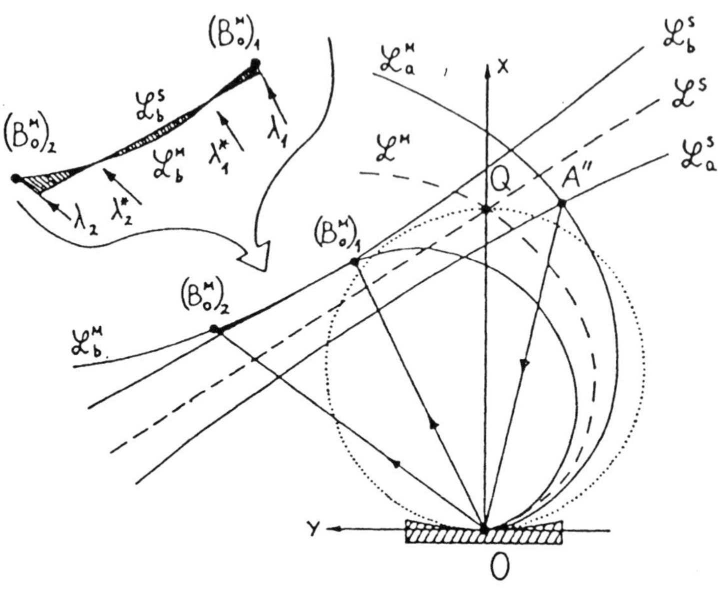

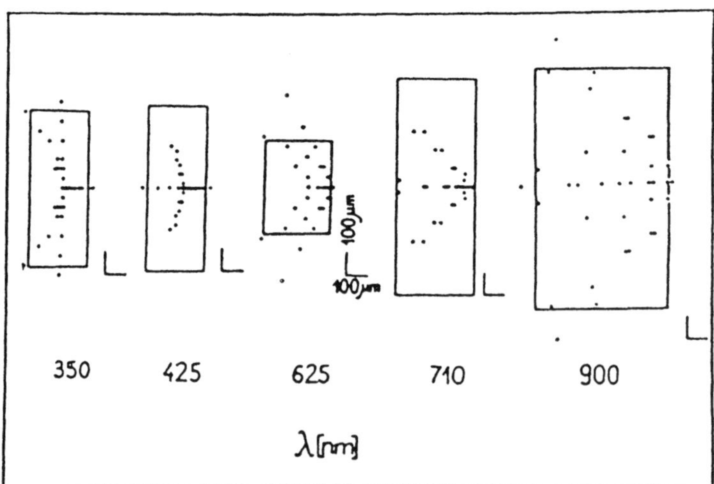

Fig.7.7 Focal curves of a flat-field spectrograph concave grating (type III according to Jobin-Yvon) and focal spot dimensions: points - ray-tracing picture; rectangles - analytical estimation according to eq.7.16 (after [7.10]).

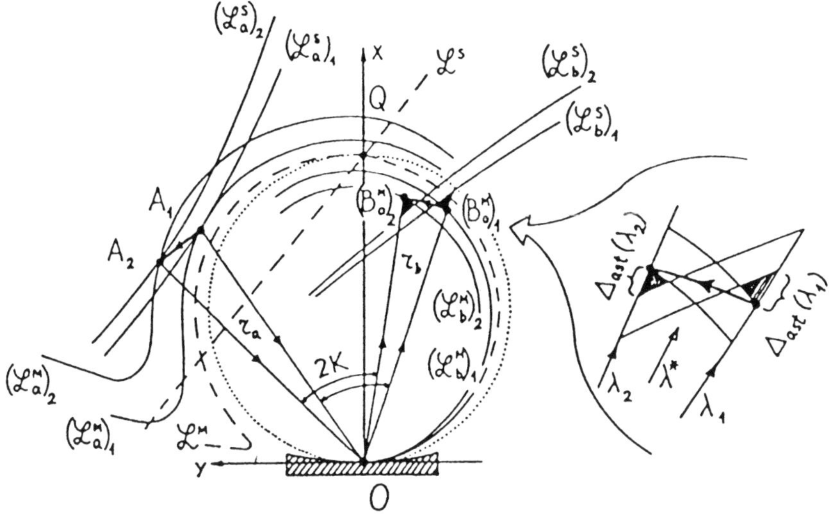

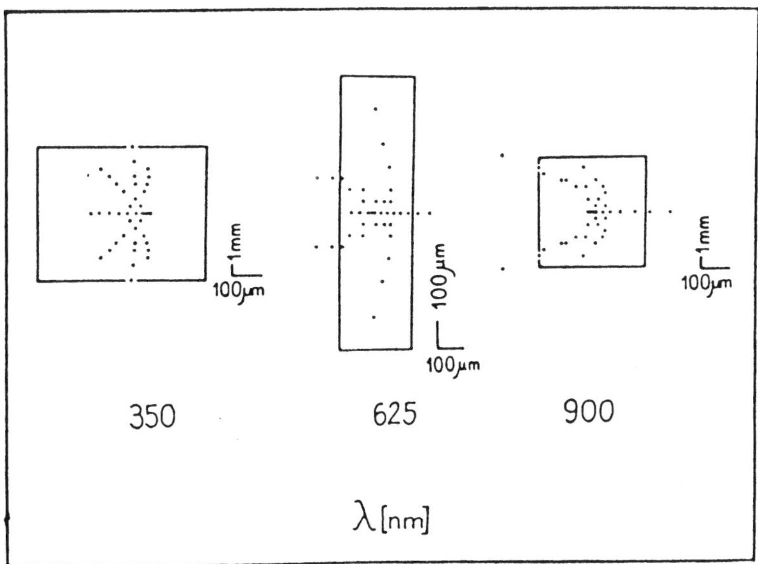

Fig.7.8 Focal curves of a concave grating for a monochromator with fixed deviation (type IV according to Jobin-Yvon) and focal spot dimensions: points - ray-tracing picture; rectangles - analytical estimation according to eq.7.16 (after [7.10]).

due to the small astigmatic difference Δ_{ast}. The divergence of the meridional focal line $\mathcal{L}_b^M$ from the detector surface, due to the slight curvature of $\mathcal{L}_b^M$, results in small *defocus* on the detector surface, expressed in small non-zero values of the meridional astigmatism F_{200}.

A *monochromator* design requires fixing the entrance and exit slits and rotating the grating in order to scan the spectrum (to focus different wavelengths at the exit slit). It means that the image position is constantly moved from one focal curve to another when the source rotates around the grating center (Fig.7.8). Contrary to the previous example, the intersection angle between the meridional and sagittal focal curves is large. Thus in the end points of the spectral interval (350 - 900nm) the registration point on the meridional curve B_M lies far from the sagittal focus and the image is extended in height due to the sagittal astigmatism (note the change in spot dimension scales at 350 nm and 900 nm in comparison with 625nm). In the middle, where the corresponding sagittal and meridional curves intersect, the image is stigmatic and its deformations are determined by the coma terms. Outside this point it is sufficient to consider only sagittal astigmatism (F_{020}) to estimate the image height (rectangles in the lower part of Fig.7.8 were obtained directly from eq.7.16).

7.5 Types of Concave Gratings

There are several different conventions for concave grating classification. In general these conventions concern specific ways of grating manufacturing and that is why they were defined by Jobin-Yvon and American Holographic Co.. It is more convenient for the user, however, to divide gratings according to their properties. There are several parameters for drawing boundary lines. The concave grating may have different meridional and sagittal radii. They can have either straight or curved grooves and constant or varying periods. There are several combinations of parameters that are most useful. Usually when the gratings are mechanically ruled the groove parameters (curvature and spacing) are known. There is also a simple way to find them for holographically recorded gratings.

7.5.1 Schemes for Holographic Recording of Concave Gratings

During ruling both groove curvature and spacing variation are well defined. Holographic recording of the grating uses two point sources (C and D). If these sources are located on the meridional plane at distances r_c = OC and r_d = OD, their position is determined by the angles γ and δ between the central rays OC and OD and the grating normal. A direct link exists between the groove radius of curvature ρ_0 and the linear coefficient of spacing variation μ with the recording parameters:

$$\mu\frac{\lambda_0}{d_0} = -\left[\left(\frac{\cos^2\gamma}{r_c} - \frac{\cos\gamma}{R_M}\right) - \left(\frac{\cos^2\delta}{r_d} - \frac{\cos\delta}{R_M}\right)\right]$$
$$\rho_0\frac{\lambda_0}{d_0} = -\left[\left(\frac{1}{r_c} - \frac{\cos\gamma}{R_M}\right) - \left(\frac{1}{r_d} - \frac{\cos\delta}{R_M}\right)\right]^{-1} , \qquad (7.17)$$

where d_0 is the spacing in the grating center and λ_0 is the recording wavelength. There are several possible recording schemes:

(1) Totally symmetrical recording (Fig.7.9). The grooves are formed at the intersections of the blank with interference planes, parallel to the grating normal. Close to the meridional plane ($z \approx 0$) the grooves are straight and equidistant. This recording corresponds to the classical ruling with constant

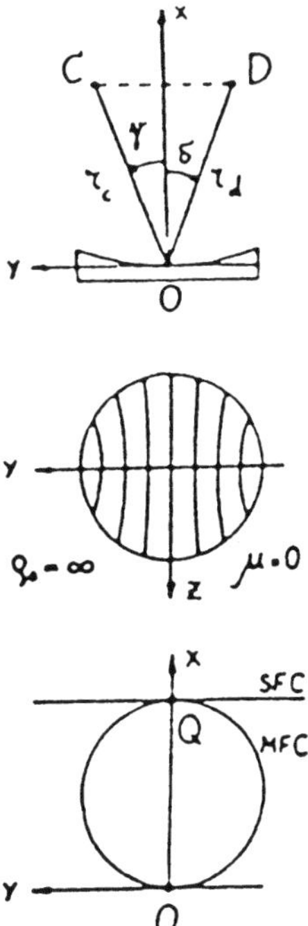

Fig.7.9 Symmetrical recording with the groove direction and spacing, and the focal curves. Q is the center of curvature.

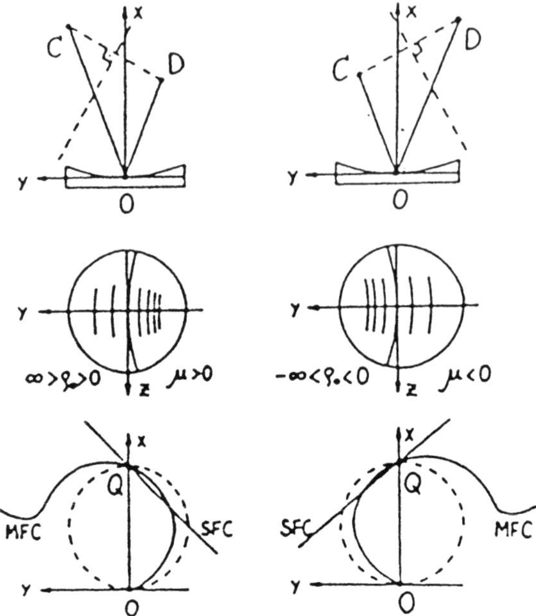

Fig.7.10 The same as in Fig.7.9 but with angular symmetry only.

spacing. Self-generated meridional focal curves coincide with the Rowland circle. Historically this is the first mounting (and grating type) used. Unfortunately, the sagittal focal curve touches the Rowland circle only in one point and the sagittal astigmatism is very large. Reduction is possible only by using a sagittal radius of curvature for the blank different from the meridional. Then the sagittal focal curve crosses the meridional in two stigmatic points. Because the angle of intersection is larger, astigmatism increases rapidly beyond these points.

(2) Partial symmetrical recording. It is possible to preserve the angular symmetry with respect to the blank normal (Fig.7.10) or to keep one of the sources on the Rowland circle (Fig.7.11). The corresponding groove shapes and spacing and their focal curves are presented in Figs.7.10 and 7.11.

(3) Completely asymmetrical recording. It is used when a full optimization of the recording conditions is required to reduce the aberrations over the working spectral region. Nevertheless, as a rule, at least one of the sources is kept close to the Rowland circle. An example of the result of such optimization is presented in Figs.7.7 and 7.8 with corresponding discussion of the properties.

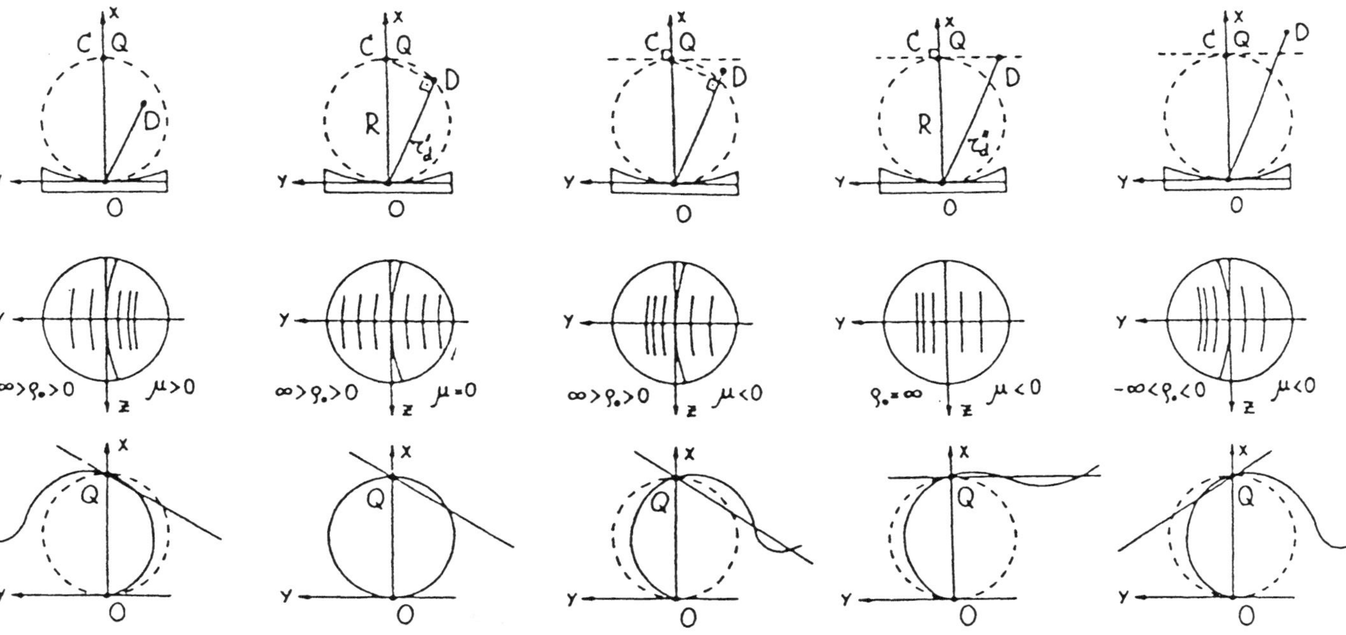

Fig.7.11 The same as in Fig.7.9, but without a symmetry of the recording sources. One recording point is on the Rowland circle.

7.5.2 Commercial Types of Concave Gratings and Their Design

A comparison between the gratings of different types with the corresponding commercial grating classification is presented in Table 7.1. Usually the last type is specified for flat-field spectrographs and monochromators and is different for each application. Moreover, special needs require special design of grating parameters, which is an interactive process between manufacturer and user (usually a designer of the device). In the case of flat-field spectrographs, the initial conditions include determination of the working spectral interval and spectral selectivity (limited also by the available detectors). These conditions, together with the maximum dimensions of the device, limit the range of the radius of curvature and the grating dispersion, as measured in its center. If vertical resolution is required in addition to the spectral one, some of the limitations can be contradictory.

Fixing the radius of the blank and the dispersion (d_0) is the most difficult part of the design because it requires experience. Their choice provides the first connection between the recording parameters leaving us with 3 free parameters. There are two possibilities, mutually complementary:

(1) To choose the type of the focal curve so that the working part of the meridional focal curve is as straight as possible and the sagittal curve is close to it. The fourth recording parameter can be used to reduce some of the coma terms (or their combination);

(2) To use ray-tracing of the image while varying the recording parameters in order to minimize either the aberration coefficients, or more usually the dimensions of the image spot. Optimization conditions depend on the grating application and require an initial choice of the working mounting. It is also most important to be able to precisely formulate the aim of the

Table 7.1

Recording source position in the holographically recorded concave gratings and their commercial classification

Recording source arrangement:	Jobin-Yvon	American Holographic Co.
Symmetrical	Type I: ruled equivalent	Classical Rowland
Angular asymmetry, both sources on the Rowland circle	Type II: aberration reduced	Circle gratings
Asymmetrical	Type III: stigmatic, and Type IV: special	Flat-field and monochromator gratings

optimization procedure. For example, it can aim to minimize the image width, or both width and height, or exit slit curvature, over an entire working spectrum, or at several fixed wavelengths, etc.

The two approaches are complementary. An initial approximation can be done using the focal curve representation and minimization of the aberration coefficients one by one and then to control the result by conducting the "final tuning" with the help of ray tracing.

7.6 Efficiency Behavior of Concave Gratings

Most investigations, starting with the paper of H. A. Rowland in 1883 [7.1], have been devoted to the geometrical focal properties of concave gratings and ignored efficiency behaviour. Compared with plane gratings the efficiency properties of concave gratings are considerably more complex. Rather than being illuminated by a uniform beam of collimated light, concave gratings usually work in conical mounting, the imaging rays being skewed out of the meridional plane. The main experimental difficulty in measuring local efficiencies is determined by the need to direct the incident beam individually for each grating part. Recording from two point sources tends to make the groove depth non-uniform over the grating surface. The typically Gaussian beam distribution is likely to overexpose the center and to underexpose near the edges of the resist-coated blank. In case of mechanical ruling, the geometrical characteristic that varies across the grating area is the blaze angle. This is the inevitable result of the interaction of a tool that has a fixed slope angle with the blank whose curvature changes progressively. For low blaze angles this leads to the introduction of bipartite or tripartite rulings in which the ruling process is interrupted once or twice to readjust the diamond tool angle.

Operating conditions usually include non-uniform incident beams with varying aperture that must also be taken into account. Conical diffraction and the variation of the groove parameters (which are never known to the necessary precision) make it difficult to extract accurate theoretical data.

Due to the limited dimensions of photodetector arrays and to the relatively narrow part of the focal curves with low aberrations, concave gratings are likely to have low groove frequencies that diffract in several orders. Thus, in view of the properties of the reflection gratings (Chapter 4), high efficiency can be expected only if the grooves have been blazed. This can include ruling with varying blazing or ion-milling, making the grating relatively expensive and reducing its advantage of being a single and "simple" optical element in the flat-field spectrograph. However, in many applications the effort is worthwhile.

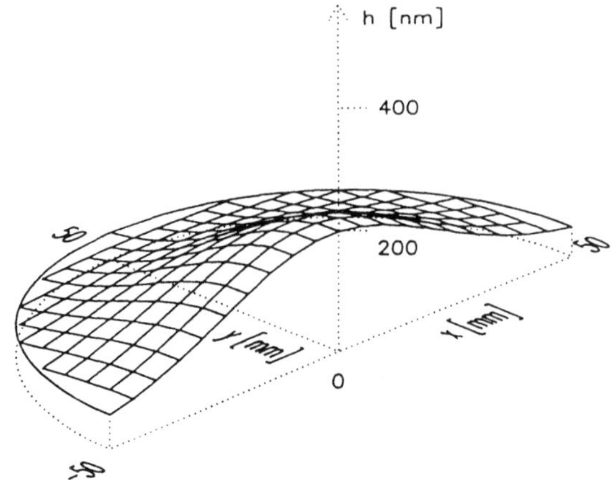

Fig.7.12 A typical two-dimensional map (distribution over the grating surface) of the groove depth of a holographically recorded grating (after [7.11]).

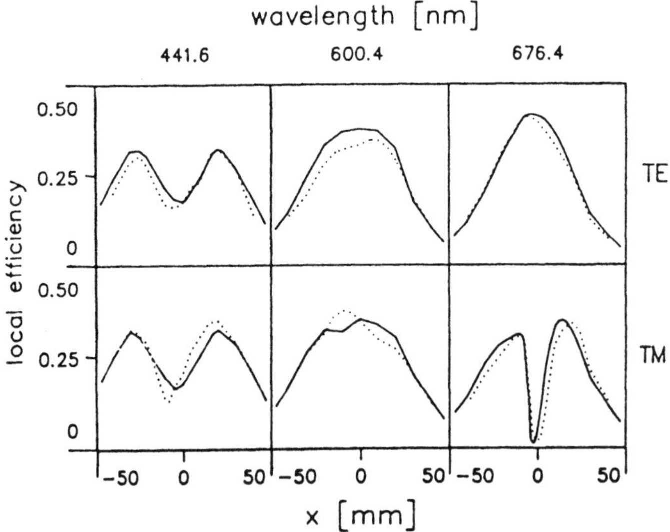

Fig.7.13 Experimental (dotted curve) and theoretical (solid curve) local efficiency distribution along the mean diameter for the two fundamental polarizations and three different wavelengths on the working mounting. Flat-field holographic concave grating with 100 mm diameter, 211 mm radius of curvature and 793 lines/mm frequency (after [7.11]).

7.6.1 Efficiency of Holographic Concave Gratings

Due to the Gaussian intensity map of the recording beams, groove depth over the grating surface reproduces its shape (Fig.7.12), unless some additional blazing has been done. In many cases, gratings operate over the resonance domain (see Chapter 4) so that there will be rapid changes in TM efficiency close to special wavelengths and angles of incidence. Concave gratings work in mountings where the angle of incidence and the wavelength vary almost independently over the surface (at least, in a specific region), so it is difficult to chose a mounting completely free of anomalies. Fig.7.13 provides an example where the local diffraction efficiency of a holographic grating is measured in the working mounting with a narrow laser beam, and where rapid changes can be observed. Fortunately, the anomalies are present only over a limited region of the surface and since the signal is averaged they are smoothed out in the large aperture incident beam (Fig.7.14). However, the outermost regions of the grating may have shallower grooves which reduce the integrated efficiency. The advice is to make gratings with the most uniform groove depths possible along the surface, although this can be difficult to achieve if recording distances are short.

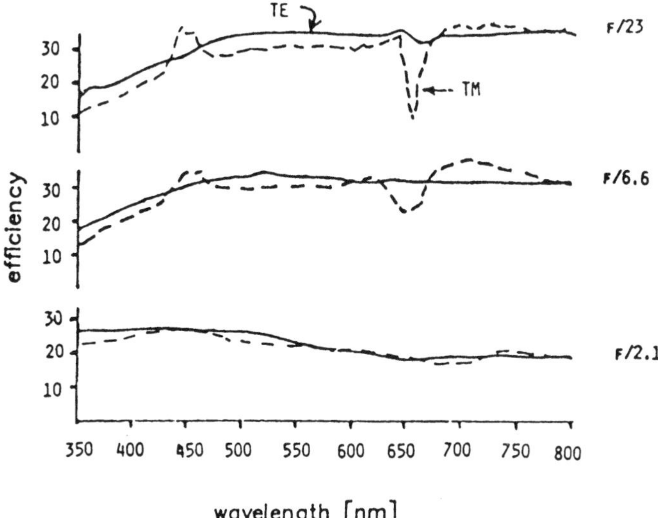

Fig.7.14 Integral efficiency curves, as measured under the working conditions of the grating of Fig.7.13, presented for different incident beam aperture F/#. Solid curve - TE polarization, dashed curve - TM polarization (after [7.11]).

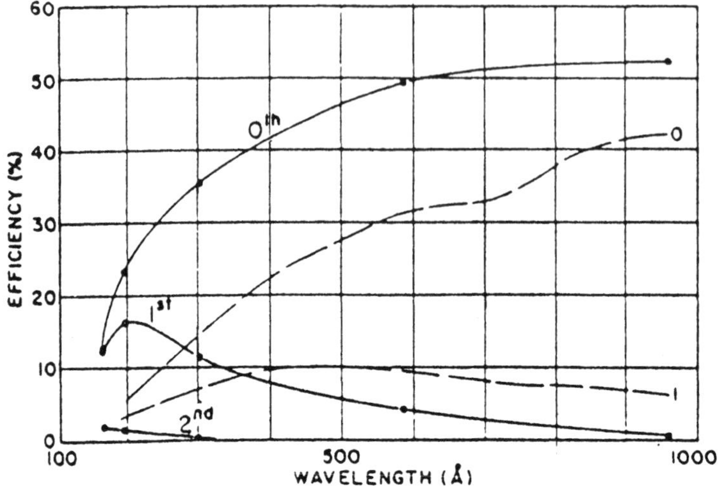

Fig.7.15 Efficiency of a conventionally ruled gold blazed grating with a 1-m radius of curvature, 1200 gr/mm and blaze angle of 2.6°. Dashed line gives efficiencies of a similar sinusoidal grating with groove depth 52 nm. Different orders are indicated in the figure (after [7.13]).

7.6.2 Blazed Concave Gratings

In some cases it is necessary to increase diffraction efficiency of the grating by blazing the grooves. This may be the case in the vacuum ultraviolet, where the natural reflection of metals is low and with it diffraction grating efficiency. Then a triangular (or quasi triangular) groove with a specific blaze angle can increase the efficiency over a particular spectral interval, reducing it outside this interval [7.12]. Fig.7.15 provides a comparison of the efficiency of a sinusoidal and echelette concave grating, where the blazed grating has efficiency in the -1st order exceeding 15% around 200Å, but lower than the efficiency of a similar sinusoidal grating above 370Å [7.13].

References

7.1 H. A. Rowland: "On concave gratings for optical purposes," Amer. J. Sci. (3) **26**, 87-98 (1883).

7.2 H. Haber: "The torus grating," J. Opt. Soc. Am. **40**, 153-165 (1950).

7.3 T. Namioka: "Theory of the ellipsoidal concave grating. I.," J. Opt. Soc. Am. **51**, 4-12 (1961).

7.4 A. Maréchal: "Optique géometrique générale" in *Handbook of Physics*, v.**24** *Foundations of Optics*, S. Flugge, ed., (Springer, Berlin, 1956).

W. Glaser: "Grundlagen der Elektronenoptik", *Teil II: Theorie der geometrischen Abbildungsfehler*, (Springer, Wien, 1952).

M. Born and E. Wolf, *Principles of Optics*, 4th edition, (Pergamon Press, Oxford 1968), chs. 4 and 5.

L. Seidel: "Ueber die Entwicklung der Glieder 3ter Ordnung, welche den Weg eines ausserhalb der Ebene der gelegenen Lichtstrahles durch ein System brechender Medien bestimmen," Astr. Nachr. **43**, no 1027, pp.289-302 (1856).

L. Seidel: "Ueber die Entwicklung der Glieder 3ter Ordnung, welche den Weg eines ausserhalb der Ebene der Axe gelegenen Lichtstrahles durch ein System brechender Medien bestimmen," Astr. Nachr. **43**, no 1028, pp.301-304, 307-320 (1856), no 1029, pp.321-332 (1856).

7.5 P. Lehmann: "Theory of blazed holographic gratings," J. Mod. Opt. **36**, 1471-1487 (1989).

P. Lehmann: "A blazed holographic concave grating for monochromators," J. Mod. Opt. **36**, 1489-1512 (1989).

C. Palmer and W. R. McKinney: "Equivalence of focusing conditions for holographic and varied line-space grating systems," Appl. Opt. **29**, 47-51 (1990).

7.6 S. Morozumi: "Abberation theory of diffraction gratings," Optik **53**, 75-88 (1979).

7.7 D.Lepere: "Monochromateur à simple rotation du réseau, a réseau holographique sur support torique pour l'ultraviolet lontain (Monochromators with single grating rotation and holographic gratings on toroidal blanks for vacuum ultraviolet)," Nouv. Rev. Optique **6**, 173-178 (1975).

T. Namioka, M. Seya, H. Noda: "Design and performance of holographic concave gratings," Jpn. J. Appl. Phys. **15**, 1181-1197 (1976).

H. Noda, T. Namioka, M. Seya: "Geometric theory of the grating," J. Opt. Soc. Am. **64**, 1031-1036 (1974).

H. Noda, T. Namioka, M. Seya: "Design of holographic concave gratings for Seya-Namioka monochromators, J. Opt. Soc. Am. **64**, 1043-1048 (1974).

7.8 L. Tsonev and E. Popov: "Focal images formed by a concave holographic grating. A corroborative investigation using three different techniques," J. Mod. Opt. **39**, 1749-1760 (1992).

7.9 W. R. McKinney and C. Palmer: "Numerical design method for aberration-reduced concave grating spectrometers," Appl. Opt. **26**, 3108-3118 (1987).

7.10 L. Tsonev and E. Popov: "Focal spot estimation for concave diffraction gratings," Opt. Commun. **90**, 11-15 (1992).

7.11 E. G. Loewen, E. K. Popov, L. V. Tsonev, and J. Hoose: "Experimental study of local and integral efficiency behavior of a concave holographic diffraction grating," J. Opt. Soc. Am. A **7**, 1764-1769 (1990).

7.12 A. V. Savushkin, E. A. Sokelava, and G. P. Startsev: "Optimization of the spectral and energy characteristics of concave diffraction gratings," Sov. J. Opt. Technol. **54**, 376-378 (1987).

7.13 W. R. Hunter: "Diffraction gratings for the vacuum ultraviolet spectral region," Nucl. Instr. Meth. **172**, 259-268 (1980).

Additional Reading

R. Bittner: "Holographische Gitter als Dioden-Zeilen-Spectrographen," Optik **64**, 185-199 (1983).

B. J. Brown, I. J. Wilson: "Holographic grating aberration correction for a Rowland circle mount. I.," Opt. Acta **28**, 1587-1599 (1981).

H. G. Beutler: "The theory of the concave grating," J. Opt. Soc. Am. **35**, 311-350 (1945).

M. Chrisp: "Abberation-corrected holographic gratings and their mountings," in *Appl. Opt. and Opt. Engineer.*, v.**10** (Academic Press, New York, 1987), ch.7.

H. Greiner, E. Schaffer: "Theorie eines Konkavgitter-Spektrometers," Optik **15**, 51-62 (1958).

H. Greiner, E. Schaffer: "Uber den Astigmatismus des Konkavgitters mit spharischer oder torisher Oberflache," Optik **16**, 288-303 (1959).

T. Harada, S. Moriyama, and T. Kita: "Mechanically ruled stigmatic concave gratings," Jpn J. Appl. Phys. **14**, suppl. 14-1, 175-179 (1974).

E. Ishiguro, R. Iwanaga, and T. Oshio: "Geometrical optical theory of diffraction gratings," J. Opt. Soc. Am. **69**, 1530-1538 (1979).

T. Kita and T. Harada: "Use of aberration-corrected concave gratings in optical demultiplexers," Appl. Opt. **22**, 819-825 (1983).

M. Koike and K. Ohkubo: "Holographic concave gratings for use with an off-plane constant-deviation monochromator," Appl. Opt. **25**, 4071-4075 (1986).

J. M. Lerner, J. Flamand, J. P. Laude, G. Passereau, and A. Thevenon: "Abberation corrected holographically recorderd diffraction gratings," SPIE, v.**240**: *Periodic structures, gratings, moire patterns and diffraction phenomena*, 72-81 (1980).

T. Namioka: "Theory of the concave grating. I.," J. Opt. Soc. Am. **49**, 446-460 (1959).

T. Namioka and M. Koike: "Aspheric wave-front recording optics for holographic gratings," Appl. Opt. **34**, 2180-2186 (1995).

M. Neviere and W. R. Hunter: "Analysis of the changes in efficiency across the ruled area of a concave diffraction grating," Appl. Opt. **19**, 2059-2065 (1980).

C. Palmer: "Absolute astigmatism correction for flat field spectrographs," Appl. Opt. **28**, 1605-1607 (1989).

C. Palmer: "Theory of second-generation holographic diffraction gratings," J. Opt. Soc. Am. A **6**, 1175-1188 (1989).

C. Palmer: "Second-order imaging properties of circular field spectrographs," Appl. Opt. **29**, 1451-1454 (1990).

M. Pouey: "New holographic grating devices for hot plasma diagnostics. Part I: Stigmatic properties of spherical holographic grating," J. Optics (Paris) **14**, 235-241 (1983).

J. Picht: *Grundlagen der geometrisch-optischen Abbildung* (Akademie-Verlag, Berlin, 1955).

T. Sai, M. Sai, and T. Namioka: "On Beutler's theory of the concave grating," Sci. of Light (Tokyo) **17**, 11-24 (1968).

M. Seya: "A new mounting of concave grating suitable for a spectrometer," Sci. of Light **2**, 8-17 (1952).

M. Singh, G. Prakash Reddy: "New mountings of concave reflection holographic grating," Optik **76**, 83-88 (1987).

G. W. Stroke: "Diffraction gratings" in *Handbook of Physics*,v.29, *Optical Instruments*, ed. S. Flugge (Springer, Berlin, 1967).

C. H. V. Velzel: "A general theory of the aberrations of diffraction gratings and gratinglike optical instruments," J. Opt. Soc. Am. **66**, 346-353 (1976).

W. T. Welford: "Abberation theory of gratings and grating mountings," E. Wolf, ed. *Progress in Optics* (Elsevier, North-Holland, Amsterdam, 1965) v. **IV**, pp.241-280.

Chapter 8

Surface Waves and Grating Anomalies

8.1 Grating Anomalies

To an instrument designer an anomaly is normally perceived as a nuisance, to a scientist it represents a challenge. It means that something is not properly understood, an interesting problem waiting to be solved. Gratings are a good example of this. Since 1902, when R. W. Wood [8.1] observed unexpected changes of diffraction efficiency by a factor greater than 10, and so sharply defined that it affected only one of the two sodium D-lines, we have followed his lead in calling this behaviour anomalous. Generations of spectroscopic scientists have investigated this phenomenon. There are several reasons for so much interest:

(1) Anomalies are connected to physical phenomena that are interesting in themselves. Moreover, sometimes the corrugation can significantly enhance this phenomenon, as is the case with surface waves and non-linear second-harmonic generation.

(2) In high speed scanning applications anomalies must be suppressed because they may lead to a rate of change in output signal that exceeds data processing capabilities.

(3) They provide a valuable stimulation for the development of the electromagnetic theory of scattering, and its numerical implementations, as well serving as a critical test of their accuracy.

Despite the large number of studies devoted to grating anomalies (or perhaps because of), there is neither a simple nor even complete definition of just what is anomalous. In general an anomaly describes something that is irregular, or unexpected from a previous level of knowledge, or unexplained by existing theory. But, in fact, even after valid explanations were found, the behavior continues to be called anomalous. A good example is that sharp dips or peaks in the spectral or angular behavior of grating efficiency are generally termed anomalous, even though they are completely explainable by electromagnetic theory.

We distinguish between three types of grating anomalies:

(1) Threshold phenomena that are connected with the energy redistribution at the cut-off of higher diffraction orders (see Chapters 4 and 6).

(2) Resonance anomalies that are due to the excitation of a guided wave along the corrugated surface, or to a resonance inside the groove. Phase matching between the incident wave and a solution of the homogeneous problem is assured by the grating.

(3) Non-resonance anomalies (sometimes called "broad" anomalies).

Lord Rayleigh [8.2] tried to explain Wood's anomalies with nothing but threshold effects. Differences in the boundary conditions for the two fundamental cases of polarization result in different efficiency behavior near the cut-off. Rayleigh's theory, while a first attempt to develop an electromagnetic theory of gratings, failed to explain later experiments. Fano [8.3] was the first to draw a clear connection between narrow anomalies and surface wave excitations, distinguishing between "sharp" and "broad" anomalies. Hessel and Oliner [8.4] provided the currently accepted formulation of the phenomenological approach to resonance ("sharp") anomalies, presented in detail in the next section. More recently Neviere [8.5] and Maystre [8.6] have contributed significantly to the physical understanding of anomalies and their classification. A microscopic picture of some peculiarities in grating behavior has been added recently [8.7].

While it is believed that resonance anomalies are usually narrow, and the non-resonance anomalies broad, there are many exceptions. Therefore, this convention is safe to use only if the fundamental reasons for specific anomalies are known. For example, the resonance anomaly in the reflectivity of a corrugated waveguide that is due to a waveguide mode excitation strongly depends on the groove depth and can become relatively broad, while a non-resonance total absorption of light by a metallic grating in grazing incidence [8.8] shows up as a very narrow dip of diffraction efficiency. Sometimes a resonance anomaly behaves like a non-resonance. A typical example is the absorption of light by deep metallic lamellar gratings [8.9]. Whereas the absorption is due to the mode resonances inside the grooves, these resonances are so weak that no field enhancement is observed, but absorption can still reach 100%.

Non-resonance anomalies include a variety of phenomena, and do not have a unified explanation. Grazing incidence total absorption of light (Fig.8.1), already mentioned above, is explained by merging of two independent zeros of the -1st and the zeroth order efficiencies in the groove depth region where no surface wave exists because of large radiation losses [8.10]. Littrow mount zero, or Bragg anomaly [8.11] is connected with the quasi-periodicity of grating properties as a function of groove depth [8.12], and appears to be due to formation of curls of energy flow inside the grooves (see Section 8.4.1).

Often anomalies of different types (as well as of the same type) merge into each other, making it difficult to distinguish between them. For example, the conditions for a surface plasmon excitation along highly conducting metallic gratings in TM polarization almost coincide with the cut-off conditions for higher orders, although these two anomalies have a different influence on the efficiency behavior. In fact, threshold phenomena alone have a minor effect in most cases, and rarely predominate, as with certain echelles (see Chapter 6). They can usually be observed as edges in the angular dependencies and appear in both polarizations, their position defined by eq.8.15. By contrast, resonance anomalies depend strongly on polarization. For highly conducting metallic gratings the cut-off anomalies almost coincide with the position for plasmon excitation and as a result they are often confused.

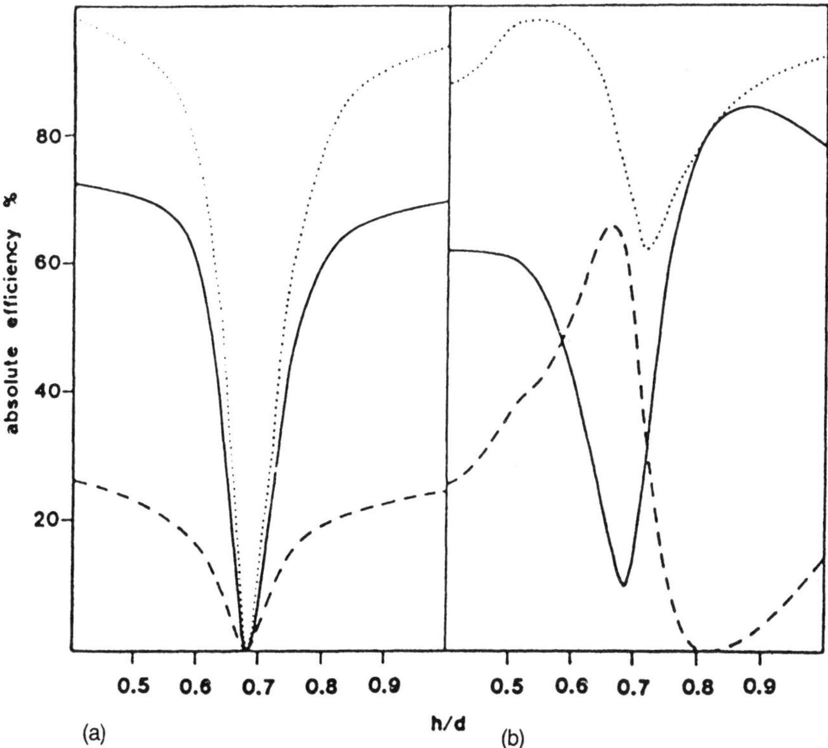

Fig.8.1 Efficiency curves as a function of modulation depth h/d for an aluminum grating with d=0.5 μm, λ=632.8 nm, θ_i=87.85°, TM polarization. Solid curve represents total diffracted energy; a) sinusoidal groove profile, b) symmetrical triangular groove profile (after [8.8]).

8.2 Phenomenological Approach

Intuitively, surface or guided waves represent a wave localized near some interface(s) between different media or inside a layer. This wave has vanishing amplitudes going away from the "guiding" structure. In general, without dissipative and scattering losses such waves can propagate without decay. Its most important characteristic is the phase velocity k_G in a direction parallel to the interface, called the propagation constant. Then, if the direction of propagation is along the x-axis, the wave harmonically varies in x as $\exp(ik_G x)$. Wave propagation equations (or, more general, Maxwell's equations) provide a link between the wave vector components k_x and k_y and the refractive index n in each of the media (see Chapter 2.1 and 2.2):

$$k_x^2 + k_y^2 = \left(\frac{2\pi}{\lambda} n\right)^2 . \qquad (8.1)$$

When we consider a surface (guided) wave, its "surface" nature requires that in the lower and uppermost media the wave amplitude decreases gradually with increasing $|y|$, i.e., the vertical component of the wave-vector has to be imaginary. In terms of diffraction orders (Chapter 2.2) this means that the surface wave is characterized by evanescent orders both in the substrate and the cladding, implying a restriction to $k_x \equiv k_G$ so that $k_G / (2\pi / \lambda)$ must be greater than the greatest refractive index of the cladding or the substrate. If we consider a dielectric waveguide with refractive index of the guiding layer higher than the refractive index of the surrounding media (Fig.8.2a), then the guided wave propagation constant has to be larger than the free-space wavevector $k = 2\pi/\lambda$, assuming the cladding to be air, whereas the incident plane wave is always characterized by $k_x = k \sin\theta_i < k < k_G$. The fundamental mode field distribution is shown in Fig.8.2a, and the plane incident and reflected wave in Fig.8.2b. In order to have a coupling between the two waves, their wave-vector components along the interface must be equal (phase-matched), which is not the case for perfect plane interfaces.

An important point is that there are always losses due to absorption inside the media as well as scattering on the surface, volume defects and roughness, so that the surface wave always decays along its propagation direction (x-axis in our case). Provided loss sources are homogeneously distributed, this decay is exponential, so that along the x-axis the entire field dependence takes the form:

$$\exp(ik_G x)\exp(-\gamma x) \equiv \exp\left[i(k_G + i\gamma)x\right] , \qquad (8.2)$$

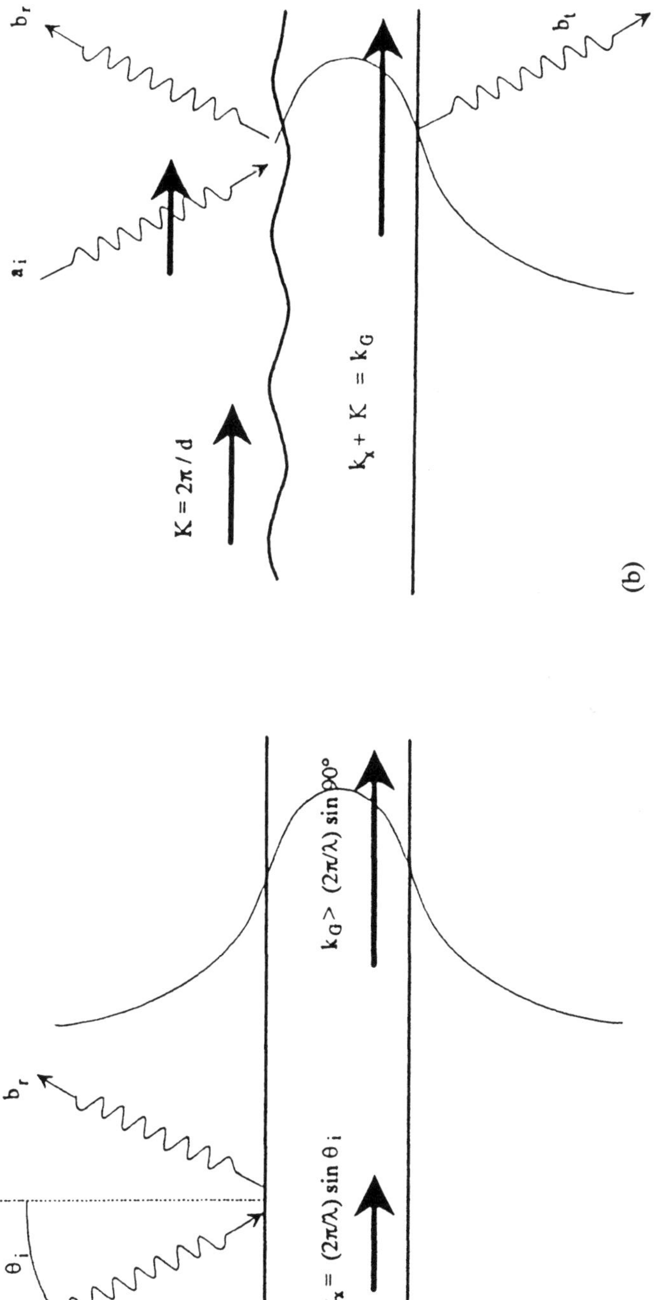

Fig.8.2 a) Schematic representation of the guided wave field distribution in a planar waveguide (right hand-side) with its wavevector along the surface. At the left a plane incident and reflected wave are presented, the horizontal component of their wavevector being smaller than k_G. b) A corrugation with wavevector **K** can lead to a coupling between the surface wave and the incident, reflected and transmitted plane waves.

so that the decay constant γ can be represented as an imaginary part of the propagation constant k_G.

A periodic perturbation of the surface (usually called a *grating*) can be characterized by a function y = f(x). Being periodic with respect to x, with a period equal to d, f(x) has the well-defined Fourier series representation:

$$f(x) = \sum_m f_m \exp(imKx) \quad , \quad K = \frac{2\pi}{d} . \tag{8.3}$$

Then a coupling between the incident plane wave and the guided wave can be carried out through the grating wave vector K, provided that phase matching is fulfilled:

$$k_G = k_x + mK , \tag{8.4}$$

where m can be any integer. Fig.8.2b represents such a case. Of course, the coupling also "works" in the opposite direction (recall the grating equation in Chapter 2): not only does the plane incident wave excite the waveguide mode, but the mode itself is coupled to the radiation orders in the cladding and/or the substrate through the profile component, so that its existence can be felt in the propagating diffraction orders. To study this influence, which is the most important consequence of the mode excitation, it is necessary to investigate in more detail the conditions for the existence of a surface wave. For an inexperienced reader, the consequences of the next section can be summarized as follows:

A surface wave requires existence of a so-called solution of the homogeneous problem of light scattering. Such a solution requires that a certain matrix M that connects the diffracted **b** and the incident **a** wave components has to have a zero determinant: An echo from matrix algebra theorems reminds us that if the equation **Mb** = 0 is to have a non-zero solution, the necessary condition is detM = 0. After that it is sufficient to realize that the scattering matrix S is the inverse of M, so that if detM has a zero when $k_x = k_G$, all the components of $S = M^{-1}$ are proportional to

$$S_{ij} \propto \frac{1}{\det M} \approx \frac{1}{k_x - k_G} . \tag{8.5}$$

That is all that we want to show in the following section, so an inexperienced reader can go directly to section 8.2.2 with the only "luggage" being eq.8.5.

8.2.1 Guided Wave and a Pole of the Scattering Matrix

Imagine a plane interface between two homogeneous isotropic linear media. It is well known that along the interface, a surface wave can only exist in TM polarization if one of the media is dielectric and the other metallic. The transverse magnetic field component has the form (see section 2.2 and Fig.8.2):

$$H_1 = b_1 e^{ik_G x + ik_{1y} y}$$
$$H_2 = b_2 e^{ik_G x - ik_{2y} y} \quad , \tag{8.6}$$

with k_x and k_y linked by (8.1). The boundary conditions require continuity of magnetic and electric field components, tangential across the interface:

$$\left| \begin{array}{l} H_1(y=0) = H_2(y=0) \\ \dfrac{1}{n_1^2}\dfrac{\partial H_1}{\partial y} = \dfrac{1}{n_2^2}\dfrac{\partial H_2}{\partial y}\Bigg|_{y=0} \end{array} \right. \Rightarrow \left| \begin{array}{l} b_1 = b_2 \\ \dfrac{k_{1y}}{n_1^2} b_1 = -\dfrac{k_{2y}}{n_2^2} b_2 \end{array} \right. , \tag{8.7}$$

where the last system can be expressed in a matrix form:

$$M\vec{\mathbf{b}} = 0 , \tag{8.8}$$

where

$$M = \begin{pmatrix} 1 & -1 \\ \dfrac{k_{1y}}{n_1^2} & \dfrac{k_{2y}}{n_2^2} \end{pmatrix} , \quad \vec{\mathbf{b}} = \begin{pmatrix} b_1 \\ b_2 \end{pmatrix} . \tag{8.9}$$

The invert of M is usually called a scattering matrix of the system:

$$S = M^{-1} \quad . \tag{8.10}$$

In order for eq.(8.8) to have a non-zero solution, it is necessary that the det(M) be equal to zero at a certain value of the horizontal component of the wavevector, namely when $k_x = k_G$. This can be fulfilled if n_1^2 and n_2^2 have opposite signs, which is true for a metallic-dielectric interface in the optical region. Such a solution is called a plasmon-polariton surface wave. In the vicinity of the zero of the determinant of matrix M it can be expanded in a Taylor series. Keeping only the first-order term $\det|M| \propto k_x - k_G$, the inverse of M has components inversely proportional to det(M):

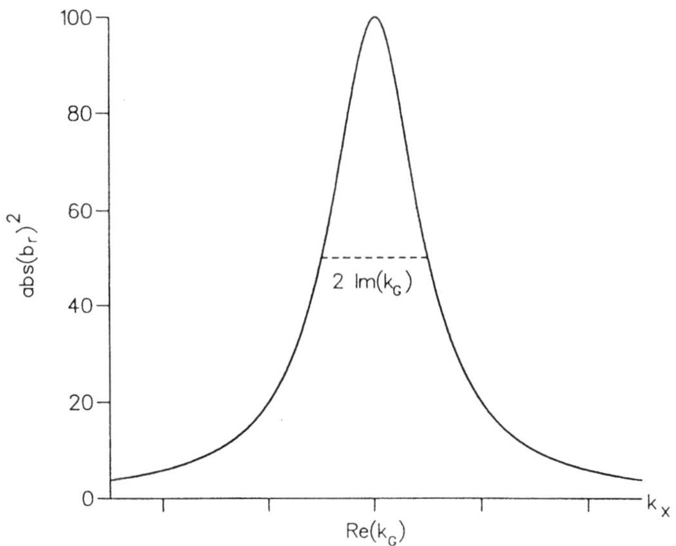

Fig.8.3 A typical Lorentzian shaped curve, representing the dependence of the scattering matrix S for real values of k_x if the expansion (8.12) holds rather than (8.13).

$$S_{ij} \propto \frac{1}{k_x - k_G}, \tag{8.11}$$

i.e., k_G is a pole of all the components of the scattering matrix, because they have a singularity (go to infinity) as $k_x \to k_G$.

Taking into account that k_G is usually complex, the dependence (8.11) for real values of k_x has a Lorentzian form (Fig.8.3).

8.2.2 Pole of the Scattering Matrix and Diffraction Efficiencies

Let us return now to our initial problem. If a plane wave is incident on a flat interface, the phase matching condition $k_x = k_G$ can not be fulfilled, so that there is no resonance anomaly in the reflectivity (or transmittivity), see Fig.8.4a, solid line. Even a slight corrugation, however, has an already well-defined wave-vector (eq.8.3). Provided the wavelength-to-period ratio is properly chosen, this can lead to coupling between the incident wave and the surface wave. Reciprocally, the same phase conditions ensure a coupling between the surface wave and all the diffracted orders, and, in particular, the specularly reflected one. Due to that coupling, its amplitude b_r exhibits a pole (eq.8.11):

$$b_r \propto \frac{1}{k_x - k_G} \tag{8.12}$$

in its angular dependence. However, as far as there is no such pole without corrugation (h=0), it is clear that this pole must be compensated by *something*, so that there is no Lorentzian-shaped curve in the orders reflected by plane interface(s). The simplest way is to introduce a numerator in eq.(8.12) which is equal to the denominator at h=0:

$$b_r \propto \frac{k_x - k^z}{k_x - k_G}, \tag{8.13}$$

with $k^z(h=0) = k_G(h=0)$. Here the upper index z stand for zero and must not be confused with the lower index z which denotes the z-component of k.

When h differs from zero there is no reason to keep k^z and k_G bound together. Then the reflected amplitude becomes zero, when $k = k^z$, which explains the meaning of the superscript z. Of course, like the pole, the zero k^z usually is complex, so that for a real angle of incidence it is rare to have zero reflectivity. With few well-defined exceptions, we can assume continuity requirements: when the system parameters (groove depth, wavelength, etc.) are varied continuously, the poles and the zeros move continuously in the complex plane. The curves they form are called *trajectories* and they can provide important information about the system behavior and the links that exist between different anomalies.

An example of the angular dependence of the reflectivity of a corrugated waveguide, given in Fig.8.4a, confirms fully the existence of pole and zero. Without corrugation reflectivity is a constant, determined by the Fresnel reflectivity of the slab. Even an infinitesimally small grating on top of the waveguide changes its response drastically in the angular interval where the system parameters provide grating-assisted coupling between the incident wave and the waveguide mode supported by the waveguide. Well-defined maxima and minima are clearly distinguished[1].

---

[1] In order to actually locate these minima and maxima for an infinitesimally shallow grating one must have a superlaser with infinitely coherent and collimated beam or other source of a real plane wave, and have an infinitely large grating.

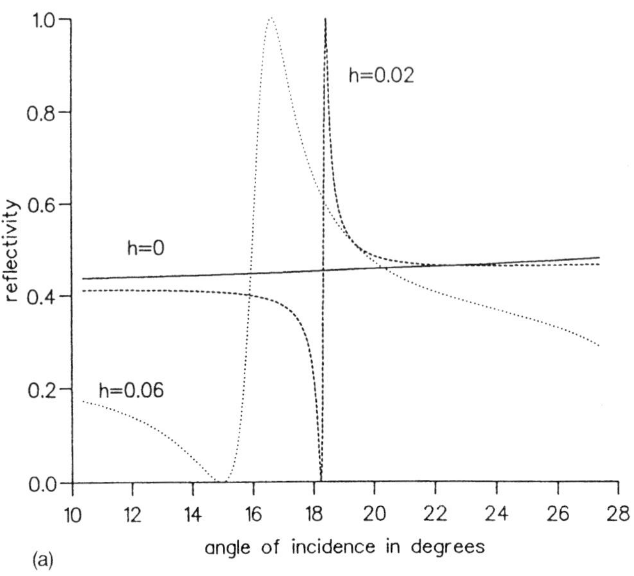

Fig.8.4 a) Angular dependence of the reflectivity of a corrugated waveguide with thickness t=0.19 μm, refractive indices: $n_1 = n_3 = 1, n_2 = 2.3$, a sin corrugation with period d=0.37 μm. TE polarization, λ=0.6328 μm. The values of the groove depth in μm are shown in the figure.

A careful look at the behavior of reflectivity in Fig.8.4, which is quite typical for resonance anomalies, leads to at least three generalities:

(1) The waveguide mode excitation has an important influence over the reflectivity (or, more general, diffraction efficiency). Without corrugation, there is no such influence, provided the system is ideal[1] .

(2) The response of the system (namely the reflection order amplitude) is non-Lorentzian. A minimum can also be observed, which corresponds to the existing zero in eq.8.13 (i.e., the rather intuitive considerations that led us to eq.8.13 are well-backed). Strictly speaking, the maximum corresponds to a zero

[1] This is not true if the surface is rough. In that case the roughness supplies a large set of phase vectors, so that usually there is at least one to ensure a coupling between surface wave and the incident wave. This is much more applicable to metallic surfaces, which are rougher than the dielectric interfaces. That is why surface roughness strongly influences the metallic grating performance in TM polarization, when a plasmon-polariton surface wave can be excited.

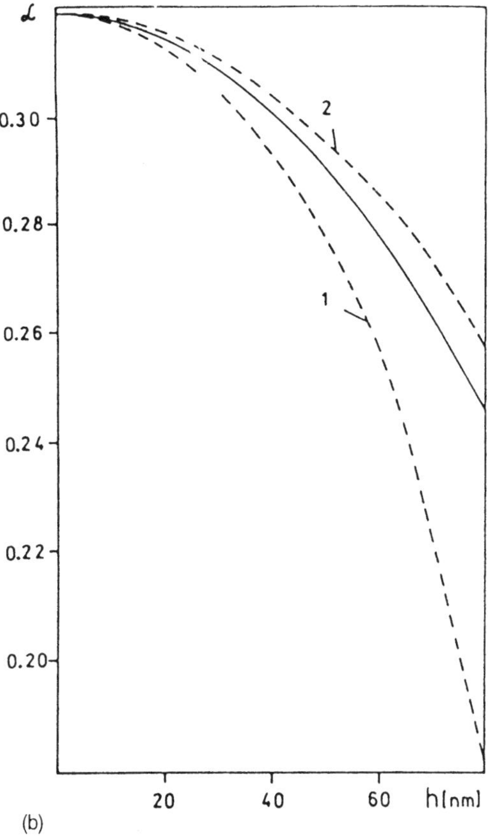

Fig.8.4 b) trajectories of the minimum (1) and maximum (2) of the reflectivity as a function of the groove depth. The solid line indicates the variation of the real part of the pole (after [8.13]).

of the transmitted wave amplitude, but, as a rule of thumb, the zero leads to a minimum, whereas the pole to the maximum. If the anomaly-free efficiency is high, the maximum can hardly be detected, as is the case with the reflectivity of a metallic grating (see section 8.4).

(3) The pole and the zero merge into each other as the groove depth decreases, and their splitting increases with h (Fig.8.4b). Thus the anomalous region is narrower for shallow gratings and wider for deep gratings.

The phenomenological formula (8.13) and, more generally, the phenomenological approach, recently called "polology", has proven to be very useful. Its "only" disadvantage is the necessity to know the positions of the poles and the zeros before efficiency behavior can be predicted. This indicates use of rigorous electromagnetic theories which are capable of finding efficiencies without pole searching. However, polology appears to be quite useful in providing physical understanding of resonance anomalies. Moreover, sophisticated codes are not always available, so the poles and the zeros can help in a fast, even if not rigorous way in explaining "anomalous" experimental data. They may on occasion also serve to analyze suspicious behavior of computer codes.

8.3 Types of Surface Waves

As we have observed, a resonance (i.e., pole) results in an anomaly in the diffraction efficiency. There are several possible sources for poles. At a bare metallic-air interface a surface wave can propagate only for TM polarization of light It is called *plasmon-polariton surface wave* (PSW). A direct consequence of eq.(8.9) is that its propagation constant is equal to:

$$k_{PSW} = \frac{2\pi}{\lambda} \frac{1}{\sqrt{1 + n_{metal}^2}} , \tag{8.14}$$

so that for highly conducting metals the real part of k_{PSW} is slightly greater than $2\pi/\lambda$ and the phase condition for excitation of PSW is fulfilled close to the cut-off conditions of the higher orders (see section 2.5), defined by:

$$\sin\theta_i = \pm 1 + m\frac{\lambda}{d} ; \tag{8.15}$$

which explains why resonance and cut-off anomalies are usually confused.

In Section 8.4 we study in detail this anomaly for various sets of grating parameters: gratings supporting a single order, where the resonance anomaly can lead to perfect absorption of the incident light combined with a large enhancement of the electromagnetic field density, and gratings having several orders.

Deposition of a dielectric layer over the metallic substrate has two direct consequences:

(1) The PSW propagating constant is shifted from its position for a bare surface, because the PSW propagates along the dielectric-metal interface, rather that along the air-metal surface. The position of the anomaly can be altered significantly, depending on the layer thickness and refractive index.

(2) Provided the optical thickness of the layer is significant, it can support another type of guided wave - *waveguide modes* (WM). They can be TE or TM polarized, usually with different propagation constants, varying over the range

$$\max(n_{cladding}, n_{substrate}) < k_{WM}\frac{\lambda}{2\pi} < n_{layer} , \qquad (8.16)$$

so that they can produce numerous anomalies over the entire spectral and angular region and in both TE and TM polarization. This has led to a lot of troubles during the last decades. Section 8.5 gives an example of these two influences, but a detailed study can be found in Chapter 5 of *Electromagnetic Theory of Gratings* [8.5].

Of course, in order to have a waveguide mode, it is not necessary to use a metallic substrate. A typical example of the resonance anomaly of a corrugated waveguide is presented in Fig.8.4 and a detailed discussion supplied.

In the last section of this chapter a multilayered dielectric grating is discussed. If its thickness is great enough to provide high reflection, it is thick enough to support a large number of waveguide modes and its properties are anomalous all over.

8.4 Influence of Surface Waves on Metallic Grating Properties

Eq.8.13 represents the well-known phenomenological formula that enables one to predict the behavior of efficiencies in the region of resonance anomalies, and has been fully confirmed by comparing numerical with experimental results. The coefficient of proportionality is a slowly varying function of the grating parameters. Pole and zero are independent of the angle of incidence and vary slowly with wavelength and profile parameters. In case of anomaly interaction when two surface waves are excited simultaneously, two (or more) terms are included in the phenomenological formula with different poles and zeros, corresponding to different surface waves.

For shallower gratings, the splitting between pole and zero is smaller (see Fig.8.4). If the substrate is perfectly conducting and the grating supports a single (specular) diffraction order, energy balance requires that the pole and the zero are mutually complex conjugated, i.e., they lie symmetrically on the two sides of the real k_x axis. Let us remember that corrugation initiates radiation losses, expressed in the growth of the imaginary part of the pole. Thus, the zero can also become complex. Because the zero and the pole are complex conjugated, no influence on the reflectivity can be detected, but the phase of the reflected wave is changed by 2π when the angle of incidence is varied across the anomaly region. And indeed, the modulus of the right-hand side term of eq.8.13 is unity, if $k^z = \overline{k}_G$, with the overbar standing for complex conjugation.

If the grating is not perfectly conducting, both imaginary and real parts of the pole and the zero are slightly modified. The small difference between their positions can have unbelievably strong effects on the reflectivity, reducing it significantly.

It must be pointed out that in grating theory another set of notations is more common. These are the normalized values of poles ($\alpha^p = k_G / k$) and zeros $\alpha_m^z = k_m^z / k$, m indicating the number of order), and the normalized wavenumber $\alpha_0 = k_x / k \equiv n_1 \sin\theta_i$.

8.4.1 Total Absorption of Light by Metallic Gratings

An unexpected theoretical discovery made by Maystre and Petit [8.14] was experimentally confirmed by Hutley and Maystre [8.15]. While a plane mirror will reflect most of the incident light, it can become a total absorber if just slightly corrugated and illuminated under certain conditions. These conditions include a set of angles of incidence, wavelength and groove depth values, that are peculiar for a specific substrate material and groove form (Fig.8.5a). This phenomenon (sometimes termed Brewster's effect in metallic gratings) has been thoroughly investigated. Its importance in grating studies can be compared with Wood's discovery of grating anomalous behavior. Its connection with the plasmon surface wave excitation was revealed and its resonant nature proved.

` Fig.8.5b represents the trajectories of the pole and the zero of the reflected wave when groove depth is varied, corresponding to the efficiency behavior seen in Fig.8.5a. The position of α_0^z is almost symmetrical to α^p with respect to their initial position at h = 0. As the corrugation depth increases, the trajectory of α_0^z crosses the real axis $\alpha_0 = \sin\theta_i$ at some value of $h = h_{B_1}$. Having real zero of the zeroth order amplitude means that when the sine of the incident angle is equal to that zero, there is no reflected light in the specular order. As it is the only propagating order, a *total absorption of incident light* is observed (or, in other words, nothing is observed to be reflected by the corrugated mirror!)[1]. For deeper grooves α_0^z moves away from the real axis (Fig.8.5b) and, according to eq.(8.13), the minimum value of reflectivity increases (Fig.8.5a).

---

[1] This has nothing to do with poor optical properties of the reflecting material (i.e., it is not just a "bad" mirror), but is purely an effect of the corrugation.

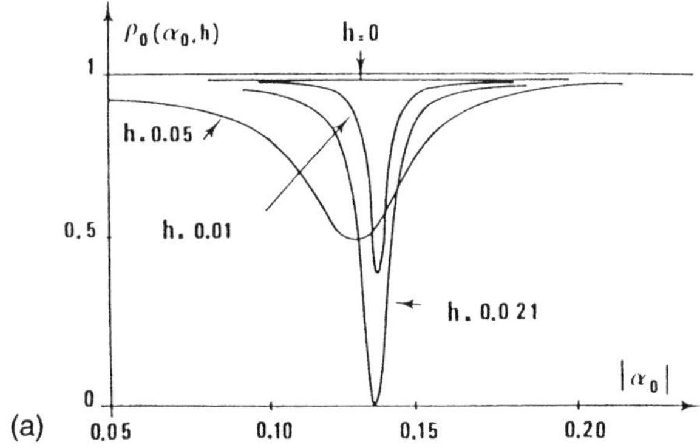

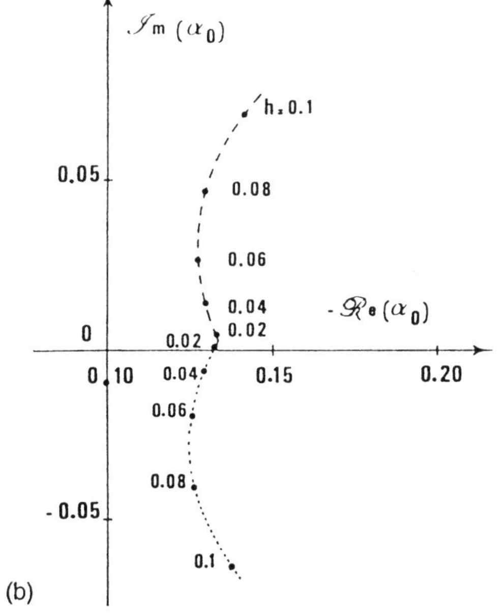

Fig.8.5 a) Zeroth order efficiency of a silver grating versus $\alpha_0 = \sin\theta_i$. TM polarization, d=1/2400 mm, λ=0.5 μm, given for different groove depth values, shown in the figure in μm. b) Trajectory of the normalized pole (dashed line) and of the normalized zero of the zeroth reflected order (dotted line) in the complex plane, as a function of the modulation depth h/d, indicated with circles. Parameters of the sinusoidal silver grating correspond to those in Fig.8.5a (after [8.16]).

Further increase of groove depth leads to stronger coupling between the surface wave and the order diffracted into air, and the diffraction (radiation) losses increase, which is expressed as an increase of the imaginary part of the pole. Above some critical groove depth value, diffraction losses become so high that the surface wave is no longer localized at the surface; the real part of α^p becomes less than unity. The pole is transferred into a zero of the zeroth order amplitude which is solitary, that is not accompanied by a pole. This zero leads to a new non-resonant total absorption of light in grazing incidence (Fig.8.1a).

With a still greater increase of the groove depth, a new pole appears, leading to the Brewster effect in deep gratings, which has a similar behavior and explanation as the total absorption of light by a shallow grating. Investigations over larger groove depths indicate repeating minima of reflectivity. A suitable choice of incidence angle can diminish any of these reflection minima to a zero value. It is remarkable that for an aluminum sinusoidal grating in the red (λ=0.6328 μm), three cases of total light absorption (for shallow, deep, and very deep grooves) almost coincide in their angular position (Fig.8.6).

The physical reason for such quasiperiodical appearance of anomalies is similar to the quasiperiodicity of grating efficiency in Littrow mount (Chapter

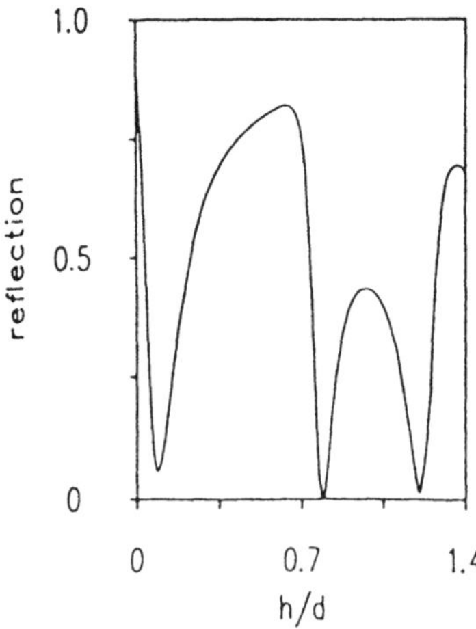

Fig.8.6 Reflectivity of an aluminum sinusoidal grating as a function of the groove depth. d=0.57 μm, λ=0.6328 μm, TM polarization, θ_i = 14.929° (after [8.8]).

4). Formation of *curls* inside the deep grooves separates the groove bottom from the energy flow above the grooves. For a suitable choice of system parameters these curls are completely hidden inside the grooves and energy flow distribution above the grooves resembles that of a shallow grating (Fig.8.7).

An interesting consequence of the resonant nature of this anomaly is the very strong field density enhancement near the grating surface. As becomes

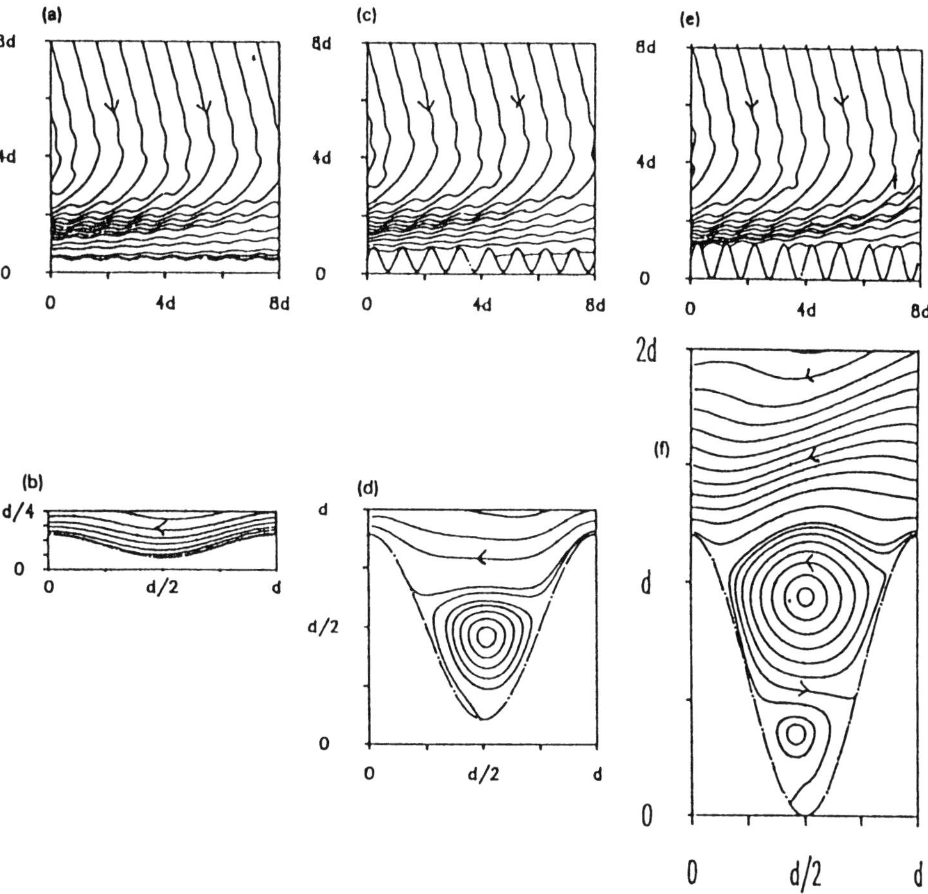

Fig.8.7 Energy flow lines corresponding to the total absorption of light by a sinusoidal aluminum grating. d = 0.57 μm, λ= 0.63287 μm, TM polarization. (a,b) h = 0.057 μm, (c,d) h = 0.3957 μm, (e,f) h = 0.67 μm (after [8.17]).

evident in Fig.8.7, energy flow lines approaching the groove tops are curved in a direction almost parallel to the grating, visualizing the excitation of a surface wave which propagates along the negative x-axis direction. When line density is increased, it represents a sharp increase in the magnitude of the Poynting vector corresponding to a strong enhancement of field density above the groove tops (Fig.8.8).

Because the energy flow above deep gratings is separated from the groove bottoms, field enhancement (FE) appears only above the groove tops (Fig.8.8b,c), while for shallow gratings (Fig.8.8a) a comparatively strong FE is also observed in the groove bottoms.

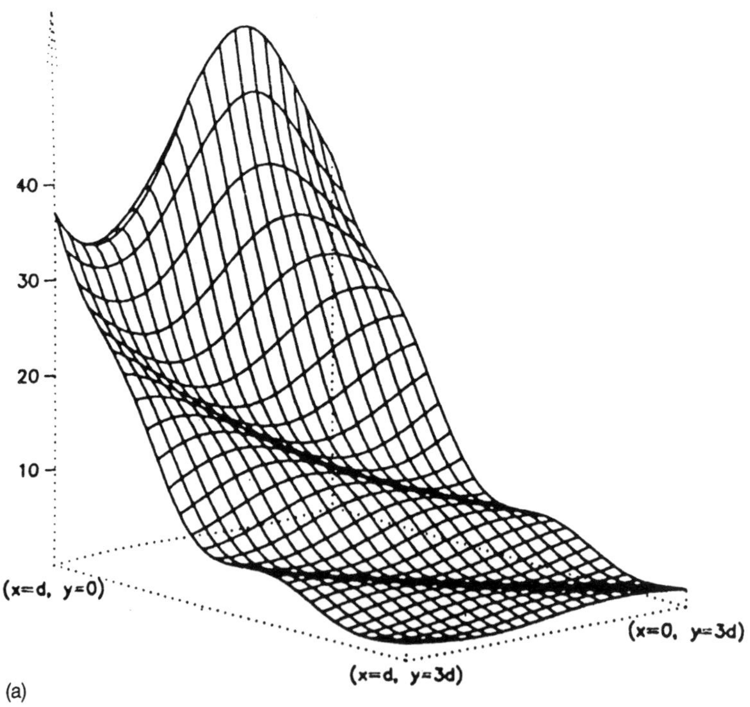

Fig.8.8 3D view of the normalized optical power density distribution corresponding to the parameters that provide total light absorption: a) h/d=0.10, θ_i=14.82°; b) h/d=0.79, θ_i=14.93°; c) h/d=1.20, θ_i=15.06° (after [8.18]).

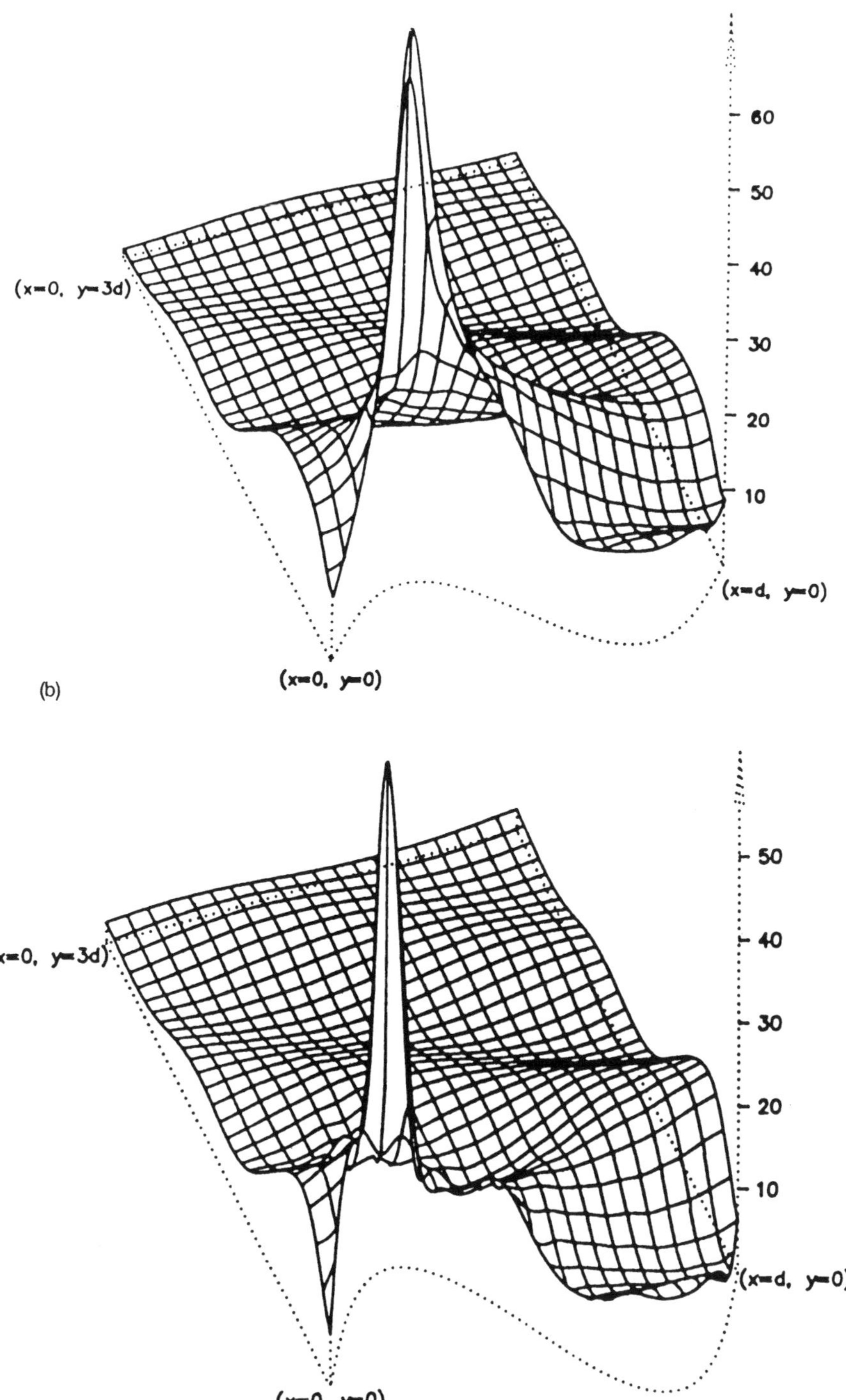
60
50
40
30
20
10
(x=0, y=3d)
(x=d, y=0)
(x=0, y=0)
(b)
50
40
30
20
10
(x=0, y=3d)
(x=d, y=0)
(x=0, y=0)
(c)

Sharp increases of electromagnetic field density near the surface can be utilized to enhance effects that otherwise are very weak:

(1) Surface enhanced Raman scattering (SERS): if a Raman-sensitive substance covers a corrugated metallic substrate, FE in the vicinity of the interface layer can significantly increase Raman scattering.

(2) Nonlinear scattering and second harmonic generation in the metallic substrate or surface plasmon luminescence in the covering layer can also be enhanced, provided the density of the electromagnetic field energy increases

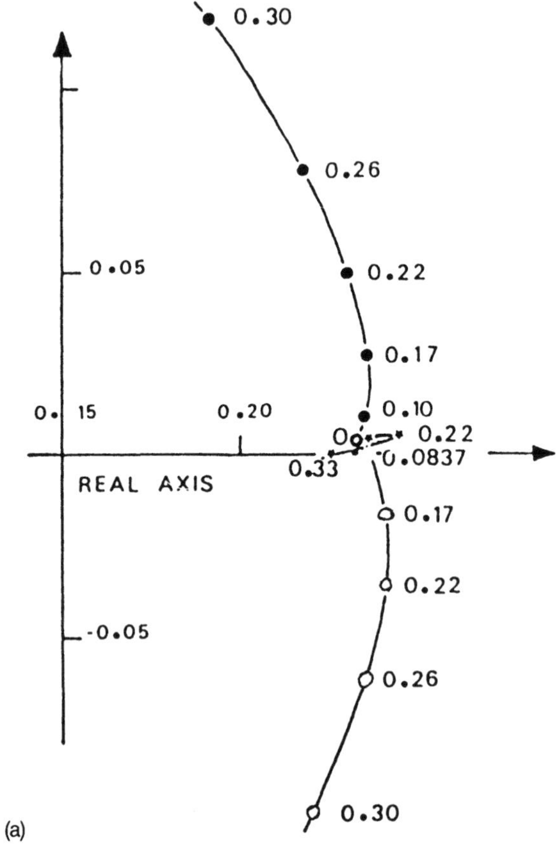

Fig.8.9 a) Trajectories of the pole $\alpha^p = k_G/k$ (black circles), -1st order zero $\alpha^z_{-1} = k^z/k$ (stars) and the specular order zero α^z_0 (crosses), when h (nm) is varied for an aluminum grating with $\lambda/d=0.8$ (after [8.19]).

sufficiently.

8.4.2 Gratings Supporting Several Orders

The main interest in diffraction gratings applications lies, of course, in utilizing their dispersive properties (i.e., use of non-specular orders). In fact, the narrow meaning of "anomaly" is connected with the existence of sharp dips or peaks especially in the -1st order efficiency. From the phenomenological point of view there is nothing unusual in this anomaly and it can easily be explained. As already discussed, resonance excitation is accompanied by zeros of diffraction order amplitudes. In general these zeros are complex and the behaviour of efficiencies is defined on the real θ_i, or α_0, or k_x axis. In the vicinity of a resonance anomaly the amplitudes of the orders are dominated by the term (8.13) and are determined mainly by the ratio of the imaginary part of the zeros and the pole and by the separation of their real parts. When the grating supports several diffraction orders, there are no simple rules guiding the behavior of poles and zeros, so that it is not easy to predict the form of the anomaly - either its depth, width, or position. As a rule of thumb, its width is proportional to the separation of the real parts of the zero and the pole, but it depends also on their imaginary parts. The smaller the imaginary part of the zero in comparison to that of the pole, the deeper is the anomaly (see again

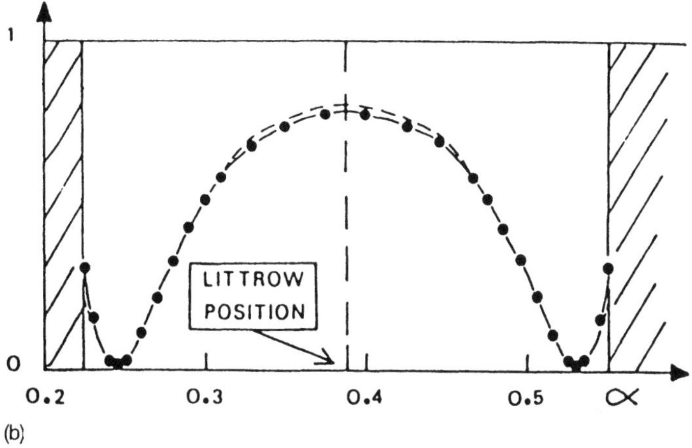

Fig.8.9 b) Comparison between the efficiency computed using rigorous theory (full line) and the reconstructed one using the phenomenological formula (8.13) (dashed line), calculated as a function of the sin of angle of incidence for h=0.227 μm (after [8.19]).

eq.8.13). The influence of the pole can be detected as a peak in efficiency, while the zero leads to a dip. If the efficiency is high the peak can hardly be observed and vice versa. Unfortunately, for highly conducting metallic gratings, the imaginary part of the -1st order amplitude α^{z}_{-1} is usually small, so that the dip in the -1st order efficiency is significant (Fig.8.9).

Many examples of resonance anomalies of reflection gratings supporting several diffraction orders can be found in Chapter 4. Over the TM spectral or angular presentation of efficiencies sharp peaks and dips are observed near the cut-off of higher diffraction orders. As already discussed, for highly conducting substrates the cut-off conditions almost coincide with the plasmon surface wave excitation conditions and that fact makes the anomalies in TM polarization much more pronounced than in the TE case. These anomalies may be annoying for grating users, but they can hardly be suppressed or removed out of the spectral region under interest. That makes the initial choice of the grating parameters quite important. For example, the best way to avoid anomalies is to choose the working point far from the cut-off anomalies, at least in TM polarization, if that is possible.

8.5 Resonance Anomalies in Dielectric Overcoated Metallic Grating

It has already been pointed out in Section 8.3 that a thin dielectric layer deposited over a metallic grating can shift the resonance anomalies in TM polarization and can cause new anomalies both in TM and in TE polarization. These new anomalies resemble waveguide modes of the optical dielectric waveguide (see Fig.8.2), but are leaky, because of the highly conducting substrate. They are Zernike-type guided waves that decrease rapidly as they propagate along the layer, so much that their propagation can hardly be observed. Nevertheless they serve as a source of poles and zeros for the diffracted order amplitudes, which show themselves as anomalies.

A detailed tracing of the poles and zeros that are due to the plasmon and leaky waves excitation, as a function of layer refractive index and thickness, is given in Chapter 5 of *Electromagnetic Theory of Gratings* [8.5]. The general conclusion to be drawn is that there is no simple *a priori* rule that guides the trajectories and, therefore the anomalies, but the grating response is always well represented and explained by the phenomenological approach, taking into account the trajectories of *both* the poles and the zeros as a function of the grating and overcoating parameters.

A representative example is supplied by Hutley *et al.* [8.20]. A typical aluminum echelette grating with 1264 lines/mm, 23° blaze and 110° apex angle when used in Littrow conditions exhibits a weak threshold anomaly in TE

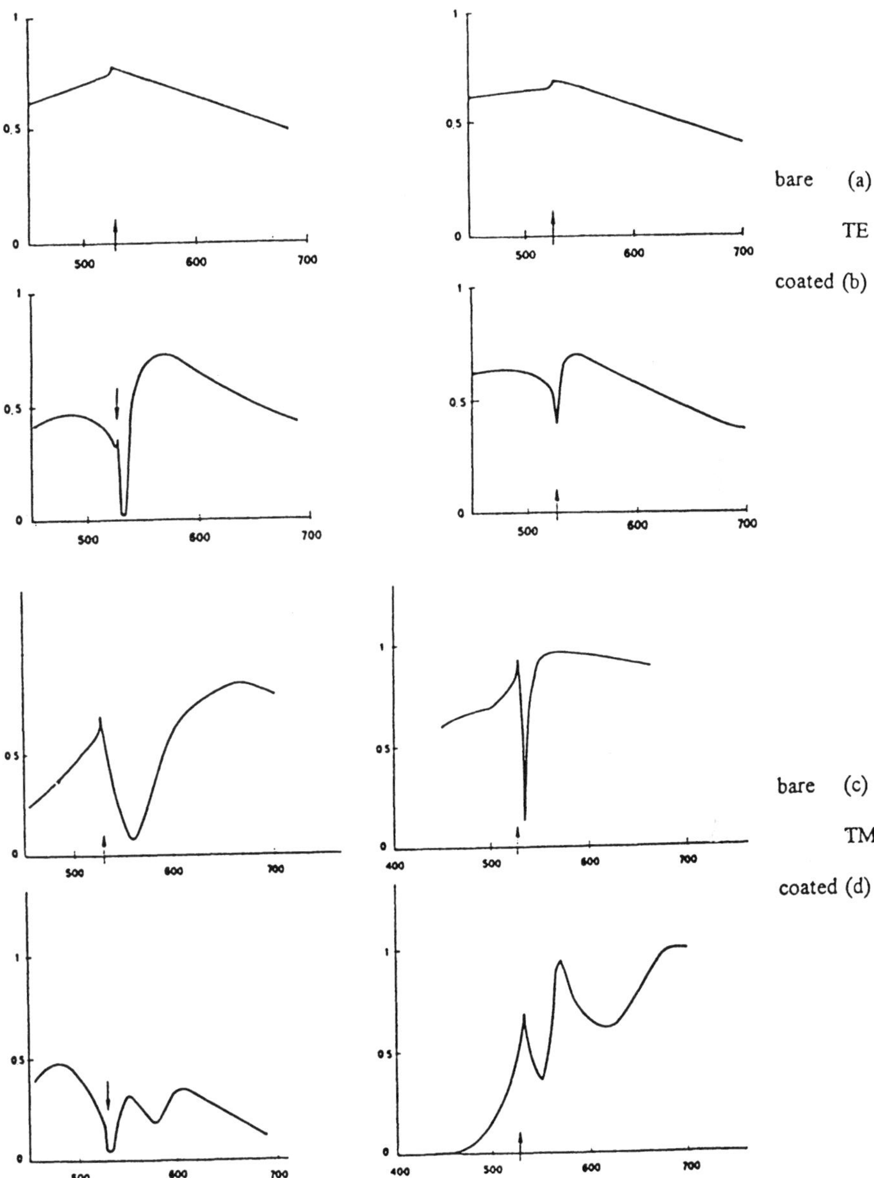

Fig.8.10 Efficiency curves for -1st order efficiency in Littrow mount as a function of wavelength in nm of a bare (a, c) and dielectric overcoated (b, d) aluminum echelette grating, used in TE (a, b) and TM (b, d) polarization. Right - theory, left - experiment (after [8.20]).

polarization at the cut-off of the -2nd and +1st orders, providing us also with a good example of a non-resonance anomaly (Fig.8.10a). Due to the surface plasmon excitation, the corresponding anomaly in TM polarization has a well-defined resonance nature, with the cut-off slightly separated from the position of very low -1st order efficiency near 560 nm, indicating a zero of its amplitude that has a small imaginary part (compare with Fig.8.9).

Deposition of a 100 nm thin dielectric layer (MgF_2) is enough to drastically change the efficiency curves. At first, it introduces a rather sharp anomaly in TE polarization so that the efficiency behavior seems more like a TM dependence. Second, the influence in TM polarization is even more dramatic. Not only has the anomaly split in two, but the longer-wavelength branch is so wide that it covers a large interval of the optical region.

The TM-like behavior in TE polarized light is due to existence of resonance phenomena inside the dielectric layer. This guided wave leads to a pole, which is accompanied by a zero of the amplitude. At specific conditions this zero can have low (or zero) imaginary part and then the amplitude becomes zero at angles of incidence, corresponding to the real part of zero. If the grating supports a single diffraction order (namely, the specular one), total absorption

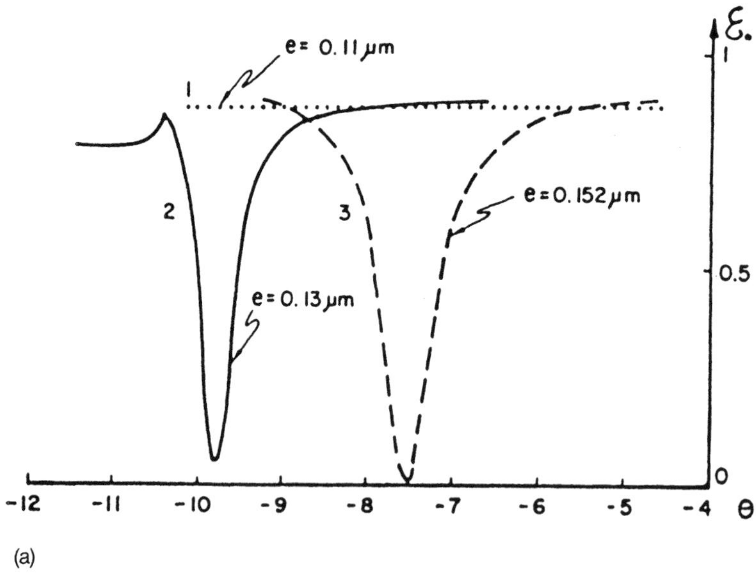

Fig.8.11 a) Efficiency in the specular reflected order for different MgF_2 layer thickness e as a function of the angle of incidence. Blazed aluminum grating with 2400 gr/mm and 10°22' blaze angle. TE polarization, λ=492 nm (after [8.21]).

of light by a dielectric coated metallic grating can be observed in TE polarization, almost identical to the corresponding effect in TM case (Fig.8.11).

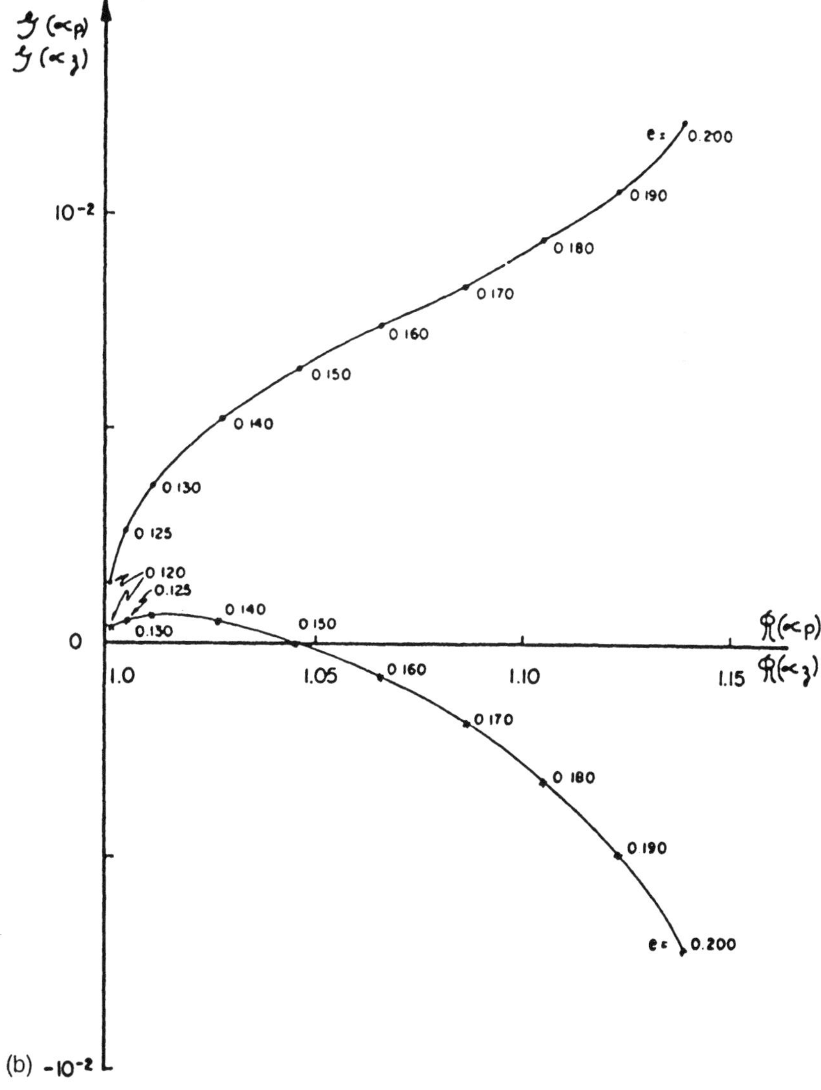

Fig.8.11 b) The trajectories of the normalized pole (upper curve) and zero (lower curve) in the complex plane when the layer thickness is varied. The zero becomes real when its trajectory crosses the real axis (after [8.21]).

8.6 Resonance Anomalies in Corrugated Dielectric Waveguides

Corrugated dielectric waveguides differ from dielectric overcoated metallic gratings in two points: (1) the reflectivity of the plane dielectric waveguide is much lower than the reflectivity of the metallic mirror, and (2) the losses of the guided wave propagating in a dielectric waveguide are much less than those of a metallic surface. These two differences determine the peculiarity of the resonant anomaly in corrugated dielectric waveguides, as shown in Fig.8.4a. A sharp peak and a dip can be observed on a relatively low non-resonant background. The theoretical value of the maximum reaches 100% if losses are negligible, when the grating profile is symmetric, and if only the specular reflected and transmitted orders propagate [8.13]. The width of the maximum is proportional to the square of the groove depth h^2, at least for shallow gratings. This immediately points to the use of the effect for narrow-band filtering. However, the resonant nature of the anomaly imposes some limitations on its usage:

(1) The higher the spectral selectivity required (i.e., the narrower the anomaly region) the more necessary it is to have better collimation of the incident beam. Divergent (or convergent) beams contain rays with different angles of incidence, which satisfy the phase matching conditions (eq.8.4) for different wavelengths and the anomaly region becomes broader with a decrease in the maximum value.

(2) The optimistic theoretical results presented in Fig.8.4a are valid for a plane incident wave, whereas in practice the beams have limited width. The narrower anomaly means weaker coupling between the incident beam and the guided wave thus longer interaction lengths are necessary and the influence of the finite beam width grows [8.5].

(3) The losses due to absorption and scatter inside the waveguide region and on the interfaces is much more important for resonance phenomena than for a non-resonance ones.

Due to these reasons experimental values of the maxima reported do not exceed 60% so far.

8.7 Multilayered Dielectric Gratings

The idea to drastically improve grating performance, or at least reduce absorption losses, is so attractive that people never give up searching for a simple solution. The natural solution is to copy the idea of dielectric coated mirrors. A stack of layers with alternating low and high optical indices turns around incident light before most of it reaches the substrate. Modern coating facilities provide dielectric layers with very small losses. Applied to gratings, this concept fails, except perhaps to enhance the performance in a very narrow spectral interval, a task better handled by simpler means. The difficulty with gratings is basic and comes from the fact that the grating is a grating: it has diffraction orders and as such, has a grating vector that can phase-match the incident wave to the numerous waveguide modes that can be supported by a multilayered dielectric stack. This never happens in dielectric mirrors, because standard polishing reduces roughness to negligible levels.

A typical example of spectral and angular characteristics of a multilayered grating is given in Fig.8.12. The grating consists of a glass substrate coated with a dielectric system $(HL)^4H$, where H denotes ZnS with index n=2.3 and L denotes cryolite (Na_3AlF_6) with n=1.35. All the layers have optical thickness $\lambda/4$, optimized to have maximum reflectivity for 632.8 nm wavelength. Whereas the corresponding flat mirror deposited under the same conditions has the properties it is supposed to have (reflectivity exceeding 98%), the corresponding grating with 1200 gr/mm comes nowhere near to achieving perfect blazing. Numerous anomalies are observed, their position

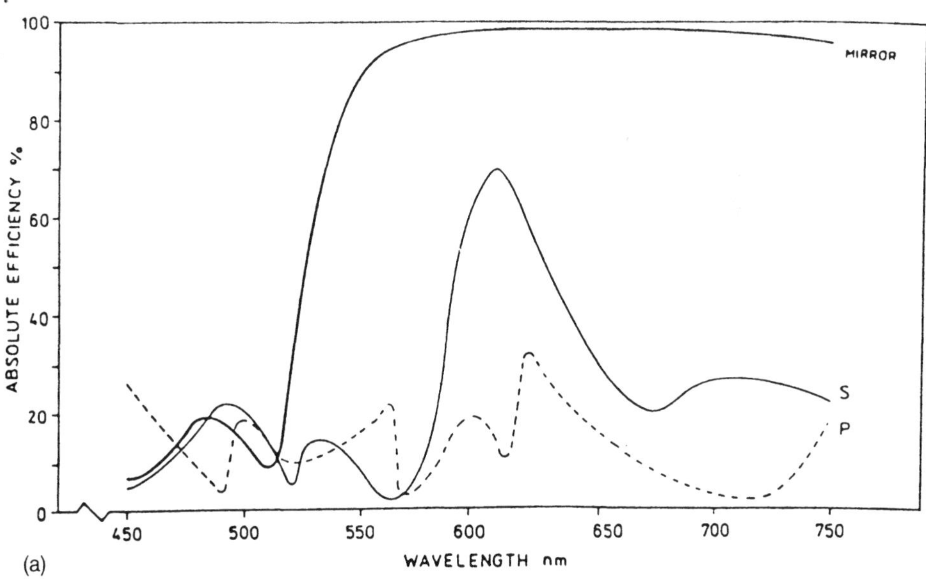

Fig.8.12 a) First order efficiency of a multilayered dielectric grating as a function of wavelength. Heavy solid line, dielectric mirror reflectivity; thin line, TM polarization; dashed line, TE polarization; border line, aluminum grating in TM polarization (after [8.22]).

well-marked in angular dependence (Fig.8.12b). All the sharp dips are due to waveguide mode excitation. And indeed, the total optical thickness of the system is more than 2λ, so that it will support four TE and three TM modes.

In the spectral dependence the anomalies are wider, and a maximum value of about 70% is observed, but only over a narrow spectral interval. If the substrate is metallic, the maximum value can be improved significantly [8.22], but again only over a narrow spectral region, surrounded by many sharp anomalies that limit the use of such gratings. (An important exception to the rule was discussed in Ch.4 in regard to the x-ray domain). There is no good reflector in the spectral region below 100 nm even less below 100Å. Fortunately, due to the naturally very low λ/d ratio in this domain the effect of

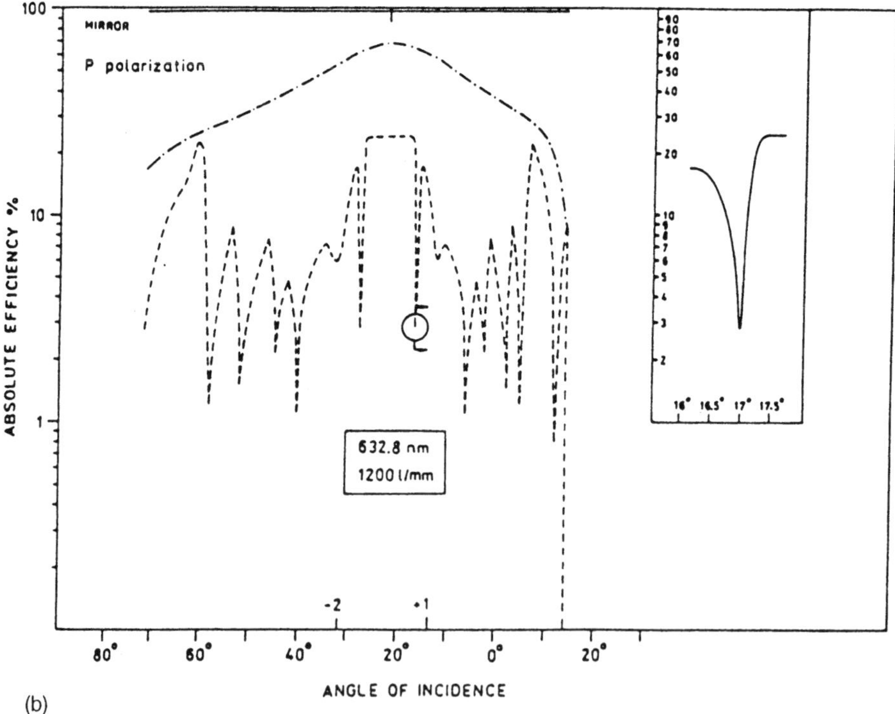

Fig.8.12 b) First order efficiency of a multilayered dielectric grating as a function of angle of incidence. Heavy solid line, dielectric mirror reflectivity; dashed line, TE polarization; border line, aluminum grating in TM polarization (after [8.22]).

guided-wave excitation is negligible and multicoated gratings are successfully used, although requiring an unusual set of materials.

The basic problem with resonance anomalies can be solved by reducing the grating period. The idea is to work in a region of λ/d where the guided waves in the multilayered system cannot be excited. The conditions are simple:

$$\sin\theta_i + \frac{\lambda}{d} > n_H \tag{8.17}$$

$$\sin\theta_i - 2\frac{\lambda}{d} < -n_H \quad , \tag{8.18}$$

where n_H is the highest refractive index of the layers. It is also necessary that the -1st order propagates, i.e.,

$$\sin\theta_i - \frac{\lambda}{d} > -1 \quad . \tag{8.19}$$

When in Littrow mount, these 3 conditions limit the useful spectral interval:

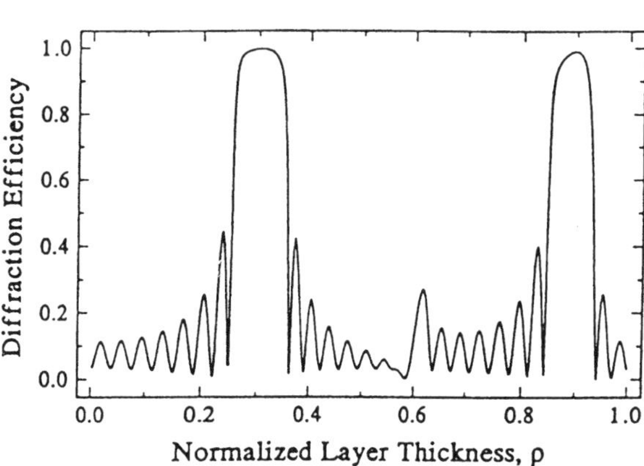

Fig.8.13 First order Littrow mount diffraction efficiency of a sinusoidal grating with 3000 gr/mm and h = 0.12 μm as a function of $\rho = \sigma/4$, eq.8.22. Wavelength 590 nm, TM polarization, substrate refractive index equal to 1.46. The reflection coating has 8 pairs of layers $H(LH)^8$ with refractive indices $n_H = 2.37$ and $n_L = 1.35$ (after [8.23]).

$$\frac{1}{3}n_H < \frac{\lambda}{2d} < 1 \quad . \tag{8.20}$$

A 3000 gr/mm grating will have an anomaly-free interval of 140 nm:

$$526.6 \text{ nm} < \lambda < 666.7 \text{ nm} \quad . \tag{8.21}$$

Lifeng Li [8.23] has investigated the optimal layer thickness. Fig.8.13 presents the efficiency in the middle of the working spectral interval, $\lambda = 590$ nm, and for groove depth h = 0.12 μm as a function of the dimensionless parameter σ, which is defined by:

$$t_{L,H} = \sigma \frac{\lambda}{4 n_{L,H}} \quad , \tag{8.22}$$

where $t_{L,H}$ are the thicknesses of the lower and higher index layers. As obvious from Fig.8.13, the highest efficiency is obtained for $\sigma = 4\rho = 1.216$.

Fig.8.14 shows that the spectral region with high efficiency is even narrower than expected from eq.(8.21), because the reflectivity of the planar system under the same incident conditions is low below 540 nm, Fig.8.15. Due to the lack of anomalies the angular dependence is much more regular (Fig.8.16) when compared with a grating of longer period (Fig.8.12b). The efficiency in TE polarization remains low for these groove depths and λ/d ratios, even for a perfectly conducting material (see Ch.4).

Due to eq.(8.20) it is possible to further expand the useful spectral region by decreasing the higher refractive index in the reflection coating, but this will require more pairs of layers to maintain high efficiency and the spectral region of high reflectivity of the corresponding mirror will be narrower.

In several limited applications the penalty of a narrow spectral interval of reflection and the expense of multilayered coatings can be accepted in view of potential virtues. This applies especially to high-power lasers where even a few percent absorption can destroy the grating (see Ch.13). Hence the desire to obtain almost perfect reflection. While theory can propose a solution, Figs.8.13 and 8.14, in practice the efficiency is always lower [8.23], probably due to profile degradation at these fine pitches. Discrepency between theory and experiment always seems to grow when groove frequency exceeds 1800 gr/mm. A different approach, which eliminates the problems with the groove profiles during the layer growth, is to build the grating on the top of a plane multylayered mirror, Fig.13.2 [8.24]. Aiming for the infrared region the period

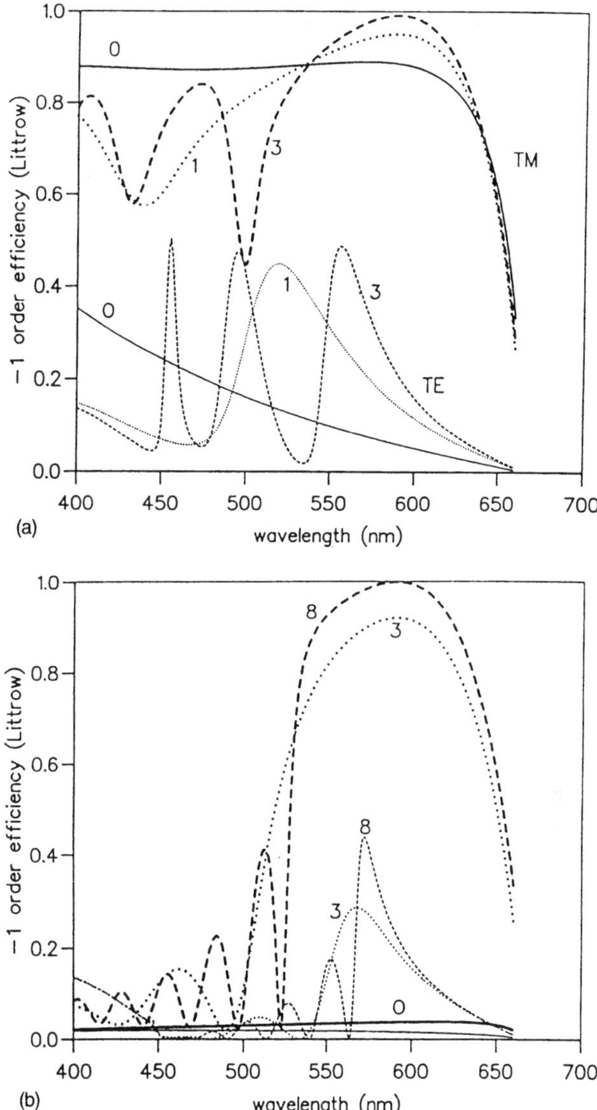

Fig.8.14 Spectral dependence of a -1st order efficiency in a Littrow mount. (a) A metallic substrate with complex index n = 1.15 + i 7.15 and a coating with a number N of pairs (LH) indicated in the figure. (b) A glass substrate coated with N pairs $H(LH)^N$. The refractive indices as in Fig.8.13. Thin lines, TE case; heavy lines, TM case.

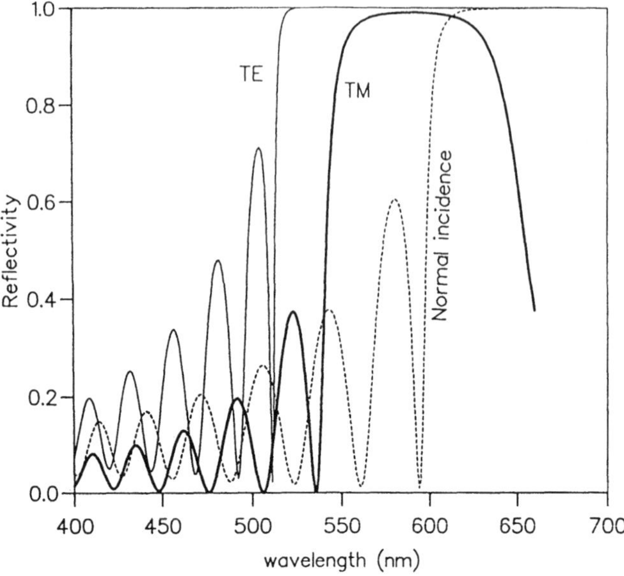

Fig.8.15 The reflectivity of a plane multilayer mirror with 8 pairs of layers and a glass substrate. Normal incidence, as indicated, and an incidence, corresponding to Fig.8.14, TE and TM case.

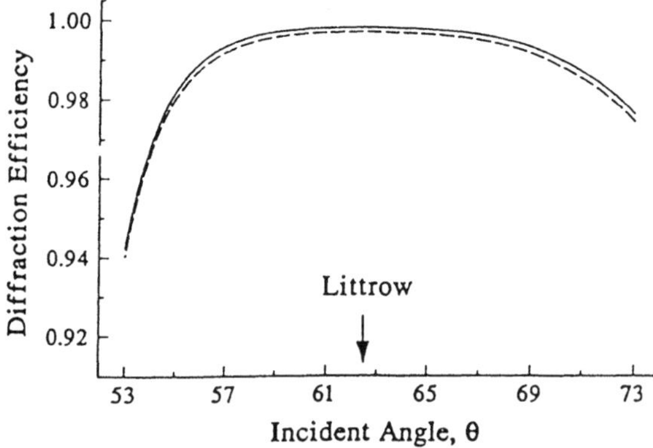

Fig.8.16 Angular dependence of the peak efficiency for a grating with parameters given in Figs.8.12 and 8.13. An Al substrate, 4 pairs of layers, dashed line; a glass substrate, 8 pairs of layers, solid line (after [8.23]).

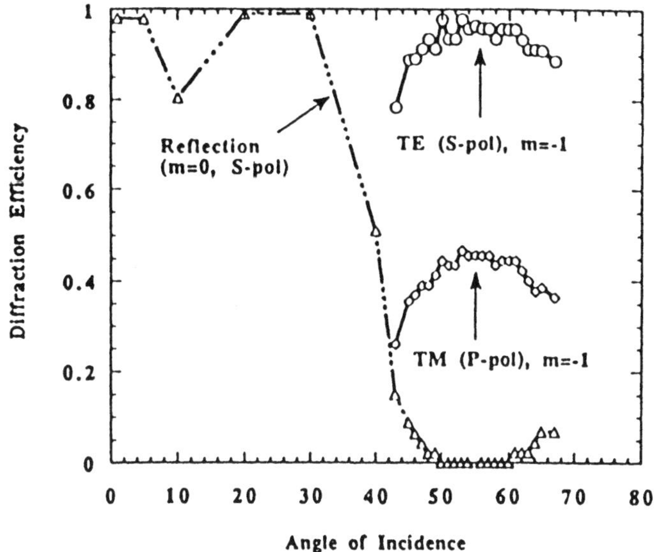

Fig.8.17 Angular dependence of diffraction efficiency in -1st order Littrow mount of a grating presented in Fig.13.2 with alternating layers of ZnS and ThF_4. Grating layer consists of ZnS with a three quarters of wavelength optical thickness and groove depth of a quarter of wavelength, 1550 gr/mm (after [8.24]).

is longer. Fig.8.17 presents the measured angular dependence of a trapezoidal grating having 1550 gr/mm, which allows for anomaly-free performance near 1.053 μm. The theoretical prediction of 98% efficiency has to be compared with the measured 96.1% peak value. The zeroth order efficiency at the -1st order maximum drops to 1% with the transmission loss of 0.5% and a net scatter of approximately 1%. Which leaves 1.4% to undetermined losses. Unmeasured scatter by itself does not cause damage, but when absorbed it can be enough to damage the structure. Unfortunately, no data on the energy damage treshold is reported in [8.24].

References

8.1 R. W. Wood: "On a remarkable case of uneven distribution of light in a diffraction grating spectrum," Phil. Mag. **4**, 396-402 (1902).

8.2 Lord Rayleigh O. M.: "Note on the remarkable case of diffraction spectra described by Prof. Wood," Phil. Mag., Series 6, **14**, 60-65 (1907).

8.3 U. Fano: "The theory of anomalous diffraction gratings and of quasi-stationary waves on metallic surfaces (Sommerfeld's waves)," J. Opt. Soc. Am. **31**, 213-222 (1941).

8.4 A. Hessel and A. A. Oliner: "A new theory of Wood's anomalies on optical gratings," Appl. Opt. **4**, 1275-1297 (1965).

8.5 M. Neviere: "The homogeneous problem," in *Electromagnetic Theory of Gratings*, R. Petit, ed. (Springer-Verlag, Berlin, 1980) ch. 5.

8.6 D. Maystre: "General study of grating anomalies from electromagnetic surface modes," in *Electromagnetic Surface Modes*, A. D. Boarman, ed. (John Wiley, 1982), ch.17.

8.7 E. Popov: "Light diffraction by relief gratings: a microscopic and microscopic view," in Progress in Optics, ed. E. Wolf (Elsevier, Amsterdam, 1993) v. **XXXI**, pp.139-187.

8.8 L. Mashev, E. Popov, and E. G. Loewen: "Total absorption of light by a sinusoidal grating near grazing incidence," Appl. Opt. **27**, 152-154 (1988).

8.9 E. Popov, L. Tsonev, and D. Maystre: "Lamellar metallic grating anomalies," Appl. Opt. **33**, 5214-5219 (1994).

8.10 L. Mashev, E. Popov, and E. G. Loewen: "Brewster effect for deep metallic gratings," Appl. Opt. **28**, 2538-2541 (1989).

8.11 D. Y. Tseng, A. Hessel, and A. A. Oliner: "Scattering by a multimode corrugated structure with application to P type Wood anomalies," U. R. S. I. Symposium on Electromagnetic Waves, Alta Freq. **38**, 82-88 (1969).

8.12 E. Popov, L. Tsonev, and D. Maystre: "Gratings - general properties of the Littrow mounting and energy flow distribution," J. Mod. Opt. **37**, 367-377 (1990).

8.13 E. Popov, L. Mashev and D. Maystre: "Theoretical study of the anomalies of coated dielectric gratings," Opt. Acta **33**, 607-619 (1986)

8.14 D. Maystre and R. Petit: "Brewster incidence for metallic gratings," Opt. Commun. **17**, 196-200 (1976).

8.15 M. C. Hutley and D. Maystre: "Total absorption of light by a diffraction grating," Opt. Commun. **19**, 431-436 (1976).

8.16 D. Maystre and M. Neviere: "Sur une méthode d'étude théorique quantitative des anomalies de Wood des réseaux de diffraction: application aux anomalies de plasmons," J. Optics (Paris) **8**, 165-174 (1977).

8.17 E. Popov and L. Tsonev: "Total absorption of light by metallic gratings and energy flow distribution," Surf. Sci. **230**, 290-294 (1990).

8.18 E. Popov and L. Tsonev: "Electromagnetic field enhancement in deep metallic gratings," Opt. Commun. **69**, 193 (1989).

8.19 D. Maystre, M. Neviere, and P. Vincent: "On a general theory of anomalies and energy absorption by diffraction gratings and their relation with surface waves," Opt. Acta **25**, 905-915 (1978).

8.20 M. C. Hutley, J. F. Verrill, R. C. McPhedran, M. Neviere, and P. Vincent: "Presentation and verification of a differential formulation for the diffraction by conducting gratings," Nouv. Rev. Optique **6**, 87-95 (1975).

8.21 E. G. Loewen and M. Neviere: "Dielectric coated gratings: a curious property," Appl. Opt. **16**, 3009-3011 (1977).

8.22 L. Mashev and E. Popov: "Diffrraction efficiency anomalies of multicoated dielectric gratings," Opt. Commun. **51**, 131-136 (1984).

8.23 L. Li: "Multilayer-coated diffraction gratings: differential method of Chandezon et al. revisited," J. Opt. Soc. Am. A **11**, 2816-2828 (1994).

8.24 M. Perry, R. D. Boyd, J. A. Britten, D. Decker, B. W. Shore, C. Shannon, and E. Shults: "High-efficiency multilayer dielectric diffraction gratings," Opt. Lett. **20**, 940-942 (1995).

Additional Reading

V. L. Brudny and R. A. Depine: "Spectral impurity in gratings due to surface plasmon scattering," Opt. Commun. **82**, 420 - 424 (1991).

J. Cowan and E. T. Arakawa: "Dispersion of surface plasmons in dielectric-metal coatings on concave diffraction gratings," Z. Physik, **235**, 97-109 (1970).

Y. J. Chen, E. S. Kotelos, R. J. Seymour, G. J. Sonek, and J. M. Ballantyne: "Surface plasmons on gratings: coupling in the minigap regions," Solid St. Commun. **46**, 95-99 (1983).

R. A. Depine, V. L. Brudny, and J. M. Simon: "Phase behavior near total absorption by a metallic grating," Opt. Lett. **12**, 143-145 (1987).

J. M. Elson, L. F. DeSandre, and J. L. Stanford: "Analysis of anomalous resonance effects in multilayer-overcoated, low-efficiency gratings," J. Opt. Soc. Am. **5**, 74 - 88 (1988).

E. N. Glytsis and T. K. Gaylord: "Antireflection surface structure: dielectric layer(s) over a high spatial-frequency surface-relief grating on a lossy substrate," Appl. Opt. **27**, 4288-4303 (1988).

G. Goubau: "Surface waves and their application to transmission lines," J. Appl. Phys. **21**, 1119-1128 (1950).

P. Halevi, and O. Mata-Méndez: "Electromagnetic modes of corrugated thin films and surfaces with a transition layer. II. Minigaps.," Phys. Rev. B **39**, 5694-5705 (1989).

A. Hessel and A. A. Oliner: "Wood's anomaly effects on gratings of large amplitude," Opt. Commun. **59**, 327-330 (1986).

M. C. Hutley: "An experimental study of the anomalies of sinusoidal diffraction gratings," Opt. Acta **20**, 607-624 (1973).

M. C. Hutley and V. M. Bird: "A detailed experimental study of the anomalies of a sinusoidal diffraction grating," Opt. Acta **20**, 771-782 (1973).

M. C. Hutley, J. F. Verrill, and R. C. McPhedran: "The effect of a dielectric layer on the diffraction anomalies of an optical grating," Opt. Commun. **11**, 207-209 (1974).

B. Kleemann and R. Guther: "Metal gratings with dielectric coating of variable thickness within a period," J. Mod. Opt. **38**, 897-910 (1991).

E. G. Loewen, W. R. McKinney, and R. C. McPhedran: "Experimental investigation of surface plasmon scatter from diffraction gratings," SPIE v.**503** *Application, Theory, and Fabrication of Periodic Structures*, 187-197 (1984).

L. Mashev and E. Popov: "Zero order anomaly of dilectric coated gratings," Opt. Commun. **55**, 377-380 (1985).

L. Mashev and E. Loewen: "Anomalies of all-dielectric multilayer coated reflection gratings as a function of groove profile: an experimental study," Appl. Opt. **27**, 31-32 (1988).

D. Maystre, J. P. Laude, P. Gacoin, D. Lepere, and J. P. Priou: "Gratings for tunable lasers: using multidielectric coatings to improve their efficiency," Appl. Opt. **19**, 3099-3102 (1980).

M. Neviere, D. Maystre, and P. Vincent: "Application du calcul des modes de propagation a l'etude theorique des anomalies des reseaux recouverts de dielectrique," J. Optics (Paris) **8**, 231-242 (1977).

I. Pockrand and H. Raether: "Surface plasma oscillations at sinusoidal silver surfaces," Appl. Opt. **16**, 1784-1786 (1977).

C. H. Palmer, Jr: "Parallel diffraction grating anomalies," J. Opt. Soc. Am. **42**, 269-276 (1952).

C. H. Palmer, Jr: "Diffraction grating anomalies. II. Coarse grating," J. Opt. Soc. Am. **46**, 50-53 (1956).

H. Raether: "Dispersion relation of surface plasmons on gold- and silver gratings," Opt. Commun. **42**, 217-222 (1982).

R. H. Ritchie, E. T. Arakawa, J. J. Cowan, and R. N. Hamm: "Surface-plasmon resonance effect in grating diffraction," Phys. Rev. Lett. **21**, 1530-1533 (1968).

A. E. Siegman and P. M. Fauchet: "Stimulated Wood's anomalies on laser-illuminated surfaces," IEEE J. Quant. Electron. QE-**22**, 1384-1402 (1986).

J. M. Simon and M. C. Simon: "Diffraction gratings: a demostration of phase behavior in Wood anomalies," Appl. Opt. **23**, 970-970 (1984).

J. E. Stewart and W. S. Gallaway: "Diffraction anomalies in grating spectrometers," Appl. Opt. **1**, 421-429 (1962).

J. Strong: "Effect of evaporated films on energy distribution in grating spectra," Phys. Rev. **49**, 291-296 (1936).

P. Tran, V. Celli, and A. A. Maradudin: "Conditions for the occurrence of k gaps for surface polaritons on gratings," Opt. Lett. **13**, 530-532 (1988).

M. Weber and D. L. Mills: "Interaction of electromagnetic waves with periodic gratings: Enhanced fields and the reflectivity," Phys. Rev. B **27**, 2698 - 2709 (1983).

R. W. Wood: "Anomalous diffraction gratings," Phys. Rev. **48**, 928-937 (1935).

Chapter 9

Waveguide, Fiber, and Acousto-Optic Gratings

9.1 Introduction

The role of microelectronics in science and common life has given birth to three fields in optics: guided wave optics, optical computing, and binary optics (or more general, photolithographic technologies). Micro-optics is a commonly used term, although with ill-defined boundaries. The influence of microelectronics was direct, although not simultaneous, in achieving device geometry in the form of planar structures, in defining new goals pointing to optical computing, and in introducing new tools using planar technology.

Planar technology, namely *photolithography*, has been so highly developed that it has led to a desire to produce all the optical elements and systems by its techniques. While applicable for elements with large characteristic lengths, diffraction drastically limits the performance when dimensions are reduced to levels close to the recording wavelengths. One solution was reinvention of holographic recording using a single grating mask, sometimes called the "*self-interference method*" or going instead to much shorter recording wavelengths, typically with electron-beams, despite their limited recording areas and resulting stitching errors (see Ch.16). While naturally tending to lamellar profiles, technology was desired to produce blazed profiles by multiple exposure, requiring a reduction of feature size, which was named *binary optics*. (It is a paradox that optics tends to limit itself from the 'bulk' to planar technologies; the only explanation is found in the great number of microchip process lines in existence.) Otherwise an experimental set-up for high-quality holographic grating recording is cheaper than even moderate-quality photolithographic equipment, unless the latter is available while the former has to be bought.

Optical computing has been the goal of a generation of scientists developing *integrated optics*. A useful argument when applying for funds, it may yet have a future in parallel processing of typically optical phenomena, like Fourier analysis, but is hardly compatible with general binary computing. In fact, optics has already taken its revenge. Most data storage makes use of optical recording. Compact disks can be found almost everywhere as a substitute for mechanical phonograph records or magnetic tapes. Magneto-optical disk access approaches that of magnetic hard disks.

Planar structures in electronics are natural, although quite undesirable. Restricted by the technology, the dream is a bulk chip or board, although a 3 to 5 layer sandwich is achievable. Using a photon of optical frequency has many advantages[1], but in most instances it is much more difficult to handle than the electron. *Integrated optics*, so promising 20 years ago, is still awaiting a major impact, although it has already resulted in several commercial devices: waveguide electro-optical modulators, in/output waveguide couplers, and DFB/DBR lasers. Recently photonic band-gap devices in guided-wave geometry emerged from the fashionable field of photonic crystals. Two main reasons limit the application of gratings to planar waveguides. The first is technological, although not basic: gratings with reasonable efficiency performance require special efforts because blazed profiles are difficult to obtain. Compared to a poor although inexpensive grating, a prism, tapered coupler or edge coupling can be much more efficient. Unfortunately, waveguide gratings inevitably have to be produced in or on the waveguide and can rarely be attached separately. The second reason is that the spectroscopic/ demultiplexing/ filtering/ sensing use of gratings calls for beam properties whose quality has to increase with resolution requirements. For example, the limited area available in planar waveguides cannot assure the degree of collimation necessary for high resolution in planar plane gratings. Use of concave planar gratings is restricted by two factors: the inherently short radius of curvature dictated by waveguide dimensions and the precision of groove direction, curvature and smoothness required. Photolithography imposes other limits on the quality of groove profiles.

The dreams of integrated optics have found a potentially wide fulfillment in *fiber optics*. Fibers possess the required collimating properties: micro-extension in two-dimensional transverse directions and macro-extension in the third or propagating dimension. Fibers, in particular the germanosilcates, can aquire index modulations, low, moderate, or high, during fiber manufacturing over lengths ranging from few microns to tens of meters. Recent applications in fiber communications use gratings for wavelength filtering (narrow or band-gap) as well as demultiplexing, Fabry-Perot fiber interferometers, mirrors for fiber lasers, and occasionally for birefringence and polarization compensation and output coupling. Fiber wavelength filters can be used for localized or distributed sensors for detecting tension, strain, or temperature, etc. Blazing properties are rarely important so that relatively simple photolithographic methods can be successfully applied, although in practice they are rarely used. More straightforward is excimer laser illumination

---

[1] Unfortunately, still mainly hypothetical, except in optical fibers.

with beam splitting (equivalent to the classical holographic recording with special care to beam coherence), or using a single beam incident upon a diffraction grating mask (similar or equivalent to the self-interference method explained in Ch.16), or by simple mechanical chirping when recording with a focused laser beam.

After the first reports in 1978 that blue-green light in fibers can lead to index changes when allowed to interfere with weak far-end reflection [9.1], it took some time to realize that Bragg conditions are not necessarily satisfied at just the writing wavelength. As in many instances of grating history, new life arose from completely different observations nine years later: frequency doubling in optical fibers found the only reasoning in 'self-organized' gratings inside the fiber when exposed to 1064 nm Nd:YAG laser light [9.2]. The true demonstration of interference grating recording was made by Meltz *et al.* just two years later [9.3], using two UV beams to form a grating perfectly designed to be used in the red, but immediately extended to the communication window around 1.5 μm by Kashyap *et al.* [9.4]. The boom of investigations, papers and, more important, many conferences and workshops that followed, gives evidence of the importance. This completely new field may expect to suffer from some of the chaos that 'new' discoveries have had in other grating applications and technologies. Although much is in flux at this writing, the field is emerging and applications sometimes precede full investigations.

9.2 Mode Coupling by Gratings

The essential property of gratings to transfer energy (i.e., provide coupling) between electromagnetic waves travelling in different directions is successfully utilized in guided wave optics for mode coupling: conversion between modes of different order, deflection and reflection of a mode, and coupling to radiated order(s) (i.e., input and output coupling). The term 'mode' has two meanings, a narrower one, naming a strictly guided wave which disappears gradually when leaving the guiding structure and a more general one, where it represents a system eigensolution that can exist without an incident wave, although not necessarily vanishing away from the guiding structure. Guided modes of a perfect fiber or waveguide belong to the narrower meaning of mode. When coupled to the radiated orders by a periodic surface corrugation or refractive index modulation these modes do not vanish far away from the waveguide. Nevertheless they are considered as modes in a general sense, corresponding to a solution of the homogeneous scattering problem. Leaky modes of non-corrugated fibers and planar waveguides are also modes in the general sense. In this chapter we shall use this more general definition.

Chapter 8 presents one of the aspects of mode coupling by gratings, namely the influence of guided wave excitation on grating performance and especially on diffraction efficiency. The other important aspect - the influence of the grating on the guided wave properties is discussed here. This influence includes both change of the guided wave characteristics (propagation constant and field distribution) and the coupling to radiated diffraction orders or to another guided wave. While coupling is the aim, the change of field distribution and the value of the propagation constant is often undesired because it leads to strong transition phenomena.

To better understand the physics of mode coupling by diffraction gratings one can use the so called *coupled wave approach* in integrated optics. It is useful as intuition, although it is never clear what exactly lies behind 'normal-mode', 'local-mode', and 'coupled-mode' approaches. In fact it is not necessary to go into such detail, because the development of numerical methods has made it easier to calculate directly the grating response and mode characteristics. Moreover, the interesting cases in practice are rarely within the limits of the approximate methods. The coupled-mode approach is capable of drawing a connection between Bragg mode coupling and photonic band-gaps without using the tools of solid state physics.

9.2.1 Coupled-Mode Approach

The essence of the coupled-mode approach is that the electromagnetic field inside the corrugated region can be adequately represented by just two components, namely the two modes to be coupled [9.5]. This is valid close to the resonance conditions as well as for weak modulation and implies that neither the field distribution nor the mode propagation constants are affected by the surface or index modulation. The modes of the planar waveguide can be represented by one of the electromagnetic field components, the transverse electric (E_z) or magnetic (H_z) component for TE or TM polarization respectively. Fiber modes can also be characterized by one field component. For simplicity, we shall discuss TE modes of a planar waveguide with a sinusoidal refractive index modulation, although the results are valid for other polarization, surface corrugation and guiding geometry (fibers), but have different field components and coupling coefficients.

The transverse electric field inside the modulated region consists of the fields of the two modes:

$$E_z(x,y) = a_1(x)\mathcal{E}_1(y)e^{i\beta_1 x} + a_2(x)\mathcal{E}_2(y)e^{i\beta_2 x} , \quad (9.1)$$

where $a_{1,2}(x)$ are mode amplitudes to be determined. They vary slowly with respect to the exponential propagation factors and with respect to the grating period. The modal fields $\mathcal{E}_{1,2}(y)$ and their propagation constants $\beta_{1,2}$ are solutions of the unperturbed homogeneous problem:

$$\left[\frac{\partial^2}{\partial x^2}+\frac{\partial^2}{\partial y^2}+k^2 n_0^2(y)\right]\left[\mathcal{E}(y)e^{i\beta x}\right]=0 \quad , \tag{9.2}$$

where n_0 is the non-modulated refractive index and k the vacuum wavenumber. The propagation constant β is equal to the x-component of the field wave-vector k_x. In general β can have an imaginary part due to absorption or radiation losses, as explained in detail in Ch.8, where a different notation (k_G) for the propagation constant is used.

A sinusoidal perturbation of the refractive index can be represented by its Fourier components:

$$n^2(x,y)=n_0^2(y)+\theta(\mathcal{M})\frac{h}{2}\left(e^{iKx}+e^{-iKx}\right) \tag{9.3}$$

with a modulation strength h. For surface relief gratings the jump of the refractive index is large and the small parameter is the groove depth h. Function θ is equal to 1 inside the modulated region $y \in \mathcal{M}$ and is 0 elsewhere. As usual K is the grating vector. Substitution of (9.1) and (9.3) into the wave equation (9.2) will result in an ordinary differential equation for the unknown amplitudes a_1 and a_2:

$$\sum_{j=1}^{2}\left(a_j''+2i\beta_j a_j'\right)\mathcal{E}_j e^{i\beta_j x}=-k^2\frac{h}{2}\left(e^{iKx}+e^{-iKx}\right)\theta(\mathcal{M})\sum_{j=1}^{2}a_j\mathcal{E}_j e^{i\beta_j x} \ . \tag{9.4}$$

Mode coupling is strong close to the resonance defined by the *phase-matching conditions*, often called *Bragg conditions*:

$$\beta_1-K=\beta_2+\Delta \ , \tag{9.5}$$

where Δ is the detuning from the resonance conditions. Eq.(9.4) is separated into two equations corresponding to the rapidly oscillating exponential terms that depend on β_1 and β_2, respectively:

$$\left| \begin{aligned} a_1' &= ic_{12}a_2e^{-i\Delta x} \\ a_2' &= ic_{21}a_1e^{i\Delta x} \end{aligned} \right. \quad , \tag{9.6}$$

where the second-order derivatives of the amplitudes are omitted because of their slow variation. The y-dependence is suppressed by an integration over the vertical co-ordinate y so that the coupling coefficients in *the coupled-mode equations* (9.6) are given by:

$$c_{ij} = k^2 \frac{h}{4\beta_i} \frac{\int_{\mathcal{M}} \mathcal{E}_j(y)\bar{\mathcal{E}}_i(y)dy}{\int_{-\infty}^{+\infty} |\mathcal{E}_i(y)|^2 dy} \quad , \quad \text{with} \begin{cases} i=1, j=2, \\ \text{or} \\ i=2, j=1 \end{cases} . \tag{9.7}$$

Solution of the coupled mode equation (9.6) is expressed in exponential form:

$$\left| \begin{array}{l} a_1(x) = \hat{a}_1 e^{(\xi - i\Delta/2)x} \\ a_2(x) = \hat{a}_2 e^{(\xi + i\Delta/2)x} \end{array} \right. . \tag{9.8}$$

Substitution of the solution (9.8) into (9.6) leads to an algebraic system which has a solution only when its determinant is zero, resulting in an equation for the exponential coefficient ξ:

$$\xi^2 = -c_{12}c_{21} - \Delta^2/4 . \tag{9.9}$$

Given the value of ξ the mode total field behavior can be easily expressed in the form:

$$\mathcal{E}(y)e^{i\tilde{\beta}x} \tag{9.10}$$

with a modified propagation constant of the incident mode equal to

$$\tilde{\beta} = \beta_1 - \frac{\Delta}{2} + i\xi \quad . \tag{9.11}$$

As shown in eq.(9.11), $\tilde{\beta}$ consists of three terms: the non-perturbated propagation constant β_1, a slowly varying oscillating term with a frequency $\Delta/2$, and a third term $i\xi$. If ξ is imaginary, the third term is added (or subtracted) to the frequency $\Delta/2$, but when real, it represents an exponential decrease of the mode amplitude.

9.2.2 Types of Mode Coupling

There are several important cases. The main one is the *contra-directional coupling* which acts as a selective waveguide mirror (fibers included). The coupling is between the same mode propagating in opposite directions:

$$\beta_2 = -\beta_1 \equiv -\beta \ . \tag{9.12}$$

Then the coefficients c_{12} and c_{21} in eq.(9.7) have opposite signs and their product is negative:

$$c_{12}\, c_{21} = -\eta^2 < 0 \tag{9.13}$$

and the coefficient ξ in eqs.(9.8) and (9.9) is real for

$$|\Delta| \leq 2\eta \ . \tag{9.14}$$

Note that Δ is the deviation from the Bragg conditions (9.5). Thus close to the resonance the amplitude of the incident mode decreases exponentially inside the modulated region with a decay coefficient ξ. Its energy is transferred to the backward propagating mode, Fig.9.1a. The amount of transmitted energy in the original direction is determined by the grating length L and the coupling coefficient ξ, the latter depending explicitly on the dephasing Δ. The rest of the incident mode energy is reflected backwards.

Co-directional couplers convert modes of a given order to modes of a different order. Then β_1 and β_2 have equal signs, although different magnitudes, and the product $c_{12}c_{21}$ is positive. In that case, independent of Δ, the coefficient ξ is imaginary, i.e., the mode amplitudes a_1 and a_2 vary sinusoidally within the modulated region (Fig.9.1b). The larger the value of Δ the faster the oscillations, thus a large Δ is outside the coupled-mode approach. The amount of energy transfered into the second mode depends on the grating length and the coefficient ξ, but contrary to the previous case, the energy which is not transferred to the mode is transmitted instead of being reflected. This type of coupling is rarely used because it implies large periods (i.e., important diffraction losses). Instead, mode conversion can be carried out using the contra-directional coupling of modes with different numbers. That is why in the following we discuss the contra-directional coupling.

It must be pointed out that the types of mode coupling neither follow from nor depend upon the coupled-mode approach and its assumptions. They are a direct consequence of the grating equation (2.2). *Reste à savoir* what happens to the grating equation when the grating region occupies only few grooves. The problem is not common to classical Integrated and Fiber Optics but emerged recently in photonic band-gap physics. It will be discussed later in this chapter.

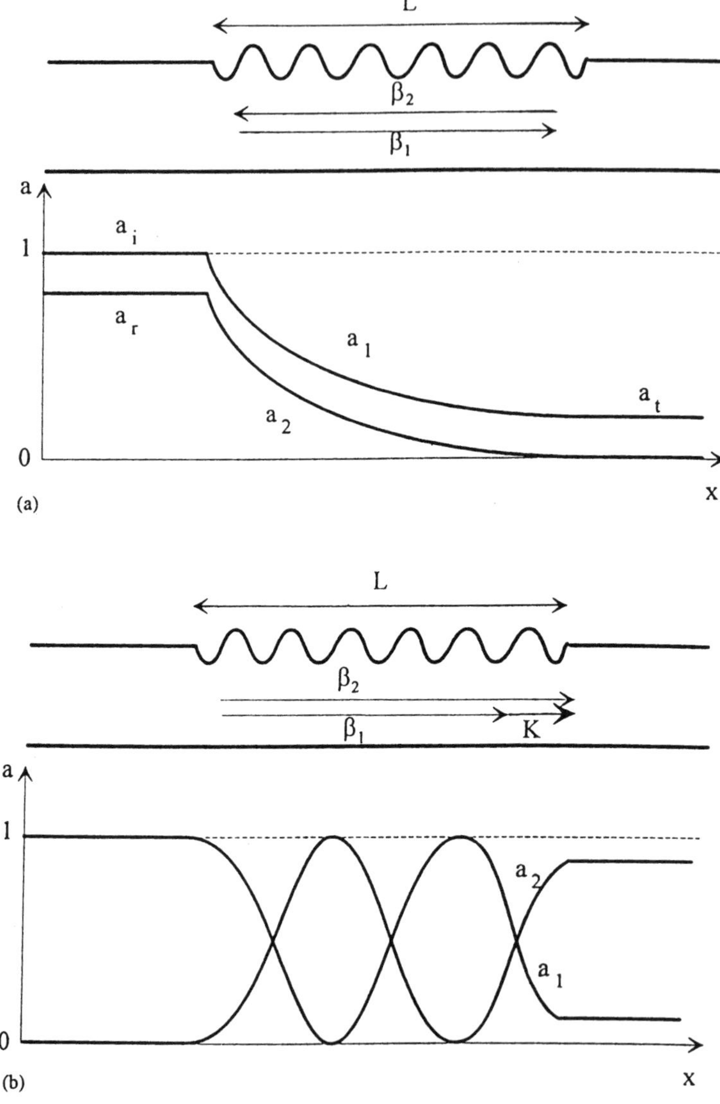

Fig.9.1 Schematic representation of contra-directional (a) and co-directional (b) mode coupling in corrugated waveguides. A mode with a propagation constant β_1 and amplitude a_1 is incident on the corrugated region from the left. Adding or substracting a grating vector K, the grating couples this mode to another mode with propagation constant β_2 and amplitude a_2. The variation of the amplitudes along the corrugation is shown at the bottom.

The coupled-mode approach seems straightforward and easy, until it is applied to a specific practical problem. It is not just the limitations and uncertainty due to modulation strength that are important. Even within the limits of validity it requires determination of mode propagation constant(s) and field overlap integrals in eq.(9.7) and their spectral dependencies. Also required are their variation when waveguide parameters are changed, plus all the other aspects. It is much easier to use rigorous theoretical methods, even though they are rarely devoted to mode coupling studies (see Chapter 10). Usually it is enough to calculate the spectral dependence of the pole of the scattering matrix, as explained in section 8.2. In the vicinity of mode coupling the imaginary part of the pole (which is nothing but one of the mode propagation constants) grows to reflect the energy transfer toward the other mode. The result is the same as by the coupled mode theory expressed in eq.(9.11). Thus it is not necessary to separately calculate the mode overlap integrals, as any rigorous theory takes them into account automatically. The problem not solved is to what extent the solution (i.e., the poles) of the infinitely long grating corresponds to the response of a finite grating. The main situation under suspicion is when the modulation is so strong that the mode is below cut-off. It then becomes necessary to consider the transition phenomena at the unperturbed-modulated regions.

Anyway, it is reassuring to note that both coupled mode theory and rigorous theories give qualitatively equal results, differing quantitatively only in the values of η and β for strong modulations.

9.2.3 Contra-Directional Coupling

Let us return to the backward mode coupling description (eqs.9.8-14). Weak modulations can be adequately described by a coupled-mode approach, whereas strong modulations require rigorous theories, giving for example the spectral dependence of the real and imaginary part of the propagation constant in eq.(9.11). In general there is a common physical behavior: Outside of the resonance (mode coupling) region, the imaginary part of β is small, at least when the absorption and diffraction losses are not too large. The latter include both scattering in the propagating diffraction orders and transition losses. Near resonance there is a region where the imaginary part of β grows rapidly due to energy transfer into the other mode. This region in the coupled-mode theory is defined by eq.(9.14). Its width depends on the mode overlapping and modulation strength (eq.9.7). Due to arguments from solid state physics this is called a '*forbidden region*', although the mode can tunnel well enough at distances inversely proportional to Δ, which can be large for weak gratings. Anyway, it is important to note that inside the 'forbidden region' the mode still exists, except that its propagation constant has a comparatively large imaginary

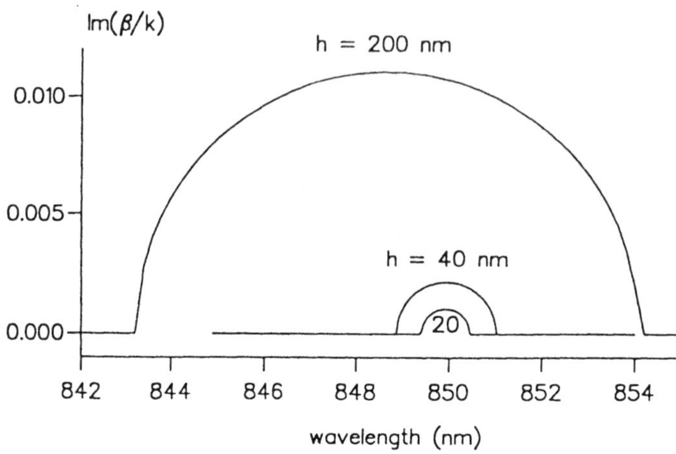

Fig.9.2 Spectral dependence of the imaginary part of the normalized mode propagation constant β/k in the vicinity of contra-directional coupling by sinusoidal grating with a full groove depth h taking three different values indicated in the figure. The waveguide region is 0.4 μm thick and has index n_g=1.7, the substrate n_s=1.5 and the cladding is air. The corrugation is at the substrate-guiding layer interface and has a period d = 0.263664 μm so that no radiation orders can propagate both in the substrate and the cladding. TE polarization.

part[1]. This is valid regardless of whether a k-gap or ω-gap is observed in the reflectivity [9.6], these gaps being connected to the behavior of the zeros of diffraction efficiencies (see Chapter 8), rather than with the pole (i.e., the mode propagation constant).

A typical example of mode propagation constant spectral dependency for weakly and moderately modulated waveguides is given in Fig.9.2 with a well-defined coupling region even for strong modulations. It is important to note that the width of this region does not depend on the grating length L (see Fig.9.1). The latter determines the total amount of the reflected energy. In fact it is the product ξL which determines the reflection, so that the length has to be compared with the characteristic length $L_c=1/\xi$. Shorter gratings lead to large transmission (Fig.9.3). Given sufficient grating length the reflection can be fully saturated inside the interaction region Δ no matter how small the

---

[1] It can be considered as a hybrid mode consisting of two coupled modes.

imaginary part of β. Thus a photonic band-gap region is formed in transmission (Fig.9.3) by longer gratings, whereas waveguide gratings with lengths shorter, or comparable with their characteristic length L_c, lead to partial mode coupling. The side lobes are due to the imaginary values of ξ which lead to a modification of the real part of the mode propagation constant, as discussed for the co-directional coupler. The saturated region for large values of L has a width exactly equal to the width of the semicircle in Fig.9.2. Shorter gratings have a well defined maximum and a half-width determined by η. The weaker the grating the narrower the resonance region but the longer the characteristic length.

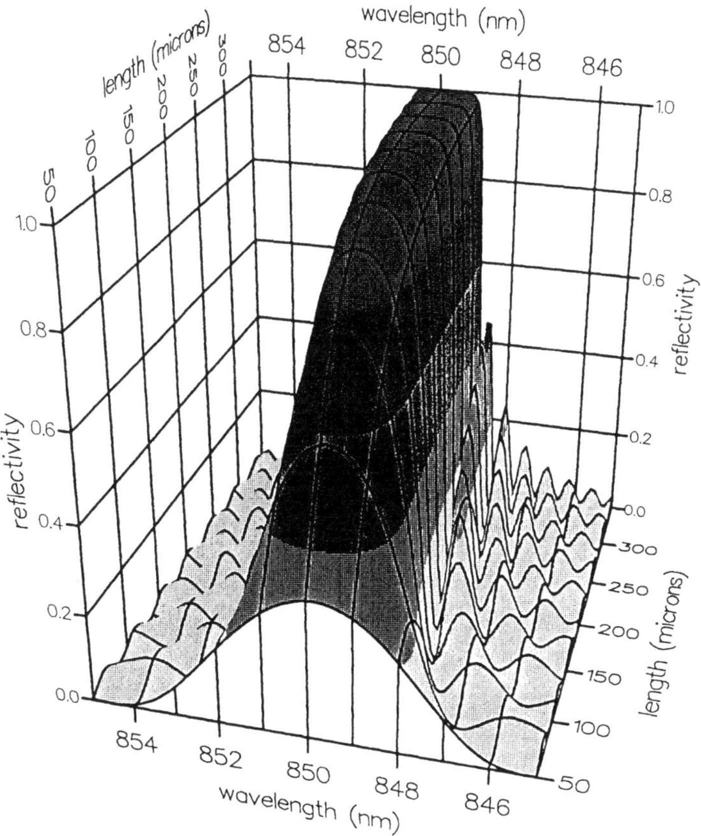

Fig.9.3 Spectral variation of the reflectivity in the back-scattered fundamental TE_0 mode of the waveguide presented in Fig.9.2 as a function of the corrugated region length. The groove depth is h = 40 nm, which corresponds to a saturated band-gap width of 2 nm.

Band-gaps in transmission do not inevitably lead to 100% reflection. Weak modulation and lack of parasitic propagating diffraction orders are necessary, but they require longer gratings and lead to narrower spectral resonances. Fortunately, they can be made and successfully used in planar waveguides and more recently in fiber gratings. A grating of sufficient length gives 100% reflectivity at the maximum and can be used as a selective planar mirror. A shorter grating will partially transmit the mode acting as a semi-transparent mirror. There are numerous possible applications, the most important being the waveguide gratings used as distributed mirrors for semiconductor waveguide lasers and wavelength filtering.

9.3 Distributed Planar Waveguide Grating Laser Mirrors

Narrow-band waveguide gratings are the natural tool for reflecting light into semiconductor lasers. The applications include both a corrugation of one of the layers inside the lasing region (DFB) or separately outside (DBR), as shown in Fig.3.2. The idea works in fibers as well, as discussed later in this chapter.

The main problems of DFB lasers are technological. Typical requirements imply shallow gratings with very short periods. When the waveguide material has an optical index close to 1.5 - 1.7, which is typical, the useful groove period is 0.2 - 0.3 μm, but refractive indices may reach 3.3 - 3.5 which would require first-order periods of 0.10 - 0.15 μm, which means that it is better to use higher order coupling. This points to deeper modulations,

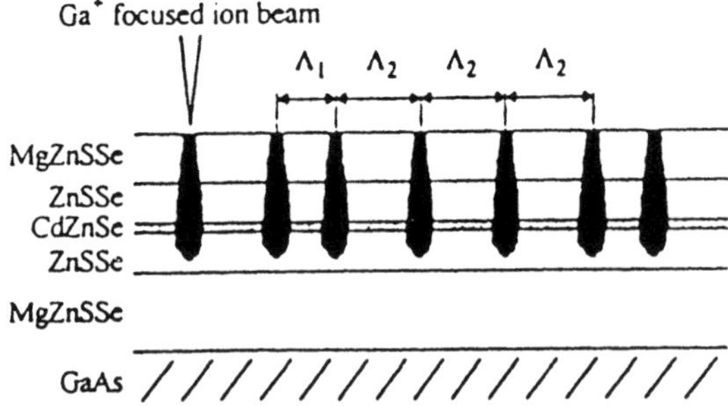

Fig.9.4 Schematic representation of gain-modulated CdZnSe/ZnSSe/MgZnSSe separate confinement laser structure. The avarage period of the distributed feedback grating is approximately 90 nm (after [9.7]).

because otherwise the coupling length becomes too large. As a result diffraction losses into lower radiated orders increase, so that blazed profiles are required to compensate. Then manufacturing is no longer compatible with the molecular beam epitaxy normally used. Such compatibility is always desirable for gratings imposed in multilayered structures and explains the desire to use classical microelectronic photolithography for the gratings as well. Classical methods of grating manufacturing cannot be used for such finc pitch gratings. The alternative is to use laser or ion beams, as discussed in Chapter 16. Two types of devices exist: DFB resonators with modulation of the index and with gain modulation. The latter is reported to allow for better control of the longitudinal mode structure. Fig.9.4 presents such a laser in ZnSe-base structures [9.7]. A focused Ga^+ ion beam with 30 nm spot size is capable of changing the defect concentration by ion implantation and therefore a periodic modulation of the optical gain is achieved with a period of about 90 nm. To provide period control with precision better than 1 nm a bi-periodic grating is used, with slightly different periods (e.g., $d_1 = 92$ nm and $d_2 = 96$ nm) obtaining an average period $d_{av.}$, which varies with the number of grooves (m_1 and m_2) contained in each subperiod: $d_{av.} = (m_1 d_1 + m_2 d_2)/(m_1 + m_2)$. With $m_{1,2}$ taking values of 1, 2, and 3, $d_{av.}$ varies between 89 and 95 nm and the laser can be tuned from 483 to 507 nm.

In turn, the separation of the grating and lasing regions (DBR lasers, Fig.3.2) provides some advantages: (1) the gratings can be longer and thus shallower and with narrower bands; grating regions several millimetres long are easily produced using either holographic or photolithographic techniques; (2) a single mode regime can be readily achieved without an artificially induced phase shift; and (3) directionally asymmetrical emission needs only two different gratings or a grating with an end-mirror, which is difficult to make for DFB lasers.

As always there is a price to be paid: low-loss waveguides for DBR are required because they lie inside the laser resonator. This demands either ion implantation [9.8] or dielectric cap disordering [9.9, 10] with a tuneable band that extends to almost 100 nm.

Both DFB and DBR semiconductor lasers suffer from simultaneous frequency hopping in the presence of even weak external optical feedback. The other principal disadvantage is the narrow region of available wavelength tuning, typically not more than 10 nm. There are proposals to increase by 10 the tuning capacity using a co-directional instead of contra-directional mode interaction [9.11]. The semiconductor sandwich includes a lasing waveguide region and a long-period grating region coupled vertically to another planar waveguide which acts as an attenuator. Its edge can be covered with a highly reflecting mirror (Fig.9.5). The composite waveguide is a multiple-mode structure. Varying electrooptically the index in the grating region changes its

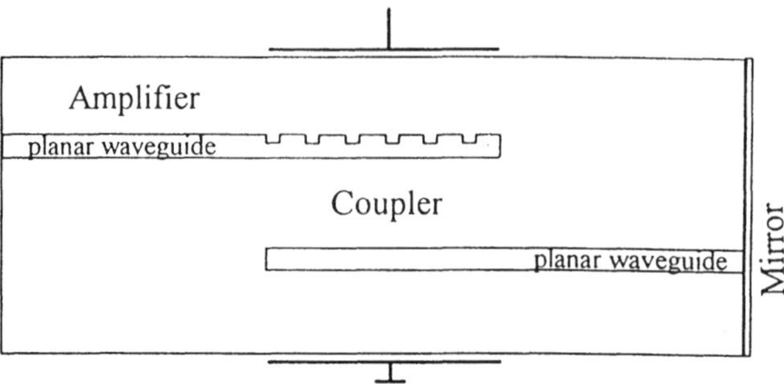

Fig.9.5 A tunable semiconducor laser based on co-directional mode coupling. The lasing wavelength varies with the applied current refractive index change which tunes the device resonance frequency.

coupling (i.e., filtering) properties. Thus wavelength selection is done by forcing the laser to operate at a selected coupled mode.

9.4 Wavelength Demultiplexing in Planar Waveguides

A classical (bulk) spectrometer can be considered a typical wavelength demultiplexer. However, the recent fiber revolution in communications has restricted the meaning of demultiplexer to a device with a fiber input and fibers/array detectors output. Field utilization imposes difficult requirements on mechanical stability and ease of handling. Optical devices are supposed to be as easily interchanged as a microelectronic chip or, at worst, as a plug-in board. Furthermore, when used in thousands, the assembly cost becomes just as important as high efficiency. The latter is typically measured in dBs, which reflects the common state of the art: it is rare to observe efficiencies greater than 50% (3 dB) as discussed further. A detailed review has become available by J. P. Laude [9.12].

For field use bulk spectrometers constructed of several separate optical elements are immediately rejected, although they are capable of high efficiency, small scattering losses, and weak aberrations (defocusing losses). Their main fault is not so much difficult manufacturing and assembling, even if it involves skilled manual operations. The problem lies in their relatively large size and especially in lack of stability and the difficulty in making operation fool-proof. The advantages of using high quality gratings, either classically ruled or holographically recorded, cannot be discarded without a loss in performance, namely efficiency, so that the most promising high-efficiency wavelength demultiplexers use waveguides as propagating media and a classical grating either attached or etched into it. Otherwise, planar waveguide gratings serve for this application mainly as an intellectual exercise or when efficiency is not important. However, low efficiency does not just mean a lower signal. Light that is not properly used always appears elsewhere. The problem is well-known to classical optical designers, but is increased by the small dimensions of waveguide demultiplexers and the consequent difficulties in shielding and baffling inside the waveguide.

These reasons are enough to deprive truly planar gratings of practical usefulness for dense wavelength demultiplexing. They can be applied safely for separating a few wavelengths, i.e., as wavelength filters (section 9.3 and 9.6). In order to efficiently separate multiple wavelengths into different channels such planar (distributed) gratings will need chirping of the period, so that different regions of the grating satisfy the phase-matching conditions (eq.9.5) for different wavelengths. More practical is to have several regions with different but constant periods, applicapble for only a limited number of channels. However, a narrow reflection band means longer characteristic lengths (see section 9.2.3), i.e., longer grating regions for high reflectivity, thus higher losses inside the waveguide and greater technological difficulties. Longer interaction regions also require very short periods to avoid diffraction losses. Another important difficulty is the resulting profile of the reflected

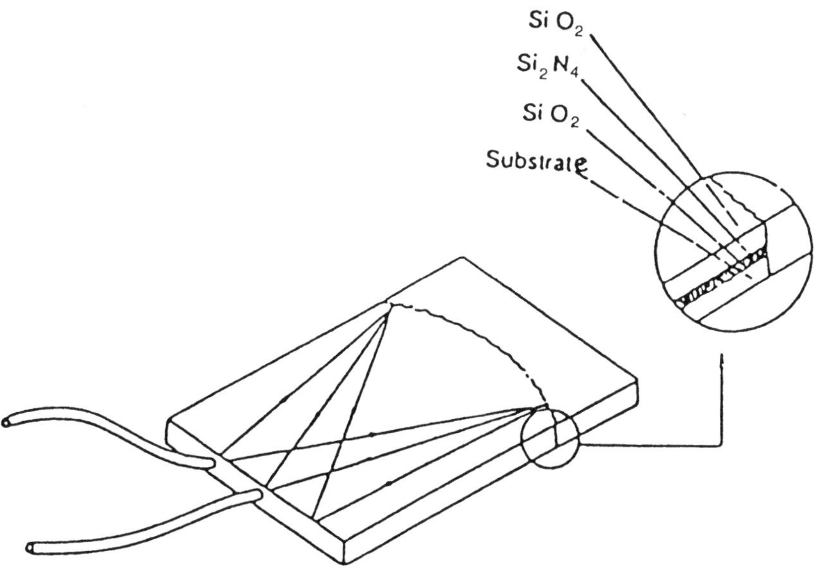

Fig.9.6 Spectrometer with a grating on a silicon substrate (after [9.12], p.92).

beams and their large transverse dimensions which require additional focusing. Thus at this writing planar-only grating demultiplexers are still mainly of academic interest.

An unusual design was recently proposed [9.13] in which the dispersion element is a transmission grating formed as microholes etched deep into the waveguide. The low efficiency (<10%) and the need for additional etched mirrors limits applications. Focusing can be made by use of geodesic lenses but the total losses of the device reported exceed 10 dB [9.14]. More practical is to use a single focusing grating instead of collimating mirrors. The grating can be either separately prepared and attached in place at the polished edge of the waveguide (Fig.9.6, [9.12]) or made in the form of an etched *niche*. The latter configuration enables aspherical grating curvature because the etching is derived from a mask produced by electron-beam lithography. Since residual polishing is impossible the cementing has to be carefully done to avoid introducing additional scatter. Another design also involves etching in the waveguide, but instead of preparing a special place, the grating itself is directly etched through a suitable mask (see Fig.9.7 [9.15]). Additional discussion on the recent design aspects can be found in section 12.9.

Unfortunately, switching to a classical grating (glued or etched) instead of a planar (distributed) does not reduce the requirements imposed on grating quality. The most difficult combination, high efficiency in non-polarized light over the entire working spectral interval requires either fine pitch deep grooves or an echelette profile (see Chapter 4). A 100% modulated grating with 1 μm

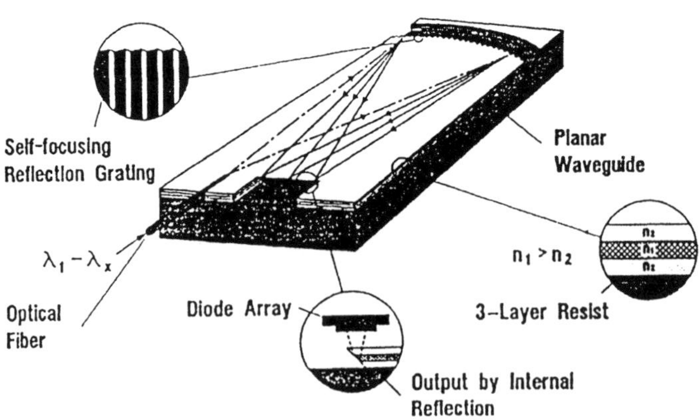

Fig.9.7 Integrated optical demultiplexer with bulk reflection concave grating. Spectral range is 400-1100 nm with resolution 7 nm and maximum efficiency of 25% (after [9.15]).

period is very difficult to replicate with a good profile form and is impossible to achieve by casting. Echelettes have problems caused by 'stitching' errors during the mask manufacturing. The etching process will always lead to rounded edges, resulting in reduced efficiency, as discussed in section 16.8.

Aberration corrected gratings prepared by other techniques can have all the virtues of the classical gratings, except for two serious problems that remain: the small characteristic dimensions need special means for aberration correction and, in addition, high efficiency requires blazing of the grooves. As a result efficiency rarely exceeds 25% [9.15].

A completely different approach uses the waveguide region only for spacial separation of different wavelengths. The waveguide is made rather thick (3.2 mm) so that it is easier to speak in terms of total internal reflection rather than about waveguide modes (Fig.9.8, [9.16]). The transmission gratings represent holograms 20μm thick. An index change of 0.02 insures diffraction efficiency of about 83% at 720 nm wavelength. Total losses of less that 3 dB are obtained within the spectral interval 700 - 740 nm with a cross-talk of -40 dB between chanels separated by 10 nm. A smaller channel separation requires more than one reflection in the substrate, increasing the device length. Channel separation of 1 nm needs a device length of 6.4 cm and propagation distance of 9 cm, which imposes stringent requirements on the mechanical and temperature stability of the system. It is better instead to increase the substrate thickness, but the preservation of a constant angle of incidence is still critical. Not known also is the influence of the incident beam divergence on the channel overlap.

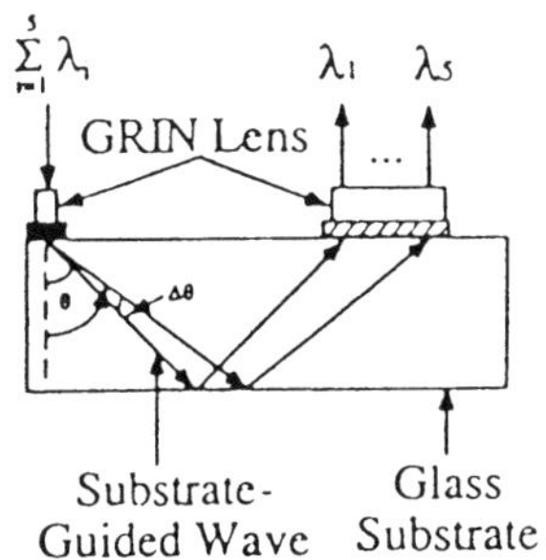

Fig.9.8 Normal incidence demultiplexer based on total internal reflection of the diffracted beams (after [9.16]).

9.5 Input/ Output Waveguide Grating Couplers

The physics of waveguide grating output coupling is simpler than the mode coupling described in section 9.2.1 because there is no feedback of radiated diffraction orders to the guided wave (Fig.9.9). The part of the guided wave energy radiated by a single groove is carried away from the waveguide and the mode 'enters' the next groove with diminished amplitude. Due to the linearity of the process the radiated part is proportional to only the mode amplitude, fixing all the other system parameters. Assuming slow change of the mode amplitude within one groove period, one immediately arrives at a single differential equation for the mode amplitude a:

$$\frac{da(x)}{dx} = -c\, a(x) \quad , \tag{9.15}$$

with the decay constant c depending on the waveguide and grating parameters, wavelength, polarization, and the number of diffraction orders which can propagate in the cladding and the substrate. The solution of eq.(9.15) is obviously an exponentially decaying amplitude:

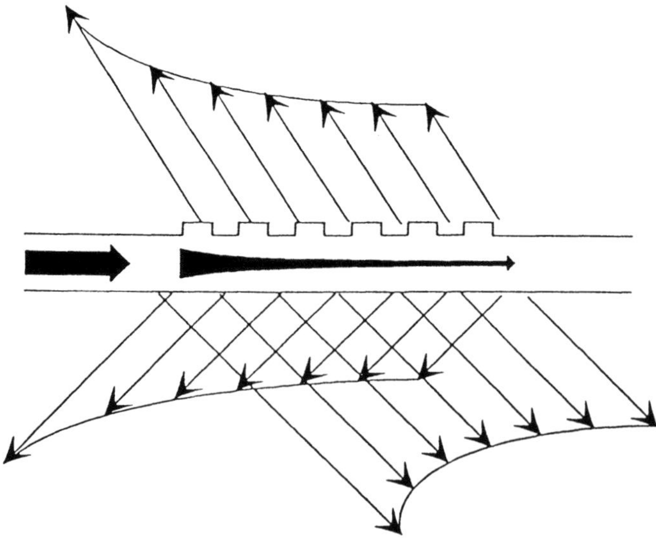

Fig.9.9 Schematic representation of output grating coupler with two orders propagating in the substrate and one in the cladding.

$$a(x) = a(0)\, e^{-cx} \quad . \tag{9.16}$$

This simple behavior can be complicated by reflectance on the substrate second surface and by transition phenomena at the interface between the corrugated (modulated) and the planar region, but the difficulty of grating coupler design and use has a different origin: There are two main problems: the exponential beam profile and the existence of multiple radiated orders. The exponential cross-sectional distribution of the radiated beams may require additional focusing. Several techniques can be used: chirping the profile can lead to focusing of the radiated beam(s) according to the grating equation (see Chapter 2) which connects the mode normalized propagation constant (effective index) β/k, the N-th diffracted order direction, the wavelength λ and the period d and the refractive index of the cladding or the substrate n:

$$n \sin\theta_N = \frac{\beta}{k} - N\frac{\lambda}{d} \quad . \tag{9.17}$$

Unfortunately, different wavelengths will be focused at different points. Beam profile can be changed during the output coupling by varying the grating strength along its length so that the coefficient c in eqs.(9.15, 16) will depend on x in such a manner as to compensate for the exponential decrease of the radiated mode amplitude, eq.(9.16). However, it is much easier to accept the penalties of the exponential beam profile than to apply complicated technologies for groove period and/or groove depth chirping.

The second and much more important problem is that of multiple radiated orders. In fact, when only one radiated order exists, the grating profile and period hardly matter: given enough grating length the entire mode energy will be radiated out. However, practice rarely permits a single order. Device designs usually require output in the cladding. As the refractive index of the substrate is higher, there is at least one order in the substrate. The second reason is technological: design can permit configurations with a single radiating order, with obligatory propagation inside the substrate, assuming it has a higher index than the cladding ($n_s > n_c$), but then the groove period must lie within the interval [9.17]:

$$\frac{\lambda}{n_s + \beta/k} < d < \frac{\lambda}{n_c + \beta/k} \quad . \tag{9.18}$$

For example, a waveguide with an effective mode index of 1.6, index of the substrate $n_s = 1.5$, and cladding index $n_c = 1$ requires grating periods in the interval $0.325\mu m < d < 0.390\mu m$ at 1 μm wavelength which is difficult but still possible to attain holographically [9.18]. Going to higher indices, for example $n_c = 3.3$ and $\beta/k = 3.4$ requires more severe limits: $0.15\mu m < d < 0.25\mu m$ at 1 μm. This region decreases linearly with the decrease of wavelength.

As with other spectral devices, the energy leakage into additional orders in grating couplers not only reduces the signal but leads to parasitic signals and increases noise. Thus the desire for effectively coupling out the light from the guided wave into a single order. Reciprocity guarantees that efficient output will more or less ensure efficient input under the same conditions. Unfortunately, having only a pair of orders (one in the cladding and one in the substrate) is not sufficient without taking special care of the groove profile. With the natural trend of holographic and photolithographic gratings to produce symmetrical grooves there is no preferable output direction. Given a symmetrical waveguide with the same cladding and substrate material, the ratio between the energy radiated in the substrate and the cladding is close to 1:1 for sinusoidal and lamellar grooves. This result is worse when the index of the substrate increases so that for a sinusoidal grating with a pair of radiating orders, the efficiency of the coupling into air is slightly above 30%.

A high quality low loss lamellar grating is reported to reach 54% efficiency (Fig.9.10) when only a single pair of orders propagate. This result is relatively high when compared to the efficiency obtained by deep blazed grooves. While theoretically the blazing of the grooves can give 100%

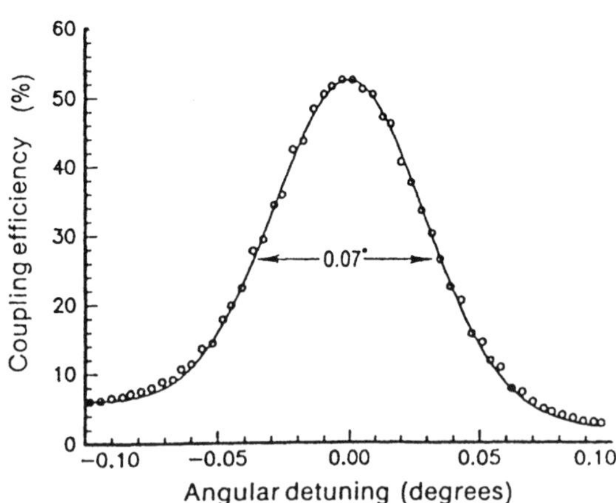

Fig.9.10 Coupling efficiency of a grating coupler consisting of a polystyrene waveguide 0.500 μm thick and a lamellar grating etched in fused-quartz. Grating period is 0.388 μm and depth 0.110 μm (after [9.18]).

efficiency in a given order [9.19], even for multiple beam output, this requires very deep grooves and a short period. Volume holograms with blazing (i.e., inclined planes) introduce unacceptable losses in the waveguide. Blazing by etching or electron-beam variable dose writing into PMMA (polymethyl-methacrylate) layers [9.20] leads to profile deformations which limit the blazing effect so that only the configurations with few radiated orders have satisfactory performance. With a stitching error of 100 nm in the electron writing beam, blazed gratings with periods between 0.8 and 2.4 μm are reported to blaze 70% into the first order in the cladding in the visible region [9.20] while only 10% of the mode energy is radiated into the substrate, where 3 orders can propagate. Fortunately, the stitching error does not have as much influence in the case of completely dielectric gratings as it appears to have in reflection.

An entirely different method for blazing is to use a pair of gratings at the lower and upper waveguide boundary. Although having symmetrical groove profile, a blazing effect is achieved through introduction of a phase shift between the gratings [9.21, 22]. Single-step manufacturing is possible when the waveguide is formed on a corrugated substrate by thermal evaporation of organic glass under a specific angle so that one obtains the required phase shift between the upper and the lower grating. The procedure requires vacuum deposition and a coupler based on a grating pair has a period of 0.347 μm and depth of 0.074 μm formed on a waveguide. With losses of 3dB/cm it reaches output blazing exceeding 98% and an input efficiency of 78%. The difference between the output and input efficiency is basic and has a simple physical background: There are three differences between output and input waveguide grating coupling. First, reciprocity requires equal transverse beam energy distribution in output and input. A similar response is expected only if the incident beam has an exponential profile, which is never the case. Moreover, its profile must vary from one coupler to another, and even within one and the same coupler, if the wavelength or angles are changed. A Gaussian beam is coupled into the waveguide with a reduction by 20% of the coupler performance, so given that other conditions are equal, the 100% output coupler will serve as an 80% input coupler. The loss is due to a re-radiation of the mode as it propagates along the corrugation *inside* the region covered by the incident beam.

The second difference is that once excited inside the coupler, the mode starts to radiate out of the waveguide. The leakage continues also *outside* the area of the incident beam. The longer the grating the larger the losses. However, the grating region (and the beam dimensions) must not be too short compared to the characteristic coupling length, which is the inverse of the coupling coefficient c in eqs.(9.15, 16). The ideal scheme of a semi-infinite grating and an exponential beam finishing exactly at the interface between the planar and the grating region has no practical meaning, so that an optimum beam width exists. A detailed analysis can be found elsewhere [9.23].

The third difference between the input and output coupling is trivial but rather annoying, although extensively used in m-line spectroscopy. Given the mode and the grating with sufficiently large period, Nature easily finds the output order direction (9.17), because these directions form a continuum. The opposite process requires coupling a discrete incident wave to a discrete mode. This is why high-efficiency input requires collimation of the beam better than the angular width of the resonance. A slight shift of the wavelength, so common with semiconductor lasers, completely alters the input conditions, which has limited the application of grating couplers as sensors, pick-up and positioning devices. Many promising and interesting configurations have been proposed and manufactured for optical pick-up and positioning [9.24], but their wide use has been limited by the lack of low cost stable laser sources. Source stabilization is too high a price to be paid in commercial devices. Compensation of the angular shift due to the wavelength drift can be achieved by using a second grating or lens (see for example Fig.9.11 from [9.18]) but this eliminates all the advantages of planar optics. It then becomes easier to use classical bulk optics [9.25].

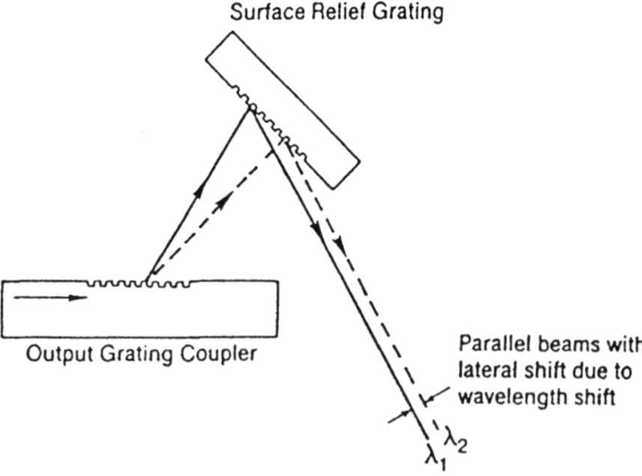

Fig.9.11 Compensation of grating coupler angular dispersion using another grating (after [9.18]).

9.6 Photonic Band-Gap in Waveguide Gratings

The main virtue of weak and moderate waveguide gratings, their strong spectral selectivity, can easily turn into a fault when wider spectral and/or angular response is required. The applications are few: a dead mirror of a tuneable waveguide laser and a photonic band-gap beam director. Wider spectral gaps in transmission need larger mode coupling coefficients η. The mode overlap integral (see eq.9.7) is maximal when identical modes are concerned, so there is a natural limit. The only 'unlimited' quantity is the modulation strength, although in practice grating studies show that there is always a limit (see Chs.4-6). Recent experiments have reached the maximum modulation for both phase and surface-relief gratings. The limit is a lamellar grating with air filling the grooves, which means a corrugation up to (or even far into) the substrate, or an index modulation as high as possible. There are

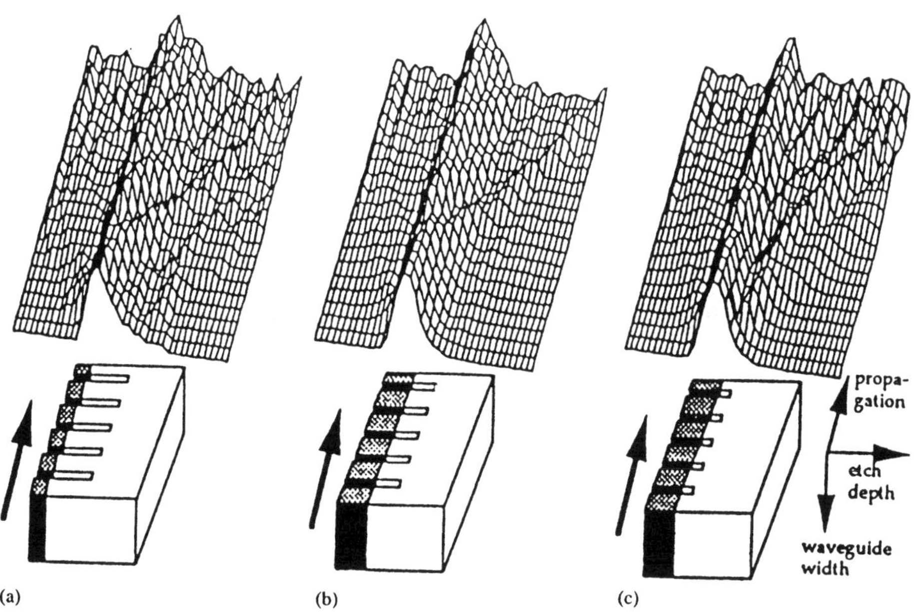

Fig.9.12 Waveguide mode transmission through a deeply corrugated region consisitng of 5 grooves with 100 nm thickness of the gap and 425 nm period. The substrate has a refractive index 3.3 and the guide 3.5. The waveguide thickness is 200 nm (a) and 400 nm (b, c) (after [9.26]).

indications that such quasi-gratings consisting of deeply etched waveguides with only few grooves (Fig.9.12) are capable of forming a band-gap in transmission 100 to 200 nm wide with out-of-band transmission as high as 40 - 70% [9.26]. Wide spectral gaps also mean weaker angular dependencies, which is important for beam directors. Etching deeply into the substrate is necessary because the modal field is stretched into the substrate due to its high index. Thus, in order to increase the coupling coefficient, the integration region $\mathcal{M}$ in eq.(9.7) must cover part of the substrate where the mode overlap is still large. However, it is not clear that the amount of light not transmitted will be reflected in the waveguide rather than lost in propagating diffraction orders and in continuous scatter due to transition phenomena at the waveguide - grating boundary. The few available experimental results give reflection in arbitrary units so that it is difficult to judge.

In order to decrease scattering losses it will be better to use waveguides with lower indices and grooves with higher indices. For example, if the waveguide index in Fig.9.12 is 1.55 and the substrate index 1.5, the grooves can have an index up to 3.3. This will maintain the high contrast of the grating and the mode field will be more strongly confined in the waveguide region.

9.7 Fiber Grating Physics

Many of the shortcomings of planar waveguide gratings which limit their wide application do not exist in fibers. The electromagnetic field is limited in two dimensions so that the main problem of planar waveguides, that of collimation, is automatically solved. Low-loss long fibers possess the required length for very narrow-band filtering. Moreover, it is not necessary to couple the fiber to a corrugated planar waveguide and back again, a process that introduces large losses. In fiber-based telecommunications it is much easier to use frequency stabilized lasers, one per entire system. Moreover, fiber gratings can be used as external resonators to stabilize the emission of otherwise drifting semiconductor lasers.

Under UV light exposure some of the Ge defects in a germanosilicate fiber core are bleached and new defects formed. The increased absorption leads to a substantial increase of the refractive index up to near-infrared, the changes reaching 10^{-4}. High-pressure low-temperature doping of the fibers with molecular hydrogen drastically increases the photosensitivity of even standard low-defect fibers by allowing almost every Ge atom to react to the UV light. Index changes of up to 0.02 are reported [9.27]. This may not seem so extreme compared to Fig.9.12 but it exceeds the index difference between the fiber core and its cladding. Fortunately, the photoinduced index change has a positive sign (refractive index increases) so that the modes stay confined in the core.

These gratings, with respectively weak and strong index changes are recently titled type I and type II fiber gratings [9.28]. Type II gratings typically have a broader reflection band. They are more lossy than type I but have much greater lifetime, especially in hostile enviroments, because the grating is formed by dosed damage of the core with a periodic structure consisting of oriented cracks [9.29].

The physics of mode coupling in fiber gratings is described directly by the coupled-mode approach (section 9.2) with the only change that the vertical coordinate y has to be replaced by the cylindrical coordinates r and φ. As in planar waveguide gratings, the practically most interesting interaction is the contra-directional coupling between one and the same mode. The coupling coefficients can be expressed as in eq. (9.7), including a two dimensional integration over the fiber cross-section. In many cases it can be assumed that the mode is more or less fully confined in the modulated region (the core) so that the two integrals (in the numerator and in the denominator) are equal when the same mode is involved. Then it is possible to write explicitly the exponential coefficient in the coupled-mode solution (9.8). The constant η in eq.(9.13) can be expressed explicitly through the mode propagation constant β and the maximum index modulation Δn_{max}:

$$\eta = \frac{\Delta n_{max}}{2\beta / k} k \ . \tag{9.19}$$

Reflection R from the grating structure is then given as a solution of the coupled-mode equations [9.5]:

$$R = \frac{\left| \xi^2 - \frac{\Delta^2}{4} \right|}{\left| \coth\left(L\sqrt{\xi^2 - \frac{\Delta^2}{4}} \right) \sqrt{\xi^2 - \frac{\Delta^2}{4}} + i\frac{\Delta}{2} \right|^2} \ , \tag{9.20}$$

where L is the grating length. ξ is a simple function of η and the detuning Δ, as given by eq.(9.9):

$$\xi^2 = \eta^2 - \frac{\Delta^2}{4} \ . \tag{9.21}$$

The reflectivity R has a complicated form, shown in Fig.9.3. At the phase matching conditions (Δ=0) the reflectivity takes a simple form:

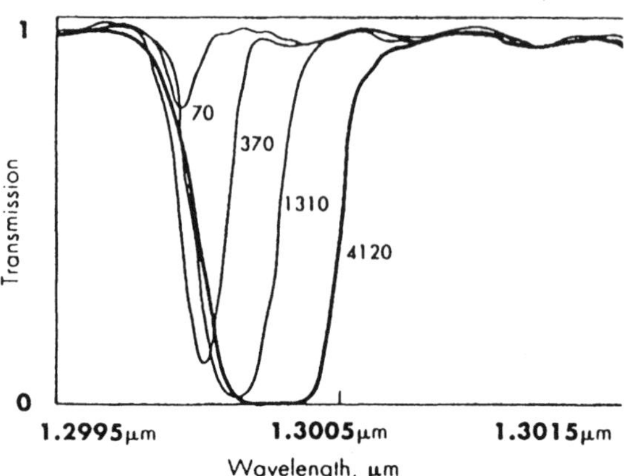

Fig.9.13 Transmission spectra of fiber gratings recorder with increasing number of pulses of 20 nsec excimer laser pulse (after [9.30]).

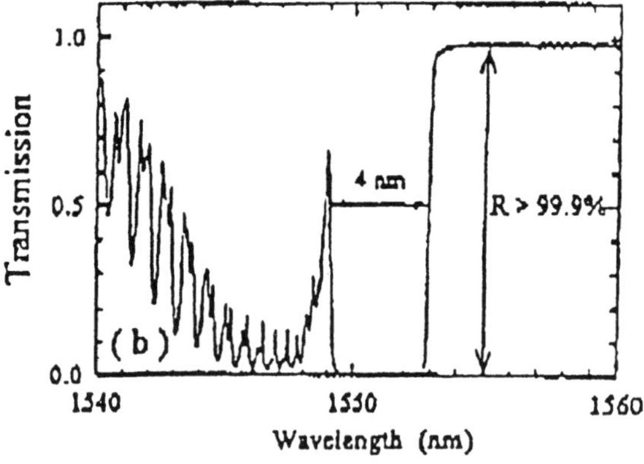

Fig.9.14 Transmission through fiber gratings patterned in a standard fiber (after [9.27]).

$$R = \left|\tanh(\eta L)\right|^2 \tag{9.22}$$

In practice, however, it is not easy to determine the photoinduced change of the refractive index Δn, so that eqs.(9.19, 22) can be used instead. Once the reflectivity is known, the coefficient η can be determined from eq.(9.22) and then Δn from eq.(9.19). The coefficient η determines the width of the band-gap region in the saturated regime, equal to 2Δ. It must be pointed out that Δ is the detuning expressed in wavenumbers and it is directly related to the spectral detuning so that the spectral width of the band-gap Δ_λ is given by:

$$\Delta_\lambda = \frac{\Delta n_{max}}{2\beta / k} n_0 \lambda \ . \tag{9.23}$$

Several different formulas can be found in the literature. They give similar results in the case when n_0 is close to β/k. The main difference comes from the mode overlap coefficient (eq.9.7) which usually lies between 0.8 and 1.

As for planar waveguide gratings (Fig.9.3) the modulation strength determines the width of the reflectivity curve, and the product ηL the maximum reflection value. The larger the photoinduced index change, the wider the minimum in transmission. Fig.9.13 presents the transmission spectra of fiber gratings formed by the standard holographic technique using exposure from 20 nsec excimer laser pulses after increasing the number of the recording pulses [9.30]. Further increase of the index up to $3x10^{-3}$ forms a well-defined stop band 4 nm wide (Fig.9.14). This width can be also obtained using eq.9.23. The losses below the gap result from light radiated into backward radiation modes.

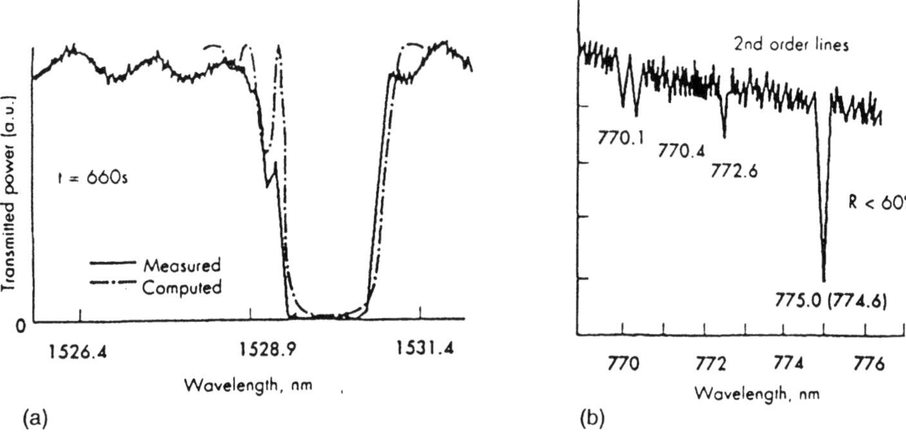

Fig.9.15 First (a) and second (b) order Bragg coupling in a 2 mm long fiber grating written in a moderately sensitive erbium and germanium doped fiber (after [9.30]).

The grating period required to couple the forward propagating mode to the same backward propagating mode is given by the grating equation:

$$2\beta / k = \frac{\lambda}{d} \ . \tag{9.24}$$

Typical values of wavelength close to 1.55 μm and a normalized propagation constant β/k close to 1.5 require periods less than 0.5 μm. Much more difficult is achieving the grating necessary to work at λ = 0.8 μm which requires periods reduced by a factor of two. It is possible to use the second-order Bragg resonance achieved through two grating vectors K but the effect is much weaker (Fig.9.15) and there is strong radiation into the cladding due to the first-order Bragg coupling to the radiation modes. The conditions for excitation of radiation modes follows immediately from the grating equation and depend on the cladding index n_c, the grating period, the wavelength and the mode propagation constant, which also depends on λ:

$$\frac{\lambda}{d} < \frac{\beta(\lambda)}{k} + n_c \ . \tag{9.25}$$

In standard fibers n_c is only slightly smaller than β/k and coupling to the radiated modes always appears at wavelengths slightly shorter than the Bragg wavelength.

The technological tools for fiber grating recording are described in Chapter 16.

9.8 Fiber Grating Lasers

Probably the most promising application of fiber gratings is in all-fiber lasers. They can be used as narrow or wider-band wavelength-selective mirrors with a variety of useful properties: pre-selected reflectivity value and working wavelength, tuneability, etc. The first fiber grating laser was made in erbium doped fiber using a single grating with 0.5% reflectivity. The other reflector was a highly reflecting mirror [9.4]. All-fiber lasers are usually pumped by semiconductor laser diodes. The main problem in fiber lasers arises from the contradictory requirements of short length and single frequency operation which leads to weak lasing powers, typically 100-300 μW and 0.1% efficiency (ratio between the emitted and the pump power). The erbium dopant concentration cannot be increased too much because this leads to self-spiking [9.31]. Fortunately, even with relatively weak gains most of the pump light is not lost, and an isolated section of an erbium fiber amplifier added to the

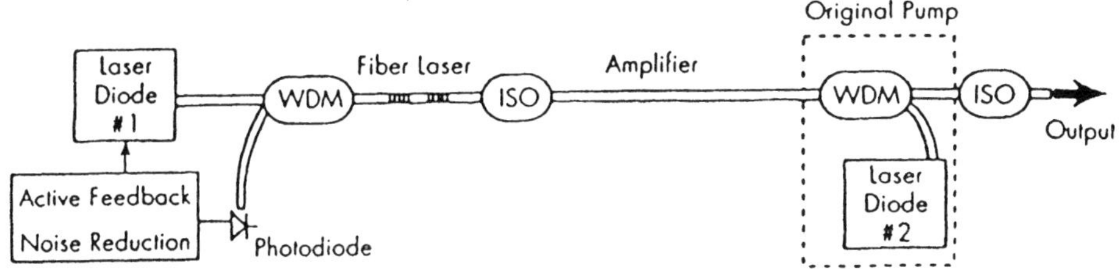

Fig.9.16 Schematic representation of a fiber laser combined with fiber amplifier. ISO indicates an optical isolator, and WDM wavelength demultiplexor (after [9.30]).

system can boost the power up to 60 mW, Fig.9.16 [9.30]. Instead of using additional amplifiers, it is possible to increase the power of fiber lasers by using $Er^{3+}:Yb^{3+}$ fibers which absorb strongly at 980 nm and can emit up to 7-8 mW in a single frequency regime with linewidth less than 2.5 kHz [9.32].

Broad band-gap fiber gratings distributed throughout the amplifying region can remove the simultaneous unwanted amplifier emission at shorter wavelengths [9.33]. Continuous wavelength tuning without mode hopping can be done by stretching or heating the fiber [9.34]. 1% strain leads to a tuning range of 10 nm, the result of change in grating period and the strain optic effect.

Mode-locked fibers lasing in 10 ps pulses with 1GHz repetition [9.35] can be combined with external electrooptic modulators and used in high-bit-rate communication systems. 2.5 Gbit/sec are reported to be transmitted over 654 km with 10^{-9} bit-error-rate [9.36]. The occasional noise burst that limits performance is due to relaxation oscillation and can be suppressed by active noise reduction - the fiber laser output is sampled out of phase to modulate the pump diode [9.37] so that a noise level of less than -140 dB/Hz is achieved, approaching that of high-power solid state lasers and DFB semiconductor lasers.

When the two gratings have slightly different periods the laser can lase simultaneously at the two corresponding Bragg wavelengths [9.38]. The output beats with a modulation frequency equal to the difference between the two resonance frequencies and can reach THz values.

9.9 Fiber Grating Filters

Fiber gratings can be used as *external selective mirrors* when combined with cw semiconductor lasers. The fiber is coupled to a semiconductor laser through a fiber lens. With an antireflection coating (Fig.9.17) the fiber grating

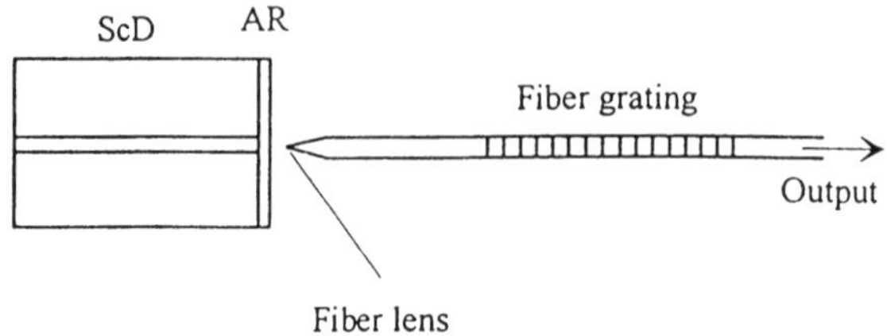

Fig.9.17 External cavity diode laser consisting of a semiconductor diode (ScD), antireflection coating (AR), fiber lens and fiber grating.

serves as a semitransparent feedback mirror for accurate determination of the lasing wavelength, set by the grating period [9.39] and tuneable by heating or stretching the fiber. If the grating is weak and there is no AR coating, it serves as a stabilizing feedback element [9.40].

The same configuration can be used to frequency control low-cost Fabry-Perot laser diodes with an upper frequency limit of 1.2 Gbits/s [9.41], determined by the round-trip length of the external cavity. A linewidth of 50 kHz can be achieved having side-mode suppression better than 30 dB.

The most direct application of fiber gratings is their use as *reflection filters*, narrow- or broad-band. The narrow-band reflection filters can have a width as small as a few pm and reflectivity can vary from 0 to 100%. There are, however, two main problems in practical applications: manufacturing and environmental stabilization. Very narrow-band reflection filters with high reflectivity require long gratings with the necessary precise control of the period and modulation strength along the entire grating length. A typical example is discussed in section 9.10.

Broad-band reflection filters can be made of strongly-modulated gratings with bandwidth up to 10 nm (see section 9.7). Several narrow-band gratings closely separated by their period can be used to form a wider gap. A similar effect is obtained by chirping the grating period along the fiber length [9.42].

Reflection filters can be combined in interferometers of different kinds to form *bandpass filters* which transmit light within a spectral interval. When two identical gratings (Fig.9.18a) reflect in phase light coming from the input port then there is no throughput. When the reflected light from the gratings is returned back to the coupler out of phase, all the signal is transmitted to the output port. Outside the reflection region, there is no reflection in either the output or the input port [9.43]. However, the path difference between the two arms has to be stabilized to within the wavelength of light, which makes the system unstable. This Michelson interferometer scheme can be improved by

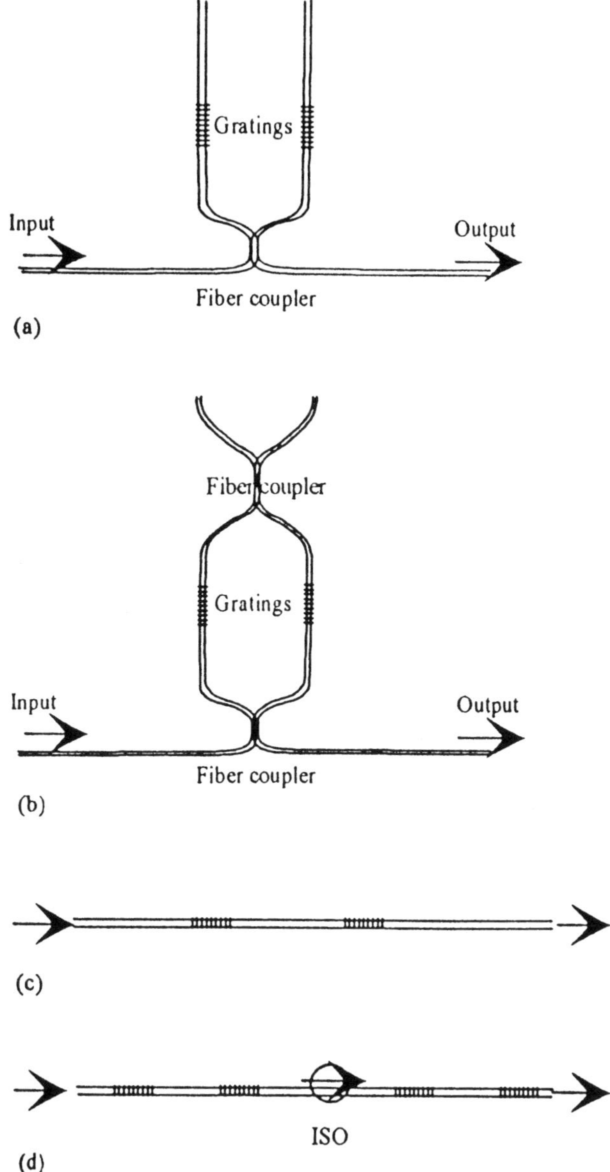

Fig.9.18 (a) Michelson, (b) Mach-Zender, (c) Fabry-Perot, and (d) Vernier interferometer schemes of fiber grating bandpass filters.

connecting the two free ports of Fig.9.18a to form a Mach-Zender interferometer (Fig.9.18b) with a switching ratio of 6:1 [9.44]. In both schemes the required phase-shift can be obtained after the gratings have been written, and the fibers coupled, by inducing an index change in one of the fibers by exposing a part of it to UV light.

The simplest method to obtain transmission band-pass filters without fiber coupling and branching is a *Fabry-Perot resonator.* By using two identical gratings in tandem separated by several millimeters (Fig.9.18c), very narrow-band filtering is obtained, combined with high peak transmission values. However, instead of a single transmitted frequency a comb-like spectrum is observed, typical of all Fabry-Perot interferometers. Its line density increases with the distance between the gratings. With a 10 cm distance between two highly reflecting gratings a transmission filter with 1.6 kHz halfwidth and 1.06 GHz free spectral range (FSR) can be made and a finesse up to 5000 is obtained ([9.30]). Increasing the FSR by reducing the length of the cavity requires a reduction of the grating length, thus increasing its strength. A FSR of 50 GHZ will require two 2 mm gratings with 2 mm distance between their centers. For a much larger free spectral range two Fabry-Perot resonators can be combined together with an optical isolator between (Fig.9.18d). The resulting *vernier filter* can have FSR many times wider than that of a single Fabry-Perot filter [9.45]. The working principle is quite simple: if the FSRs of

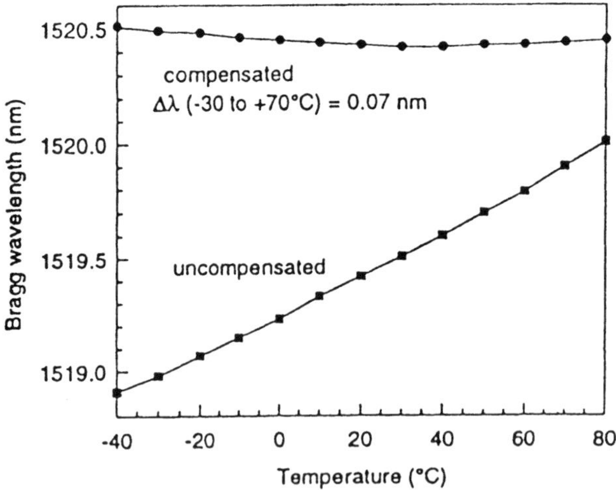

Fig.9.19 Variation of Bragg wavelength with temperature for an uncompensated fiber grating (lower curve) and a fiber grating in compensating package (after [9.47]).

the two Fabry-Perot resonators are related as (N+1)/N, and N is an integer larger than 1, the positions of the maxima in the two 'combs' of each of the resonators coincide only for each N-th maximum so that the FSR of the vernier filter is increased N times. This result is obtained by using a total of 4 gratings and an optical isolator. A FSR of 250 GHz can be obtained with an increase of the bandwidth so that the finesse maintains almost the same value as the single Fabry-Perot filter. A simpler configuration is the double-cavity Fabry-Perot resonator with only three fiber gratings and no isolator [9.46].

The main problem of fiber grating filters is the same as for all high-precision spectral devices: maintaining stability. As discussed in the next section, environmental changes cause a shift in the position of resonance. Attempts have been made to stabilize the filter response, for example by compensating the thermal expansion of the grating. The modulated region of the fiber is mounted in a holder with an initial stretch so that the expansion of the metallic holder with temperature reduces the stretch of the grating by just the amount necessary to cancel the shift in the Bragg wavelength. The compensated device response is almost independent of temperature: it presents a 0.07 nm shift over 100° when compared with 1 nm shift of the uncompensated fiber (Fig.9.19). However, such performance requires special mounts and compensates only for temperature variations.

9.10 Fiber Grating Sensors

Fiber gratings can be used for constructing sensors both in the detecting and in the probing part of the device. The simplest role in the detector part is for wavelength filtering, in that a slight change of the incoming wavelength can lead to large change in transmission intensity, often with a contrast of 100%. In the probing part the grating changes its resonant frequency with environmental conditions. *Temperature* changes modify the grating period through fiber thermal expansion as well as change in refractive index. Typically the integral shift of the Bragg wavelength is 12 pm/°C at 1.55 μm light wavelength (see lower curve in Fig.9.19). Fiber longitudinal *stretching* also influences the resonant frequency through both the expansion and contraction of the groove period and the strain optical effect. A 1% strain, almost the supportable limit, has an integral effect of about 12 nm shift in the resonance wavelength.

There are many possible applications of localized or distributed sensing of temperature and strain. To distinguish 1° temperature difference, the fiber grating filter with 5 pm spectral width and, say, 30% reflectivity has to be 22 mm long with relative error of the period not greater than 10^{-6}. This is a difficult requirement for the grating manufacturer and needs great stability of working conditions. Detection of strain is important in construction. A single fiber with several longer gratings can act as a distributed sensor of construction deformations. However, there must be some method by which one can distinguish between the effect of temperature changes and strain. A wavelength modulation technique for detection of zero-crossing of the derivative of the signal can be applied to drastically increase the sensitivity. A shift of 0.3 pm of the maximum wavelength position could be measured due to 0.3 μstrain in 30 Hz bandwidth [9.48]. By exciting two polarization modes in a birefringent cavity it is possible to avoid wavelength detection and to use electronic monitoring of the beat-tone of the signal amplitude [9.49]. Strain distribution along the fiber can be detected by a simultaneous measurement of amplitude and phase of the spectrum reflected by a grating with uniform or pre-chirped period [9.50].

9.11 Mode Conversion by Fiber Gratings

Under conditions given by eq.9.25 the fiber mode is coupled to lossy core modes or to radiated modes. It hardly makes sense to use this as an output coupler, even with slanted grooves for obtaining blazing direction. Neither input coupling can have reasonable efficiency. However, the effect is successfully used to filter out light of shorter wavelength, if necessary. As already discussed, such a broadband absorption filter (*tap*) can be used to suppress the spontaneous emission of a fiber laser amplifier. Large grating plane tilt (blazing) with respect to the core axis makes the output coupling polarization sensitive and can be used as an 'absorbing' polarization filter.

Gratings several meters long with larger periods (e.g. 500 - 600 μm) can be used for a co-directional coupling between two modes propagating in the same direction. The applications are in compensating for the group-velocity dispersion of the fiber in long-distance communications [9.42, 51].

9.12 Acousto-Optic and Electro-Optic Gratings

Although quite different in nature another type of grating can have behavior and use quite similar to that of waveguide and fiber gratings. These are the so called dynamic gratings that can be switched on and off. They are phase gratings with low modulation level of refractive index due to an applied electric field or presence of an acoustic wave. In general, there is also a periodic deformation of the surface accompanying the index modulation but it plays a minor role.

The common feature for both electro-optic and acousto-optic gratings is that periodic modulation is excited through a comb-like electrode structure. Both gratings exist in bulk and waveguide version. The electrodes can be either a bi-polar counter-directional couple or a plane electrode as a substrate and a single comb-like electrode as a cladding. While the latter structure allows for shorter periods, the counter-couple can excite bulk waves and in general requires lower voltages. While electro-optic gratings are formed directly due to the electro-optic effect, acousto-optic gratings are a secondary effect due to the propagating (or standing) acoustic waves which have been excited using an electric field by, typically the piezo-effect. Therefore the main difference is that the electro-optic grating is confined to the electrode region, while the acousto-optic grating can occupy a much larger area. This difference determines limitations in application: in the first years of *integrated optics* numerous devices were proposed ([9.52]) which used electrooptically induced gratings for high-speed modulators, optical correlation and computing. However these devices later found only limited practical applications. The reason is that with a relatively weak phase grating, effective coupling needs long coupling lengths, i.e., larger electrode regions, which imposes stronger requirements on technology and on the tolerances. In addition, larger electrodes limit the time response and therefore the upper frequency. In addition, phase matching in waveguides requires equal optical and microwave velocities by means of complicated techniques and device geometries ([9.53]). Thus electro-optical grating-based devices until recently have had limited applications except where grating use was inevitable, such as large-angle beam-deflectors [9.54] and pulse compression systems, capable of producing 2.1 ps pulses at 9.35 GHz repetition rate [9.55]. Otherwise electro-optic modulators and switches manage quite well without a grating structure [9.56], or at least without electrooptically induced gratings. However, gratings serve an important role as feedback DFB ([9.57]) or DBR ([9.58]) mirrors, combined with multiple-quantum-well structures.

Contrary to electro-optic gratings, acousto-optic gratings have found many applications [9.59]. A narrow electrode region generates through the piezo-effect an acoustic wave which can propagate along several centimeters without beam divergence [9.60]. This provides the necessary length for Bragg diffraction by an otherwise weak phase grating, so that the entire incident light is reflected into another direction. Another feature of the acoustic wave determines its potential for applications. Its speed is about 10^5 times less than the light velocity, which allows for a perfect match between the GHz region and the optical spectrum. For example, a 2GHz acoustic wave induces an index modulation grating with a period of approximately 1.8 μm, depending on the sound velocity. This grating leads to Bragg diffraction at angle θ_B, independent of whether planar or bulk geometry is used, as described by the grating equation:

$$2n_M \sin\frac{\theta_B}{2} = \frac{\lambda}{d_{a.w.}} \quad , \tag{9.26}$$

where n_M is the material index, $d_{a.w.}$ the acoustic wavelength and λ the light wavelength. In fact, eq.(9.26) describes first-order Littrow mount in transmission, and although diffraction is possible in other mountings, this regime is usually prefered. In $LiNbO_3$ for example, a longitudinal sound wave of frequency 10 GHz along its trigonal axis causes deflection of red He-Ne laser beam at about 11° [9.61]. Higher diffraction angles at lower frequencies can be reached using a paratellurite (TeO_2) because there is a direction in which the acoustic wave travels with a very low velocity (v = 615 m/s [9.62]), which increases the angle of diffraction (see eq.9.26'). A sufficiently wide acoustic wave can lower to zero the light transmitted into the specular direction when switched on (Fig.9.20) and can totally transmit light when off. A more convenient form of eq.(9.26) connects the diffraction angle with frequency f and speed v of the acoustic wave:

$$2n_M \sin\frac{\theta_B}{2} = \frac{\lambda f}{v} \tag{9.26'}$$

When the width of the acoustic wave is narrower, or its amplitude smaller, the transmitted beam is only partially diffracted. Thus its amplitude can be modulated by modulating the amplitude of the acoustic wave.

In general, in the practical frequency region below 10 GHz several (often many) diffraction orders are generated by the acoustic wave. As far as the index modulation caused by the acoustic wave is weak, higher order efficiencies are negligible when compared with the first order, but sometimes they can cause problems when allowed to interfere with the working orders. A simple method for filtering off all but the zeroth and -1st orders is to use acousto-optic gratings in a planar waveguide. The approach presented in section 9.2 is directly aplicable. In particular, the phase matching equation (9.5) can be written in a vector form:

$$\vec{\beta}_d = \vec{\beta}_i + \vec{K}_{a.w.} \tag{9.27}$$

where $\left|\vec{K}_{a.w.}\right| = 2\pi f/v$ is the acoustic wave vector and $\vec{\beta}_i$ and $\vec{\beta}_d$ the incident and diffracted light modal wave vectors. In a monomode waveguide eq.(9.27) can be satisfied for only a single direction of the deflected beam (Fig.9.20) given by:

$$\sin\theta_i \left(\equiv -\sin\theta_d \equiv \sin\frac{|\theta_B|}{2} \right) = m\frac{\lambda f}{2v\beta_G / k} \tag{9.28}$$

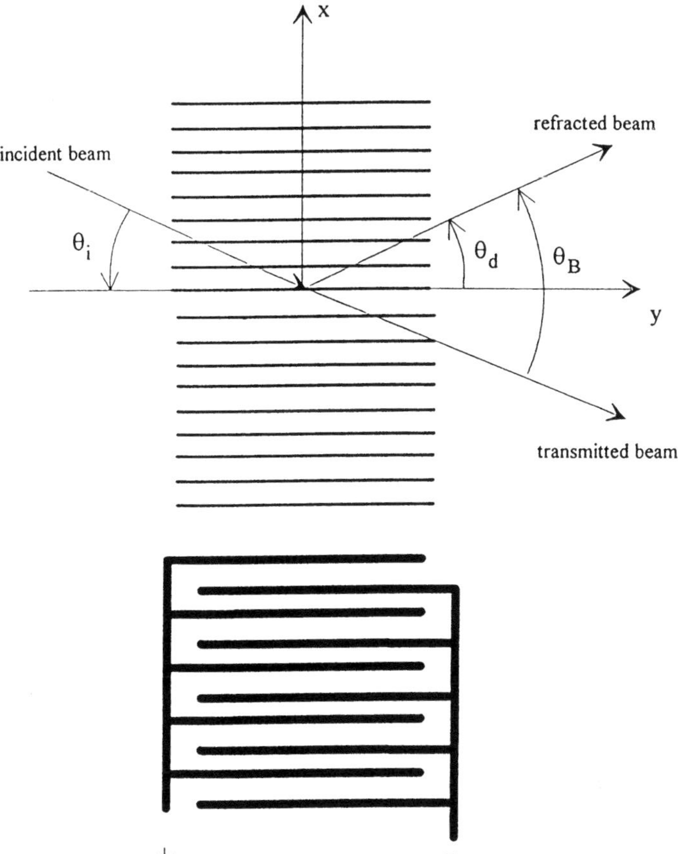

Fig.9.20 Schematic presentation of light beam interacting with an acoustic wave propagating along the x-axis.

where β_G/k is the light mode effective index. Due to the discrete existing values of β_G/k eq.(9.28) can be satisfied only for discrete angles of incidence. Then the total diffracted light is distributed among the transmitted beam and the single diffracted order, allowing for precise beam splitting, deflection and modulation.

By varying the acoustic wave frequency the deflection angle can be varied and by changing or modulating its amplitude the light beam intensity can be modulated. There are many useful applications in laser physics where acousto-optic beam deflectors and modulators are used instead of mechanical analogs. More sophisticated applications use specific grating properties to obtain more specific results. For example, as discussed in section 2.12, due to the grating dispersion pulse broadening appears so that when used in pulse delay lines, a double-pass configuration is required ([9.63]). Another application involves shifting of the frequency of a monochromatic laser used in optical gyroscopes ([9.64]). In addition to changes in the beam direction, interference between the acoustic and electromagnetic waves leads to a shift of frequency of the optical wave by an amount equal to the frequency of the acoustic wave. This allows for a frequency shift of several MHz with an absolute precision of 1 Hz. If the angle of diffraction is small, typical for lower frequencies, this can give rise to difficulties in detection due to the heterodyne effect when both the zero and the first order enter the detector. Two acoustic waves propagating under an angle to each other can be used then to increase the angle of diffraction [9.65].

References

9.1 K. O. Hill, Y. Fujii, D. C. Johnson, and B. S. Kawasaki: "Photosensitivity in optical fiber waveguides: Application to reflection filtre fabrication," Appl. Phys. Lett. **32**, 647-649 (1978).

9.2 R. H. Stolen and H. W. K. Tom: "Self-organized phase-matched harmonic generation in optical fibres," Opt. Lett. **12**, 585-587 (1987).

9.3 G. Meltz, W. W. Morey, and W. H. Glenn: "Formation of Bragg gratings in optical fibres by a transverse holographic method," Opt. Lett. **14**, 823-825 (1989).

9.4 R. Kashyap, J. R. Armitage, R. Wyatt, S. T. Davey, and D. L. Williams: "All-fibre narrowband reflection gratings at 1500 nm," Electron. Lett. **26**, 730 (1990).

9.5 H. Kogelnik: "Theory of Dielectric Waveguides," ch.2 of *Integrated Optics*, ed. T. Tamir (Springer, Berlin, 1975).

9.6 E. Popov: "Plasmon interactions in metallic gratings: ω and k minigaps and their connection with poles and zeros," Surf. Sci. **222**, 517-529 (1989).

9.7 D. Eisert, G. Bacher, N. Mais, J. P. Reithmaier, A. Forchel, B. Jobst, D. Hommel, and G. Landwehr: "First order gain and index coupled distributed feedback lasers in ZnSe-base structures with finely tunable emission wavelengths," Appl. Phys. Lett. **98**, 599-601 (1996).

9.8 J. H. Marsch, S. I. Hansen, A. C. Bryce, and R. M. De La Rue: "Applications of neutral impurity disordering in fabricating low-loss optical waveguides and integrated waveguide devices," Opt. Quant. Electron. **23**, S941-S954 (1991).

9.9 T. Miyazawa, H. Iwamura, and M. Naganuma: "Integrated external-cavity InGaAs/InP lasers using cap-annealing disordering," IEEE Photonics Technology Letters **3**, 421-423 (1991).

9.10 S. G. Ayling, J. Beauvais, and J. H. Marsch: "A distributed Bragg reflector laser using dielectric cap disordering and strontium fluoride masking," Proc. 6th European Conf. Integr. Opt., April 1993 (Neuchatel, Switzerland), pp.7_10-7–11.

9.11 S. Illek, W. Thulke, and M.-C. Aman: "Codirectional coupled twin-guide laser diode for broadband electronic wavelength tuning," Electron. Lett. **27**, 2207-2208 (1991).

9.12 J.-P. Laude: *Le Multiplexage de Longueurs d'Onde* (Masson, Paris, 1992).

9.13 G. H. B. Thompson, S. M. Ojha, and S. Clements: "Fabrication of a lowloss integrated InGaAsP/InP demultiplexer using Ch4/H2/CO2 reactive ion etching," Proc. 6th European Conf. Integr. Opt., April 1993 (Neuchatel, Switzerland), pp.2–12-2–13.

9.14 M. Takami: "Wavelength division system transmit five wavelengths by integrated optics technology," JEE (Japan), v.**23**, no. 74.7, p.99 (1986).

9.15 J. Göttert, J. Mohr, A. Müller, and C. Müller: "Fabrication of microoptical components and systems by the LIGA technique," Proc. "Procèdes et Modèles pour la Micro-Optique Passive", Georgia Tech Lorraine, Metz, April 1995, pp.31-38.

9.16 M. M. Li and R. T. Chen: "Five-channel surface-normal wavelength-division demultiplexer using substrate-guided waves in conjuction with a polymer-based Littrow hologram," Opt. Lett. **20**, 797-799 (1995).

9.17 R. Ulrich: "Efficiency of optical-grating couplers," J. Opt. Soc. Am., **63**, 1419-1431 (1973).

9.18 T. A. Strasser and M. C. Gupta: "Grating coupler dispersion compensation with a surface-relief reflection grating," Appl. Opt. **33**, 3220-3226 (1994).

9.19 K. C. Chang and T. Tamir: "Simplified approach to surface-wave scattering by blazed dielectric gratings," Appl. Opt. **19**, 282-288 (1980).

9.20 R. Waldhäusl, P. Dannberg, E. B. Kley, A. Bräuer, and W. Karthe: "Asymmetric triangular (blazed) grating couplers in planar polymer waveguides," Proc. 6th European Conf. Integr. Opt., April 1993 (Neuchatel, Switzerland), pp. 4_14-4_15.

9.21 I. A. Avrutsky, A. S. Svakhin, V. A. Sychugov, and O. Parriaux: "High-efficiency single-order waveguide grating coupler," Opt. Lett. **15**, 1446-1448 (1990).

9.22 J. C. Brazas, L. Li, and A. L. McKeon: "High-efficiency input coupling into optical waveguides using gratings with double-surface corrugation," Appl. Opt. **34**, 604-609 (1995).

9.23 M. Neviere: "The homogeneous problem," in *Electromagnetic Theory of Gratings*, R. Petit, ed. (Springer-Verlag, Berlin, 1980) ch. 5.

9.24 H. Nishihara, M. Haruna, and T. Suhara: *Optical Integrated Circuits*, (McGraw-Hill, New York, 1989).

9.25 H. Iwaoka and K. Akiyama: "A high-resolution laser scale interferometer," SPIE **v.503**, 135-139 (1984).

A. Teimel: "Technology and application of grating interferometers in high-precision measurement," *Progress in Precision Engineer.*, eds. P. Seyfried, H. Kunzmann, P. McKeown, and M. Weck (Springer, 1991), pp.15-30, Proc. 6th Intern. Precision Engineer. Seminar, Brauschweig, Germany, 1991.

9.26 T. F. Krauss and R. M. De La Rue: "Exploring the two-dimensional photonic bandgap in semiconductors," Proc. NATO ASI *Photonic Bandgap Materials*, Elounda Crete (June 1995).

T. F. Krauss and R. M. De La Rue: "Optical characterization of photonic microstructures," Opt. Photon. News, v.**6**, n.12, pp.19-20 (Dec. 1995).

9.27 T. Erdogan and V. Mizrahi: "Fiber grating technology grows," IEEE LEOS Newsletters, 14-18 (February, 1993).

9.28 J.-L. Archambault, L. Reekie, and P. St. J. Russel: "100% reflectivity Bragg reflectors produced in optical fibres by single excimer laser pulses," Electron. Lett. **29**, 453-455 (1993).

9.29 P. C. Hill, G. R. Atkins, J. Canning, G. C. Cox, and M. G. Sceats: "Writing and visualization of low-threshold type II Bragg gratings in stressed optical fibers," Appl. Opt. **33**, 7689-7694 (1995).

9.30 W. W. Morey, G. A.. Bail, and G. Meltz: "Photoinduced Bragg gratings in optical fibers," Opt. Photon. News, pp.8-14 (February 1994).

9.31 P. LeBoudec, M. Le Flohic, P. Francois, F. Sanchez, and G. Stepkan: "Self-pulsing in Er^{3+}-doped fiber laser," Opt. Quantum Electron. **25**, 359-367 (1993), "Influence of ion pairs on the dynamical behavior of Er^{3+} doped fiber lasers, " *ibid*, v. **25**, 501-507 (1993).

9.32 J. T. Kringlebotn, P. Morkel, L. Reekie, J.-L. Archambault, and D. N. Payne: "Efficient single-frequency erbium:ytterbium fiber laser," Proc. 19th European Conf. on Opt. Commun., Montreux 12-16 Sept. 1993, (Swiss Electrotechn. Assoc., Zurich, 1993), vol. **2**, paper TuC3.2, pp. 65-68.

9.33 P. St. J. Russell, J-L. Archambault, and L. Reekie: "Fiber gratings," Physics World, pp.41-46 (Oct. 1993).

9.34 G. A. Ball and W. W. Morey: "Continuously tunable single-mode erbium fibre laser," Opt. Lett. **17**, 420-422 (1992).

9.35 R. P. Davey, K. Smith, R. Kashyap, and J. R. Armitage: "Mode-locked erbium fibre laser with wavelength selection by means of fibre Bragg grating reflector," Electron. Lett. **27**, 2087-2088 (1991).

9.36 J. L. Zyskind, J. W. Sulhof, W. D. Megill, K. C. Reichman, V. Mizrahi, and D. J. DeGiovanni: "Transmission at 2.5 Gbit/s over 654 km using an erbium-doped fiber grating laser source," Electr. Lett. **29**, 1105-1107 (1993).

9.37 G. A. Ball, G. Hull-Allen, and W. Morey: "Low noise single frequency linear fiber laser," Electr. Lett. **29**, 1623-1625 (1993).

9.38 S. Chernikov, R. Kashyap, P. F. McKee and J. Taylor: "Dual frequency all fiber grating laser source," Electron. Lett. **29**, 1089-1091 (1993).

9.39 G. D. Maxwell, R. Kashyap, and B. J. Ainslie: "UV written 1.5 μm reflection filters in single mode planar silica guides," Electron. Lett. **28**, 2106 (1992).

9.40 C. R. Giles, T. Erdogan, and V. Mizrahi: "Reflection-induced changes in the optical spectra of 980-nm QW lasers," IEEE Photonics Technol. Lett. **6**, 903-906 (1994).

9.41 D. M. Bird, J. R. Armitage, R. Kashyap, R. M. Fatah, and K. H. Cameron: "Narrow line semiconductor laser using fibre grating," Electron. Lett. **27**, 1115-1116 (1991).

9.42 F. Ouellette: "All-fibre filter for efficient dispersion compensation," Opt. Lett. **16**, 303-305 (1991).

9.43 W. W. Morey: "Tunable narrow-line bandpass filter using fibre gratings," Opt. Fiber Commun. Conf., San Diego, 18-22 Febr. 1991, Technical Digest OFC'91, paper PD20-1, Conference edition (Opt. Soc. Am., Washington, 1991).

9.44 R. Kashyap, G. D. Maxwell, and B. J. Ainslie: "Laser trimmed four-port bandpass filter fabricated in simple mode photosensitive fibre," Phot. Tech. Letts. **5**, 191 (1993).

9.45 Y. H. Ja: "Optical vernier filter with fiber grating Fabry-Perot resonators," Appl. Opt. **34**, 27, 6164-6167 (1995).

9.46 A. A. Saleh and J. Stone: "Two-stage Fabry-Perot filters as demultiplexers in optical FDMA LAN's," J. Lightwave Technol. **17**, 323-330 (1989).

9.47 G. W. Yoffe, P. A. Krug, F. Ouellette, and D. A. Thorncraft: "Passive temperature-compensating package for optical fiber gratings," Appl. Opt. **34**, 6859-6861 (1995).

9.48 A. D. Kersey, T. Berkoff, and W. Morey: " Multiplexed fiber Bragg grating strain-sensor system with a fiber Fabry-Perot wavelength filter," Opt. Lett. **18**, 1370-1372 (1993).

9.49 G. A. Ball, G. Meltz, and W. W. Morey: "Polarimetric heterodyning Bragg-grating fiber-laser sensor," Opt. Lett. **18**, 1976-1978 (1993).

9.50 S. Huang, M. M. Ohn, and R. M. Measures: "Phase-based Bragg intragrating distributed strain sensor," Appl. Opt. **35**, 1135-1142 (1996).

9.51 L. S. Tamil, Y. Li, J. M. Dugan, and K. A. Prabhu: "Dispersion compensation for high bit rate fiber-optic communication using a dynamically tunable optical filter," Appl. Opt. **33**, 1697-1706 (1994).

9.52 Y.-K. Lee and R. P. Kenan: "Four-wave theory of electrooptic gratings with a simple expansion for use in design," Appl. Opt. **28**, 74-81 (1989).

9.53 J. H. Schaffner, W. B. Bridges, and A. E. Popoa: "Electro-optic modulator," U. S. Patent No. 5.005.932 (April 1991).

9.54 L. B. Aronson and L. Hesselink: "Integrated-optical switch arrays in GaAs based on electrically controlled dynamic free carrier gratings," Appl. Phys. Lett. 62, 449-451 (1993).

9.55 T. Kobayashi, H. Yao, K. Amano, Y. Fukushima, A. Morimoto, and T. Sueta: "Optical pulse compression using high frequency electro-optic phase modulation," IEEE J. Quantum Electron. **QE-24**, 382-387 (1988).

9.56 S. L. Chuang: *Physics of Optoelectronic Devices* (Wiley, New York, 1995).

H. Nagata, K. Kiuchi, S. Shimotsu, J. Ogiwara, and J. Minowa: "Estimation of direct current bias and drift of Ti:$LiNbO_3$ optical modulators," J. Appl. Phys. **76**, 1405-1408 (1994).

9.57 M. Aoki, M. Takahashi, M. Suzuki, H. Sano, K. Uomi, T. Kawano, and A. Takai: "High extinction-ratio MQW electroabsorption-modulator integrated DFB laser fabricated by in-plane bandgap energy control technique," IEEE Photonics Techn. Lett. **4**, 580-582 (1992).

K. Sato, I. Katoka, W. Wakita, Y. Kondo, and M. Yamamoto: "Strained In-GaAsP MQW electroabsorption modulator integrated DFB laser," Electron. Lett. **29**, 1087-1089 (1993).

M. Suzuki, Y. Noda, H. Tanaka, S. Akiba, Y. Kushiro, and H. Isshiki: "Monolithic integration of InGaAsP/InP distributed feedback laser and electroabsorption modulator by vapor phase epitaxy," J. Lightwave Techn. **5**, 1277-1285 (1987).

9.58 J. Maserjian, P. Andersson, B. Hancock, J. Iannelli, S. Eng, F. Grunthaner, K.-K. Law, P. Holtz, R. Simes, L. Coldren, A. Gossard, and J. Merz: "Optically addressed spacial light modulators by MBE-grown *nipi* MQW structures," Appl. Opt. **28**, 4801-4807 (1989).

9.59 E. Gordon: "A review of acoustooptical deflection and modulation devices," Appl. Opt. **10**, 1629-1639 (1966).

9.60 A. A. Oliner, editor: *Acoustic Surface Waves*, v. **24** of Topics in Applied Physics (Springer-Verlag, Berlin, 1978).

9.61 V. N. Mahajan and J. D. Gaskill: "Diffraction of light by sound waves according to the vector wave equation," J. Opt. Soc. Am. **64**, 400-401 (1974).

9.62 Y. Ohmachi, N. Uchida, and N. Nüzeki: "Acoustic wave propagation in TeO_2 crystal," J. Acoust. Soc. Am. 51, 164-167 (1972).

9.63 R. Piyaket, S. Hunte, J. Ford, and S. Esener: "Programmable ultrashort optical pulse delay using an acousto-optic deflector," Appl. Opt. **34**, 1445-1453 (1995).

9.64 R. Cahill and E. Udd: "Phase-nulling fiber-optic laser gyro," Opt. Lett. 4, 93-95 (1979).

C. Lee: "Optical-gyroscope applications of efficienc crossed-channel acoustooptic device," Appl. Phys. B **35**, 113-118 (1984).

P. Wysocki, M. Digonnet, and B. Kim: "Broad-spectrum, wavelength-swept, erbium-doped fiber laser at 1.55 μm," Opt. Lett. 15, 879-881 (1990).

9.65 M. G. Gazalet, M. Ravez, F. Haine, C. Bruneel, and E. Bridoux: "Acousto-optic low-frequency shifter," Appl. Opt. **33**, 1293-1298 (1994).

Additional Reading

G. P. Agraval: *Fiber-Optic Communication Systems* (John Wiley, New York, 1993), ch.7.3.

G. A. Ball and W. Glenn: "Design of a single-mode linear-cavity erbium fiber laser utilizing Bragg reflectors," J. Lightwave Technol. **10**, 1338-1343 (1992).

G. A. Ball and W. W. Morey: "Continuously tunable single-mode erbium fibre laser," Opt. Lett. **17**, 420-422 (1992).

T. Erdogan and J. E. Sipe: "Tilted fiber phase gratings," J. Opt. Soc. Am. A **13**, 296-313 (1996).

M. Hatori: "Light beam deflector," U.S. Patent No. 5,048,936 (1991).

C.-X. Shi: "Fabri-Perot resonator composed of a photoinduced birefringent fiber grating," Appl. Opt. **33**, 7002-7008 (1994).

See also the reference lists of Chapters 12 and 16.

Chapter 10

Review of Electromagnetic Theories of Grating Efficiencies

10.1 Introduction

It is well known that diffraction of light by corrugated structures is a complex problem that requires sophisticated tools for its solution[1]. Analytical solutions, at least for metallic gratings, are so restricted that even in the simplest cases they are of little interest. Approaches to rigorous solutions, largely developed in the two decades since 1970, generally go through the following steps:

1. Formulation of the physical problem
2. Mathematical interpretation
3. Choice of the model
4. Algorithm for solution
5. Numerical (or, rarely, analytical) procedure for finding the solution
6. Verification of results.

In each of the steps some assumptions have to be made. Listed below are the formulations most commonly used in the different steps, together with the corresponding assumptions. Not mentioned in the above simple scheme is the procedure of Error Searching, which can easily take as much time and efforts as all the other steps combined.

10.2 The Physical Problem

Most generally, the physical problem of light scattering by a grating can be defined as follows: An electromagnetic wave is incident on a corrugated surface. Due to the corrugation periodicity, the scattered field is concentrated

---

[1] However, there seem to be two categories of scientists who insist on being able to solve the grating problem easily. The first type resembles inventors of perpetual motion: they claim that just another (always the next) approximation to *their* theory will be enough to determine all the details of grating efficiency behavior. The second type calls their theory "*Maxwell solver*", that is not restrticted to simple gratings, but involves 3D rough surfaces, extended non-periodic objects, etc. Whether they are right, only the future can tell.

into specific directions, called diffraction orders (see Chapter 2). The aim of the theory is to determine the distribution of the diffracted field in each of the orders (i.e., diffraction efficiency), as well as the near field distribution (in special cases), although a "good" theory must, by definition be able to determine the connection between the far and near field. Sometimes the incident field can be quite specific, for example in the case of surface wave scattering. There are several natural assumptions:

(1) The incident field is considered monochromatic. This assumption works well for media with linear response, because a beam with spectral structure is diffracted as a linear superposition of its monochromatic components, which is not the case for non-linear media. However, the device(s) within which the grating is incorporated, or which serve to measure the efficiencies, may have specific spectral functions that must be taken into account. The monochromatic assumption is also connected (at the upper limit of precision), with the limit size of the grating and finite aperture of the other optical elements, with the surface roughness scattering, etc.

(2) The corrugated region is considered to be infinitely long and perfectly periodic. This assumption greatly simplifies the next steps of solving the problem, and is not a handicap in practice. Some recently developed theories of scattering omit this assumption, but the resulting complications are severe, without adding important insight, except in the study of the effect of surface roughness on grating behavior.

(3) Plane wave assumption. It resembles the monochromatic assumption in its consequences, limitations and applications. As a rule of thumb, if the beam width exceeds many wavelengths, its finite width can be neglected. In cases of high resolution, the collimation of the beam becomes as significant as its monochromaticity.

(4) Regularity of the grating media: with few exceptions it is assumed that the optical properties of the grating surface are known. This may lead to some slight uncertainty as to whether the thin aluminum coating (or not so thin optically), backed up with a resin in replication, can be considered identical with one deposited on glass (or the bulk properties) from which the complex index values are determined. Deviations from nominal behavior show up mainly in regions with anomalies, which are especially sensitive to them.

(5) The next assumption is to take the case of classical diffraction, with the plane of incidence perpendicular to the grating grooves (see Chapter 2). The more general case of conical diffraction does not introduce fundamental difficulties, but complicates the mathematical and numerical treatment and also is not widely used in practice. Any theory that can deal successfully with the two fundamental cases of polarization can be generalized, if necessary, to deal with the conical case.

(6) It is usually assumed (with just a few exceptions) that the grating surface and the surrounding media are isotropic, and will not change the state of polarization. Then the two fundamental cases can be treated independently within the assumption (5).

To summarize, the most generally accepted formulation of the physical problem is that a monochromatic plane, linearly polarized wave is incident from a lossless (upper) media, usually air), onto an infinitely extended corrugated perfectly periodic structure (linear, homogeneous, and isotropic) with known optical constants, the direction of incidence being perpendicular to the grooves. The solution is supposed to fulfill the following conditions:

1. Maxwell's equation everywhere - in the cladding and the substrate, and inside the corrugated region.

2. Boundary conditions along each of the boundaries involved in the scattering.

3. Radiation conditions: the assumption is that only one incident wave carries energy towards the grating. All the diffraction orders must be either evanescent or propagating away from the corrugation structure (Chapter 2).

These three conditions have a straightforward mathematical formulation: in almost all the methods the field outside the corrugated region (above the line that connects the top of the groove and below the lines connecting their bottom) is represented as a sum of diffraction orders: propagating and evanescent, and the incident wave (in the upper media):

$$\begin{pmatrix} E_z \\ \text{or} \\ H_z \end{pmatrix} = \sum_m b_m \, e^{ik_{m_x}x + ik_{m_y}y} + a_i \, e^{ik_{i_x}x - ik_{i_y}} \quad . \tag{10.1}$$

The horizontal components of the wavevector of the different components in eq.(10.1) (usually called diffraction orders) are connected through the grating equation (Chapter 2). The link between them and the vertical components is given by the Maxwell's equations:

$$k_x^2 + k_y^2 = \left(\frac{2\pi}{\lambda} n \right)^2 \quad . \tag{10.2}$$

In fact, the representation (10.1), where the x and y-components are connected by (10.2), is not unique. The grating periodicity and the radiating conditions require that at infinity ($y \rightarrow \pm \infty$) the solution of the diffraction problem can be represented as a sum of finite (propagating) orders, whereas at a finite distance from the corrugated surface evanescent orders can be detected (see Chapter

2.2). Nevertheless the form of eqs.(10.1) and (10.2) is the most natural one, as far as it directly matches radiation conditions and Maxwell's equations.

Differences between theoretical models begin with the third step: the choice of the model, usually the model of the field representation in the vicinity (and inside) the corrugated region. There are two basic choices: rigorous and approximate. The latter make an approximation at this step, while rigorous methods make approximations only at the numerical solution stage. For example, all the methods based on the Rayleigh hypothesis assume that plane wave expansion of the field is valid over the whole space, and in particular inside the grooves [i.e., an assumption is introduced in the field representation (choice of the model)]. While this hypothesis is valid for relatively shallow and smooth profiles, it is not rigorous for deep grooves, or for profiles with sharp edges.

In another type of approximation, often used in waveguide theory, the geometrical corrugation of the surface is transferred into a periodicity of the optical properties of the media (i.e., the periodicity is moved from the boundary conditions into Maxwell's equations, see section 9.2.1). While simple in formulation, and in its mathematical treatment, the disadvantage of approximate methods is that their results are valid only over a limited region. Furthermore, it is never possible to tell *a priori* whether in a specific case they are valid or not.

With rigorous methods no approximations are made in the choice of the model and in the setting up the solution algorithm. In the subsequent numerical treatment some approximations cannot be avoided, for example the finiteness of the computer word, which leads to truncation and round-off errors and consequently to the loss of precision. Secondly, there is always a limitation based on memory and calculation time, which requires truncation of the field expansion series. However, this hardware truncation limit becomes less and less of concern with each new processor type. Fortunately, the results of such limitations can be easily checked and controlled and often a change from double to quadruple precision can at least double the regions of application[1].

The only disadvantage of the rigorous methods is their greater complexity, although that varies significantly from one method to another. As a rule of thumb, the more complex the theory, the better the performance of the resulting code. For example, the integral method can deal with almost any grating problem, a development that took twenty years. The classical differential method, after a similar period, also approaches "the ideal", by incorporating some recent numerical techniques. Simpler methods are also useful. While being more restricted, they are in general more efficient and can

[1] Of course, the other limitation could be the lack of programming skill or a bad choice of the numerical method, two subjective disadvantages that never appear separately.

be used on low-performance personal computers. It must be pointed out that these restrictions are known *a priori* and are not due to some approximations, but these simpler methods are more specialized. For example, the modal method is the most efficient approach for lamellar gratings but is not so suitable for other profiles.

A detailed review of the different theoretical methods and their mathematical treatment may be found elsewhere [10.1]. In this text we present the basic principles of the different methods, and discuss the advantages and the difficulties of their available numerical implementations. We will concentrate on the most commonly used methods, since they cover the widest range of grating parameters. This can serve only as a guide to draw attention and in no case should be considered as a complete presentation of the theory and be used for direct programming. From the various sets of approximate methods we have chosen the ones based on the Rayleigh hypothesis and the scalar theory, because they sound more "electromagnetic", have a well-defined approximation basis and their disadvantages and region of application are more or less known.

It is difficult to make a full classification of the rigorous methods. All of them start from Maxwell's equations in partial derivatives. In general, there are differential methods that integrate these equations numerically in one or two dimensions. Today, most of the differential methods use one-dimensional integration or other numerical methods of solving a system of ordinary differential equations. The integral method represents Maxwell's equations as integro-differential equations and solves them numerically. There are a variety of other methods, that can be classified as integral or differential, but we should try to avoid such terminology. We keep the terms "differential" and "integral" for the methods, classically known under these names, so as to avoid being a source of confusion to the inexperienced (or, even to the experienced) reader. Anyway, it is useful to draw a boundary line between the integral and differential methods: Integral methods normally use as unknowns the components of the field (and its normal derivative) on the grating surface and evaluate the field outside the corrugated region using the integral Green theorem, thus leading to integral equation(s). The differential methods start the integration from some surface outside the corrugated region and make numerical (or analytical when possible) integration over the grating region. This usually implies crossing the profile which can bring huge numerical difficulties in TM case. Conformal mapping, the method of Chandezon and the classical modal method, although considered as being differential, do not cross the profile (or, rather, use a different basis at the two sides of the grating surface), but they can deal only with limited types of profiles.

Fig.10.1 presents a systematization of recently used rigorous methods. Hatched regions draw the connection between common physical and mathematical backgrounds, although numerical implementations can differ significantly. The proper choice of basic functions and coordinate system, for example, can avoid numerical integration for several methods, grouped in Fig.10.1 as 'analytical integration', however limited by the profile form. The other group of differential methods requires numerical integration, as do all the integral methods. The latter need either two-dimensional integration or evaluation of curvilinear integrals, while the former imply for a numerical solution a system of ordinary (predominantly) differential equations. The classical integral method differs from the approach of finite-element of surface by the method of numerical integration, the latter using the finite-difference scheme. There are also hybrid methods, for example transformation of coordinate system, as in the method of Chandezon, combined with numerical integration as in the classical differential method, made to deal with anisotropic and/or inhomogeneous media.

The name convention used is now fairly well accepted and clear, even for the so-called method of Chandezon, which has become quite fashionable recently. The method of Moharam and Gaylord is usually classed as a modal

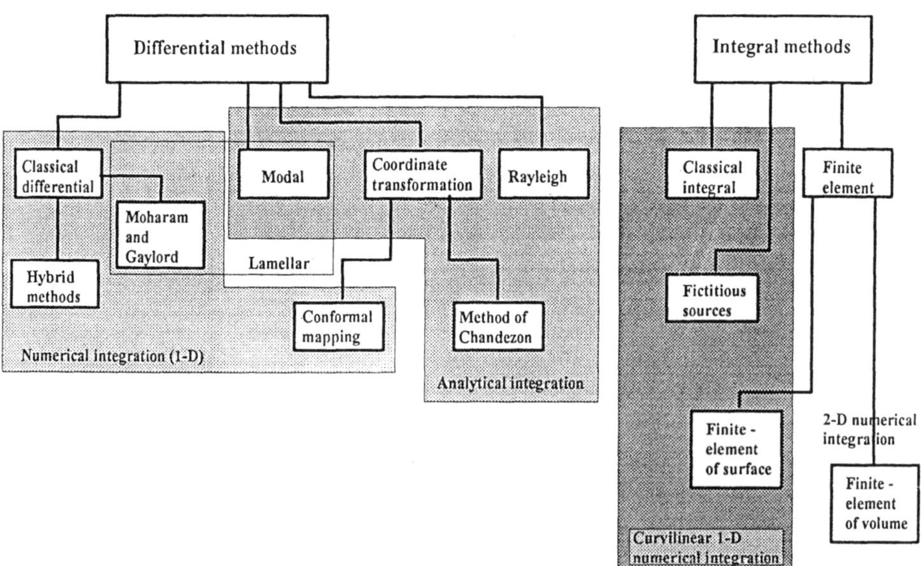

Fig.10.1 Classification of rigorous methods for grating modelling. Boxes represent different methods, together with their mutual links. Hatched region point out to some common properties linked to the method of numerical solution. The Rayleigh method is included for completeness.

method, but we prefer to consider it as a sub-class of the classical differential method, because in the classical modal method *all* the tangential field components are a priori continuous, which is not the case in TM polarization for the classical differential method and the method of Moharam and Gaylord.

10.3 The Rayleigh Hypothesis

The availability of elegant mathematical software, such as MATHEMATICA(TM)[1] is not sufficient to attack complex problems. In fact, the less time is devoted to the basis of the problem, the greater the difficulties that follow. No mathematics can help if the choice of the basic function is wrong. A typical example is the Rayleigh hypothesis [10.2]. It is assumed that the expansion (10.1) together with the link (10.2) (i.e., the plane wave expansion) can be used not only above the groove tops, but also inside the corrugated region. This assumption has a great advantage: it means that the boundary conditions on the corrugated surface can be expressed in a simple formula that connects the amplitudes of the diffracted orders above and below the grating. There are several ways of mathematically expressing this formula into a system of linear algebraic equations, a dream of all the theories. The main limitation of the Rayleigh hypothesis is that it is not rigorous. Its validity has been the subject of great deal of study, and a detailed review can be found in [10.3]. It has been shown that the plane wave expansion is not rigorous for a groove profile with edges, and even a smooth one should not exceed a certain depth. For the most commonly found sinusoidal groove shape the theoretical limit is a modulation depth $h/d = 0.142$.

There is a simple intuitive argument why the Rayleigh hypothesis cannot be valid in the general case. Eq.(10.2) can be satisfied for two opposite signs of k_{m_y}, each solution corresponding to the components propagating "upward" and "downward" to the grating surface. The expansion (10.1) is obviously valid above the groove tops. Then the outgoing radiation conditions require that only the "upward" components contribute to the solution (except, of course, for the incident wave). When this expansion is continued to the surface, it follows that as a consequence of the Rayleigh hypothesis, inside the grooves all the diffracted field components propagate "upwards". Provided the groove is deep enough, however, there are always scattered waves inside the grooves that can meet the grating surface below the scattering point (Fig.10.2, wave $\mathcal{A}$). Reciprocally, for each point there are waves (wave $\mathcal{B}$), incident from "below", while the plane wave expansion includes only a single incident wave inside the grooves. Although this argument is expressed in ray-tracing terms, it serves to

[1] MATHEMATICA is a trade mark of the Wolfram Research, Inc.

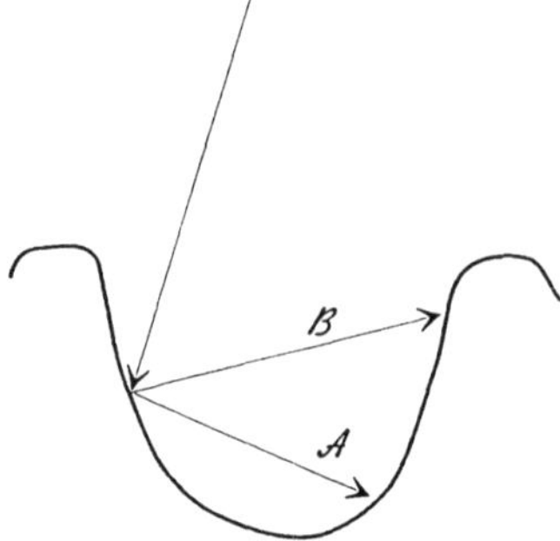

Fig.10.2 The grating groove with an incident wave and two scattered wave going downwards ($\mathcal{A}$) and upwards ($\mathcal{B}$).

give an intuitive understanding why the plane wave expansion is not adequate *inside* the grooves.

The interest devoted to the Rayleigh hypothesis is not only occasioned by its simplicity, but because some of the numerical codes based on it work surprisingly well even for deeper gratings. For example, the so-called Rayleigh-Fourier method[1] deals successfully with sinusoidal gratings five times deeper than the theoretical limit [10.4]. The *variational formulation* of the Rayleigh hypothesis deals much better with triangular groove profiles. It is almost impossible to determine why this happens: a detailed analysis on diffraction efficiency convergence and of the near field properties showed that the convergence of the results beyond the theoretical groove depth limit is not obtained for the boundary conditions, but only for the far-field efficiencies [10.5]. This phenomenon has a direct consequence in the theory of non-linear second harmonic excitation: at the second harmonic frequency the efficiency of the propagating orders diverge above the theoretical limit of validity of the Rayleigh hypothesis, long before the divergence of efficiencies on the pump frequency is observed, because the evanescent orders play an important role in the second-harmonic excitation.

This brief discussion reminds us how dangerous it can be to use approximate methods without knowing (or, often, without accepting) where the limits are. While codes can be easily developed on the basis of the Rayleigh hypothesis, they must be used with care. The programming and computation time saved can easily be lost in the need to search for convergence in each case, unless real time comparison with rigorous methods is available.

---

[1] The boundary condition equations are projected on a Fourier series basis.

10.4 Scalar Theory

Another typical example of a widespread approximate theory of gratings is the scalar one. It is the simplest (and the oldest) method that can lead to analytical solutions in limited cases, especially when the wavelength is much shorter than the period and when incidence is near normal. It happens that these conditions apply to the optics of CD players and similar devices for optical data storage, although recent designs tend towards periods getting shorter and approaching the wavelength of use.

There are several formulations of the scalar theory of gratings, starting with Fermat's principle and the Kirchhoff approximation, the results differing usually in the coefficient before the integral in the scalar theory formula (see for example [10.1]). The simplest approach is illustrated here in order to show the method. The first assumption is the Rayleigh hypothesis, i.e., the validity of the plane wave expansion (10.1 and 10.2) inside the corrugated region. The second is that the groove profile function f(x) varies slowly with respect to the wavelength, so that the reflected (E_R) and transmitted (E_T) fields *on* the corrugated interface are assumed equal to the corresponding fields that exist when light is incident on a plane interface. Some of the approaches can take into account the inclination of the surface, while others neglect that and evaluate only the phase difference between components diffracted at different points of the surface. Anyway, uniqueness of the field on the surface enables expressing the diffracted field as a sum of diffraction orders, and their amplitudes can be evaluated in a closed form:

$$b_m = \frac{R}{d}\int_0^d e^{-imK-i\left(k_{i_y}+k_{m_y}\right)f(x)}dx \ , \qquad (10.3)$$

where the coefficient R depends on the refractive indices and on the angle of incidence and, unfortunately, on the choice of theory. Near normal incidence, it can be taken equal to the reflection coefficient of the corresponding plane interface:

$$R = \frac{n_1 - n_2}{n_1 + n_2} \ . \qquad (10.4)$$

For a sinusoidal grating with a total groove depth equal to h, the integral in eq.(10.3) is equal to the m-th order Bessel function - the well known case of Raman-Nath diffraction is easily obtained:

$$b_m = R\, J_m\left[h(k_{i_y} + k_{m_y})/2\right]. \tag{10.5}$$

Another important result that follows directly from the scalar theory is the blaze wavelength for blazed gratings: Provided the blaze angle φ_B is not too large, the integration in eq.(10.3) may be performed just over the large facet, whose geometry is described by $f(x) = x \operatorname{tg} \varphi_B$. Maximum efficiency is expected when the factor of the exponent in eq.(10.3) is null, and it is easily shown that this is obtained when the direction of diffraction of the m-th order coincides with the direction of the light reflected by the facet.

The main advantages of scalar theory are the simplicity of the calculations (if any), the ease of interpretation of the results, and the physical insight it gives. It is also useful in treatment of transmission gratings and should remain a useful "zeroth-order" tool, as long as its limitations are understood.

Scalar theory does not take into account the vector character of the electromagnetic field, and the diffraction process is assumed independent of polarization. While the reflection coefficient in eq.(10.3) can vary with polarization for larger angles of incidence [which is not the case with the simplest assumption in eq.(10.4)], this may not be sufficient. In fact, when the incidence is no longer normal, or the profile is more complicated, even small λ/d ratios are not sufficient to use scalar theory safely. There are two main problems: shading of different parts of the profile and the effect of secondary scattering. Geometrical optics considerations are used to solve the first problem, although often they are ambiguous. Secondary scattering can also be treated, but then the method becomes so complicated that it looses its main attractiveness: simplicity and clarity. Moreover, it often happens that higher order approximations are even worse. And in no way can the scalar theory deal with cut-off anomalies[1].

10.5 Classical Differential Method

From a mathematical point of view Maxwell's equation can be expressed naturally as partial differential equations. In the grating case, within the assumptions of section 10.2, they have constant coefficients. Strictly speaking, they can be integrated numerically in two dimensions. Just as with the approximate methods, the simplest solution is the worst - two-dimensional numerical integration of the equations exhibits numerical instabilities (at least in the numerical implementations that have been made so far), so that

[1] It also cannot deal with resonance and non-resonance anomalies, but we stress the cut-off anomalies, because they can be important in long-wavelength cases (see Chapter 6).

applications are limited to shallow gratings. Only recently techniques originating from the integral method appear to directly integrate the two-dimensional differential equations (see section 10.8).

Using grating periodicity, the two-dimensional equations can be expanded on some basis, which reduces the problem to a system of ordinary differential equations [10.6]. After the truncation, its integration can be performed quite efficiently using well developed numerical algorithms. The natural basis that is usually used is the Fourier series representation of the field F(x,y) and permittivity ε(x,y) inside the corrugated region:

$$F(x,y) = \sum_m F_m(y)\, e^{imKx}\ , \quad \varepsilon(x,y) = \sum_m \varepsilon_m(y)\, e^{imKx}\ . \tag{10.6}$$

The main feature of the differential method(s) is that it goes "across" the profile (Fig.10.3). And indeed, in the representation (10.6) a Fourier transform of the electromagnetic field components (F) and of the material characteristics (ε) is performed for a constant value of y.

It is important to know that the expansion (10.6), although resembling in its x-dependence the plane wave expansion, is not equivalent to it and is rigorous if the number of terms goes to infinity. Although not a complete proof,

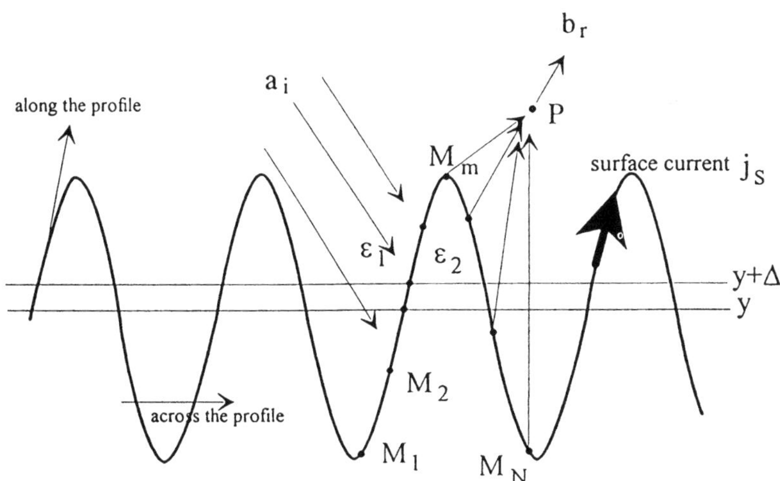

Fig.10.3 Schematic representation of the grating profile with the notations used in the text.

the following argument may be used to understand the difference intuitively: In the Rayleigh hypothesis both projections on the x and y axis in the field expansion (10.1) are determined from Maxwell's equation in a homogeneous medium (eq.10.2), whereas the form of the y-dependent set of functions in the field expansion (10.6) is determined numerically, depending on the profile form and the refractive index values. On a more naïve level, the field representation in the differential method (eq.10.6) is not restricted to outgoing waves only, but permits solutions going up and down inside the modulated region, their amplitudes matching the plane wave expansions outside this region.

In the TE case the Maxwell's equations, when projected on the Fourier basis (eq.10.6), lead to a system of ordinary differential equations with non-constant coefficients:

$$\mathbf{F}''(y) = V(y)\,\mathbf{F}(y) \ , \tag{10.7}$$

where the vector $\mathbf{F}\ (\equiv E_z)$ contains the field components $F_m(y)$ and the matrix V(y) is expressed through the permittivity components, wavelength, angle of incidence and grating period:

$$V_{mn}(y) = \varepsilon_{m-n}(y) - \delta_{mn} k^2_{m_x} \ . \tag{10.8}$$

Solution of the system of equations (10.7) is performed most effectively by using a fixed integration step Δ along the y-axis (Fig.10.3) using standard numerical techniques, for example the Numerov algorithm. A convergence of the results reducing Δ and increasing the number N of Fourier harmonics is a usual test for the method. Even for highly conducting substrates, for which the jump of the permittivity when crossing the profile at a constant y (Fig.10.3) cannot be well defined by a small number of Fourier components, the method gives rigorous results for relatively low values of N, with groove modulations that may exceed 100% in the resonance domain. The explanation can be easily found: In the TE case the unknown functions are E_z and H_x, both of which are continuous across the grating surface for non-magnetic media. Thus the convergence of their Fourier representation is relatively fast, although the Fourier representation of the jump of the refractive index is converging slowly.

For the TM-polarized case the situation is not so favorable. The system of ordinary differential equation then becomes of the first order, but requires doubling the number of the unknown functions. However, this is not so critical and the classical Runge-Kutta procedure is applicable. In order to save computation time, this predictor-corrector method can be used as a self-starting integration procedure, followed by the less time-consuming Adams-Moulton scheme. The same procedure can be applied with great success in TE polarization, unifying the codes for the two fundamental polarization cases into one. However, in TM polarization, due to the different boundary conditions, the y-derivative E_x of the field $\mathbf{F}$ ($\equiv H_z$) undergoes a jump when crossing the corrugated surface. This requires a lot of Fourier components in expansion of the field (10.6) for highly conducting substrate materials. The large number of equations to be integrated can lead to numerical instabilities if the groove depth is not shallow, and sophisticated procedures must be applied, such as orthonormalization during the integration, or dividing the integration region in several parts, connecting them by the so called R-matrix method [10.7]. Otherwise, provided the substrate is not highly reflecting, the classical differential method is fast and successful all over the spectral region, even in the XUV domain (where all the "optical" indices are close to unity) and for gratings supporting a large number of orders.

10.6 Modal Methods

10.6.1 The Method of Moharam and Gaylord

Sometimes equation (10.7) (or the corresponding equation for the TM case) can be integrated in a closed form. The simplest case corresponds to lamellar profiles, where the permittivity inside (and, of course, outside) the corrugated region does not depend on the vertical y-coordinate. More complicated cases can include step-like profiles, which can serve as a good approximation for any profile form, provided the slices are thin enough [10.8]. However, with numerous thin slides, we arrive again at the classical differential method.

In each separate slice the solution of (10.7) is found as a sum over 'modes' which is nothing but the solution of the system of differential equations with constant coefficients:

$$\exp(i\sqrt{\xi_m}\, y) \ , \qquad (10.9)$$

where ξ_m are the eigenvalues of matrix V. The modal coefficients are determined by matching the field expansions at the slice boundaries. In fact this method is a direct application of the classical differential formalism to lamellar profiles: the differential equations and the basic functions are the same for the two methods, the difference lying in the numerical method of solution. However, historically the method of Moharam and Gaylord was called modal method, a convention still in use.

A scheme was proposed for reaching very deep grooves (more than 300% modulation depth), but without numerical examples to include highly conducting gratings. For the latter case, difficulties similar to the classical differential formalism are observed. They have the same origin: In the TE case and non-magnetic media both E_z and H_x (which are the unknown functions) are continuous across the profile, so that their Fourier representation rapidly converges. In the TM case, however, not only is the refractive index discontinuous (as in the TE case as well), but so is one of the field components (E_x). Then its Fourier spectrum converges as slowly as 1/N, N being the order of the component. Taking a sufficient number into account (more than 200 for metallic gratings, where the jump of the field is large) requires long computation times and large memory. More critically, it leads to numerical instabilities, like loss of precision, which implies special techniques, such as orthonormalization or the R-matrix algorithm, as is the case for the classical differential method.

The other problem appears at short wavelength-to-period ratios for metallic gratings with profiles other than lamellar: The essence of the method implies a staircase-like representation of the profile, which can lead to large errors in the efficiencies if the deviation from the real profile are comparable with the wavelength, as happens with echelles.

Otherwise, the method is highly applicable for dielectric gratings, where a relatively small number of Fourier harmonics (10.6) of the field (as well as of permittivity) is enough to determine the grating efficiency correctly and the staircase-like representation of the profile is not critical.

10.6.2 The Classical Modal Method

For a step-like (lamellar, rectangular) profile, it is not even necessary to use the Fourier series expansion (10.6) with respect to the x-axis. A solution of the Maxwell's equations can be found in closed form in each of the grooves and lamellae in Cartesian coordinates:

$$F_m(x,y) = u_m(x)\, e^{i\mu_m y} , \tag{10.10}$$

so that the function u(x) has a different form inside the grooves and inside the lamellae, determined by the optical index of the media. The boundary conditions are then applied on the vertical groove walls. Due to the periodicity, there is a discrete set of values for μ, called modal constants. The total field is represented as a sum over all the modes of the corrugated system and the coefficients in the modal expansion are determined from the boundary conditions on the interfaces between the corrugated and the homogeneous media. Unfortunately, for highly conducting materials (again!), the modal constants are spread over the complex plane and cannot easily be located. Several different techniques have been proposed [10.9-10.11].

The main disadvantage of the modal method is that it is too narrowly specialized and can be applied to profiles other than the lamellar one only if it can be represented as a few rectangular steps. The main advantage is that each of the modes represents a solution of Maxwell's equation *and* the boundary conditions inside the different media of the corrugated region, so that it becomes possible to evaluate the electromagnetic field characteristics inside the grooves with great precision. Its main difference in comparison to the classical differential method and the method of Moharam and Gaylord is that it does not require a Fourier representation of the permittivity *and* of the field components at both sides of the corrugated interface (Fig.10.3) and so can deal with highly reflecting surfaces.

10.7 The Integral Method

This is historically one of the first rigorous methods and it is based on the idea that the solution of electromagnetic scattering problems can be readily found by simple integration, provided that the field and its normal derivative on the surface of the scattering object (and at infinity) are known [10.12, 10.13]. Using the well-known Green's theorem, the field in each volume point can be represented by an integral over the surrounding surface, which is the grating and a hemisphere with infinite radius.

Intuitively, the incident plane wave (or limited beam) generates a surface current in the grating material (Fig.10.3, points M_1, M_2, ... M_m). Then the diffracted field is a field generated by the surface current j_S along the grating surface. As each point of the grating (points M_m ... M_N in Fig.10.3) radiates a field, proportional to the local surface current density, the total diffracted field in each outside point (P) can be evaluated as a sum (i.e., integration) over all the components radiated by the surface points. Due to the grating periodicity, components radiated by points identically situated on different grooves add to form the diffraction orders with amplitudes b.

Unfortunately, this pastoral picture is disturbed by the fact that each surface point radiates a field, which acts as an incident (source) field for the other points of the surface (see again Fig.10.2), so that the surface current density j_S depends not only on the incident field, but on its values at the other points of the surface. The total diffracted field is then obtained by integrating a function which is unknown and depends on the diffracted field. Thus the simple integration is transformed into an integral equation. For example, in the simplest case of a perfectly conducting grating in TE polarized light, this integral equation has the form:

$$\Phi(x) = 2\Phi_0(x) + \frac{1}{d}\int_0^d N(x,x')\Phi(x')dx' \ , \qquad (10.11)$$

where the unknown function $\Phi(x)$ is directly linked with the surface current density:

$$\Phi(x) = -i\omega\mu_0\sqrt{1+f'(x)^2}\ j_S(x) \ . \qquad (10.12)$$

The incident wave determines the term

$$\Phi_0(x) = -i\left[k_{i_y} + k_{i_x} f'(x)\right] e^{i\left[k_{i_x}x - k_{i_y}f(x)\right]} \ , \qquad (10.13)$$

and the Kernel N is given by

$$N(x,x') = \sum_m \left\{ \text{sign}\left[f(x)-f(x')\right] - \frac{k_{m_x}}{k_{m_y}} f'(x) \right\} e^{i(x-x')k_{m_x} + [f(x)-f(x')]k_{m_y}} \ . \qquad (10.14)$$

For finitely conducting gratings it is necessary to introduce two unknown functions (the field and its normal derivative) in order to express the diffracted field everywhere in the space, but then two coupled integral equations are written instead of eq.(10.11) and their solution gives rise to severe numerical problems. Fortunately, an approach proposed by Maystre leads to a single integral equation.

The solution of eq.(10.11), or of similar integral equations, can be performed by several methods. The most straightforward one is to project the unknown function on the grating profile, taking its values over a limited number of profile points (namely, M_1, M_2, ..., M_N, in Fig.10.3) as unknowns Φ_1, Φ_2, ...,Φ_N. The integral equation is projected over this basis (usually substituting the integral by a trapezoidal rule summation). Thus the integral equation is reduced to a linear algebraic system of N equations for the N unknown quantities Φ_1, Φ_2, ...,Φ_N. The main difficulties arise from the singularities of the kernel. This sounds quite complicated, but has a simple physical background: The surface current density, generated in point M_i of the profile by the radiated field from the neighbouring point M_{i-1}, tends to infinity when the points tends to each other. The more complicated profiles and deep grooves require more points per wavelength, thus the coupling between the neighbouring points can grow unlimited. Fortunately, it is possible to evaluate these singularities analytically and to extract them from the numerical integration. In fact, this procedure for eliminating the singularities, as well as the proper choice of the discretization points along the grating surface, is the most important for success of the generated code [10.1, 13].

The greatest advantage of the integral formalism, when compared to the differential method, is that it "follows" the profile, without crossing it. As a result, it is not necessary to develop in Fourier series some quantities that exhibit a jump over the surface. Sometimes, however, this can be a disadvantage, as in the case of non-linear dielectrics, characterized by volume distributed sources, which are difficult to include in the integral equation. In addition, working in real space, instead of the transformed one, requires numerical derivatives, if all the field components are searched near the surface.

As a result of its generality, the integral theory is able to deal with practically any kind of grating, including some limiting cases where it is the only available method. An example here is the case of echelle gratings (used in 50, 100, or even higher orders, and at high angles of incidence), or highly conducting very deep gratings with arbitrary profile, etc. This advantage is obtained at the cost of more complex mathematics, larger codes, and longer computation times, as well as larger memory storage requirements.

10.8 The Finite-Element Method

As in many other instances, the world of gratings rarely accepts easy solutions. Standard numerical techniques may fail for many reasons. Only for shallow grooves can the electromagnetic field close to the grating surface be divided into rapidly oscillating exponential terms with slowly varying amplitudes. This causes enormous numerical problems for deep gratings if a direct two-dimensional integration of Maxwell's equations is tried. Only recently [10.14, 10.15] has it became possible to apply an appropriate meshing procedure well known in radar applications. The entire field is represented as a sum of elementary functions ϕ_i over the mesh cells number:

$$E = \sum c_m \phi_m(x, y) \ , \qquad (10.15)$$

with c_m being the unknown amplitudes. ϕ_m is assumed to be different from zero only in the m-th cell and the simplest form is the pyramidal function:

$$\phi_m(M) = \begin{cases} 0, & \text{for point M outside the cell} \\ |M\,M_m|\, e^{ik\sin\theta_i x}, & \text{inside the cell} \end{cases} \ , \qquad (10.16)$$

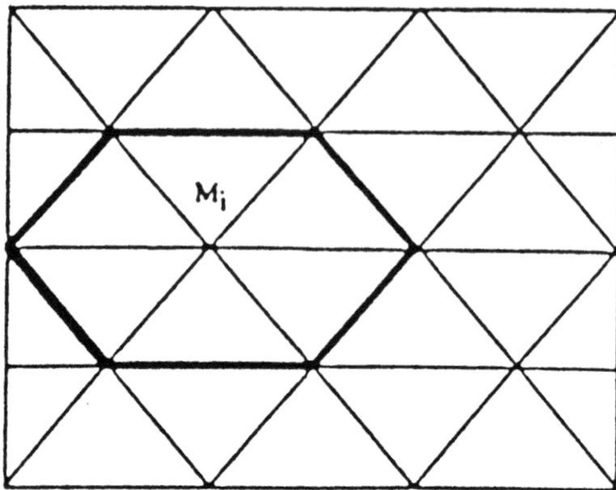

Fig.10.4 An elementary cell used to represent the field in the finite-element method.

where M_m is the central cell point (Fig.10.4). The explicit choice of such a simple basis allows analytical integration of the differential equations and the problem is reduced to a linear algebraic system of equations for the unknown amplitudes a_i..

The simplicity of the idea however has to be paid for by the complex meshing algorithm and extended calculation times. For example, highly conducting gratings need meshing that takes into account the rapid changes of the field inside the metal (Fig.10.5a). When compared with the other methods, advantages of such "direct" treatment appear for complex non-standard gratings, such as anisotropic, chiral, or non-homogeneous materials, multilayered structure with varying thickness, aperiodic objects, etc.

10.9 The Method of Fictitious Sources

A peculiar way of removing the singularities in the integral equation without taking them into account is to go above or below the profile (or, usually, simultaneously below and above) [10.16, 10.17]. Imagine (Fig.10.6) a fictitious surface $\tilde{\mathcal{P}}$ placed below the grating surface. There are at least two possibilities to evaluate the field above the real surface $\mathcal{P}$: (1) to consider a continuous surface current along the fictitious surface $\tilde{\mathcal{P}}$ which is generated by the incident field and which acts as a source for the diffracted field *above* $\mathcal{P}$; (2) similarly, to consider a set of points (or wires, or any multiple source) on $\tilde{\mathcal{P}}$ that act as a source of the diffracted field *above* $\mathcal{P}$.

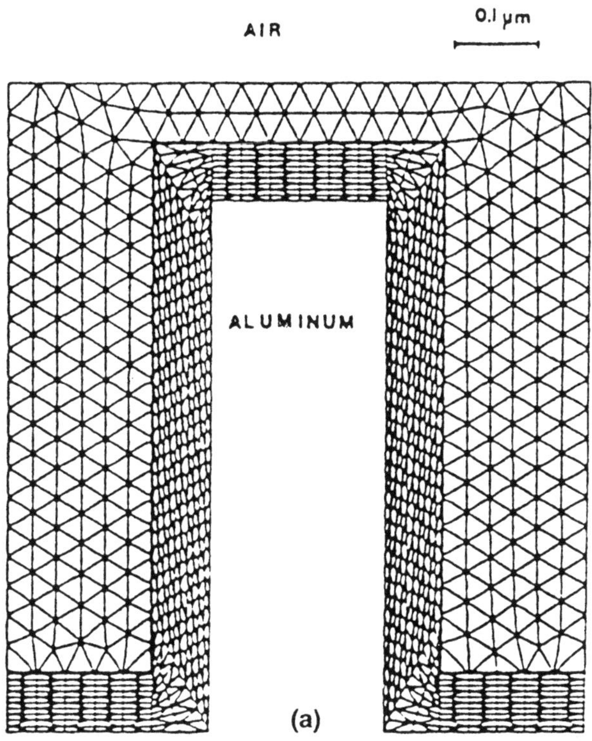

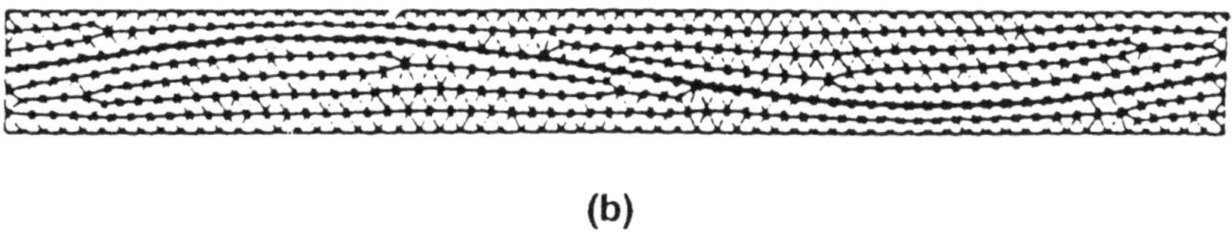

(b)

Fig.10.5 Mesh for applying the finite element method for (a) a deep metallic lamellar grating and (b) dielectric sinusoidal grating (after [10.15]).

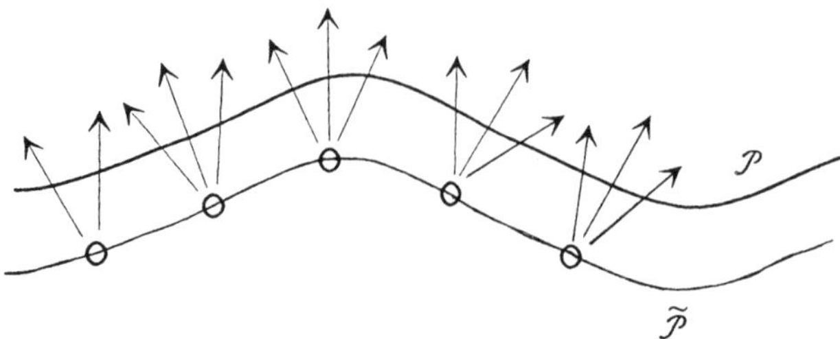

Fig.10.6 The real and fictitious grating surfaces and the fictitious point sources with the field radiated by them.

Because the region where the diffracted field is evaluated is separated from the fictitious source region, singularities that come from the source and the viewing point approaching each other do not exist. Then the scattering problem is reduced to evaluating the amplitudes of the fictitious sources, usually by least square fit of the field, radiated by the entire set of fictitious sources and of the incident field. The condition to be fulfilled is that the total electric field inside the perfectly conducting substrate vanishes. If the substrate is not perfectly conducting, the diffracted field below the grating surface is represented as if being radiated by another set of fictitious sources situated above the grating, i.e., at a finite non-vanishing distance from the substrate.

The method of fictitious sources seems, at a first glance, to avoid the disadvantages of both the differential method (passing "across" the profile) and the integral method (the complicated way of isolating the singularities). However, a price is paid by the user of the code: there is no reliable general algorithm for choosing the position of the fictitious surfaces and the position and the type of the fictitious sources. From general considerations, one can expect that a single source per groove is enough, having a complex field such that it corresponds to the field diffracted by the grating. However, in order to find the source it is necessary to know the solution. The user of such code must possess good intuition and a lot of experience of how to move the sources when changing the profile, a task that can be quite challenging.

Otherwise, despite its subjectivity in utilization, the method of fictitious sources is a relatively simple tool that can serve to determine the diffracted fields in very complicated cases.

10.10 The Method of Coordinate Transformation

Another approach is to use the differential method without crossing the profile, i.e., not necessarily representing the jump of the material refractive index in a Fourier expansion. This is possible if a suitable transformation of the coordinate system can "straighten" the corrugated boundary so that the integration of the Maxwell's differential equation to be carried out in each of the media separately, matching the two solutions at the interface *after* the integration, as it is done in the integral method and in the classical modal method (see Section 10.6.2) and, of course, in the approximate methods, based on the Rayleigh hypothesis. The idea sounds simple, but it leads to a complication with the Maxwell's equations which, even for homogeneous media, become partial differential equations with *varying* coefficients [10.18].

There are two useful transformations: The first one, called *conformal mapping* is applied exclusively to infinitely conducting gratings, where the classical differential methods fail by definition. Conformal mapping maps the grating into a layer with periodically varying permittivity, deposited on a perfectly conducting substrate, with optical properties tending to those of the cladding at an infinite distance from the grating. Numerical integration, identical to the classical differential method, is applied to a finite layer thickness.

The main difficulty of this method lies in the complicated way of determining the conformal mapping, especially for deep grooves. Going to finitely conducting gratings, two conformal mappings become necessary, adding new difficulties.

The second method uses the most natural transformation of coordinates to straighten the profile:

$$\left| \begin{array}{l} u = x \\ v = y - f(x) \\ w = z \end{array} \right. \qquad (10.17)$$

As mentioned, Maxwell's equations in the new curvilinear non-orthogonal cooordinate system have non-constant coefficients. Using the grating periodicity, however, they can be simplified to a system of ordinary differential equations with *constant* coefficients, which can be expressed in a matrix form:

$$\mathbf{F}'(v) = T\,\mathbf{F}(v) \quad , \qquad (10.18)$$

where T is a matrix that depends on system parameters, but not on the coordinates. That makes a great difference with systems of equations of the type (10.7). While the latter require numerical integration, the solution of eqs.(10.18) can be expressed through the eigenvectors and eigenvalues $\{r_m\}$ of matrix T, i.e., in the basis:

$$\{\exp(ir_m v)\} . \tag{10.19}$$

The expansions in this basis (eq.10.19) above and below the corrugation are matched on the *flat* boundary v=const, taking into account the appropriate outgoing wave conditions. The diffraction order amplitudes can be obtained by backward transformation of the basis (10.19) after the amplitudes of the field expansion are determined. As far as this method does not require crossing the profile during a search for the solution, but only when matching the different expansions, it is able to deal with deep gratings independent of polarization and refractive index, and including multilayered gratings. It has relatively simple computer codes and short computation times, comparable to those of the differential method. The ease of incorporating several overcoating layers that follow the initial profile may be stressed. The main limitation concerns the types of profile that it can deal with. This limitation comes from the nature of the coordinate transformation (10.17), which does not allow non-continuous functions f(x). Strictly speaking, it requires that the derivative of f(x) be a continuous function, so edges are excluded, but this concerns all the electromagnetic methods and, fortunately, nature does not allow edges. Practically, the slope of the steepest part of the profile is much more important: when $f'(x)$ becomes large not only does its Fourier representation converge slowly, but the transformation of the coordinate system degenerates, the derivative with respect to the first and second coordinates tend to each other $\left(\frac{\partial}{\partial v} \to \frac{\partial}{\partial u}\right)$. However, numerically this limitation is not so severe as it appears: sinusoidal gratings with h/d > 10 have a very steep slope, but only along a limited part of the profile and can be successfully treated, while lamellar gratings are excluded by definition[1], and triangular gratings with limited asymmetry of the profile can be treated without difficulties.

10.11 Theory of Waveguide Gratings

There has been a long and strong belief that approximate theories, namely the so called coupled-mode formalism, are sufficient to adequately determine the properties of waveguide grating couplers and mode converters. It

[1] From this point of view, this method seems complementary to the classical modal method, which is valid only for rectangular profiles.

is difficult to deny the role of coupled-mode theory in physical understanding and even design of such devices. It has proven capable of predicting the efficiency of output and input couplers and performance of mode converters (waveguide filters, demultiplexers, etc.) when the modulation is low (see Chapter 9). Fortunately, practical applications do not often require high modulation depths, mainly because this imposes change of mode parameters and strong transition phenomena. However, the situation can change: practice requires high-efficiency couplers and wide-band filters (e.g., photonic band-gap devices for semiconductor lasers) with blazed profiles and with modulation depth reaching 100%. The fault of all the approximate theories is that they (or, namely, their users) are always tempted to go beyond limits of validity. As a result coupled-mode theory is cloned into normal-mode and local-mode approaches. These are accompanied by numerical methods based on the Rayleigh expansion. Unluckily for it, practice requires deep blazed profiles for in-/output and deep lamellar grooves for mode conversion and photonic band-gap (see Chapter 9).

Rigorous grating theories are *a priori* capable of solving the problem. As far as Maxwell equations and proper boundary conditions, including conditions at infinity, are solved, a resonance solution must inevitably emerge, including mode excitation and coupling. Usually missing is a demonstration of how rigorous theories can be used for solving the waveguiding problem, because there are several peculiarities.

There are three main physical problems when guided waves are involved. The main concern of rigorous grating theories is to determine diffraction efficiencies (i.e., the ratio of the energy carried away by propagating order and the incident wave intensity). This implies existence of an incident wave, which is expressed mathematically as a solution of the *inhomogeneous* problem. Grating input couplers fall directly within the limits of that problem. Thus any numerical implementation of a rigorous grating theory can be used to solve the grating input coupling, by simply evaluating the amplitude of the evanescent order, corresponding to the waveguide mode. This brings information on the coupling efficiency, after proper evaluation of the mode field cross-sectional distribution. The position and width of the resonance response (see Ch.8) point to the real and imaginary part of the mode propagation constant. Limited beam response can be evaluated as a superposition of the response to its plane-wave components or, more directly, using the phenomenological approach (see Chapter 8).

The other two problems, output coupling and mode interaction are linked with the solution of the *homogeneous* problem: existence of scattered (and guided) waves without incident ones. The output coupling can be obtained from the input coupling using the reciprocity theorem (see Ch.2) by multiple solutions of the inhomogeneous problem with different incident waves each corresponding to the radiated diffraction order. But the inhomogeneous problem cannot lead to a solution of mode interaction. Rigorous methods are capable of solving this problem as well, but their numerical implementation is usually not adapted to it. Solution of the homogeneous problem implies that the determinant of the matrix M, which is the inverse of the scattering matrix of the system, has a zero (representing a *pole* of the filed) for some system parameters (see Ch.8). Mathematically this means that all the scattered field amplitudes (i.e., diffraction orders) depend on a single scattered diffraction order amplitude, if the zero is single. If this 'principal' order is the guided wave amplitude, then the other diffracted orders amplitudes are just their output coefficients. In the vicinity of another mode excitation, or of the same mode propagating in the opposite direction, the dependence of the amplitude of the latter on the amplitude of the 'principal' order gives the mode coupling coefficient, normalized using the mode field distribution. The behavior of the resonance curves of the mode propagating constant (i.e., the pole of the determinant of the system scattering matrix) carries information on the mode scattering properties, namely its attenuation constant and optical band-gap width. There is a direct link between the pole of the scattering matrix, eqs.(8.2)-(8.11), which represents the complex propagation constant $k_G + i\gamma$ of the guided wave and the mode coupling coefficient in eqs.(9.9), (9.13) and (9.20-22). Comparison of eqs.(9.10) and (8.2) directly leads to the relation:

$$\gamma = \xi_{|\Delta=0} \equiv \eta \quad , \tag{10.20}$$

i.e., the imaginary part of the pole γ is equal to the mode coupling coefficient.

To successfully deal with the homogeneous problem, any numerical method must be equipped with tools for solving singular problems: once the system scattering matrix (or rather its inverse) is obtained, instead of solving a linear system of algebraic equations with a source term corresponding to the incident waves, it is necessary to invoke a root-finding procedure of searching for singularities (i.e., zeros of the determinant). Once a singularity is found there are standard tools to find out the matrix column dependencies (i.e., to find the solution of the homogeneous problem).

Then comes the nontrivial problem of mode normalization. Unresolved by the *common* grating theories are also the transition phenomena at the plane-corrugated waveguide region. The latter can be quite important for strongly modulated waveguides. A theory capable of dealing with these problems has to avoid the periodicity which is so useful in the grating theories. The only available numerical codes [10.19, 20] are based on the integral method, namely the finite-element method (section 10.8). They consider a finite system consisting of several grooves loaded on an infinite plane surface. The incident mode corresponds to the mode of the unperturbed system and is considered as an incident wave with a known field distribution. All the transition phenomena, mode coupling and radiation effects, are taken rigorously into account. Contrary to the coupled mode approximate theories, here the main problem, long computation time and large memory required, arises for weak coupling due to the large physical dimensions of the system.

10.12 Conclusions

The following conclusions are based on a snap-shot of the recent state of the art: it happens, although more rarely than desired, that in applying a new algorithm (or an old one, well known in some other science), some of the limitations of a given method can be significantly reduced. For example, just by changing the mathematical formulation of the differential equations, a considerable improvement can be obtained in the method of Moharam and Gaylord for highly conducting gratings in TM polarization ([10.21]), a case causing numerical problems till now. This process usually works in two directions: the regions of applications of the rigorous methods are extended, while more new examples that limit the applications of the approximate methods are found.

These conclusions are intended only to serve a user of the theory, or a future grating-code-programmer looking for some simple guidelines. Otherwise it must be stressed, there is no answer to the common question "What is the best theory?" Or the answer is that there is no "best" theory. The term "best" is rather subjective: it depends on the aims you follow: simple, general, or cheaper? There are, however, two basic concepts, that should be noted:

(1) The grating theory and its numerical implementations is a powerful and sophisticated tool that must be treated with the same respect as any complicated and reliable experimental equipment. Fool-proof solutions exist but their efficiency is doubtful.

(2) Grating methods are complementary: what one gains from the simplicity of one method is lost due to its limitations, whereas the gain in generality can be lost in the large computation time (or just a change in computer!). Request for fast (efficient) algorithms for a variety of cases, requires a variety of methods, specialized for each case.

(3) Skill in programming has to be necessarily combined with understanding of the mathematical background of the theory. One of the authors was amazed that one day of code optimizing could reduce computation time by 90%, and combined with half a day of simple formula tricks could lead to a reduction of computation time by a factor of 125 (worse for the initial code!).

Finally, these conclusion can hardly be drawn from just the context of the current chapter. Behind them is the authors' experience and the discussions with many of the members of The Grating Society.

Most general is the integral method. It covers almost all of the cases, provided the background mathematics and the code itself are well done, which requires significant effort and time. Simplified and restricted versions can be run on computers such as PC-AT; 80386 based computers (or equivalent) are the minimum requirement, but recent workstations are preferable.

The differential method is very good for dielectric gratings of any profile form and reasonable groove depth. It works quite well for metallic gratings in the TE case (again for experimentally achievable groove depths). Only recently have algorithms been created to deal with highly conducting gratings in the TM case. Most complicated profiles can be treated, as well as multilayered gratings and inhomogeneous and anisotropic media. With gratings having a large number of orders, problems can appear unless energy is localized near the zeroth order. The method of Moharam and Gaylord can deal with arbitrary profiles, but it is especially efficient for dielectric gratings with few propagating orders.

Modal method is very efficient for lamellar gratings, provided the period is not too large compared to wavelength.

The method based on the transformation of the coordinate system is applicable in the resonance domain, independent on the polarization and grating material, with the main limitation lying in the existence of very steep parts of the profile.

The method of fictitious sources can be applied to almost the entire variety of grating problems, provided an experienced user is able to locate the sources in each particular case.

The finite-element method requires, in general, longer computation times and should be used mainly when the grating material has some exotic properties.

Approximate methods are useful because of their simplicity, being easy to implement numerically (sometimes even on a pocket calculator), and requiring short computation time. They are not safe to use in a general case, unless they are checked against one of the rigorous methods. Higher approximation terms hardly enlarge the region of validity, and can even lead to divergent results.

Table 10.1 is an attempt to summarize these conclusions. Several criteria can be applied and we have chosen those important for a débutante: universality and difficulty, else a specialist will need at least several. A useful combination depends on the aims and possibilities. Although not obvious, often it is cheaper to acquire a code or ask for consulting than to develop it.

Table 10.1 Comparative analysis of mostly used numerical methods in grating theory

Method	polarization		Special media (inhomogeneous, anisotropic, chiral)	Computation time and memory requirements	Problems
	TE	TM			
Volume finite-element	independent		easy	very large	meshing
Surface finite-element	independent		difficult	moderate	programming, singularities
Classical integral	independent		difficult	moderate	programming, singularities
Fictitious sources	independent		difficult	moderate	meshing
Classical differential	O. K.	problems	easy	short for TE and/or dielectrics	TM case due to jump of E_x.
Moharam & Gaylord	O. K.	problems	more difficult but realistic	short for TE and/or dielectrics	TM case due to jump of E_x. Staircase representation of the profile for short λ/d.
Modal	independent		difficult	short	Staircase representation of the profile for short λ/d.
Conformal mapping	independent		difficult	short	Difficult mapping.
Method of Chandezon	independent		difficult (has to be combined with numerical integration)	short	Differentiable profile (not very steep). Not too many diffraction orders.
Rayleigh method	independent		easy	short	Shallow grooves without edges.

References

10.1 D. Maystre: "Rigorous vector theories of diffraction gratings," E. Wolf, ed., *Progress in Optics* (Elsevier, North-Holland, Amsterdam, 1984) v.**XXI**, pp.2-67.

10.2. Lord Rayleigh O. M.: "On the dynamical theory of gratings," Proc. Royal Soc. (London) A **79**, 399-416 (1907).

10.3. R. F. Millar: "The Rayleigh hypothesis and a related least-squares solution to scattering problems for periodic surfaces and other scatteres," Radio Sci. **8**, 785-796 (1973).

M. Neviere and M. Cadilhac: "Sur la validite du developpement de Rayleigh," Opt. Commun. **2**, 235-238 (1970).

10.4. A. Wirgin: "Sur la théorie de Rayleigh de la diffraction d'une onde par une surface sinusoidale," C. R. Acad. Sc. Paris, v.**288**, Serie B, 179-182 (1979).

10.5. P. M. Van den Berg: "Reflection by a grating: Rayleigh methods," J. Opt. Soc. Am. **71**, 1224-1229 (1981).

10.6. M. Neviere, G. Cerutti-Maori, and M. Cadilhac: "Sur une nouvelle méthode de résolution du problème de la diffraction d'une onde plane par un réseau infiniment conducteur," Opt. Commun. **3**, 48-52 (1971).

M. Neviere, P. Vincent, and R. Petit: "Sur la théorie du réseau conducteur et ses applications a l'optique," Nouv. Rev. Opt. **5**, 65-77 (1974).

P. Vincent: "Differential methods," in *Electromagnetic Theory of Gratings*, R. Petit, ed. (Springer-Verlag, Berlin, 1980), ch. 4.

10.7. F. Montiel and M. Neviere: "Differential theory of gratings: extension to deep gratings of arbitrary profile and permittivity via the R-matrix propagation algorithm," J. Opt. Soc. Am. A **12**, 2672-2678 (1995).

10.8. M. G. Moharam and T. K. Gaylord: "Rigorous coupled-wave analysis of planar-grating diffraction," J. Opt. Soc. Am. **71**, 811-818 (1977).

M. G. Moharam and T. K. Gaylord: "Rigorous coupled-wave analysis of dielectric surface-relief gratings," J. Opt. Soc. Am. **72**, 1385-1392 (1982).

10.9 J. R. Andrewartha, G. H. Derrick, and R. C. McPhedran: "A general modal theory for reflection gratings," Opt. Acta **28**, 1501-1516 (1981).

10.10 J. Y. Suratteau, M. Cadilhac, and R. Petit: "Sur la Détermination Numérique des Efficacités de Certain Réseaux Diélectriques Profonds," J. Optics (Paris) **14**, 273-282 (1983).

10.11 G. Tayeb and R. Petit: "On the numerical study of deep conducting lamellar diffraction gratings," Opt. Acta **31**, 1361-1365 (1984).

10.12 M. A. Wirgin: "Considérations théoriques sur la diffraction par réflexion sur des surfaces, quasiment planes, applications à la diffraction par réseaux," Compt. rend. Acad. Sci. **259**, 1486-1488 (1964).

R. Petit: "Etude théorique de la diffraction d'une onde plane et monochromatique par un réseau métallique infiniment conducteur," Appl. Opt. **4**, 1551-1554 (1965).

10.13 D. Maystre: "Integral methods," in *Electromagnetic Theory of Gratings*, R. Petit, ed. (Springer-Verlag, Berlin, 1980), ch.3.

10.14 S. D. Gedney, J. F. Lee, and R. Mitra: "A combined FEM/Mom to analyze the plane wave diffraction by arbitrary gratings," IEEE Trans. Microwave Theory Tech. **40**, 363-370 (1992).

10.15 T. Delort and D. Maystre: "Finite-element method for gratings," J. Opt. Soc. Am. A **10**, 2592-2601 (1993).

T. Abboud: "Etude mathématique et numérique de quelques problèmes de diffraction d'ondes électromagnétiques," Thése de Doctorat de l'Ecole Polytechnique, Paris, 1992.

10.16 V. D. Kupradze: "On the approximate solutions of problems in mathematical physics," Russian Mathematical surveys **22**, 58-108 (1967).

10.17 C. Hafner: *Generalized multipole technique for computational electromagnetics*, (Artech, Boston, 1990).

G. Tayeb:" The method of fictitious sources applied to diffraction gratings," in *Generalized Multipole Technique (GMT)*, Appl. Comput. Electromag. Soc. (ACES) Jl. **9**, no.3, 90-100 (1994).

10.18 J. Chandezon, D. Maystre, G. Raoult: "A new theoretical method for diffraction gratings and its numerical application," J. Optics (Paris) **11**, 235-241 (1980).

10.19 F. Pincemin and J.-J. Geffet: "Propagation and localization of a surface plasmon polariton on a finite grating," J. Opt. Soc. Am. B, to be published.

10.20 A. Sentenac and J.-J. Greffet: "Scattering by 2D particles deposited on a dielectric planar waveguide: a near-field and far-field study," Waves in Random Media **5**, 145-155 (1995).

10.21 P. Lalanne and G. M. Morris: "Highly improved convergence of the coupled-wave method for TM polarization," J. Opt. Soc. Am. A **13**, 779-784 (1996).

Additional Reading

Reviews on Theoretical Methods

M. Cadilhac: "Some mathematical aspects of the grating theory," in *Electromagnetic Theory of Gratings*, R. Petit, ed. (Springer-Verlag, Berlin, 1980), ch.2.

R. Petit and D. Maystre: "Application des Lois de l'Electromagnetisme a l'Etude des Reşeaux," Rev. de Phys. Appliquée, **7**, 427-441 (1972).

R. Petit, ed. *Electromagnetic Theory of Gratings*, (Springer-Verlag, Berlin, 1980).

G. I.. Stegeman and D. G. Hall: "Modulated index structures," J. Opt. Soc. Am. A **7**, 1387-1398 (1990).

General Theoretical Problems

G. Bouchitte and R. Petit: "Homogenization techniques as applied in the electromagnetic theory of gratings," Electromagnetics **5**, 17-36 (1985).

D. Maystre: "A rigorous theory for problems of scattering: the delta boundary operator method," J. Mod. Opt. **34**, 1433-1450 (1987).

R. Petit and M. Cadilhac: "Electromagnetic theory of gratings: some advances and some comments on the use of the operator formalism," J. Opt. Soc. Am. A **7**, 1666-1674 (1990).

A. Roger: "Generalized reciprocity for gratings of finite conductivity," Opt. Acta **30**, 575-585 (1983).

P. Vincent and M. Neviere: "The reciprocity theorem for corrugated surfaces used in conical diffraction mountings," Opt. Acta **26**, 889-898 (1979).

A. Maréchal and G. W. Stroke: "Sur l'origine des effets de polarisation et de diffraction dans les réseaux optiques," C. R. Ac. SC. **249**, 2042-2044 (1959).

Differential Methods

M. C. Hutley, J. P. Verrill, R. C. McPhedran, M. Neviere, and P. Vincent: "Presentation and verification of a differential formulation for the diffraction by conducting gratings," Nouv. Rev. Opt. **6**, 87-95 (1975).

P. Vincent: "Differential methods," in *Electromagnetic Theory of Gratings*, R. Petit, ed. (Springer-Verlag, Berlin, 1980), ch.4.

Modal Methods

J. R. Andrewartha, G. H. Derrick, and R. C. McPhedran: "A modal theory solution to diffraction from a grating with semicircular grooves," Opt. Acta **28**, 1177-1193 (1981).

I. C. Botten, M. S. Craig, R. C. McPhedran, J. L. Adams, and J. R. Andrewartha: "The dielectric lamellar diffraction grating," Opt. Acta **28**, 413-428 (1981).

I. C. Botten, M. S. Craig, R. C. McPhedran, J. L. Adams, and J. R. Andrewartha: "The finitely conducting lamellar diffraction grating," Opt. Acta **28**, 1087-1102 (1981).

E. N. Glytsis and T. K. Gaylord: "Three-dimensional (vector) rigorous coupled-wave analysis of anisotropic grating diffraction," J. Opt. Soc. Am. A **7**, 1399-1420 (1990).

M. G. Moharam and T. K. Gaylord: "Rigorous coupled-wave analysis of metallic surface-relief gratings," J. Opt. Soc. Am. A **3**, 1780-1787 (1986).

D. M. Pai and K. A. Awada: "Analysis of dielectric gratings of arbitrary profiles and thicknesses," J. Opt. Soc. Am. A **8**, 755-762 (1991).

Conformal Mapping Methods

M. Neviere, M. Cadilhac, and R. Petit: "Contribution a l'Etude Theorique d'Influence d'Une Couche Dielectrique sur l'Efficacite d'un Reseau Infiniment Conductreur," Opt. Commun. **6**, 34-54 (1972).

M. Neviere, M. Cadilhac, and R. Petit: "Applications of conformal mappings to the diffraction of electromagnetic waves by a grating," IEEE Trans. Ant. Propag. AP-**21**, 37-46 (1973).

Transformation of Coordinate System

J. Chandezon, M. T. Dupuis, G. Cornet, and D. Maystre: "Multicoated gratings: a differential formalism applicable in the entire optical region," J. Opt. Soc. Am. **72**, 839-846 (1982).

L. Li: "Numerical modeling of multilayer-coated gratings," Tech. Digest Series Opt. Soc. Am. v.**11**: *Diffractive Optics: Design Fabrication, and Applications*, Rochester, 264-267 (1994).

E. Popov and L. Mashev: "Conical diffraction mounting: generalization of a rigorous differential method," J. Optics (Paris) **17**, 175-180 (1986).

Integral Methods

H. A. Kalhor and A. R. Neureuther: "Numerical method for analysis of diffraction gratings," J. Opt. Soc. Am. **61**, 43-48 (1971).

D. Maystre: "Sur la diffraction d'une onde plane par un reseau metallique de conductivite finie," Opt. Commun. **6**, 50-54 (1972).

D. Maystre: "A new general integral theory for dielectric coated gratings," J. Opt. Soc. Am. **68**, 490-495 (1978).

D. Maystre: "A new theory for multiprofile, burried gratings," Opt. Commun. **26**, 127-132 (1978).

R. P. McClellan and G. W. Stroke: "Solution of the non-homogeneous Helmholtz equation for optical gratings with perfectly conducting boundaries," J. Math. Phys. **45**, 383-390 (1966).

A. R. Neureuther and K. Zaki: "Numerical methods for the analysis of scattering from nonplanar periodic structures," U. R. S. I. Sypm. Electrom. Waves, Alta Freq. **38**, 282-285 (1969).

J. Pavageau, R. Eido and H. A. Kobeissé: "Equation intégrale pour la diffraction électromagnétique par des conducteurs parfaits dans les problemes à deux dimensions. Application aux réseaux," C. R. Ac. Sc. Paris **264**, Serie B, 425-427 (1967).

J. Pavageau and J. Bousquet: "Nouvelle méthode pour la détermination de la répartition angulaire de l'energie diffractée par un réseau conducteur," C. R. Ac. Sc. Paris **268**, Serie B, 776-778 (1969).

J. Pavageau and J. Bousquet: "Diffraction par un réseau conducteur nouvelle méthode de résolution," Opt. Acta **17**, 469-478 (1970).

R. Petit: "Etude théorique de la diffraction d'une onde plane et monochromatique par un réseau métallique infiniment conducteur," Appl. Opt. **4**,, 1551-1554 (1965).

J. L. Uretsky: "The scattering of plane waves from periodic surfaces," Ann. Phys. **33**, 400-427 (1965).

M. A. Wirgin: "Considérations théoriques sur la diffraction par réflexion sur des surfaces, quasiment planes, applications à la diffraction par réseaux," Compt. rend. Acad. Sci. **259**, 1486-1488 (1964).

Fictitious Sources Methods

A. Boag, Y. Leviatan, and A. Boag: "Analysis of two-dimensional electromagnetic scattering from non planar periodic surfaces using a strip current model," IEEE Trans. Ant. Propag. **37**, 1437-1449 (1989).

G. Tayeb, R. Petit, and M. Cadilhac: "The 'synthesis method' applied to the problem of diffraction by gratings; the method of fictitious sources," SPIE vol.**1545** Intern. Conf. on the Applic. and Theory of Periodic Structures, 95-105 (1991).

Rayleigh Methods

N. R. Hill and V. Celli: "Limits of convergence of the Rayleigh method for surface scattering," Phys. Rev B **17**, 2478-2481 (1978).

J. P. Hugonin, R. Petit, and M. Cadilhac: "Plane-wave expansion used to describe the field diffracted by a grating," J. Opt. Soc. Am. **71**, 593-598 (1981).

B. A. Lippmann: "Note on the theory of gratings," J. Opt. Soc. Am. **43**, 408-408 (1953).

D. Maystre and M. Cadilhac: "Singularities of the continuation of fields and validity of Rayleigh's hypothesis," J. Math. Phys. **26**, 2201-2204 (1985).

W. C. Meecham: "Variational method for the calculation of the distribution of energy reflected from a periodic surface. I," J. Appl. Phys. **27**, 361-367 (1956).

R. Petit and M. Cadilhac: "Sur la diffraction d'une onde plane par un réseau infinitment conducteur," C. R. Acad. Sci. Paris, Serie B **262**, 468-471 (1966).

E. Popov and L. Mashev: "Convergence of Rayleigh-Fourier method and rigorous differential method for relief diffraction gratings," Opt. Acta **33**, 593-605 (1986).

G. W. Stroke: "Diffraction gratings" in *Handbook of Physics*,v.**29**, *Optical Instruments*, ed. S. Flugge (Springer, Berlin, 1967).

Yasuura Method

H. Ikuno and K. Yasuura: "Improved point-matching method with application to scattering from a periodic surface," IEEE Trans. Ant. Propag. AP-**21**, 657-662 (1973).

T. Matsuda and Y. Okuno: "Computer-aided algorithm based on the Yasuura method for analysis of diffraction by a grating," J. Opt. Soc. Am. A **7**, 1693-1700 (1990).

Approximate Methods

H. J. Gerritsen, D. K. Thornton, and S. R. Bolton: Application of Kogelnik's two-wave theory to deep, slanted, highly efficient, relief transmission gratings," Appl. Opt. **30**, 807-814 (1991).

H. Kogelnik and C. V. Shank: "Coupled-wave theory of distributed feedback lasers," J. Appl. Phys. **43**, 2327-2335 (1972).

S. T. Peng, T. Tamir, and H. Bertoni: "Theory of periodic dielectric waveguides," IEEE Trans. Micr. Theor. Techn. MTT-**23**, 123-133 (1975).

G. I. Stegeman, D. Sarid, J. J. Burke, and D. G. Hall: "Scattering of guided waves by surface periodic gratings for arbitrary angles of incidence: perturbation field theory and implications to normal mode analysis," J. Opt. Soc. Am. **71**, 1497-1507 (1981).

A. Wirgin: "Scattering from sinusoidal gratings: an evaluation of the Kirchkoff approximation," J. Opt. Soc. Am. **73**, 1028-1041 (1983).

A. Yariv: "Coupled-mode theory for guided-wave optics," IEEE J. Quant. Electron. QE-**9**, 919-933 (1973).

Chapter 11

Testing of Gratings

11.1 Introduction

Three attributes of diffraction grating quality generally determine their spectrometric usefulness. Therefore, measuring them with sufficient accuracy becomes an important concern, but one that varies from simple to complex. In some instances there are difficulties with establishing an adequate independent definition. The relative importance of the criteria depends on the application. In critical situations all three must be respected, while in other instances one of them may dominate.

Spectral purity, or the absence of stray light, is nearly always of interest because it largely describes the signal to noise obtainable. It is strongly related to the uniformity of groove spacing.

Efficiency describes the fraction of the incident light that is diffracted into the order of use at the wavelength band of interest. It is a function of the macrogeometry of the optics, but is especially sensitive to the microgeometry of the grooves and the nature of their surface.

Resolution (resolving power) of a grating is the third major attribute. It is of special interest in atomic and astronomic spectrometry, but there are many instruments, especially in molecular spectrometry, that do not require even 10% of what is theoretically attainable. It is a function of both the ruled width of a grating and the uniformity of its groove spacing.

In the case of concave gratings an additional criterion enters the field, in the form of *image aberration*. This arises from the fact that these gratings are required to perform both as imaging and dispersing devices. Sometimes one has to choose between minimizing astigmatism (for energy considerations) or coma (for resolution considerations).

Occasionally other grating properties must be considered as well. An example might be the ability to *survive adverse environment*, such as encountered in Synchrotron or high power laser applications. *Cosmetic appearance* is a classic source of concern, because the human eye is such a superb detector of minute local blemishes, even though the effect on instrumental behavior may be negligible.

Since most gratings are used in the form of replicas it is important to appreciate which of the attributes are likely to be modified in the process and which are not. There is no point in wasting time making measurements that always give the same result.

11.2 Spectral Purity

An ideal grating is one of infinite size with perfectly uniform groove spacing (perfect in the sense of 1 part in 100,000 of the groove spacing), and uniformly illuminated with collimated light. It is designed to diffract only in the directions dictated by the grating equation. In practice the finite size of real gratings will generate small secondary maxima between the orders due to Fraunhofer diffraction. Fortunately this rarely becomes a detectable problem in practice. Any light diffracted over and above the Fraunhofer diffraction is, by definition, a spectral impurity and considered undesirable. It can originate from grating imperfections, our concern here, but also from the auxiliary optics that make it into an instrument, and without which a grating cannot be tested. This already points to one of the difficulties of accurately evaluating the stray light behavior of gratings.

An experimental demonstration of Fraunhofer diffraction was shown in Fig.6.4. for a 316 gr/mm 63.5° groove angle echelle grating.

11.2.1 Effects of Grating Deficiencies on Spectral Purity

With classically ruled gratings it was long taken for granted that given the extreme sensitivity of gratings to periodic errors (see Chapter 14) no grating would fail to generate the multiplicity of faint secondary slit images, called ghosts because they are exact but low intensity replicas of the principal image. At one time ghosts were used to advantage as low intensity references, when spectroscopy was largely the domain of photography, thus making a virtue of necessity. Largely for historical reasons we distinguish between two types of ghosts, with the names Rowland and Lyman attached. Most important are those traceable to the inevitable defects in the lead screw and its mounting, which have long periods compared to the groove spacing. The inverse relationship between period and the corresponding diffraction angle leads to ghosts that are closely spaced and usually symmetric about either side of the parent line. These are the Rowland ghosts, easily found by simply observing the immediate vicinity of a narrow intense spectral line image, as formed in a high quality spectrometer with at least 1m focal length. Intensities can be as high as a disastrous 1%, occasionally even more, but with the help of modern interferometric feedback control ghosts have become almost extinct, or at least negligible. It is rare today to find a grating whose performance is seriously compromised by the presence of Rowland ghosts, even at high angles or fine pitches. For routine purposes we can consider that a 600 gr/mm grating used at

20° angle of diffraction should never have a ghost that exceeds 0.05% of the parent.

To claim that a ghost is undetectable merely indicates that its value does not exceed the background noise, and is therefore a statement that must be viewed with caution. The 'cleaner' the grating, i.e., the lower its background noise, the more even a small ghost becomes measurable. To put it in context, it is useful to measure the first order ghost as a fraction of the parent line, i.e., I_{Gh}/I_p, and note that to a good approximation (i.e., neglecting higher harmonics and phase effects), the amplitude of the periodic error ε is given by

$$\varepsilon = \frac{\lambda}{2\pi \sin\theta_i}\sqrt{\frac{I_{Gh}}{I_p}} = \frac{d}{m\pi}\sqrt{\frac{I_{Gh}}{I_p}}, \qquad (11.1)$$

or the inverse relationship

$$\frac{I_{Gh}}{I_p} = \left(\frac{2\pi\varepsilon \sin\theta_i}{\lambda}\right)^2 = \left(\frac{m\pi\varepsilon}{d}\right)^2, \qquad (11.1')$$

where λ is the wavelength, θ_i the angle of incidence and diffraction (Littrow conditions) and m is the order number. This equation assumes a simple harmonic error. On older gratings one may encounter higher harmonics, especially the second, caused mainly by eccentricity of the lead screw bearings. It will thus vary in amplitude as well as phase along a ruling, giving rise to complex ghost patterns. With strong monochromatic light and low background noise one can sometimes see as many as 100 ghosts, even though most of them will have intensities $< 10^{-6}$. From eq.(11.1) it is evident that for a given error amplitude ε the ghost ratio will increase with the $\sin^2$ of the angle of diffraction, for a given wavelength, and with λ^{-2} for constant angle. This explains why high-angle diffraction gratings, echelles in particular, are especially sensitive to periodic error. To locate the first order Rowland ghost one notes the number of grooves ruled for each turn of the lead screw, g. The first order ghost order will be found in the first diffraction order at a wavelength of λ/g removed from the parent line. The second order ghost will be located twice as far, etc. If the grating is used in higher orders m, the first order ghost is located λ/mg from the parent. The illustration in Fig.11.1 is for a 76° 31.6 gr/mm echelle. The 632.8 nm wavelength corresponds to the 97-th order, so that the screw ghost will be located 0.081 nm at either side of the parent wavelength. This corresponds to an angle of 0.068°, as recorded. An error amplitude of ~1 nm is easily detected, due to the 10^{-5} background noise.

The phase difference between parent and ghost lines is capable, under

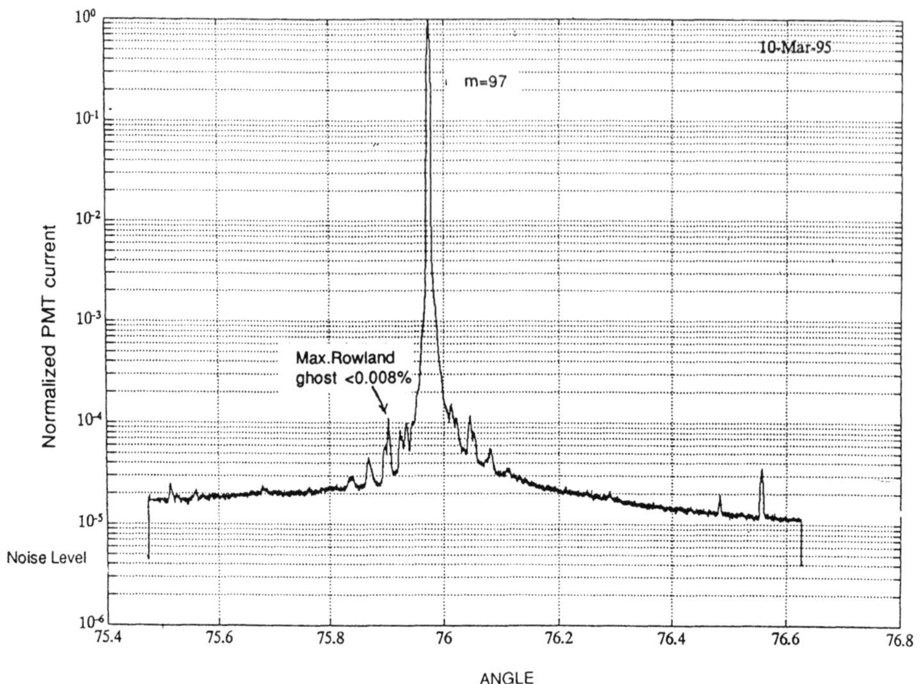

Fig.11.1 Ghost trace at 632.8 nm of a 31.6 gr/mm 76° blaze echelle grating at order 97. Maximum ghost of 0.008% corresponds to the 2.54 mm pitch of the lead screw and an error amplitude of 0.93 nm. Background stray light is about 10^{-5}. Beam diameter 200 mm (courtesy Spectronics Instrument Co.).

special conditions, of displaying bands generated by interference effects, which spoil the uniformity of the diffracted field. This has been found especially disturbing for high power laser pulse compression applications. The only solution is to replace such gratings with alternatives ruled under better control.

Lyman ghosts are somewhat more difficult to eliminate completely in mechanical ruling because they are the result of short term periodicities, which are not so easily suppressed by control systems because of integration used in the feedback loop. In addition they are not nearly as easily detected because their short period (a few grooves) places the ghosts far from the parent line. In fact it is easy to miss them entirely, unless the region between two orders is completely and slowly scanned, while given intense monochromatic input. The presence of a weak Lyman ghost line may easily be mistaken for a true spectral emission line. In practice this rarely happens, especially given the low residual

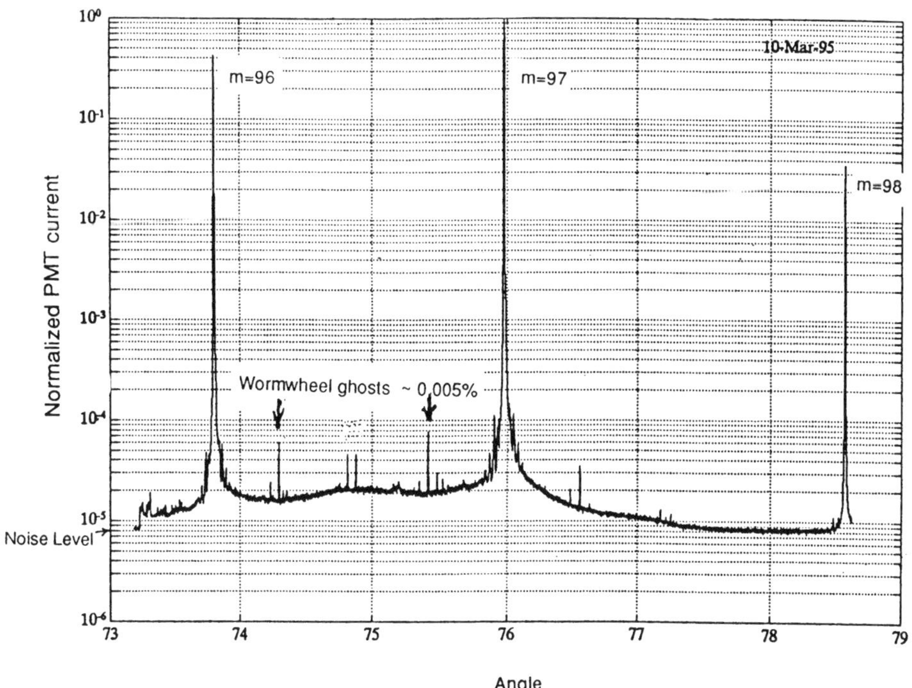

Fig.11.2 Lyman ghost trace of a 31.6 gr/mm 76° echelle grating, in order 97 at 632.8 nm, as seen on a 10 m Czerny-Turner test bench. Entrance slit 0.3x10 mm. Note logarithmic recording of intensity. Wormwheel ghost intensity of 0.005% corresponds to an error amplitude of ~ 1.5 nm (courtesy Spectronic Instruments Co.).

values seen with modern ruling engines. Worth mentioning is that care is indicated in using eq.(11.1) to estimate the Lyman error amplitude, because it makes no allowance for the fact that intensity will not only be a function of the parent line, but may be further reduced by being located away from a blaze peak. For example if the ghost efficiency is half that of the parent the error amplitude will be underestimated by a factor of 1.4. In absorption spectrometry they play no role.

The source of Lyman ghosts is usually traceable to less than perfect gears or belts in the final portion of the lead screw drive system. Especially vulnerable is the commonly used worm-wormwheel combination, with the latter attached directly to the leadscrew. As an example we can take a ruling engine (see sec.14.5.5) with a screw of 2.54 mm pitch, driven in turn by a wormwheel having 180 teeth. Since one revolution of the worm will advance the carriage by 14.11 μm, this becomes the primary period of Lyman ghosts. One conclusion is that any groove spacing that is an integral multiple of this figure will lead to a grating free of this particular ghost. To predict the location of the corresponding Lyman ghost is simple. For the case above, if the groove spacing is 31.6 μm the fractional order position where Lyman ghosts will be found is given by 14.11/31.6, or 0.446. A spectral scan of a grating that that demonstrates this is shown in Fig.11.2. The Lyman ghost is clearly visible 2.1 Å from the exciting line, despite its low intensity (0.005%), and corresponds to a ruling error amplitude of about 1.5 nm.

For comparison we may note the presence a single 0.002% Rowland ghost, whose error amplitude is no more than 0.5 nm. It is interesting that the noise intensity of this grating is so low that ghosts become invisible only when the error amplitude is < 0.3 nm (3Å).

11.2.2 Non–Periodic Groove Position Errors

There was such intense preoccupation with periodic errors over the long interval between Fraunhofer (1821) and Harrison (1949) that much less thought was given to other types of groove mislocations. One likely reason is that they do not lend themselves to elegant mathematical analysis, and thus are less interesting to academics. It is certainly not because they are incapable of causing spectrometric mischief, or that they are easy to eliminate.

Random Errors

Random errors in the location of individual grooves are, by definition, difficult to suppress. They arise on a ruling engine almost spontaneously through such effects as small changes in critical oil film thicknesses or mechanical deformations caused by variable load or friction effects, since many of them are outside the error feedback control loop. Another source may be the occasional adherence of a small particle of aluminum to the face of the diamond tool, or a small spit mark on the metallic ruling coating.

The ability of small random errors to generate optical noise, is just as great as that of periodic errors (see sec.14.4.3), should their RMS values be similar. Luckily that is usually not the case. Nevertheless they are the reason why mechanical rulings can approach, but never quite match the spectral purity of interference (holographic) gratings. Their presence is simple to detect, requiring no more than a laser in a dark room. Pointing the laser at the grating produces spectral order images on the wall, although a piece of white paper will improve visibility. Random errors show up as a fuzz between two orders, easily seen by eye, even when local intensities are $< 10^{-7}$. In addition, should there be local spots of more concentrated light they will be recognized as Lyman ghosts.

To see Rowland ghosts with the naked eye takes more resolution.

To keep random errors at a minimum is one of the true challenges for ruling engine builders, especially in the diamond carriage drive system and oil film control, usually outside the control loop. Equally important is the quality and uniformity of the relatively thick metal coatings in which ruling takes place, and the quality of the diamond tools. It is interesting to compare the ease with which random errors can be observed with the great effort required to suppress them (scc Ch.14).

The faint fuzz between orders is sometimes called *grass*, because before lasers such testing was usually performed in the green light of a filtered mercury source (546.1 nm), at which the human eye has its maximum sensitivity. The resemblance to a strip of lawn is obvious to anyone who has made the observation.

One can consider the random errors as a Fourier sum of local periodic functions, varying in amplitude, phase and period. They can be thought of micro–gratings, generating their 'private' ghost patterns [11.1].

Satellites

When there are groove location errors that cover a significant portion of a ruling, without being periodic, they will deform the diffracted wavefront in such a way as to destroy the symmetry of the spectral line image. The result is close–in satellites, as in Fig.11.3. In high resolution spectroscopy, where line shapes are important, and especially in absorption spectroscopy, such satellites are undesirable because they tend to fill in the line just where the measurement is to take place. To keep more than one percent of the incident light from being converted into satellites it is necessary to maintain RMS errors below 0.028 λ, where λ is the wavelength of use. At 300 nm wavelength that would be 8 nm, and although partially suppressed with feedback control, is never completely removed.

This points to one of the tests for the presence of satellites, which is to scan the spectrum with white light input to a high resolution system, and interpose an absorption filter. This may take the form of a piece of Didymium or Holmium oxide glass or an iodine vapor cell. Any light detected in the center of an intense absorption line must be the result of scatter. The amount will be a function of the spectral pass band, i.e., of the specific set-up, rather than a simple well defined number, but serves well for comparing similar gratings. The test set-up needs careful baffling to prevent light reflected from mounting hardware reaching the detector.

An indirect test for satellites, that may be easier to perform, is to examine the uniformity of a wavefront interferogram, from which satellite shapes can be derived by Fourier transform [11.3]. It is especially useful for

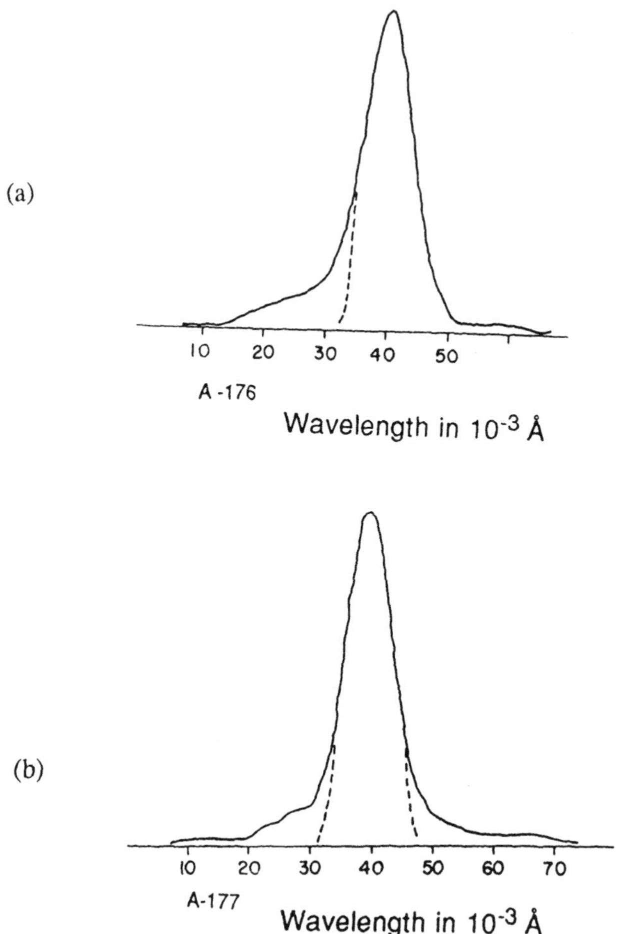

Fig.11.3 Spectral line profiles of ^{198}Hg source at 541 nm obtained with an 11m scanning spectrograph of two vintage 300 gr/mm, 63° blaze echelle gratings. Both have excessive satellites, but (a) shows much higher degree of asymmetry than (b). Dotted lines are for ideal scans (after [11.2]).

predicting behavior at wavelengths different from the one under test, particularly shorter ones (see section 11.4.3).

Roughness Induced Scattering

Roughness in UV – VIS gratings will be at subwavelength levels in amplitude and period. This generates scatter in all directions, not limited to the meridional plane where one normally observes. It is thus easily recognized, especially with the help of laser illumination, and a suitably mounted screen. The shorter the wavelength, for a given roughness, the greater will be the effect. It is therefore of special concern in the UV region. In theory it should vary with λ^{-4}, rather than λ^{-2} as is the case for ghosts. In view of the small influence this is hard to prove [11.4]. In the x-ray region wavelengths are so much shorter that roughness values may even exceed the wavelength, which puts great pressure on manufacturing skills.

Effect of Variations in Groove Depth

In mechanical ruling we have come to be concerned with small variations in groove *spacing*. Not often appreciated is that there is a similar effect that arises from small variations in groove *depth*. This lack of concern can be explained by the fact that they become obvious only under unusual conditions. These occur when interaction between the elasticity of the tool mount of the ruling engine and the plastic flow properties of the ruling surface results in every other, or even every third or fourth groove being ruled deeper than the rest. The effect is to get two gratings superimposed on one another, highly undesirable. More likely in practice are small random variations in groove depth, whose diffraction effect is mixed with that of spacing variations. Both are too small to be individually measurable, and to get a separate measure of them they must be deconvoluted from stray light measurements. Complicating the picture still more is the need to simultaneously deconvolute the effect of scatter traceable to residual roughness. Since its effect is three dimensional, it can be identified by noting the degree to which scatter increases when the length of entrance and exit slits is doubled.

In order to perform the deconvolution the only method proposed so far is that of Sharpe and Irish [11.5]. It takes advantage of the fact that scalar analysis of the problem predicts that the response to the different sources of stray light depends on wavelength. Using this approach they analyzed a typical spectrophotometer grating, with 1276 gr/mm, 8° blaze angle, at two wavelengths, 313 and 633 nm. The conclusion was that the random groove spacing variation had 0.8 nm standard deviation from the nominal 784 nm (i.e., 0.1%), while the groove depth showed a 0.3 nm standard deviation from the nominal 120 nm (i.e., 0.25%). The surface roughness calculated to have a 10 nm correlation distance, enough to provide about 10% of the total stray light at 313 nm, but virtually none at 633 nm.

11.2.3 The Measurement of Grating Stray Light

To test for overall stray light behavior in an instrument the most effective tool is a set of glass cut–off filters. These transmit virtually no light at wavelengths below their designated value. Thus any light detected at wavelength settings below that value must represent stray light. Results taken from the above paper are shown in Fig.11.4. The conclusion is that at a 500 nm setting stray light is almost solely traceable to groove spacing variations, while at 200 nm the three sources contribute about equally. The key result is that one cannot judge grating stray light behavior by testing at a single wavelength, and that the shorter the wavelength the more severe and complex the problem.

Stray light measurements became much easier with lasers available to supply highly intense monochromatic light. Recording with photographic plates is nearly obsolete, since most spectrometry is done with photoelectric detection. Photomultipliers are the ideal detector for stray light tests because of their large dynamic range ($>10^6$), even though spectrometric instruments are more likely to use array detectors with a dynamic range of about 10^4. The grating is set up as part of a standard mount, usually a Czerny-Turner, the entrance slit illuminated with incoherent laser light. Light fibers are a convenient means of feeding the light and also supply incoherence. A typical measuring cycle starts by rotating the grating around an axis lying in its face, parallel to the grooves, until the diffracted beam in the blaze order hits the exit slit, and when maximum is reached the output reading is set to 1.0. The grating is then slowly rotated, or the exit slit detector combination may be slowly translated. A chart recorder will typically record the signal as a function of time, so needs to be correlated with the wavelength change. Because of the wide dynamic range it is highly desirable to use a logarithmic amplifier to display the data, as in Fig.11.2, but, if not available, one can change the amplifier attenuator settings and accept a stepwise chart [11.6, 7].

The size of the exit slit makes a difference to what is recorded. With Rowland and Lyman ghosts there is no problem, because their images are always identical in size to the primary image from which the initial setting was made. However, grass extends over the entire region between the two orders, so that flux reaching the detector will increase with the width of the exit slit, presuming that its width is sufficient to capture at least the entire primary image. The height of the slit will make no difference here. On the other hand in looking for diffuse scatter we take advantage of the fact that it increases with the height of the slit pair. This is because diffuse scatter takes place in all directions.

Such tests can become confusing when comparing gratings of different specifications, especially ruled and interference gratings. The former will have most of their stray light in the meridional plane while the latter will have most

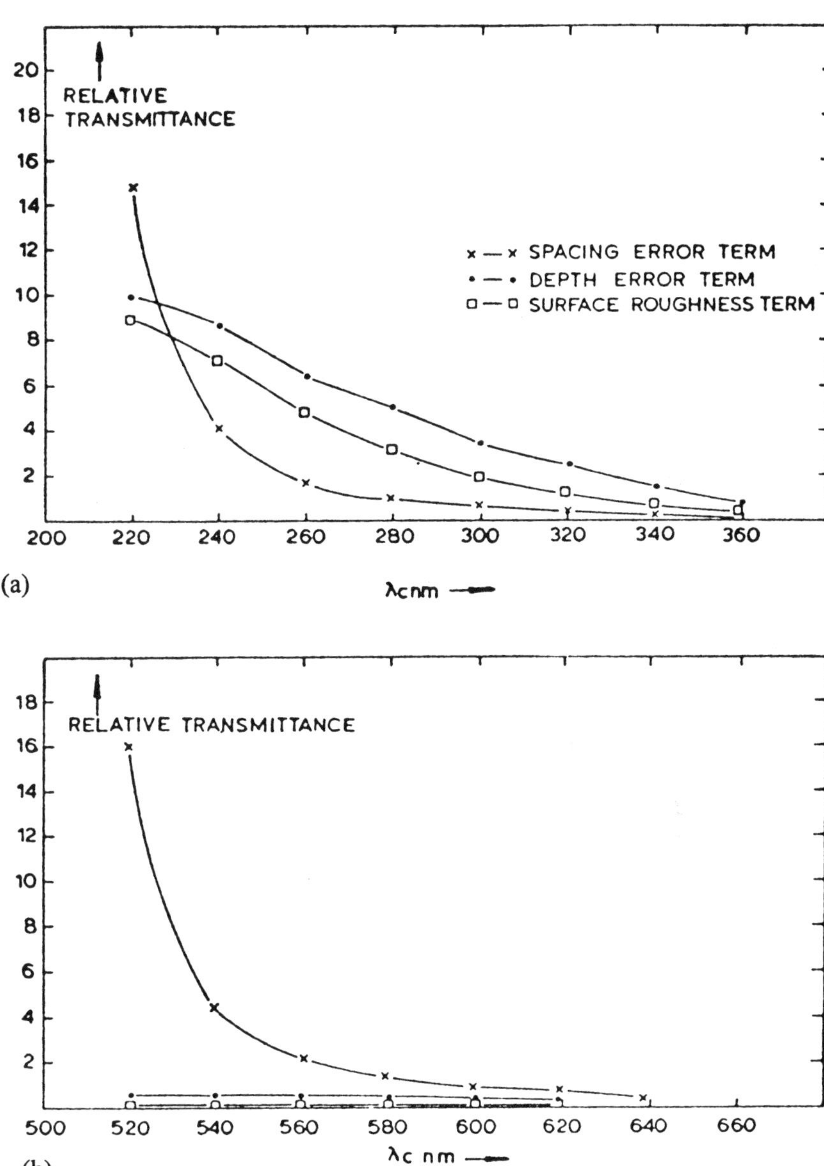

Fig.11.4 Computed transmittance of glass cut-off filters in a Pye-Unicam SP 1800 spectrophotometer with a 1276 gr/mm grating, plotted as a function of the filter cut-off wavelength λ_c and split into the contributions from the three types of grating imperfection, with the monochromator wavelength set (a) 200 nm and (b) 500 nm (after [11.5]).

of it outside, and will look worse, relatively, when slit height is increased. It is easy to make interference gratings look worse than ruled ones for stray light performance, when we know that is not really true.

One conclusion is that slits should always be as short as possible, the limit being set by the need to get enough flux for an adequate signal. Since there is no accepted standard, the reporting of grating interorder scatter is never well defined, except for a specific instrument.

The situation becomes still more confused when attempts are made to compare stray light behavior of ruled gratings with interference gratings of a differing groove frequency, or dispersion. For instance, the single most commonly used ruled grating has 1200 gr/mm (because it yields optimal energy distribution from the UV to the IR). For efficiency reasons it is sometimes replaced by an 1800 gr/mm interference grating. If the spectral band width is to be maintained constant the slits can be widened by a factor of 1.5, and thus increase throughput. If left constant it will appear to reduce the amount of stray light transmitted. In either case the comparison may not be fair, a point easy to overlook.

In most instruments, especially those intended for use in the UV, the stray light situation is more complex than indicated above. This is because the grating is not illuminated by a single wavelength, so convenient for analysis, but by a continuum of varying intensity, typically from a tungsten lamp. The stray light picked up by a detector will be the *integrated sum* generated by all the input wavelengths, convoluted by their relative intensities, as well as grating efficiency, into values that will be *different for every instrument*. This is often a source of frustration and is difficult to pin down, but for practical applications is tamed by the use of bandpass filters.

For example, take a spaceborne UV spectrometer designed to study relatively faint spectral lines in the vacuum UV (120 – 160 nm). This is impossible if there is any stray light transmitted from the intense flux delivered over the entire visible region. The result is that it would make little difference if the grating could have had its stray light performance improved 100 times. The only valid solution is to make use of detectors that are solar blind, such as CsI, which do not respond to any radiation above 160 nm, thus acting as ideal filters.

The conclusion to be drawn is that stray light represents a mix of phenomena, with a complex convolution of grating, slits, source and detector characteristics, plus the optics of the instrument walls. While in most cases the grating contribution is the most important, it is often observed that the largest single source of stray light is none of the factors discussed above, but consists of the spectrometrically useless zero order bouncing from the instrument walls. Instrument designers must keep such light from reaching the detector with the help of suitable baffles and absorbers.

11.2.4 Locating Stray Light Sources on a Grating Surface

A problem that has diminished greatly with modern rulings, but used to preoccupy many experimenters, is how to locate areas on ruled gratings that produce more than their share of ghosts or other stray light. The idea was that once identified such areas could be masked off. Since 'disturbed' areas manifest their misbehavior by diffracting in a slightly different direction a technique is required to detect it. The Foucault test is ideally suited to such testing, and is described in section 11.4.2.

11.3 The Measurement of Efficiency

There is no grating attribute that is more carefully and more frequently measured. The reason is easily appreciated: efficiency determines throughput of any spectrometric device that uses gratings, in the case of high power lasers it may even determine whether the grating will survive. While grating suppliers may pride themselves on the quality and uniformity of their replication process, it is often observed that there are variations in the performance of the final product. In many cases they are too small to worry about, in others they are so critical that every grating has to be evaluated. The principal source of variation lies in the behavior of replicating resins. There are none that do not shrink slightly during polymerization. To make matters worse, there can be slight variations in the degree of shrinkage, which leads to variations in groove geometry. In some cases as little as 5 nm variation in groove angle over its length is sufficient cause for concern, so it is easy to appreciate the need for testing (see Chapter 17).

In defining efficiency we distinguish between relative and absolute efficiency, largely for historical reasons and convenience. For the grating user it will be the *absolute efficiency* that matters, the fraction of incident monochromatic light that is diffracted into a given order under specific incidence conditions. The bandpass requirement for the incident light will vary with the degree of dispersion involved. Since the input monochromator as well as the grating under test will often have polarization properties it is usually advisable to take complete sets of data in both planes of polarization, with the electric vector of the incident light parallel to the grooves (TE or P plane) and perpendicular (TM or S plane). Unpolarized incidence can either be generated by taking the arithmetic average of the two, or measured directly with an input polarizer set at 45°. The latter saves time, but provides no information about the polarizing properties of a grating.

Relative efficiency is defined as the ratio of the diffracted light in a given order compared to that reflected by a mirror of the same metallic surface under the same angle of incidence. The 100% setting is obtained from the beam reflected from a reference mirror, instead of by the incident beam itself. There is a technical advantage in that it reduces instrumental manipulations. It also gives the appearance of measuring the groove efficiency itself, not handicapped by lack of perfect reflectance. There is the implication that absolute efficiency is easily derived by multiplying the relative readings by the known reflectance of the reference mirror. This holds true in most instances, but there are special cases (in particular 1800 to 2400 gr/mm gold or copper gratings) where there is significant departure from this simple rule at wavelengths < 600 nm. There will also be discrepancies when there are localized absorption effects, as in the anomaly region of TM polarized light (see Chs.4 and 8). It should also be clear that the angle between incident and diffracted beams is not usually the same as with a mirror.

11.3.1 Efficiency Measurement Systems — Plane Gratings

The classical efficiency measuring system consists of a monochromator supplying light to a second monochromator mount containing the test grating. The supply monochromator should cover a wide range of wavelengths, typically 190 nm to 2.5 μm, which implies at least two light sources and three interchangeable gratings. A Deuterium lamp is ideal for the 190 to 400 nm range, while a tungsten lamp covers from 300 nm to 2.5 μm. Two 1200 gr/mm gratings, one blazed in the UV the other in the visible, and a 600 gr/mm for the IR region are good choices. Often a low pressure mercury lamp is also supplied. A polarizer such a Glan-Thompson prism as well as a selection of order sorting cutoff filters may be interposed in the beam. They are necessary to prevent high orders of short wavelengths superimposing themselves onto longer wavelengths, Fig.11.5.

The test grating must be illuminated in collimated light, which the concave mirrors serve to collimate and de-collimate. The test grating rests on a traversing stage so that it can have its properties explored along its length. A light chopping wheel allows the use of lock-in amplifiers, which greatly improves signal to noise ratio, especially important with IR detectors. The latter are typically PbS types, while photomultipliers are used over the rest of the spectrum. The grating under test rests on an accurate rotary indexing table (RT), which in turn is located co-axially on a swinging arm (SA) that carries the detection optics, and whose angle is also tracked with a built–in transducer.

The detector arm should be able to rotate at least 180° so that it can operate in the absolute mode for setting the output amplifier reading to 100% for the incident light at each wavelength. This position is also necessary for evaluating transmission gratings. If the entrance slit and exit detector are both located in the meridional plane they will interfere mechanically at some minimum angle, which then becomes the angular deviation angle closest to

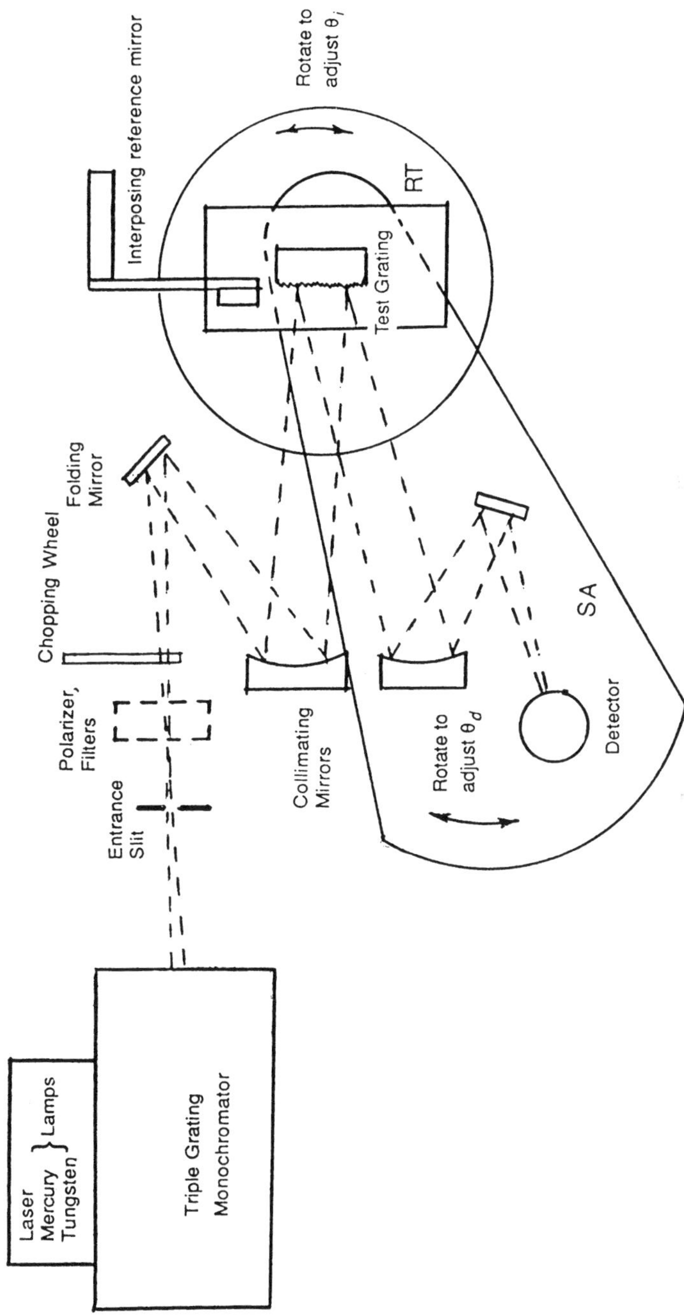

Fig.11.5 Two monochromator system for measuring grating efficiency.

Littrow. Alternately, one can chose to work slightly outside the meridional plane (i.e., in conical diffraction) by an angle of say 2° above and below, in which case one can work in Littrow with A.D. = 0. In most cases the effect of such a small conical angle on efficiency can be considered negligible. If the demands are sufficiently critical the grating can be tested at the A.D. used in a specific instrument, typically from as low as 8° to as high as 60°. The corresponding effects can be judged by the families of efficiency curves shown in Chapter 4.

The measuring sequence usually involves programmed stepping through an appropriate sequence of wavelength intervals, interposing the reference mirror after the grating (and mirror) incidence angles have been set, rotating the detector arm to pick up the reference signal from which the 100% reading is set or recorded. With the reference mirror retracted the detector arm is rotated to pick up the diffracted order under test. Depending on the wavelength the appropriate cutoff filters must be inserted, and even changed if the wavelength band is wide enough to require that. The entire sequence is then repeated in the alternate plane of polarization, after resetting the polarizer. This becomes quite a lengthy sequence, but lends itself to control by a small computer, that can also collect the output information and print or plot it. When the supply voltage and temperature are sufficiently well stabilized, it is practical to shorten the sequence by recording monochromator output values as a function of wavelength just once or twice a day. Only tests can show whether this can be done safely. If the opposite should be true the input light intensity can be continuously monitored with a beam splitter to a reference detector. The effect of small fluctuations can then be introduced into the data reduction chain.

A look at the efficiency curves in Chapter 4 shows that efficiency varies rapidly in some portions of the spectrum and quite slowly in others. The regions are quite well defined by the ratio of λ/d. Thus, when this ratio is <0.9 wavelength steps should be taken in units of 0.025 λ/d, while above that value steps of 0.1 will be quite adequate. Much time can be saved by taking this into account.

When efficiency testing is used to monitor production uniformity it is not necessary to generate complete efficiency curves. It will be quite adequate to monitor just two or three wavelengths, one near the peak value, and do so only in the TM plane of polarization, since it is always more sensitive to small variations than the TE plane.

If gratings are to be used in a spectrograph mode the angle of incidence will be fixed, and only the detection arm moved with change in wavelength. In this instance it will save time to measure the 100% settings for each wavelength in a single sequence before testing the grating, providing there is adequate stability for the accuracy desired.

Although the procedure just described is intended for plane gratings it works quite well for concave gratings with radii > 1 m. While imaging may not be perfect that does not matter as long as the detector accepts all the diffracted light. This is a fortunate circumstance since it avoids building large test equipment for every radius, especially long ones.

11.3.2 Efficiency Measurement Systems - Concave Gratings

To measure the efficiency of concave gratings with radii < 1 m requires modification of the system described above, because we must respect the image forming conditions of such gratings, at least to some extent. The input monochromator as well as the detection system can be made simpler because such gratings are rarely used in the IR region; a typical system is shown in Fig.11.6.

Light from an instrument monochromator covers the 190 to 900 nm region that is usually of interest, and is focussed into an entrance pinhole of the second monochromator. An order sorting filter wheel and polarizer are interposed as above. The whole monochromator is located on a simple slide so

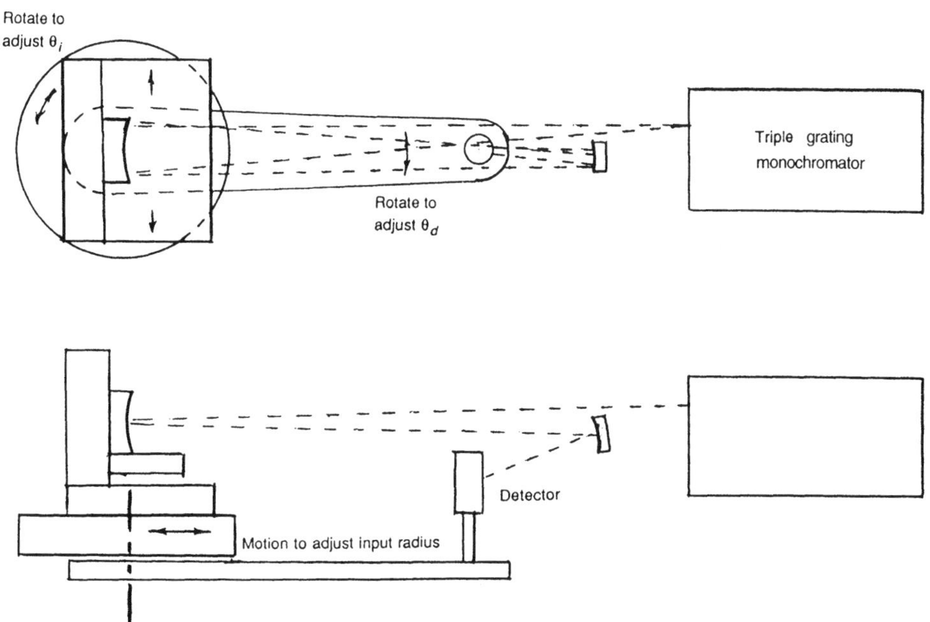

Fig.11.6 Schematic Diagram of system for measuring efficiency of concave gratings (courtesy Spectronic Instruments Co.).

that the entrance aperture will be at the correct distance from the concave test grating. The latter must be on a rotating stage, its axis coinciding with the grating pole, which allows setting the incidence angle. Underneath the stage is a detector arm, as before, whose axis of rotation matches the grating stage, and whose detector assembly and associated preamplifier can be moved during set–up to accommodate the imaging distance. This is not a critical adjustment, as long as the detector receives all the diffracted light.

There is no longer a good method for making this method absolute, so that relative measurement will be the norm. It is done by taking a concave mirror of the same radius in place of the grating, except that now there is little choice but to rapidly scan the entire desired spectrum first with the mirror and then with the grating, recording the results graphically or numerically, and then extracting the efficiency ratios. Given the usually small diffraction angles there will be only a small discrepancy when converting relative efficiency to absolute by multiplying by the known metal reflectance. There may be discrepancies in the TM plane because in the anomalous diffraction region there can be local absorption effects. The method presumes that the input monochromator maintains an output sufficiently constant over the combined scanning and blank changeover time or that appropriate compensation is made for fluctuations.

11.3.3 Efficiency Measurement Systems — Echelle Gratings

Echelle gratings (see Chapter 6) present two special challenges in measuring their efficiencies. This is because they operate at high dispersion and with a small free spectral range. Both of these attributes call for input light of extreme spectral purity. For example, at low wavelength and high order a complete free spectral range may be as small as 2 nm, so that to scan through one order with enough resolution to be useful requires that bandwidth of the incident beam be no more than 0.1 nm, and yet contain enough flux for a good measurement. Even when the FSR is larger than 2 nm the bandwidth must still be small, to avoid dispersion large enough for the spectral image to exceed the detector area, leading to incorrectly low values for efficiency. Even if a large high resolution monochromator is utilized its slit size would have to be so narrow that the product of flux times bandwidth may be too small, even with an intense tungsten lamp. The only solution is to use a light source of high luminosity over a narrow band, in other words a spectral source, such as a mercury lamp or laser.

A dye laser of sufficient purity and wavelength control (0.01 nm) would perhaps be ideal, but is relatively expensive. A low pressure mercury source is adequate and so are gas lasers, but they provide only a limited number of wavelengths. If the object is to determine accurately the efficiency in the center of a free spectral range the wavelength needs to fall within 10% of its width. It takes considerable luck to find among the limited selection a wavelength that happens to qualify, so that alternative approaches need to be found.

The second attribute of an echelle, important to instrument designers, is to determine with an unusual accuracy of 0.1° the diffraction angle at which maximum efficiency will be attained, preferably near the key wavelength of use, because this angle changes somewhat with wavelength (see section 6.4.6). An interesting technique for doing so requires a long focal length test spectrometer with a small A.D., for which the light source needs to be no more than a relatively simple, wide bandpass monochromator with a tungsten light source. Bandpass of the monochromator is set wide enough to encompass as many as six or more orders so that the flux will be high and there will be good averaging. The measurement consists of slowly scanning the test grating over

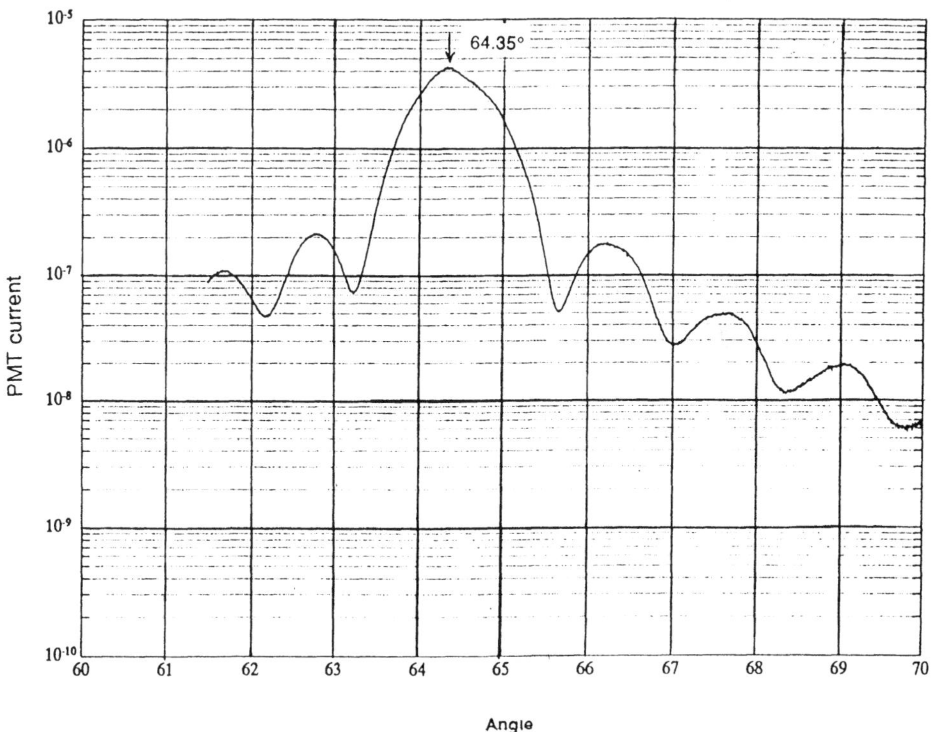

Fig.11.7 Spectral scan of 31.6 gr/mm echelle grating with 63.5 nominal blaze angle, using quasi-monochromatic light at ~543 nm with pass band of ~ 52 nm. Peak observed at 64.3° (courtesy Spectronic Instruments Co.).

an angular range from 3 degrees below to 3 degrees above the nominal blaze angle, and chart recording the output of the detector. This test has to be repeated until, by trial and error, a monochromator wavelength setting is found which produces a scan as symmetrical as that in Fig.11.7. The chart drive needs to be coordinated with the grating angle setting to within 0.05° so that the location of the blaze peak can be identified. This method has been compared with the result from a laser check that happened to be in perfect coincidence, and results obtained in the center of an order agreed with the method above to within the experimental uncertainty of 0.1°. The unexpected conclusion is that without knowing either the exact wavelength or the spectral pass band of the input monochromator such a fussy measurement can still be accomplished. Once completed it is a simple matter to use the grating equation and the approximate wavelength to determine the central order (105 in the case of Fig.11.7).

There is an alternate method for determining the effective groove angle of echelles, that is somewhat simpler, but gives the same result. The echelle is illuminated with the direct beam of a gas laser, such as the HeNe in the available 633 or 543 nm wavelengths, and the relative efficiency η_1 and η_2 is measured at the two brightest orders m_1 and m_2, as in Fig.6.8. Given the echelle groove spacing and a rotary table on which the grating can be placed to measure the diffraction angle (operating as close to Littrow as possible) there will be no doubt about which orders are involved. If we know the order number m and groove spacing d, the exact angles of diffraction θ_1 and θ_2 can be obtained from the grating equation

$$\sin\theta = m\lambda/2d \quad . \tag{11.2}$$

The exact blaze direction φ_B is obtained from

$$\sin\varphi_B = \frac{\eta_1 \sin\theta_1 + \eta_2 \sin\theta_2}{\eta_1 + \eta_2} \quad . \tag{11.3}$$

11.3.4 Checking Blaze Specifications

Ruled gratings are usually specified by the groove or blaze angle φ_B, and sinusoidal ones (interference or holographic) by their groove depth modulation h/d. A natural question is how to test for conformance. As in other instances this turns out to be a multifaceted problem.

The most obvious answer is to use direct measurement, for example in the form of a diamond stylus dragged across perpendicular to the grooves. The Talystep™ surface measuring instrument has been used for this purpose, but despite its low measuring pressure can still leave a visible mark [11.8]. More disturbing is that the finite cone angle of the stylus tip prevents reaching the bottom of the groove, leading to under-reporting of the groove depth. For sinusoidal grooves this problem vanishes, until groove spacing becomes so fine that the scanning rate can no longer be reduced to match it.

Modern atomic force microscopes (AFM) perform better in all respects, and deliver area information instead of being limited to a single line trace. However, while horizontal magnification is self-checking from observing the groove spacing, this does not hold for the vertical. A separate calibration artifact must be supplied, which could be a diffraction grating whose behavior is accurately known. Scanning electron microscopes (SEM) give a visual insight into the grating grooves, but are not much use for directly supplying depth information, see Fig.6.2 for an example.

In contrast the transmission electron microscope (TEM) is an excellent tool, although it requires considerable effort to prepare appropriate carbon film replicas, using asbestos fibers to cast slant shadows and calibrating the shadow

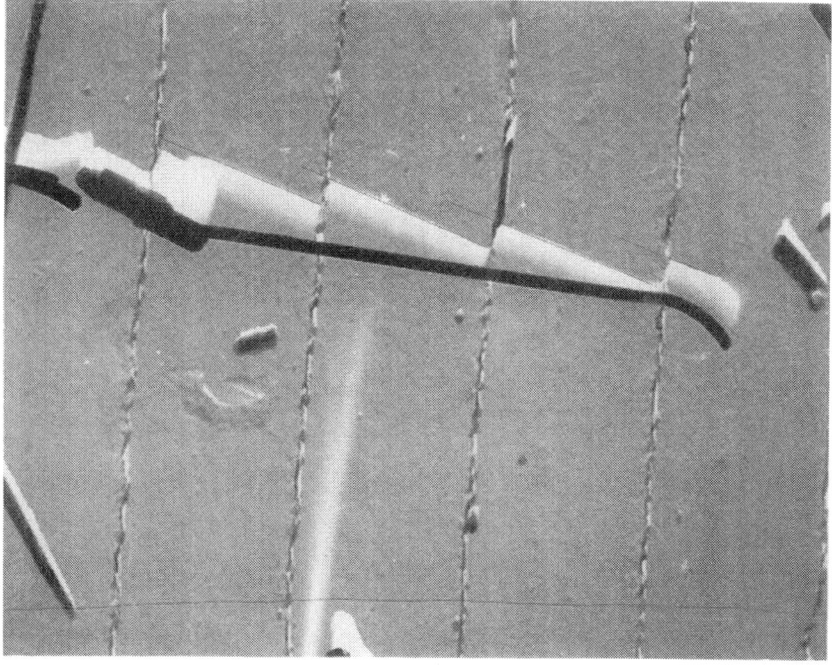

Fig.11.8a Electron micrograph of carbon replica given asbestos fiber shadowing: 300 gr/mm gold coated grating. Blaze angle 4°, groove step height 233 nm. Due to slant shadowing vertical magnification of asbestos fiber shadow is24x horizontal (courtesy Spectronics Instrument Co.).

angle with styrofoam balls to get vertical calibration [11.9]. Its principal use is for very fine pitch or low blaze angle rulings, as in Fig.11.8. Another possibility is to take a pair of TEM stereo photographs and evaluate them with photogrammetric viewers, hardly a routine process.

Other microscope techniques have been tried. One technique is to prepare a cross-sectional sandwich, in which a thin gold overcoating is removed from an aluminum grating by simple plastic film replication. A second plastic film is applied to the free surface of gold, the combination sliced with a microtome. This method is self-calibrating, and merely calls for a microscope of adequate resolution, even an SEM.

The simplest method for determining blaze conformance is to measure and plot the efficiency curves according to the methods outlined in sections 11.3.1 and 11.3.2. As long as the grating behaves in a sufficiently scalar fashion the wavelength of peak efficiency in unpolarized light will be λ_B, the blaze wavelength. It is related to the Littrow blaze angle by the grating equation

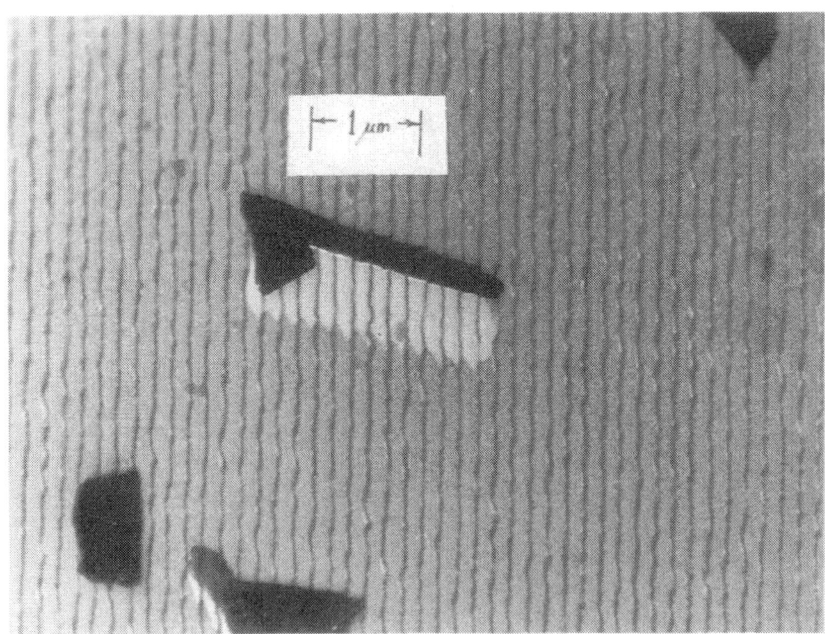

Fig.11.8b Electron micrograph of carbon replica given asbestos fiber shadowing: 5880 gr/mm gold coated grating. Blaze angle 24°, groove depth 67 nm (courtesy Spectronics Instrument Co.).

$$\theta_B = \sin^{-1}(m\lambda_B/2d)\ . \qquad (11.4)$$

This approach works well, even at angles well into the electromagnetic domain (i.e., up to about 30°), because of the relative shifts of both the TE and TM efficiency curves, even though that may not be evident from a quick glance. At greater angles it is better to make measurements in higher orders. The limit is reached with echelles, as described in section 11.3.3.

With sinusoidal groove gratings our concern is to determine the groove modulation h/d. There is no equivalent to equation (11.4). The alternative is to plot the efficiency curves in the TE and TM planes and compare their shapes with those theoretically derived. The necessary familiy of reference curves is found in Chapter 4, specifically Figs.4.28 to 4.47. If the groove frequency does not match those provided there we take advantage of the fact that efficiency is basically a function of the λ/d ratio, so that values in the abscissa are simply scaled in proportion. For example, if the grating has 600 gr/mm, the wavelength values compared to the 1200 gr/mm scale are doubled. While changes in λ do modify the curves somewhat the effect is unlikely to seriously affect judgment by more than a few percent.

11.4 The Measurement of Resolution

Spectral resolution, also commonly known as resolving power, has been an important characteristic of gratings from the earliest days. It is a difficult quantity to measure with accuracy, because any direct determination involves the optical system with which it is used. The more a grating approaches diffraction-limited performance the more important it becomes for the imaging optics to be of maximum quality, and this includes not only collimating optics but slits, the light source and the environment as well. Detectors and their noise levels also play a role.

The well known Rayleigh diffraction limit also applies here. Two spectral lines of equal intensity are considered to be separate if the peak of one matches the foot of the adjacent one, and this condition implies that the wavefront deviates from perfection by no more than $\lambda/4$ at the observing wavelength. Although necessary, it turns out that this is not always sufficient, because nothing is said about the possible presence of unsymmetric satellites at the foot of each line (see section 11.2.2). Traditionally astronomers have guarded themselves against this problem by asking for higher resolution than demanded by the Doppler width of stellar spectra. For the same reason the traditional FWHM (full width at half maximum) or line width at the 0.4 of peak value, as measured with light of maximum purity, such as a frequency stabilized laser, is not always sufficient.

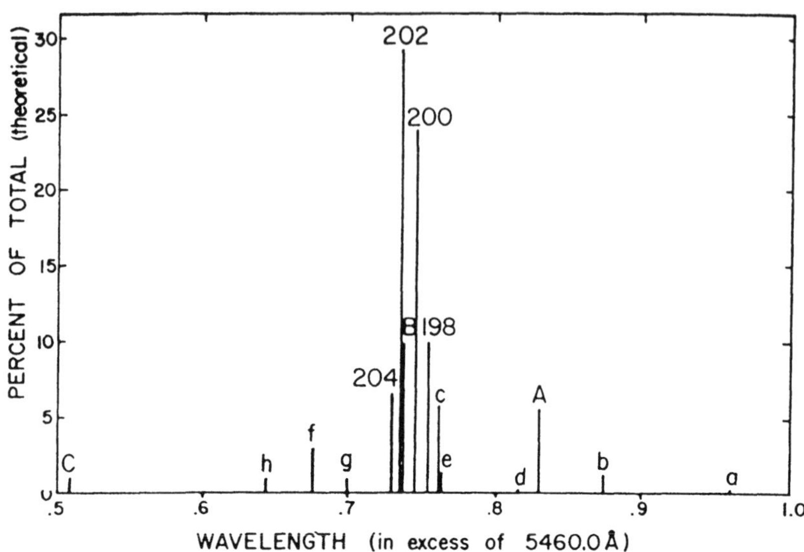

Component	Wavelength (Å)	Resolving power	
		Interval	Thousands
198	5460.753	199*C*–201*h*	40.5
199*A*	5460.831	201*h*–201*f*	165.0
199*B*	5460.737	201*f*–201*g*	260.0
199*C*	5460.509	201*g*–204	182.0
200	5460.745	204–202	910.0
201*a*	5460.961	202–199*B*	1820.0
201*b*	5460.875	199*B*–200	685.0
201*c*	5460.762	200–198	685.0
201*d*	5460.816	198–201*c*	606.0
201*e*	5460.763	201*c*–201*e*	5461.0
201*f*	5460.677	201*e*–201*d*	103.0
201*g*	5460.698	201*d*–199*A*	364.0
201*h*	5460.644	199*A*–201*b*	124.0
202	5460.734	201*b*–201*a*	63.5
204	5460.728	—	—

Fig.11.9 Spectral structure of the Hg 546.1 nm green line used for resolution testing. Resolution table is from theoretical values and does not take into account differences in intensity (after [11.10]).

The alternative is to measure interferometrically the shape of the diffracted wavefront, rather than the spectrum. Given enough data from an interferogram, it is possible to construct the corresponding line shapes expected with ideal optics, and do so for wavelengths other than the test one (see section 11.4.3). A second alternative is to observe the derivative of the wavefront with the aid of a Foucault knife-edge test (see section 11.4.2).

Probably the simplest test of all is to observe visually the spectrum of the blue or green hyperfine line set that is produced by a simple low cost, low pressure mercury source. It does not give exact numbers, but readily provides a minimum value that the grating must have (see section 11.4.1).

11.4.1 Testing with the Mercury Spectrum

Testing resolution by observing the hyperfine lines of an air-cooled Hg lamp requires a spectrometer with a focal length and mirror quality that matches the resolution of the test grating. Smaller gratings are readily checked with a 3 m focal length, but the largest gratings (400 mm) are better tested with 10 m optics. Mirrors or lenses can serve as collimators, as long as they are of diffraction-limited quality.

The two most useful spectral lines are at 435.8 and 546.1 nm, the latter being preferred for visual work. Both wavelengths have a large number of lines, because mercury happens to have not only 4 even numbered isotopes but 2 odd ones as well with many atomic transitions. They are shown in Fig.11.9, from which it is apparent that the lines are not only distributed irregularly but have widely differing intensities. Since no two neighboring lines have equal intensities any separation observed means that the resolution must exceed the simple Rayleigh limit. As an example, it takes a nominal 260,000 resolution grating to observe the f–g pair, whose wavelength difference is only 0.022 Å. In the center of the spectrum are a set of 5 closely spaced, intense lines from the 198, 200, 202, 204 isotopes plus the 199B line [11.10]. Only large gratings (>300 mm) at high angles of diffraction can separate the components, as photographed in Fig.11.10. Even at a 10 m focal length the 200–202 pair lines are separated by only 40 μm linearly and must be magnified. Residual vibrations and air currents always cause some smearing out during the long exposure necessary. An observer with good, dark adapted eyes will see the lines more clearly, and in particular can tell how clean (dark) the spectrum is between lines. Observers usually underestimate the resolution they can see and report by comparing with the numbers of Fig.11.9. An alternative is to replace the eye by a slow scan through the spectrum, best by moving the exit slit, and recording the output, as shown for laser light at 632.8 nm in Fig.11.1 and for the 453.8 nm line in Fig.6.46.

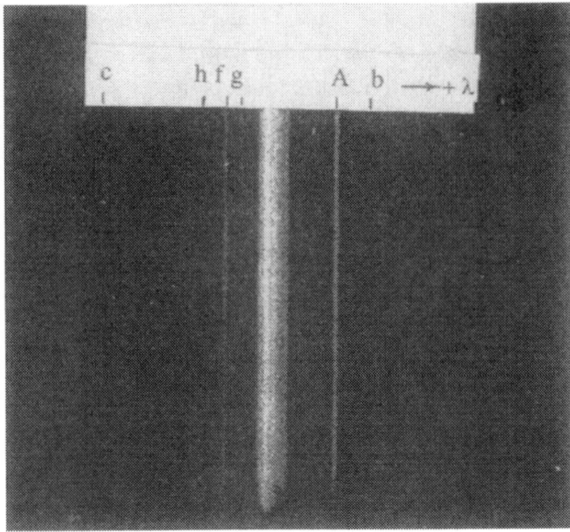

Fig.11.10 Photograph of Hg 546.1 nm hyperfine spectrum taken with a 300 mm wide r–2 echelle grating on a 10 m spectrometer. The 5 central components are just separated on the photograph, from which one concludes that resolution exceeds 10^6, or 90% of theoretical (courtesy Spectronic Instruments Co.).

11.4.2 The Foucault Knife Edge Test

Using the set–up shown in Fig.11.11 an experienced observer can quickly detect whether the knife edge cut–off is sufficiently uniform to give the desired resolution. The reason why experience is required is that the test is a measure of wavefront slope so that the observer is required to perform a kind of mental integration. It works surprisingly well but is hard to describe in words. Naturally, the closer a grating is to perfection the easier the judgment becomes, because no photometric decisions need to be made. The test has the special virtue of being fast, once the grating has been adjusted. It is important that the illuminating pin hole be exactly in the focal plane of the collimator, because otherwise coma effects are introduced into the wavefront. One way to make the absolute adjustment is to place a reference flat normal to the collimated beam.

The same set–up can be used to determine with fair accuracy the level of astigmatism that might be present in a grating. This can be done numerically by noting the position of the pinhole slide micrometer when set for the flat and when looking at the grating first in zero order and then in the diffraction setting. The zero order readings will tell whether the blank or its coating were not

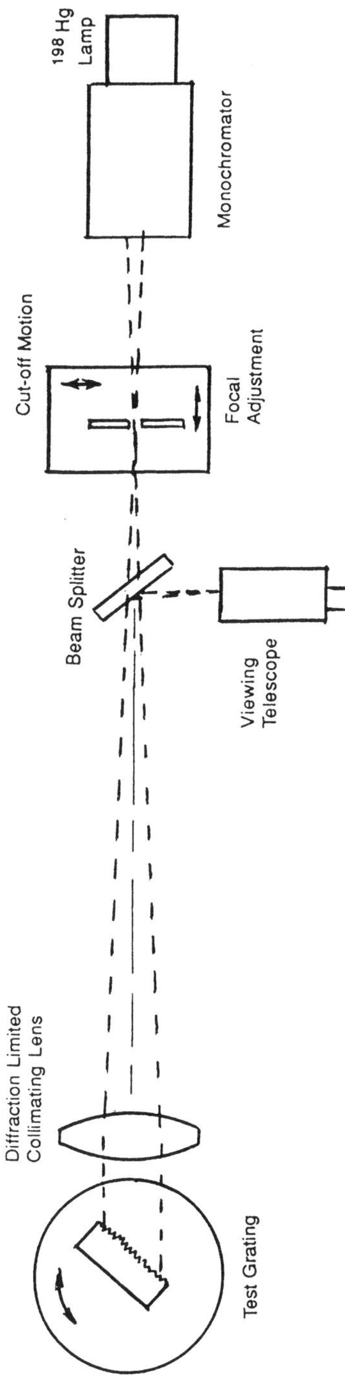

Fig.11.11 Schematic diagram for Foucault test (courtesy Spectronic Instrument Co.).

uniform, and in the diffraction direction there will be superposed the effects of errors of run and groove curvature.

11.4.3 Resolution Testing by Wavefront Interferometry

As with any optical element that is part of an imaging system, the diffracted wavefront contains all the information necessary to judge the imaging quality of a diffraction grating. In principle, therefore, nothing more than an accurate wavefront interferogram is needed to derive from it every bit of information needed (except efficiency). However, this is not how it always works in practice. While Rowland ghosts can be detected by a zig-zag pattern across the fringes, by the time that is visible to the eye the amplitude is too great to consider the grating useful. While sensitivity can be improved by using heterodyne interferometry, ghosts are easier to analyze by slow speed spectral scans such as in Fig.11.1. Lyman ghosts are even harder to find this way because interferograms rarely have enough horizontal resolution to detect high frequency variations, even should the amplitude be large.

However, interferograms do serve a highly useful purpose, especially for high angle gratings. Fringe patterns that are completely uniform over the entire grating give instant proof that diffraction limited performance can be expected, and invariably correspond to perfect Foucault knife edge cut–offs. Slight curvature of the fringes gives an indication of astigmatism that can be focussed out, as long as the axis is parallel or normal to the groove direction.

The other important application is that for the ruling engine operator nothing reveals so much and so quickly what might have gone wrong during the ruling cycle, and just where and when disturbances occurred. If the control system skipped one fringe there will be a parallel displacement of the fringe pattern. If the temperature control system misbehaved enough to shift the effective zero position, the fringe pattern will be seen to drift and change angle. Should there be error of run its amount and location is immediately visible by fringe spacing changes. If the barometric pressure correction system failed the fringe pattern will become a good facsimile of the weather record. In one instance the passage of regular summer thunderstorms was related to visible wavefront fringe shifts on the grating just ruled that were baffling because they always showed up about an hour or two after the storm had passed, and then returned to normal. After much puzzlement the problem was finally traced to the fact that the engine rested on a wooden support structure that absorbed moisture, expanded, and caused the engine base casting to bend slightly, thereby shifting the relationship between the optical and mechanical zeros. Once identified the problem was easily eliminated by supplying an improved kinematic support system. In another instance the brief presence and timing of unexpected visitors in the ruling lab was revealed by the interferogram.

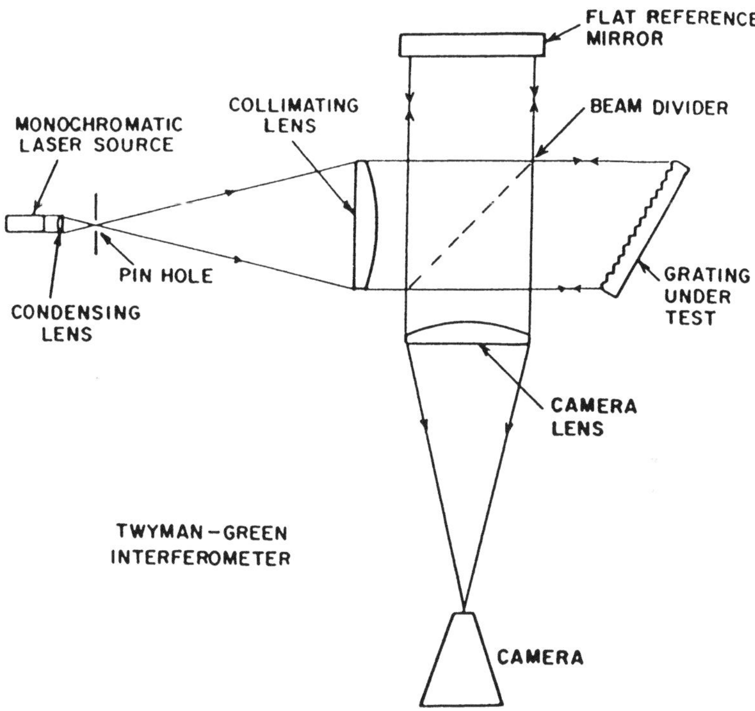

Fig.11.12 Schematic representation of Twyman–Green interferometer, as set up for testing deviation from ideal diffracted wavefront. Grating is in autocollimating or Littrow position (courtesy Spectronic Instruments Co.).

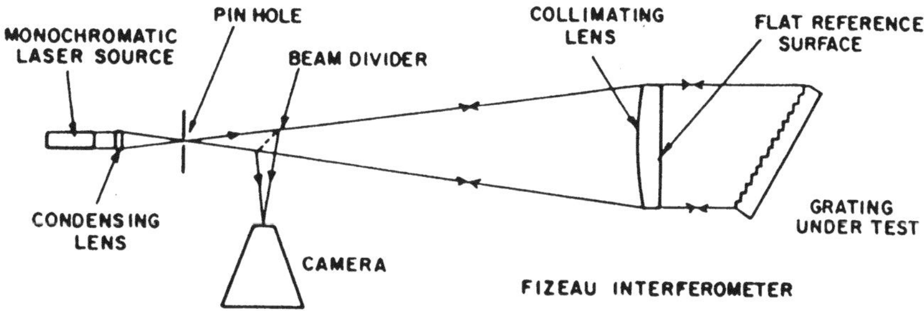

Fig.11.13 Schematic of Fizeau interferometer, as used for testing diffracted wavefront from grating set up in autocollimating or Littrow condition (courtesy Spectronic Instruments Co.).

Both Twyman–Green and Fizeau interferometers are used for wavefront testing. Both operate with the necessary collimation, and require high quality lenses for that purpose, Figs.11.12, 11.13. In the Twyman–Green the other critical component is the beamsplitter, which must both reflect and transmit beams with minimal disturbance to the wavefront (i.e., λ/20), and requires a diameter roughly equal to the aperture of the largest grating to be tested as tilted to the blaze angle. A laser makes the best light source, because its intensity allows taking exposures short enough to eliminate concern for air currents and vibrations. However, its coherence produces secondary fringes from parallel reflecting surfaces, which disturb the primary fringe pattern. If

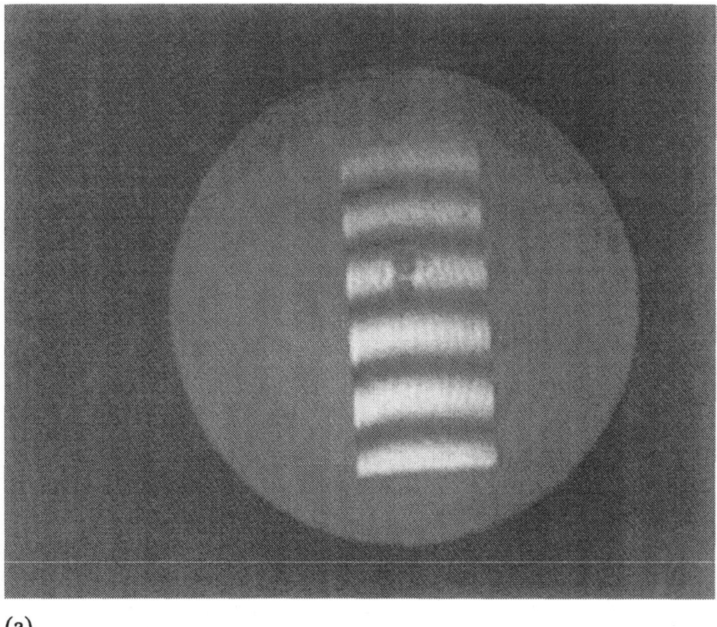

(a)

Fig. 11.4 Typical interferograms of high angle gratings obtained in HeNe light with equipment of Fig.11.12. Fringes perpendicular to grooves on left, parallel on the right. (a) 79 gr/mm 63 blaze grating with near perfect wavefront. (b) Same as (a) except fringes are parallel to the grating grooves. (c) Defective 316 gr/mm echelle ruling a 63 blaze. Zig-zag fringe deviations correspond to Rowland ghost of about 40 nm error amplitude. Local disturbances indicate areas of slightly varying groove spacing, producing close in satellites. Long period changes in fringe curvature at right end are due to error of run (progressive change in spacing), which broadens line image, i.e., loss of resolution (courtesy of Spectronic Instruments Co.).

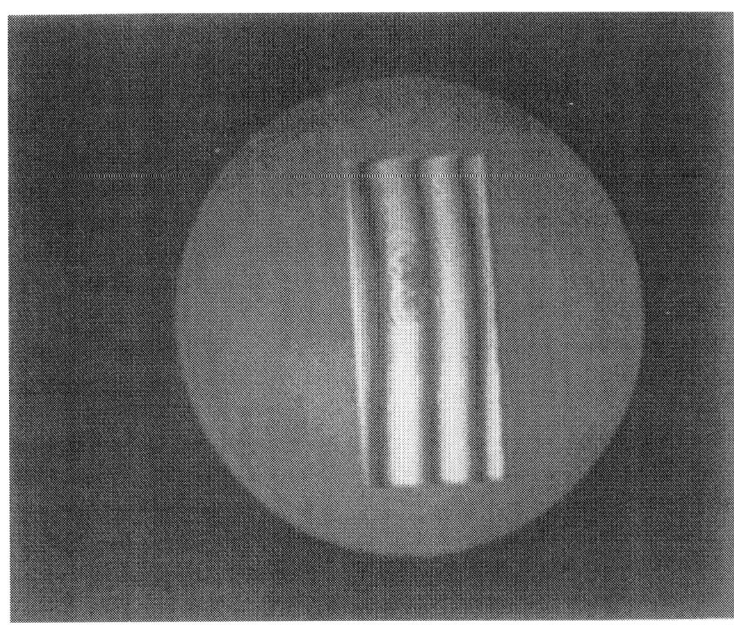

(b)

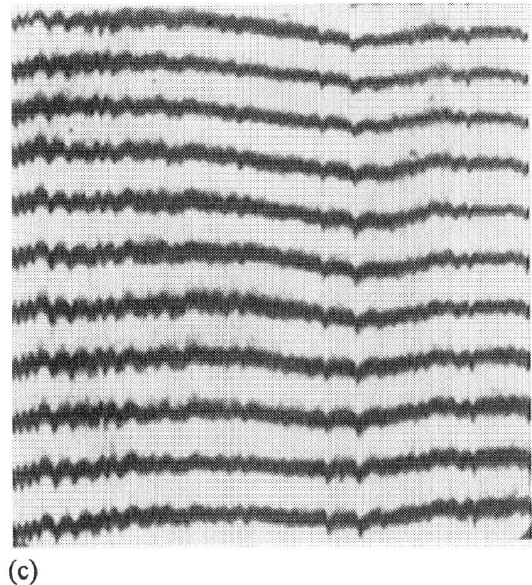

(c)

coherence is reduced too much to suppress this effect it may prevent seeing the extreme ends of a ruling. The operation consists of comparing the test wavefront, generated at the diffraction angle, with that of a near perfect reference mirror.

In making the adjustments the grating will be tilted slightly around two axes to produce two interferograms. On one the fringes are made parallel to the grooves, and will display any fanning error or lack of straightness of the grooves. More important is the other setting, with fringes perpendicular to the grooves, because this shows any variation in spacing. Examples are shown in Fig.11.14.

11.5 Testing of Concave Interference Gratings

Efficiency measurement of concave gratings has been described in section 11.3.2. A key facet to recall is that, contrary to plane gratings, the illumination is largely with conical rays from a pin hole entrance aperture, which undergo diffraction at varying compound angles. In addition the groove modulation may not always be uniform. There is also the additional convolution with the intensity profile of the illuminating beam, typically Gaussian. The result is a rather complex picture, from which one can conclude that measurements are not likely to exhibit the simple concurrence with efficiency theory that is routinely expected from plane gratings [11.11].

11.5.1 Measurement of Imaging Properties

One of the inherent properties of concave gratings is that, like concave mirrors, they have significant imaging aberrations. This handicap limited the use of concave gratings in routine instruments until the development of interference (holographic) gratings which could be made with aberrations reduced to levels that make them far more more useful. Chapter 7 describes the design approaches needed to achieve this, which are not simple.

Since in many cases image quality only barely meets design demands, it should be tested in the final grating. This demands that the grating be set up in the exact design and use configuration, which usually requires special fixturing. In particular the entrance pinhole must be in the meridional plane (unless the design calls for a slight departure), located at the correct distance from the grating pole and inclined at the correct angle. Since for a low f/ratio grating the imaging is sensitive to even a 0.1 mm shift in radius, this is not a trivial matter. For example, a 100 mm diameter f/2 grating will show an image growth of about 50 μm if the pin hole is moved 0.1 mm from optimum. Some relief is available in that there are conjugate properties between entrance and exit conditions, so that adequate judgment of image quality can still be made

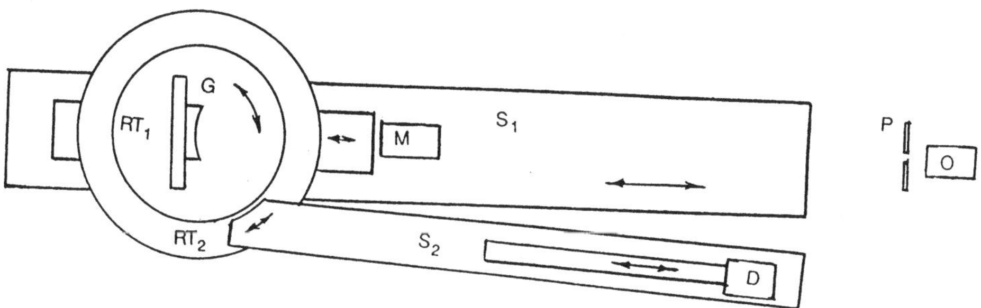

Fig.11.15 Schematic of a universal test fixture for determining the imaging properties of grating (G).

provided excess incident radius is matched by an equal reduction in the imaging radius, or vice versa. Image size and shape can be detected most easily with a high resolution 2-D CCD, although photographic film can be used by making a succession of exposures at appropriate changes in radius. To get meaningful results the angular aperture of the illuminating beam should be similar to that expected in the instrument for which the grating is intended.

To test gratings of different radii requires a universal fixture with provision for proper alignment even if blanks are of different shape. The input source should be either a laser for monochromatic testing or a monochromator for varying the wavelength; it should have resolution equal to the highest value of any grating likely to be tested. To adjust the incident radius it is probably better to arrange for moving the grating mount rather than the bulkier input system. One such system is sketched in Fig.11.15. It shows grating G mounted on a rotary table RT_1, which in turn is mounted on a slide S_1 with enough motion to cover all the desired radii for quick setting, but with a short travel micrometer motion M superimposed for fine adjustment. It will be convenient to support the grating by one of a set of V-cradles (not shown), one for each shape of blank. This automatically assures that the grating center will be on the same plane as the input beam. The input light is focussed by an objective O through an entrance pinhole P, the pair being made interchangeable for different input requirements. A second rotary table RT_2 carrying a second slide S_2 allows the position of image detector D to be adjusted for the desired radius and angle, to pick up the optimal image.

In setting up for measurement the first step after mounting the grating is to adjust the radius. This can readily be done by moving the radius slide, first by hand and then with the micrometer fine adjust screw, until the image returned to a small screen next to the input pin hole remains stationary when the grating is rotated about its axis with the rotary table. This requires that the backing wall against which the grating is placed is distanced from the table axis by the thickness of the grating, as measured at the pole. If not, appropriate shims must be used. The next step, which takes advantage of the fact that the entrance condition is rarely moved very far from the center of the radius, is to move the grating with the micrometer from the initial condition by the exact amount specified, and then rotate the grating to the incidence angle specified for the wavelength of test. The output detector D, is then positioned at the exit radius, and the optimal imaging conditions searched for by observing the CCD display, or its equivalent.

This type of tooling can be made more elaborate by equipping both slides with 0.05 mm resolution displacement transducers, whose readings can be zeroed against a calibrated concave mirror, so that absolute radius determinations can be carried out.

11.6 Role of Replication

Fortunately most of the important grating properties are replicated with such fidelity that there is no need to repeat them once the master has been certified. In the case of smaller gratings (<100 mm), efficiency is the only attribute that needs monitoring. For larger gratings it is a good idea to check the wavefront as well, by one of the methods described in this chapter.

11.7 Cosmetics

It is rare that under intense illumination the human eye will fail to detect some sort of blemish on the grating surface, especially reflection gratings. If used to judge high grade mirrors, this may give rise to needless alarm on part of the inspector. R.W.Wood once observed that in his experience the most useful gratings were never the ones with the best appearance, although he did not claim the corollary. The reason is simple: the spectral image is the result of integration of light from the entire illuminated area. Even an entire missing groove, which looks bad, represents a tiny fraction of the total area. Stray light would hardly be affected. Of course, if whole families of grooves were missing the grating should be rejected.

References

11.1 M. C. Hutley: *Diffraction Gratings*, Chapter 5: Spectroscopic Properties of Gratings (Academic Press, London, 1982).

11.2 R. C. M. Learner: "Notes on gratings ruling, tetsing and test equipment," AURA Techinical Report No.**39**, Kitt Peak National Observatory, Tucson, AZ (1972).

11.3a G. W. Stroke: "Diffraction gratings" in *Handbook of Physics*,v.**29**, *Optical*

Instruments, ed. S. Flugge (Springer, Berlin, 1967).

11.3b G. W. Stroke: "Attainment of high resolution gratings by ruling under interferometric control," J. Opt. Soc. Am. **52**, 1321-1339 (1961).

11.4 M. R. Sharpe: "Stray light in UV-VIS spectrometers," Anal. Chem. **56**, 339-356 (1978).

11.5 M. R. Sharpe and D. Irish: "Stray light in grating monochromators," Opt. Acta **25**, 861-893 (1978).

11.6 J. F. Verrill: "Specification and measurement of scattered light from diffraction gratings," Opt. Acta **25**, 531-547 (1978).

11.7 A. W. S. Tarrant: "Optical techniques for studying stray light in spectrographs," Opt. Acta **25**, 1116-1174 (1978).

11.8 J. F. Verrill: "A study of blazed diffraction grating groove profiles using an improved Talystep stylus," Optica Acta, **23**, 425-432 (1976).

11.9 W. Anderson, G.Griffin, C.F.Mooney, and R.Wiley: "Electron microscope method for measuring diffraction grating groove geometry," Appl. Opt. **4**, 999-1003 (1965).

11.10 D. Richardson: "Diffraction gratings," in Appl. Optics and Opt. Engineer., R. Kingslake, ed., v. **V**, ch.2 (Academic Press, London, 1969).

11.11 E. G. Loewen, E. K. Popov, L. V. Tsonev, and J. Hoose: "Experimental study of local and integral behavior of a concave holographic diffraction grating," J. Opt. Soc. Am. A **7**, 1764-1769 (1990).

Additional Reading

J. Calatroni and M. Garavaglia: "New analysis of the theory of Rowland ghosts," Appl. Opt. **12**, 2298-2301 (1973).

G. Dunning and M.Minden: "Scattering from high efficiency diffraction gratings," Appl. Opt. **19**, 2419-2425 (1980).

M. Garavaglia and C. Massone: "False spectra from a plane grating produced by laser illumination," Appl. Opt. **7**, 1443-1445 (1968).

G. Geikas: "Stray light from diffraction gratings," SPIE, **675**, 140-151 (1986).

R. Göhring: "Bestimmung des Apparateprofils eines Gitterspectrographen mit Hilfe eines Lasers," Zeitschrift für Astrophysik, **69**, 403-417 (1968).

D. Hammer, E. Arakawa, and R. Berkhoff: "A simple grating calibrator for the visible and vacuum ultraviolet," Appl. Opt. **3**, 79-81 (1964).

E. Inglestam and E. Djurle: "The study of diffraction grating characteristics by simplified phase contrast methods," J. Opt. Soc. Am. **43**, 572-580 (1953).

W. Kaye: "Stray radiation from holographic gratings," Anal. Chem., **55**, 2018-2021 (1983).

H. Kondo, Y. Chiba, and T. Yochida: "Automatic apparatus for measuring veiling glare distribution," Opt. Acta, **27**, 939-947 (1980).

H. Kondo, Y. Chiba, and T. Yochida: "Veiling glare in photographic systems," Opt. Engineer. **21**, 343-346 (1982).

D. H. Rank: "Theoretical resolving power of diffraction gratings," J. Opt. Soc. Am. **42**, 279-281 (1952).

D. H. Rank, G. Skorinko, D. P. Eastman, G. D. Saksena, T. H. McCubbin Jr., and T. A. Wiggins: "Hyperfine structures of some HgI lines," J. Opt. Soc. Am. **50**, 1045-1052 (1960).

J. H. Schroeder and B. P. Ramsay: "The optical properties of the grating interferometer," J. Opt. Soc. Am. **30**, 355-361 (1940).

M. L. Scott: "Diffraction grating evaluation," SPIE v. **171**: *Opt. Components: Manufacture and Evaluation*, 57-63 (1979).

G. W. Stroke: "Interferometric Measurements of wave-front abberations in gratings and echelles," J. Opt. Soc. Am. **45**, 30-35 (1955).

J. F. Verrill: "The limitations of currently used methods for evaluating the resolution of diffraction gratings," Optica Acta **28**, 177-185 (1981).

J. F. Verrill and E. W. Palmer: "A coma free spectrometer with interferometer attachment for testing diffraction gratings, Optica Acta **28**, 169-175 (1981).

J. Walz: "Fast measurements of straylight in photographic cameras," in *Research in Optics*, Tech. Rep. Inst. Opt. Research, 7-11 (Stockholm, 1982).

T. Woods, R. Wrigley III, G. Rottman, and R. Haring: "Scattered-light properties of diffraction gratings," Appl.Opt., **33**, 4273-4285 (1994).

Chapter 12

Instrumental Systems

12.1 Introduction

The purpose of spectrometric instruments is to separate incoming light into its various frequency components whose strength and value provides information either about the source or the intervening medium. All systems share as basic components the source of radiation, a dispersive medium, and a detector so that instruments can be defined by these features.

The simplest imaginable system is probably a rainbow. The source is the sun, bright and reasonably well collimated, and a vast array of raindrops constitutes the dispersing device. The eye of the observer naturally represents the detector. With an entrance "slit" so large (the sun subtends an angle of 0.5°), and a dispersion that is rather low, we cannot discern any solar physics, but are free to enjoy the beauty of nature. A polarizer in front of the eye will increase the contrast and perhaps whet an interest in optics.

From a physics point of view a key decision is usually whether the concern is for atomic transitions or molecular ones. Atomic transitions are characterized by narrow frequency bands, too high in frequency to be directly measured. Instead we measure the wavelengths to which they are related via the speed of light. Unfortunately the speed of light is constant only in a vacuum, but for most cases this effect is ignored. Typically we need to measure in units of 1 Å or less. Molecular transitions, in solutions at least, tend to be more broadly defined and as a result require instruments that seldom have resolutions below 10Å (1 nm).

Detection devices have been important because the human eye is poorly equipped for obtaining quantitative information. It is a notoriously poor judge of intensity, and in addition sensitivity varies strongly with wavelength. The range of the eye covers barely one octave of the spectrum, so it is easy to appreciate why photography played such an important role in spectrometry from the moment of its discovery. In various forms it can cover the spectrum from X–Rays at 0.1 Å to the infrared at 10^5 Å. The expense, the non–linear response, the nuisance of wet processing, and the rather tedious methods of data extraction were all accepted, at least as long as there was no alternative.

A major shift in viewpoint occurred in the 1930's, when improved photoelectric detectors offering far greater dynamic range were developed. In later years they were complemented by solid state multiple arrays such as CCDs. These not only started to supply parallel detection, but worked at improved quantum efficiencies, and established the concept of scanning without moving parts. This combination was irresistible and photography has gradually become relegated to special purpose applications. An example are the spectrometric instruments aboard the Skylab satellite that flew four missions during 1972 and 1973. Here the enormous amount of parallel information that could be encoded in a single exposure (up to 10^8 bits) provided the incentive, but was practical only because astronauts were on hand to retrieve the films. Modern unmanned satellites rely solely on electronic detection with data telemetered to ground.

Developments in electronic signal processing had a profound effect on instrumentation. It began around 1955 with analog output fed into standard recorders, but much information still had to be processed by hand. This satisfied demands until the advent of digital electronics. Now there is no limit to the amount of data manipulation possible. Some colorimeters, for example, display almost instantaneously the color coordinates of a sample, and print instructions on how to mix paint to duplicate the color of a sample swatch!

As long as photography dominated data recording, some 70 years, most instruments were in the form of *spectrographs.* They tended to be large, some in the form of enormous fixed installations, whose main task was the measurement of wavelengths. In the 1930's, thanks to electronics advances, they were gradually superseded by *monochromators* where spectral bands, detected behind the exit slit, were defined by rotation of the dispersing element, prisms, until gratings largely replaced them in the 1960's. Following the seesaw that one often observes in technology, the development of silicon–based array detectors in the 1980's regenerated interest in spectrographs of different designs. Progress was slow because it took time to produce arrays with a large number of elements, enough sensitivity (especially in the blue), and at the same time were low enough in cost. A concurrent and important development was the discovery that concave gratings could be modified to produce image fields flat enough, and with sufficient resolution, to just match that of their detectors. This was accomplished by slightly modifying the uniformity of both groove spacing and straightness, either mechanically, or more often, by interference (holographic) methods. The advantage lies in being able to make, for the first time, scanning instruments with no moving parts. A number of texts have been published that cover the field of instrumentation [12.1-6]

12.2 Terminology

Since there is no international standard for instrument terminology we describe what seems common usage:

Spectrometric: adjective describing measurements whose purpose is to separate physical properties by their wavelength or frequency.

Spectrometer: Classically an assembly of two telescopes mounted on a pair of concentric rotating graduated tables, and whose axis intersect the axis of rotation, containing a fixed central support for the dispersing device. One telescope serves as collimator for light from an entrance slit, the other as the viewing device.

Spectroscope: A viewing tube containing collimating optics, slit, dispersing element, and optics for viewing by eye the spectrum of luminous sources or corresponding absorption.

Spectrograph: A system for delivering multiple images of an illuminated entrance slit onto a photosensitive surface, whose location is a function of wavelength. Normally characterized by absence of moving parts.

Monochromator: A device that delivers an image of the entrance slit onto a normally fixed exit slit, and where the single wavelength band transmitted is controlled by *rotation* of the dispersing element.

Spectrophotometer: A device for measuring and recording the degree of absorption of a gas or liquid as a function of wavelength. Can also be configured to work in reflection.

Colorimeter: A specialized form of spectrophotometer, usually working in reflection, capable of calculating the color coordinates of a sample.

12.3 Classification of Instruments

There is a strong temptation to classify instruments in some fashion, with the thought that this will help to select the best solution to a given problem. In some aspects this turns out to be quite simple, but in others the large number of choices leads to complex decision making.

The most obvious feature to consider is the wavelength region to be studied. It often dictates critical design aspects, for example the use of concave gratings at very short wavelengths or special detectors in either the UV or the IR regions. If the wavelength range is unusually wide, the choice of dispersing device can become the critical item.

Another way to look at criteria is to consider the field of application. Biomedical instruments may have to meet government standards to be acceptable and demonstrate accuracy, reliability and special means for calibration. In some circumstances it may be critical to have instruments without moving parts. Industrial instruments sometimes need athermal designs in order not to loose calibration under widely varying environment. Instrumentation for spacecraft must have extraordinary reliability to survive untended for years and may also need ability to be reprogrammed from distances that are measured in light–minutes. Raman spectroscopy requires high resolution, and depends especially on filtering out the strong exciting wavelength. Colorimeters need careful attention to sample illumination and processing of the information.

The type of dispersion system may be a primary consideration, especially when one type has a well established reputation in a given field. For example, Fourier transform systems are well established in the IR, although not to the total exclusion of alternatives. The choice of gratings over prisms used to be controversial, and even now it is not always obvious whether concave gratings are to be preferred over plane grating systems or vice versa. It may not always be clear whether mechanically ruled or holographic gratings are to be preferred.

Much study has been given to the many optical mounts that are possible. The choice between plane and concave gratings is basic. If a plane grating design has been picked a designer must choose between the popular Czerny–Turner mount, the Ebert–Fastie where a single mirror replaces a pair, or the Monk–Gillison mount that sacrifices image quality for maximum simplicity. There are a considerable number of concave grating mounts based on ruled gratings, named after their originators, Rowland, Eagle, Wadsworth, Namioka, and are discussed in many texts [12.1-5]. With holographic methods entering the picture, the number of possibilities has climbed, which may be important for specific applications, but actually does not constitute anything basically new. The major motivation has been the desire to obtain flat fields as well as high apertures together with acceptable aberrations.

The resolution desired has much to do with the size of spectrometric instruments. Originally the relationship was direct and led to 6 and even 10-meter focal length spectrographs, but with echelle based two-dimensional imaging concepts overall dimensions could be scaled down. Both monochromator and spectrographs have long been classified by their focal lengths, and it has become customary to think of commercial instruments as 'subcompact' (100 mm or 1/8 m), 'compact' (1/4 m), 'regular' (1/2 m), 'full size' (3/4 to 1m), and 'jumbo' (1.5 m or greater). Larger instruments have been built, but these tend to be specials for research applications.

In addition to all these ways of classifying instruments there are a number of additional concepts, such as those where glass or silica fibers are used to make the optical connection between instrument and the light source being monitored. Others specialize in ultra fast scanning of spectra to observe fleeting events. There are interesting applications that have almost nothing to do with spectral analysis. A prominent example is wavelength division multiplexing (known as WDM), an important tool for increasing the information carrying capacity of optical fibers. A totally different application takes advantage of the efficiency behavior of gratings in the anomaly region, using it to detect small changes in refractive index of a film under study. The fluid sample is placed on top of a reflection grating, allowed to dry, and illuminated at the appropriate angle and wavelength to give a sensitive response. Thanks to laser illumination, gratings are small and unusual in that they are made disposable.

12.4 How to Choose a Design

As one can easily appreciate from the above, it is not a simple process to make the many choices involved in designing an instrument for a given application. In many cases there will be two or more that appear to serve equally well, and the decision must then be based on experience, availability, or perhaps even personal prejudice. Selection is likely to be based simply on what is most economical, where it is necessary to balance cost of components against the cost of assembly. The latter is often strongly affected by whether it is left to a skilled technician or whether quantities lead to well-designed tooling that converts this into a simple procedure. Should an instrument be required to feature interchangeable gratings, for example in order to cover an exceptional range of wavelengths, the gratings must be supplied in carefully aligned kinematic mounts with wavefronts matched to maintain parfocality, or mounted in sets on an indexing turret.

The first consideration will usually be the wavelength range because it dictates so many other decisions. An extreme example is X–ray instruments, because no standard materials reflect in normal incidence. The classical choice is to adopt grazing incidence, despite its large aberrations and low resolution. A modern alternative is to use supermultilayers, although they are expensive.

Changes in available technology can alter perceptions greatly, as in the above X–ray example. In the IR region, where detection is inherently afflicted with a noise problem, there is a great advantage in abandoning direct dispersion in favor of Fourier transform approaches. These in turn became practical only with the advent of high speed digital computers and efficient data reduction algorithms.

The visible and near UV spectrum contains much of the information of interest in standard instruments. The first choice to be made is instrument focal length or size, established largely by the desired resolution, which in turn depends on what kind of analysis is to be performed. Atomic transitions and Raman lines demand high resolution while most molecular, biological or color analysis is served by instruments with resolution typically one or two orders less. Catalog instruments are found largely in the 50 mm to 1 m focal length range, with a center of gravity near 100 or 200 mm. Focal ratios typically vary from about 15 to 4, but occasionally extend to 30 on the high side and 1 on the low end. Obviously the lower the ratio the greater the luminosity, but there is invariably a trade-off with increased imaging aberrations, whose effects increase when wavelength coverage is wide. The skill required to make optimal decisions is quite similar to that required in lens design, where, despite a degree of design automation that far exceeds that for spectrometers, much experience is still needed. The kind of trade-offs can be appreciated by noting that it is in the design of concave holographic gratings that very low f/numbers have been realized, but they have reduced peak efficiency and suffer from aberrations that can be fully minimized only in the vicinity of the generating wavelength. To rule concave gratings mechanically at f/numbers below 9 is a difficult feat.

Modern developments in concave gratings have made more difficult the choice between optical systems based on their use compared with plane gratings. Often the decision will be one of economics. Plane gratings are significantly less expensive, but complete mounts require one or two mirrors and their assembly and alignment into mounts. In principle a concave grating needs no focussing optics. However, many designs require plane steering mirrors in order to have light enter and leave in convenient directions. Plane gratings are available in a wider range of groove frequencies and blazing, and are much more readily interchanged.

Signal to noise is sometimes a critical factor, for example in measuring Raman or absorption spectra, or in spectral regions where sources are weak and detectors have poor sensitivity. Since the grating is often the principal source of optical noise this points to the use of interference or holographic gratings.

Increased throughput can be achieved by reducing aberrations, especially astigmatism, or by increasing slit sizes made possible by going to higher dispersion, or alternately by adopting lower f/numbers even though they increase coma aberrations and reduce resolution. One way to escape imaging restrictions is to depart from spherical surfaces on either grating or mirrors, whenever their increased cost can be justified.

To remind the reader of the many aspects that must be considered to varying degrees the following list is presented here. Not all apply to every design, and many are in plain opposition.

Summary of spectrometer design considerations

1. Choice of plane gratings or concave grating systems.
2. Dispersion and resolution required.
3. Wavelength range.
4. Methods of wavelength drive for slew or scan and their accuracy concerns.
5. Methods of wavelength readout or recording and their calibration.
6. Rate of wavelength scanning.
7. Slit controls and mechanisms, possible coupling to wavelength drive.
8. Wavelength coupled filters to control order overlap.
9. Location and direction of entrance and exit beams.
10. Ability to gang two units for conversion to a double system.
11. f/number or speed capability and focal length (i.e., size).
12. High resolution compared to high flux transmission requirements.
13. Imaging aberrations (coma, astigmatism) and methods to control them.
14. Stray light and spectral purity requirement.
15. Ability to function in special environment: temperature, vibration, humidity, shock, vacuum.
16. Flexibility to permit switch from monochromator to spectrograph modes.
17. Design of entrance optics to fill the dispersion optics.
18. Interchangeability of components.
19. Ability to make adjustments and provide service.

12.5 Plane Grating Mounts

Despite the remarkable advances made in the manufacture of concave gratings and instruments built around them, there are still many applications where plane gratings are preferred. This may be because of cost, imaging qualities (i.e., high resolution), interchangeability, wavelength range, or stigmatic imaging. The plane grating mounts described here are all based on reflection gratings. With one exception, described below, all plane grating mounts direct collimated light to the grating and focus the diffracted light onto an image plane.

The concept that measurement accuracy would be enhanced by operating in collimated light was slow to be appreciated. The first publication to mention it specifically was by Babinet in 1839 for an instrument built by Arago [12.7], although Fraunhofer had used it 15 years before. The importance of this concept was not properly understood, and overlooked for years, only to be re-invented several times, a fate not unusual for an idea ahead of its time.

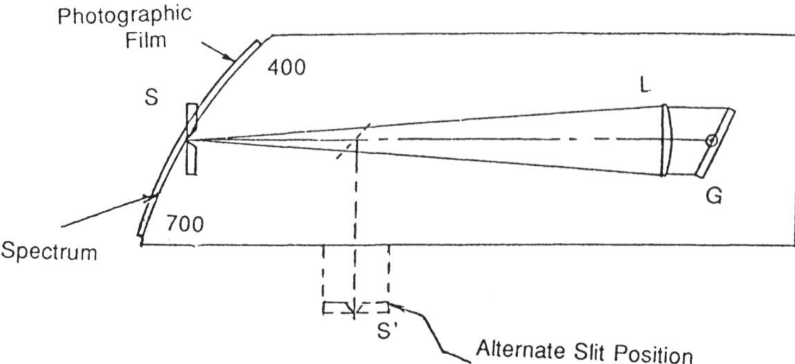

Fig.12.1 Schematic diagram of Littrow spectrograph. S is the entrance slit, L the collimating lens, G the grating that rotates around the axis indicated for wavelength tuning, E the exit slit. For plate and S to be physically separated they are usually located above and below the central plane (after [12.9]).

The classical design is the Littrow spectrograph. While generally regarded as obsolete it survives in two forms. One is a subminiature version, useful for fiber optics multiplexing and demultiplexing [12.8]. The other is as the wavelength tuning device for lasers (see section 12.11). Its basic characteristic is the use of a single lens that serves both to collimate and focus the light so as to operate in auto–collimation, as shown in Fig.12.1. Originally it was designed to work with one or a whole set of prisms with light reflected back either by a mirror or by metallizing the back face of the last prism [12.9]. In such an instrument it is a simple matter to substitute a plane reflection grating for the prism, as first described by Lippich in 1884 [12.10] and later by Ebert [12.11]. (Both used quartz collimating lenses). It is interesting that Lippich felt that mirrors were to be preferred because they would lead to more compact instruments, as well as avoiding chromatism, but he failed to follow through. This had actually been done a few years earlier by Abney, who was driven by difficulties of working in the IR [12.12], but whose contribution has been almost completely ignored.

In order to separate the entrance slit from the imaging field they are usually located just above and below the central plane respectively. A major limitation is the lens which is impossible to make sufficiently achromatic unless the spectral band is small. It is obviously limited in range by the spectral pass band of its component lenses. Mirrors have the great advantage of having no wavelength restrictions, no chromatic aberrations, and low cost, although the latter applies only to spherical mirrors which unfortunately are never quite free of aberrations. Much effort has been devoted to the study of how these effects can be minimized and the most obvious solution is to limit the focal ratio, which rarely goes below 4.

Only three commonly used plane grating mounts are separately identifiable. In order of their popularity they are the Czerny–Turner, Ebert–Fastie, and the Monk–Gillison. The first was a joint development by M.Czerny in Berlin and his graduate student Francis Turner, neither expecting their concept to sweep the field [12.13]. Ebert and Fastie never met as they were several generations apart. Ebert's publication in 1889 [12.11] was pushed into obscurity because H. Kayser, dean of the spectrometric world at that time, pronounced it unworkable [12.14]. It was rediscovered by W. Fastie over 60 years later but with some important features added [12.15]. The Monk–Gillison mount was developed by two inventors whose paths also never crossed [12.16, 17]. Despite some limitations it enjoys a certain popularity because it is the simplest and least expensive system imaginable.

12.5.1 The Czerny - Turner Mount

The Czerny–Turner mount is by far the most popular grating mount used in instrument designs. It took a long time from its first description in 1930 for designers to fully appreciate its advantages. Probably the major reason was that low-cost high-quality plane grating replicas were not available until about 1958. Its basic characteristic is to use two identical off-axis concave spherical mirrors as the collimating and focusing (camera) elements, with the important property of canceling the coma aberration that is inherent with spherical mirrors and which otherwise inhibit resolution. Its continuing popularity rests in the low cost of the three basic optical elements, two mirrors and a plane grating. While

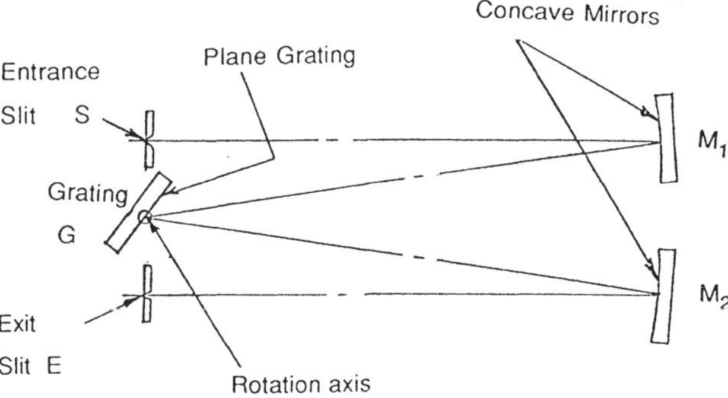

Fig.12.2 Czerny–Turner monochromator. Entrance and exit slits designated S and E, G is the grating, M_1 and M_2 are the collimating and focusing (camera) mirrors (symmetrical mount).

most often configured as a monochromator it can also function as a spectrograph, especially with array detectors, even though they may may use only a fraction of the total angular range at one time. Some instruments are configured to work in both modes with two output ports that can be selected by a steering mirror. Czerny and Turner do not give any references, specifically none to Abney, who used the identical two mirror system 45 years earlier [12.12]. Abney was clearly happy with the imaging quality attained, but did not make a big deal of it. None of the authors mention coma by name, but the pictures in the Czerny–Turner paper make clear that their mount eliminates it, although the paper's title refers only to astigmatism.

The basic Czerny–Turner mount consists of an entrance slit, collimating and focussing mirrors, and a plane diffraction grating mounted so that it can be rotated about an axis through the center of the grating surface, parallel to the grooves (Fig.12.2).To function as a monochromator there will be an exit slit in the image plane with a detector behind it, or a photodetector array such as a CCD when used as a spectrograph. The diameter of the mirrors and their focal length (1/2 of their radius) will determine the size of the instrument, while their ratio fixes the f/number, which largely determines the luminosity or throughput. This assumes that the beams fill the grating aperture. If the grating is the defining element for throughput it has been customary to use the diameter that has the same area as the usually square- or rectangular-ruled area. The lower the ratio (or the larger the numerical aperture) the greater the throughput, but the larger any aberrations. Either longer focal lengths or higher grating groove frequencies will increase the linear dispersion and allow for wider slits for a given resolution. The latter alternative has the advantage of not increasing the size of an instrument, but in return limits the maximum wavelength that can be diffracted.

In order to take full advantage of the potential for eliminating coma – essential for attaining maximum resolution – it is useful to note the conditions that will do so. According to Schafer [12.18] coma due to the two concave mirrors will cancel completely only when the equation below is satisfied:

$$\frac{\sin\beta}{\sin\alpha} = \frac{R_2^2}{R_2^2} \frac{\cos^3\beta \cos^3\theta_i^3}{\cos^3\alpha \cos^3\theta_d^3} \quad , \qquad (12.1)$$

where α and β are the angles between the central ray from the slits to the normals at the center of the two mirrors, and θ_i and θ_d are the angles of incidence and diffraction measured from the grating normal, Fig.12.3. The radii of the two mirrors are designated R_1 and R_2, although usually made equal. When $R_1 = R_2$ and the off axis angles α and β are small enough so that $\cos^3\alpha = \cos^3\beta \approx 1$, eq.(12.1) reduces to

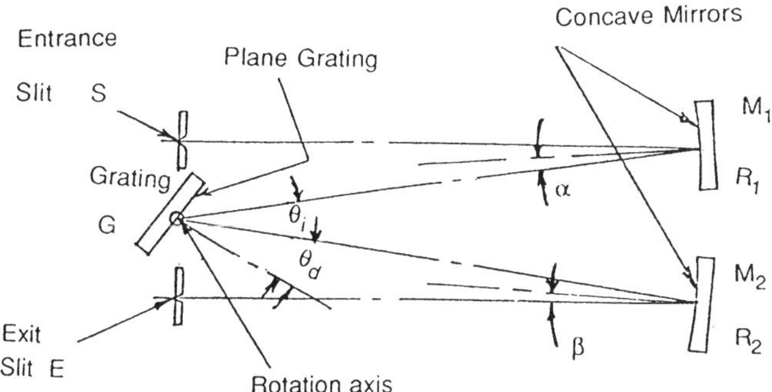

Fig.12.3 Same as Fig.12.2, with angular designations (after [12.18])

$$\beta/\alpha = \cos^3\theta_i \,/\, \cos^3\theta_d\,, \tag{12.2}$$

with the obvious conclusion that there is only one wavelength where the condition of eq.(12.2) can be exactly maintained. Fortunately, this is not a serious problem in practice. Spherical aberration is typically small, which is helpful because, given the symmetry of the system, the amount contributed by each of the mirrors is additive. Astigmatism due to the off–axis mirrors, which is not negligible, is similarly additive and hence may be a cause for concern. One solution is to replace the spherical mirrors with toroids, but this adds greatly to their cost, with the possible exception of small systems (about 100 mm) where ophthalmic toroidal blanks sometimes serve this purpose.

In most spectrometric instruments stray light is a concern. A common source is imperfection of the grating itself, but often there are reflections of undesired wavelengths towards the exit slit. Careful baffling becomes an important issue, since there is no zero reflectance material with which to paint instruments walls. Frequently dangerous is that the zero order is always reflected by a grating, and sometimes special traps must be provided to swallow it. On no account should it be possible for stray light to be focused onto the exit slit.

In order to make sure that no light from the entrance slit can fall directly onto the camera (focusing) mirror it has been suggested that the standard Czerny–Turner W–configuration be changed to a crossed or X–version, as shown in Fig.12.4. It is not a popular concept because, even though this goal is met, it leads to a rotation of the exit image and also increases astigmatism [12.19].

Choice of slit width is determined by the desired bandpass, in turn set by the type of analysis being done. In surprisingly few instances is it necessary for

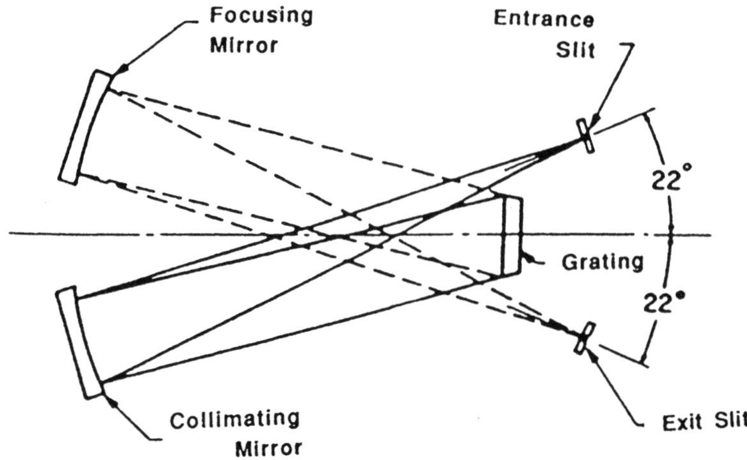

Fig.12.4 Crossed beam version of the Czerny–Turner mount (after [12.19]).

the full diffraction resolution capabilities of a grating to be exploited, in which case the angle ψ subtended by the slit is given by

$$\psi = 1.22\,\lambda / W \quad , \tag{12.3}$$

where W is the effective width of the grating. For a 100 mm wide grating at 500 nm wavelength, ψ equals 1.2 arc seconds, or a 6.1 μm slit width for a 1 m focal length system. A slit smaller than this cannot increase resolution, merely reducing throughput, but it helps to remind us that the slit can be a critical component. A typical spectrophotometer might operate at a 1 nm bandpass, which for an 1/8 m focal length corresponds to slit width of 300 μm, or about 1% of theoretical resolution. It implies that often the grating need not be illuminated with light coherent over its whole width, and that such slit design belongs to the realm of geometrical optics. The anamorphic effect derived from the angular conditions of the grating means that the exit image typically will be somewhat magnified.

When the Czerny–Turner mount is to be used in the spectrograph mode the detector should be located at a distance from the camera mirror where the image field most closely approximates a plane. This distance is equal to $\left(1 - 1/\sqrt{3}\right)R_2$, where R_2 is the radius of the focussing mirror as measured along a line passing through the center of curvature of the mirror and the center of the grating, as shown in Fig.12.5 [12.20]. It is possible to optimize the location still

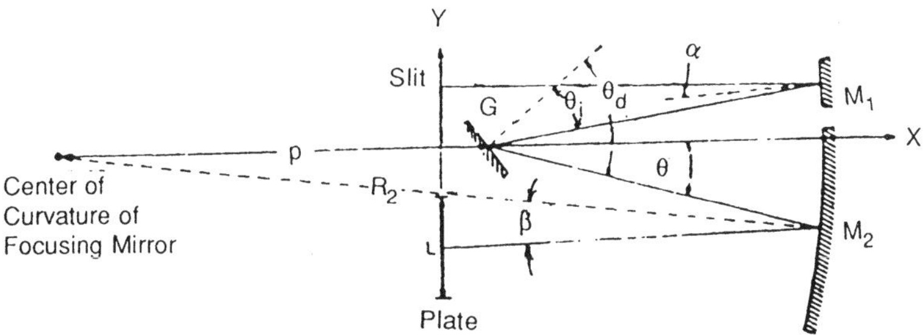

Fig.12.5 Czerny–Turner spectrograph configuration showing location of imaging field for photographic plate (after [12.20]).

further by a small tilt and slight displacement, in which case the field remains flat over a width that can be as great as 1/6 of the focal length [12.20].

One can readily imagine improving the imaging of a Czerny–Turner mount by replacing the spherical mirrors with equivalent off–axis parabolas. However, this is not necessarily a good idea, because coma aberrations, instead of subtracting will now add. Although spherical aberration and astigmatism will indeed be removed, the trade–off may not be worth it. On the other hand if the standard W–configuration is changed to a U–type, coma errors will again subtract and all three aberrations will be at a minimum. The design becomes even better if the angular deviation is made large enough (e.g., 13° for an f/6 system) so that multiple dispersion that might lead to stray light is no longer possible [12.21]. Applications of this concept have been limited because of the relatively high cost of aspheric mirrors.

12.5.2 The Ebert - Fastie Mount

Ebert described his concept of a spectrograph in which a single mirror performs both the collimating and camera functions in 1889, with a sketch reproduced here as Fig.12.6. The term 'camera' is used in classical literature because spectral images were routinely recorded photographically. Ebert pointed out the advantages of his design, its simplicity, the ability to be configured as a monochromator, even the ease of putting it into an evacuateable case, but never followed it up [12.11]. Kayser felt disdain for use of a single mirror to perform both functions leading him to dismiss the concept [12.14], his authority sufficient to bury the idea. He failed to see the advantage of having to adjust only a single mirror. Given the number of degrees of freedom involved in the adjustment of a mirror this is a matter that becomes important whenever resolution needs to be maximized.

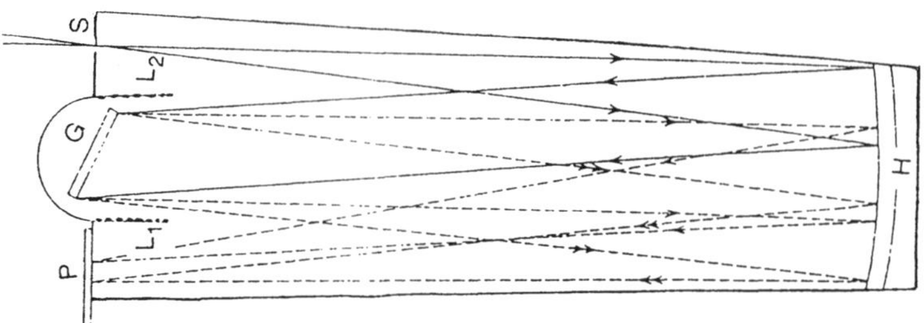

Fig.12.6 Schematic of single mirror Ebert spectrograph. S is entrance slit, H the collimating mirror, G the grating, and P the photographic plate in image plane. L_1 and L_2 are the important baffle plates (after [12.11]).

Fastie's re-invention of the concept in 1948 was made with the idea that it would be useful to avoid dealing with two mirrors that could be misaligned with respect to one another [12.22]. While this simplification in assembly found limited acceptance in the general instrument industry, it proved invaluable in the design of a whole series of instruments for spectrometry from satellite spacecraft, such as the Mariner missions to Venus, Mars, and Jupiter [12.23].

Fastie points out that an instrument should be made as large as necessary to obtain the desired resolution. This is done by fixing the ratio of slit width to focal length (i.e., the angle subtended). Once this ratio is fixed, the signal output will be a function of the ratio L_s/F, where F is the focal length and L_s is length of the slit. This demonstrates the importance of making slits as long as possible and in turn emphasizes the role played by astigmatism, which normally limits their useful length. Fastie showed that the situation could be greatly improved if the slits were made curved, specifically with their center of curvature located on the central axis of the large mirror. This is because each point on the entrance slit forms a short astigmatic image, tangent to the curved exit slit. That the center of curvature should be on the central axis could be deduced from the alternate version of the grating equation

$$m\lambda = 2d \sin\theta \cos\phi \quad , \tag{12.4}$$

where θ is the angle between the grating normal and the central axis of the mirror, and ϕ is the half–angle between incident and diffracted rays, m the order number, λ the wavelength, and d the groove spacing. When the slit is centered as prescribed, the angle ϕ is constant at all points on a slit, which means that it is the only configuration in which the diffracted image falls exactly where it should, *regardless of wavelength*. This concept of curved slits also holds for

Czerny–Turner mounts, and is used whenever the demands for throughput justifies the higher cost of such slits.

12.5.3 The Monk - Gillison Mount

The unique characteristic of the Monk–Gillison mount is that a plane grating is illuminated in convergent light, coming to a focus after diffraction. The number of optical elements is thus reduced to the absolute minimum for a plane grating mount (i.e., a reflection from a single concave mirror to a grating as the sole link from entrance to exit slit, Fig.12.7). Imaging qualities are not exactly ideal and optimum focus is attainable at only one wavelength. It is thus advisable to maintain a high f/number. But there are instruments and monochromators where a 20 nm bandpass is entirely adequate and the simplicity of this system quite desirable [12.16, 17].

The thought that led to the design was the well-known Wadsworth mount (see Section 12.6.5) which delivers stigmatic images with a concave grating (the only one to do so) provided it is illuminated with collimated light in such a way that the converging diffracted beam is normal to the grating (θ_d = 0). In the Monk–Gillison mount the roles are reversed. The grating is plane and the illumination is in converging light from a concave mirror. The astigmatism inherent in the imaging of a concave mirror can be at least partially offset by the anamorphic behavior of the grating (i.e., the diffracted beam is smaller in width than the incident). The astigmatism due to the mirror, Z_1, is given by

$$Z_1 = \frac{2v_2}{R} \sin\Phi \tan\Phi \quad , \tag{12.5}$$

where v_2 is the second conjugate distance from the mirror (usually equal to r), r is radius of curvature of the mirror, Φ is the angular deviation of the diffracted beams (θ_d - θ_i), and the astigmatism due to the grating, Z_2, is given by

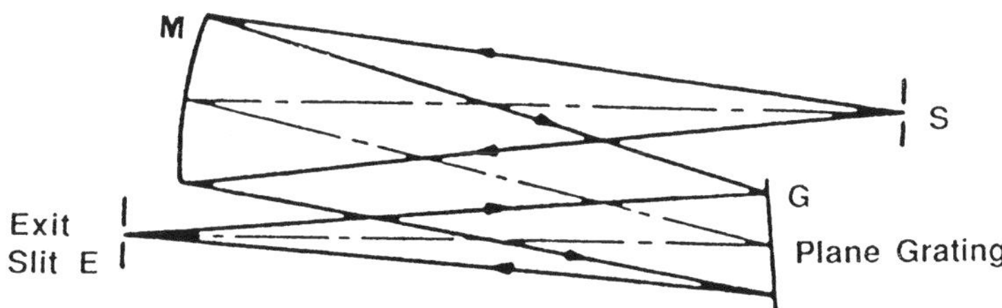

Fig.12.7 Schematic diagram of the Monk–Gillison mount. S is the entrance slit, M the focusing mirror (after [12.16]).

$$Z_2 = L\frac{\cos^2 \theta_d}{\cos^2 \theta_i} \quad , \tag{12.6}$$

where θ_i and θ_d are the angles of incidence and diffraction with respect to the grating normal and L is the length of the grating grooves. Since the combined astigmatism is the sum of the two components this can be reduced to zero if the following equation is satisfied:

$$\frac{\cos^2 \theta_d}{\cos^2 \theta_i} = 1 + \frac{2v_2}{R}\sin\Phi \tan\Phi \quad . \tag{12.7}$$

Once again it is clear that this cancellation can hold only for one set of angular conditions, or one wavelength. There is no way to cancel the residual coma except to replace the spherical concave mirror by an ellipsoid of revolution – a costly complication that defeats the main purpose of this mount. Astigmatism will usually become a limiting aspect if the focal ratio is made less than 10.

12.5.4 Grating Drives

In a monochromator the angle ϕ is fixed in eq.(12.4) so that the wavelength scanned is a function purely of θ. To obtain a readout linear in wavelength the mechanical rotation must be a function of sin θ. In many

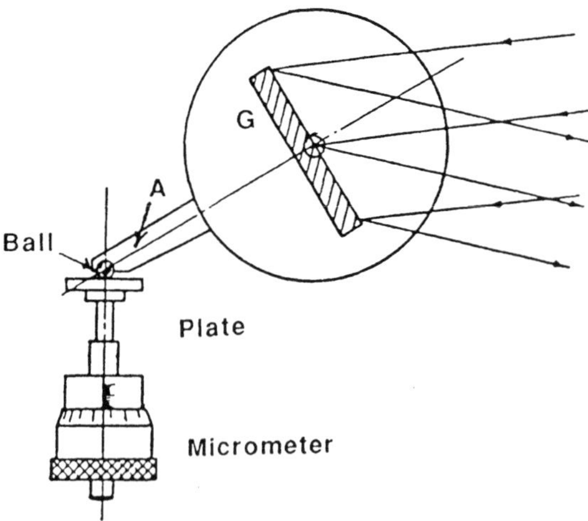

Fig.12.8 Design concept of sine drive. Grating G rotates around axis indicated under control of sine arm A with precision ball at tip. Micrometer M drives a precise flat planeperpendicular to its axis.

instruments this is done with a mechanical sine drive, Fig.12.8, in which the linear motion of a high precision screw is transformed into rotation via a sine function, provided that motions are straight, the bearing surfaces flat, and the contacting sphere sufficiently accurate. Alternate mechanisms have been constructed that provide for a readout linear in wavenumber. A more recent development has been to drive the grating with either a fine pitch stepping motor, or its equivalent in the form of a servo motor with feedback from a suitable encoder. In either case the transformation from rotation to wavelength is derived electronically with a microprocessor that generates whatever mathematical function is needed. An obvious advantage is not only high operating speed, but avoiding the cost of an accurate slide and screw mechanism.

12.6 Concave Grating Mounts

Concave grating mounts have played a major role ever since their invention by Rowland in 1882. A number of them, now regarded as classical, were developed over a twenty year period and are described below. New mounts were not developed for the next several decades because there was neither the need nor appropriate technology. As soon as need arose, solutions were found.

Applications also changed significantly over time. Initially the primary preoccupation was accurate determination of as many wavelengths as possible from every element of the periodic table, in order to set the foundation for analytical spectrometry. Some of it was later used to identify details of atomic structure. The ideal instruments were room–sized spectrographs. Length of photographic exposure was not a great concern so that astigmatism, which was responsible for doubling or tripling exposures, was simply accepted as a given. What was important was the large linear dispersion available. It is doubtful whether many of these giant spectrographs are still in use, and they are too large to fit into a museum.

When photoelectric flux detection systems became available in the 1930's, new possibilities opened up, especially with photomultipliers and their large dynamic range. Behind strategically located exit slits in the image plane they could record the specific wavelengths of elements to be monitored, providing data that not only was more accurate but was also displayed rapidly. The wavelength research data of previous decades could now be put to practical use, in the metallurgical industry for example. The instruments were called direct reading spectrographs, or simply direct readers. They were much smaller than the original instruments, but with grating radii of 1 to 2 m, were not exactly desk top size.

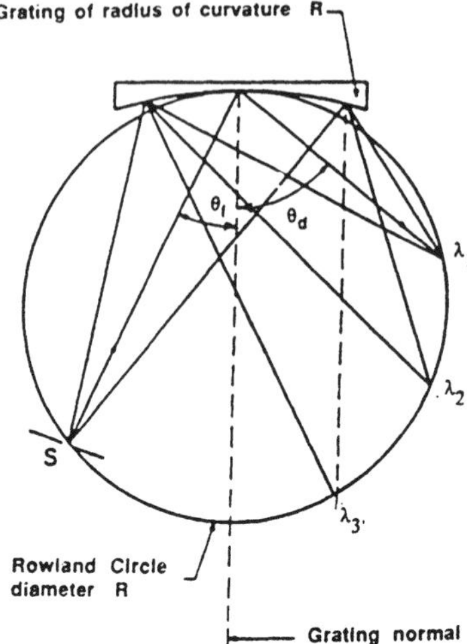

Fig.12.9 Rowland Circle. A grating with radius R is located tangent to a circle with radius R/2. Entrance slit S is imaged on circle shown for different wavelengths λ (after [12.3]).

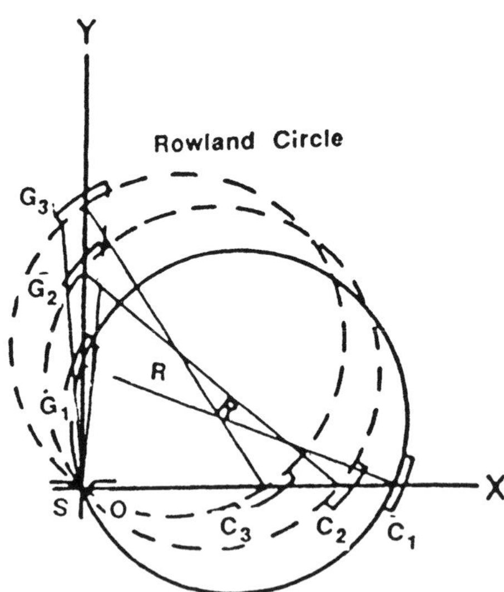

Fig.12.10 Rowland mount. Grating G is constrained to move along track OY, which carries entrance slit S at its origin. A perpendicular track OX defines the location of camera C. Grating and camera are connected by linkage bar labeled R, whose length must equal grating radius (after [12.3]).

Small concave grating instruments, although popular today, were not of much interest at first, until research began into wavelengths below the oxygen cut-off around 185 nm. Since low reflectance of mirrors pointed to the use of single optic grating systems there was no choice but to use concave gratings in mounts that could be placed inside vacuum chambers. When interest developed in the soft X-ray region (4 to 50 nm) another complication was added because gratings could reflect only under grazing incidence conditions.

When protected aluminum coatings were developed that reflect well to 110 nm or even below, concave gratings soon were replaced by plane grating equivalents in this spectral region. The reluctance to use concave gratings (unless there were truly compelling reasons) was based in part on their greater cost, but even more on their large imaging aberrations, especially astigmatism. This picture changed in the 1970's when it was found that aberrations could be significantly reduced by slightly modifying the uniformity of groove spacing as well as departing from perfect groove straightness. While this has been done with special computer controlled ruling engines it is much simpler to use holographic techniques (see Ch.7). Applications that had effectively been forbidden suddenly made sense. Two applications in particular stand out. First is that of scanning monochromators and the second is flat field spectrographs. The latter would not have been of much interest were it not for the simultaneous development, for other reasons, of solid state photodetector arrays.

12.6.1 The Rowland Mounting

Basic to most of the classical concave grating mounts is the Rowland circle. It follows Rowland's initial discovery that horizontal focussing of vertical slit images is assured when the entrance slit and detector are both located on a circle whose radius is 1/2 that of the grating. The grating is tangent to the circle, as shown in Fig.12.9.

Rowland's response was to design a mount that has been given his name [12.24]. Rowland maintained the 'camera' (in the form of a photographic plate) on the grating normal, since this would deliver spectra with dispersion sufficiently linear to allow him to determine wavelengths with linear interpolation of his plates. The mechanical arrangement was rather cumbersome, but allowed him to change angles of incidence, i.e., entrance slit location, for different wavelength ranges, while maintaining the Rowland circle conditions, Fig.12.10. It is no longer used in this form.

12.6.2 The Abney Mount

Abney, just four years after Rowland's announcement of the concave grating, maintained the same geometry through somewhat simpler means

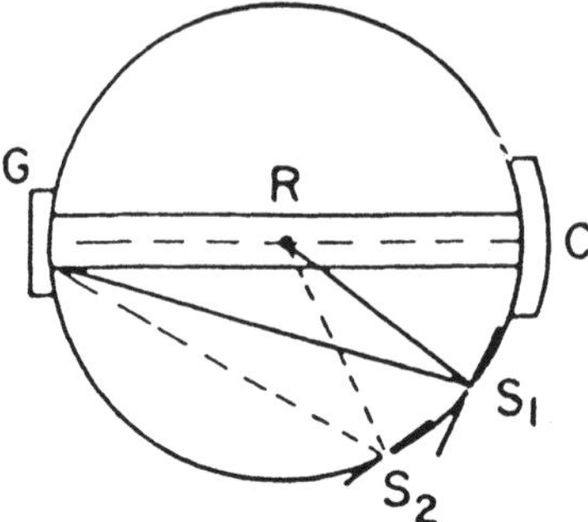

Fig.12.11 Abney mounting. Both grating G and camera C are fixed on single bar, separated by R, the radius of the grating. Input selection controlled by locating entrance optics and slit S on second bar of radius R/2 rotating about the center of the first bar, also rotating on it to face grating (after [12.25]).

[12.25]. The grating and the plate holder were at opposite ends of a suitably rigid bar. By definition, the center of the bar is the center of the Rowland circle. Thus if a second bar is made to rotate about this center and carries on top a slit whose distance from the center is 1/2 the grating radius, the Rowland circle conditions are fulfilled. A complication is that the slit *and its associated entrance optics* must be able to rotate around an axis through the slit, in order that the entrance beam always points to the grating, Fig.12.11. Like the Rowland mount, it has long served its purpose and is now considered obsolete.

12.6.3 The Paschen - Runge Mount

The Paschen–Runge mount was developed about the same time as the Abney [12.26]. It is a relatively simple and stable design and as a result is still used today. One characteristic is that both the grating and entrance slit are fixed in position, which means that the angle of incidence is kept constant, Fig.12.12. In some instances versatility has been added by giving the instrument a second entrance slit, complete with a second set of illuminating fore-optics. In the original design one or more curved plate holders were located around the Rowland circle to cover a wide range of the spectrum. Only the one normal to the grating picked up the desirable linear dispersion. When wavelength measurements were a major preoccupation these instruments were large, with gratings of 6 and 10 m radius. Obviously vibrations and constant temperature are important for obtaining coherent data, and as a result such instruments were invariably located in special basement labs.

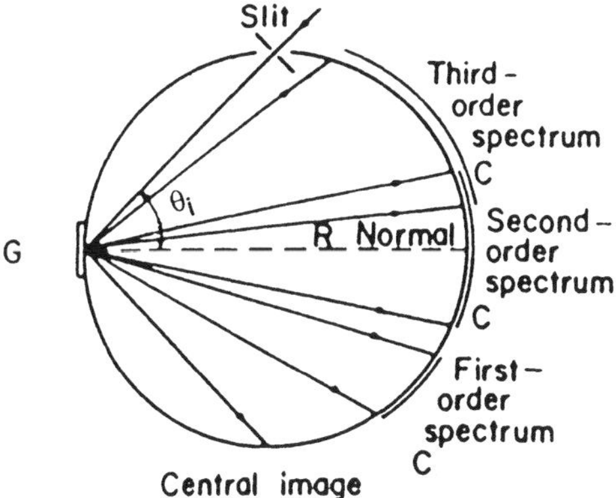

Fig.12.12 Paschen–Runge mounting, maintains fixed position for grating G and entrance slit S. One or more cameras C are accurately located around Rowland circle to record different spectral orders (after [12.4]).

The Paschen–Runge mount was the first to be converted from a spectrograph into a photo–electric direct reader, or polychromator, because its large focal curve left plenty of room for detectors. However, should two of the desired lines be too close together to leave room for the detectors that can easily be remedied with steering mirrors. Astigmatism is seldom a concern here, as long as the slit image does not exceed the size of the detector. Calibration is important, because detector sensitivity is not always uniform over the whole wavelength region. The first such direct reader to be made commercially was by Hasler [12.27].

Paschen–Runge mounts have also been used in the vacuum region, 30 to 200 nm, where 30 nm is the limit below which normal incidence cannot be used with standard gratings.

12.6.4 The Eagle Mount

The Eagle mount can be considered as the concave grating equivalent of the Littrow mount for plane gratings. There is only a small angular deviation between incident and diffracted beams [12.28]. Just like the Littrow mount it is used in two versions. The first has entrance slit and image field separated by placing them above and below the Rowland circle plane, so that in the center they will be near auto-collimation. The second is to put them side by side in the Rowland circle plane (Fig.12.13).

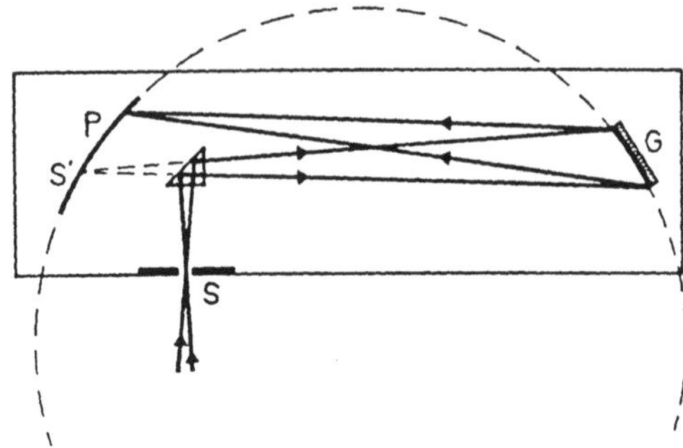

Fig.12.13 Eagle mounting with 45° prism in path from entrance slit S to grating G. Plate P shown located on Rowland circle, indicated in dashed form (after [12.4]).

If the spectral range covered by the image plane is insufficient the grating may be moved along a set of rails along the optical axis, towards or away from the entrance slit. At the same time it must be rotated around a vertical axis through the grating pole. The effect is one of rotating the Rowland circle about an axis through the slit. To stay on the Rowland circle also requires that the plate holder be rotated about the same axis. It is evident that complex and accurate mechanisms are required to increase the wavelength range–explaining why the concept was rarely put to use.

A simple trick is capable of greatly reducing astigmatism. It calls for inserting a cylindrical lens in the imaging beam, with an axis 90° to the slit. If made of fused silica it will transmit over a wide enough band (180 to 2500 nm) for most practical purposes [12.29]. Actually this was not a new idea, since Rowland mentions it in his first paper [12.24].

12.6.5 The Wadsworth Mount

The Wadsworth mount owes its high standing to the fact that it is the only concave grating mount capable of stigmatic imaging [12.30]. To achieve this performance requires departing from the Rowland circle, illuminating the grating in collimated light, and operating under angular conditions, which lead to diffraction taking place normal to the grating (i.e., $\theta_d = 0$), Fig.12.14. It is clear that there is another 'penalty' in that with a need for collimating optics one usually has to give up the luxury of operating with a single active optical element.

Although Wadsworth used a lens to collimate the input beam this was soon replaced by a mirror in order to simplify the system and, above all, avoid

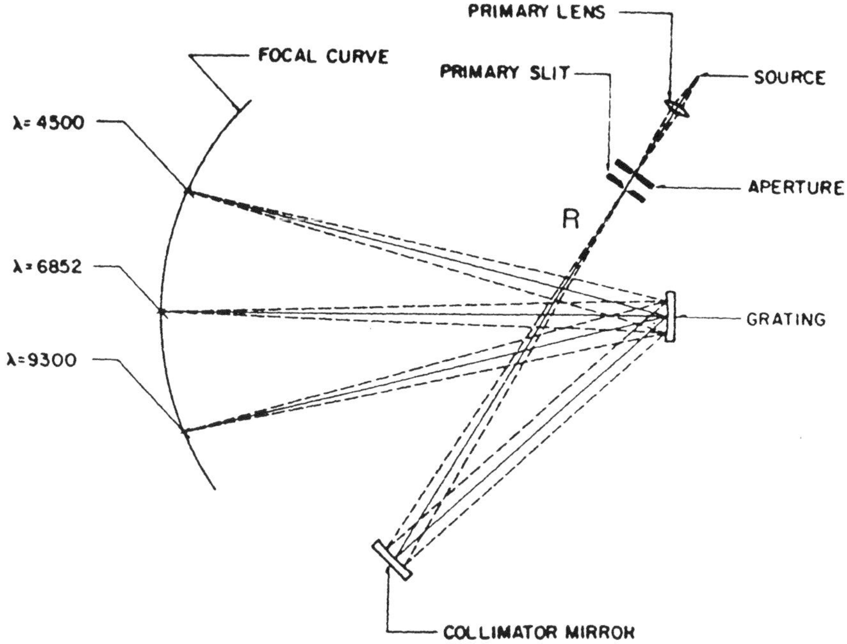

Fig.12.14 Wadsworth mount schematic. Light from source is focussed on primary slit, expands to collimating mirror which directs it to concave grating. Slit images are focussed on focal curve located 1/2 R from the grating pole, R is grating radius (after [12.4]).

chromatic changes in focus. Since the image field lies half the distance from the grating, compared to the Rowland circle, dispersion will be 1/2 that of the same grating used in any of the other mounts. However, this leads to four times brighter images, an effect enhanced still more by the lack of astigmatism.This usually more than makes up for light lost through use of a collimating mirror.

Many commercial instruments have been built since the first one appeared in 1942 [12.31]. An exceptional version of a Wadsworth was built by Bartoe and Brueckner, who needed a compact spectrograph for a rocket experiment to study solar radiation over a relatively wide range (120 to 270 nm) [12.32]. This meant using diffraction angles which at the ends of the range no longer delivered a sufficiently low degree of astigmatism, as can be appreciated by noting that the distance between horizontal and vertical focus, i.e., astigmatism $|A|$, is given by

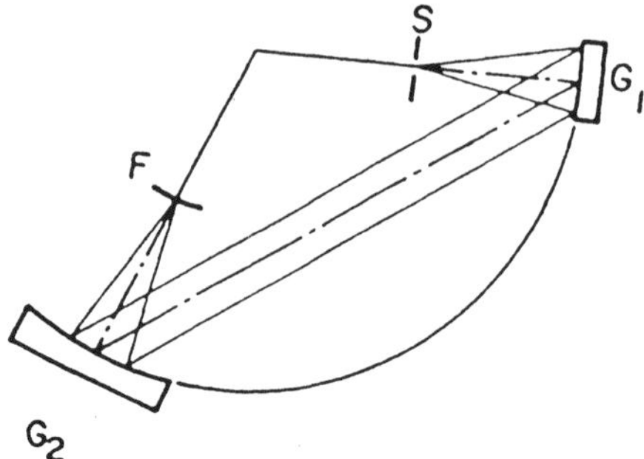

Fig.12.15 Bartoe–Brueckner stigmatic coma free spectrograph. Grating G_2 illuminated with collimated light diffracted by grating G_1, which is illuminated by light from normal entrance slit at center of curvature (inverse Wadsworth). Final Wadsworth image produced on focal surface F. Both gratings have same radius and groove frequency (after [12.32]).

$$|A| = 2\frac{1-\cos^2\theta_d}{1+\cos^2\theta_d} \quad , \tag{12.8}$$

where θ_d is the angle of diffraction. For small values of θ_d this reduces to

$$|A| = 2\theta_d^2 \quad . \tag{12.9}$$

Obviously astigmatism is reduced to zero only for small values of $\theta_{d,}$ an option not available, given the dispersion necessary for the experiment. The elegant solution was to arrange for predispersing the light incident on the concave grating in such a way that the light bundle of each wavelength falls on that portion of the grating that will diffract it normal to the grating. One consequence is a much larger grating, so that different portions can handle different wavelength regions. Proper control over this incidence can be done in two ways. One is with a plane grating of the same grove frequency, receiving light from a collimating mirror. The other is to combine these two elements into a single concave grating, again with the same groove frequency (Fig.12.15). With such a tandem design not only is astigmatism kept to low values over a wide range of angles, but coma is reduced nearly two orders of magnitude compared to a standard Wadsworth mount. The two gratings in this instance both had 2400 gr/mm frequency and a radius of 850 mm. The larger grating was 90x270 mm, or f/3, and could only be produced by holographic methods. This made it interesting to compare performance with what could have been expected from a single aberration reduced holographic grating: It would have had astigmatism 1 to 2 orders of magnitude greater than the more complex tandem mount, and thus not suitable for this experiment.

12.6.6 The Seya - Namioka Mount

All the previous concave mounts were designed basically for use as spectrographs. However, there was a gradually expressed need for a monochromator based on a concave grating in order to have a single element system. Initially demand arose from interest in the vacuum UV region (<180 nm) where no mirror has sufficient reflectance to consider using two of them in series, one for collimation and one for focussing. The Rowland circle, considered basic to all single optic designs, does not lend itself to wavelength tuning with respect to fixed entrance and exit slits. It remained for Seya to discover that if one starts with entrance and exit slits on a Rowland circle and gives them an angular separation of 70.25°, the image degradation resulting from wavelength tuning by grating rotation was at a minimum, and acceptable for many applications [12.33]. As the grating is rotated for wavelength selection the slits will in effect move away from the Rowland circle, as shown in Fig.12.16. The exit image will be slightly curved so that resolution is poor when long straight exit slits are used. Some improvement occurs when the exit slit is curved, but is much better if the spherical grating is replaced by a toroidal one. A detailed analysis by Namioka can be found in [12.34].

The role of the Seya-Namioka mount was reduced with the 1959 development of UV-enhanced aluminum [12.35]. Fast fired aluminum, rapidly

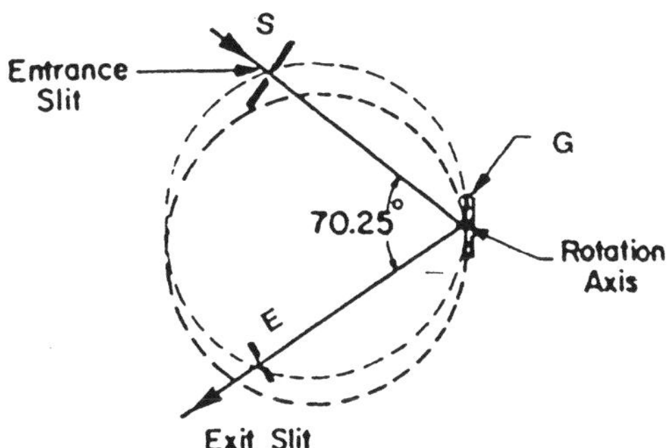

Fig.12.16 Seya–Namioka concave grating monochromator. Entrance and exit slit S and E separated by angle of 70.25° and located on Rowland circle define by normal position of grating G.Wavelength tuning by rotation of G around axis indicated (departing from Rowland circle) (after [12.33]).

coated with a thin layer of MgF_2, provides an oxide–free surface that reflects well down to 120 nm. As a consequence Czerny-Turner mounts, with much better imaging properties, gradually replaced the Seya-Namioka for wavelengths > 120 nm.

In the visible there had been no need to seriously consider the use of the Seya-Namioka mount. The advantage of a single optic was not enough to overcome imaging limitations. However, this picture changed dramatically with the development in the 1970's of holographic concave gratings whose imaging properties, particularly astigmatism, could be improved just enough to become really useful, and led to a whole new class of single optic monochromators (see Chapter 7). Actually most of these monochromators added a pair of beam steering mirrors in order to have entrance and exit slits in line, with a minimal optical penalty [12.36]. In the near UV and visible region these instruments, with 100 and 200 mm radii gratings, could provide a modest resolution of about 2 nm – adequate for many applications. It may be noted that optimal imaging now required departing from the 70.25° angular deviation to smaller angles, 61 and 49° being typical [12.37]. Later designs expanded this approach to shorter wavelengths in the vacuum UV, where with a switch to a toroidal shaped grating, 0.5 nm resolution could be obtained at 160 nm, the angular deviation increasing to 142° [12.38].

12.6.7 Flat Field Concave Grating Spectrographs

Strictly speaking there is no such thing as a flat field spectrograph with a single concave grating as the optical element. However, it turns out that it is

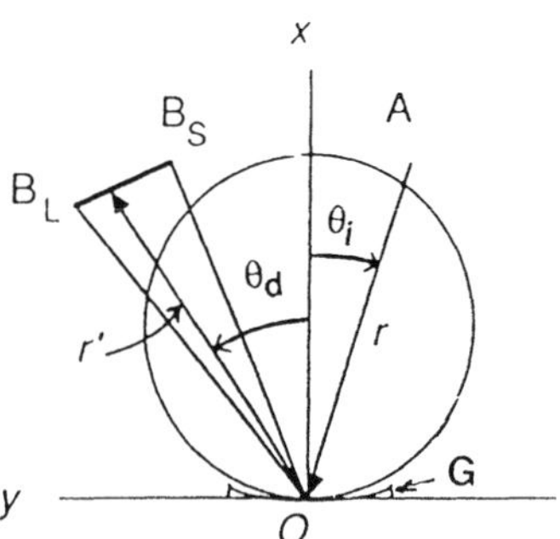

Fig.12.17 Principle of concave grating flat field spectrograph, Departure from Rowland circle is exaggerated. B_S and B_L designate long and short wavelength end of spectrum, r and r' represent distances from pole of grating to entrance slit and image plane respectively, γ is the inclination of the image plane (after [12.39]).

possible to design holographic concave gratings that produce an image field sufficiently flat and with resolution adequate to match typical photodetector arrays [12.39]. A typical example is shown in Fig.12.17. There is no standard design or mount, with a name attached, to describe these systems. They are usually designed to order. The larger the aperture (or the lower the focal ratio) the more difficult it will be to design the grating with an image field sufficiently flat and with acceptable aberrations. Typically the angle subtended by the spectral image will determine image quality and in general should be made as small as possible, with limits that are set by the wavelength range and size of the detector elements, convoluted with the desired resolution.

Since the limited number of degrees of freedom available in design have to concentrate on maintaining the focussing conditions over the required range (see Chapter 7.4), there may not be enough left to control remaining aberrations, especially coma. Palmer has shown that while absolute astigmatism can be attained, absolute focus cannot [12.40]. A perhaps obvious approach to improved imaging behavior is to add degrees of freedom by replacing the spherical waves normally used to generate such holographic concave gratings by equivalent aspheres [12.41]. Some noticable improvement is attainable, but the aspheric terms are able to influence only the higher order aberrations. The unfortunate result is that while one can improve resolution by reducing coma, one cannot reduce astigmatism at the same time. Conversely, reducing astigmatism prevents coma reduction [12.42]. For example, in one design the image heights of a flat field spectrograph were reduced by a factor of five, compared to first order design, but image curvature increased, indicating that coma was not reduced. In another example, the line width of a Seya-Namioka monochromator was reduced by at least a factor of two, due to better focusing and reduced coma, but with a small increase in astigmatism.

12.6.8 Grazing Incidence Mounts

Grazing incidence mounts with their large diffraction angles are afflicted with large imaging aberrations, primarily astigmatism, and would never be used if there were not compelling reasons. These are not hard to discern, since at sufficiently short wavelengths (< 30 nm) metallic surfaces no longer reflect unless light is incident at large angles (i.e., near grazing). The shorter the wavelength the closer one must approach grazing incidence. Not discussed here are mirrors and gratings whose normal reflectance has been extended to the X-ray domain through the use of a large number of quarterwave layers, for example alternating beryllium and molybdenum.

Applications of EUV (< 50 nm) and soft X-ray (< 10 nm) regions of the spectrum are found in satellite instruments and research in fields of plasma and Synchrotron light sources. None of them is likely to be a catalog item.

In order to reduce astigmatism to acceptable levels it has become common practice to replace spherical blanks with toroids even though they are difficult to make and expensive. To reduce aberrations still further most of the gratings will have non-standard rulings. These may be derived from holographic methods even though the degree of improvement is inherently limited. A greater degree of control can be exercised by taking advantage of modern digitally controlled ruling engines (see Chapter 14) [12.43]. They allow for much larger variation in groove spacing and can be programmed to follow any mathematical function. One proposal is to use a plane VLS (variable line spacing) grating in a monochromator whose wavelength selection drive (that rotates the grating) is coupled to a grating translation designed so that the grating is always illuminated at its optimal spacing [12.44]. In some cases the most important level of improvement is obtained by giving the grooves an appropriate degree of curvature. The Hitachi approach [12.43] is to provide an inclined plane on the ruling engine whose angle controls the lateral excursion of the diamond tool as it traverses the concave blank (see Chapter 14). Many other possibilities may be found in the literature [12.45].

12.7 Tandem Monochromators

When the spectral purity obtainable from a monochromator is insufficient for the task at hand there are some interesting solutions. Since the principal source of stray light is likely to be the grating, the first search should be for one where it is minimized. Normally this means replacing a mechanically ruled grating by an interference or holographic grating. This may reduce interorder stray light by as much as an order of magnitude (to levels of about 10^{-5} at a distance of 25Å), but even that is not always sufficient. There are applications, such as Raman spectrometry where in addition to interorder scatter (far field) we are also concerned with near field scatter due to a strong exciting line. Even a theoretically perfect grating used by itself will be inadequate. The classical solution is to make use of two or more monochromators in series. The basic concept is a simple one: any light coming through the exit slit of the first monochromator (which becomes the entrance slit for the second monochromator), will contain stray light of other wavelengths. In transmission through the second system they will be diffracted away from the exit slit. In special cases it has been found useful to repeat the process by adding a third monochromator. It is interesting to note that with a triple system it no longer matters whether the gratings have special low ghost properties – only that they have high efficiency – while for double systems clean gratings are still desirable. Background levels can decrease to 10^{-10} for a double system, and 10^{-15} for a triple, although that is no longer measurable directly. The power of this approach can be judged by the fact that with a triple system one can observe a weak Raman line that is only 1 or 2 Å away from a strong laser exciting line. In less demanding applications Raman analysis has sometimes been made possible with a single monochromator due to the development of holographic notch filters that remove light from the exciting line instead of depending on a second monochromator.

In a double monochromator the dispersion of the second system is additive to the first, so that one can obtain higher resolution for the same throughput or double the throughput for the same resolution, through the use of wider slits. An important concern is that tandem systems can operate properly only when each of the units tracks the wavelength of the other over the entire scan. This does not happen by itself. For example, it is insufficient to have matching gratings rotating in exact synchronism if the focal lengths of the multiple mirror pairs are not exactly alike. A typical system is sketched in Fig.12.18.

It is not always essential to build tandem systems around two or three grating systems. For example, it is possible to use a single grating in double pass [12.46]. An obvious advantage is that only one grating needs to be wavelength tuned and the system is more compact. The mirror that reflects the dispersed light back into the monochromator must invert the wavefront, which means that it takes a pair of mirrors or a Porro prism to reverse the beams.

In some instances the second monochromator is designed to operate in a subtractive mode, (usually termed a *dual* monochromator) which at first seems counterintuitive. The second monochromator has no influence on spectral

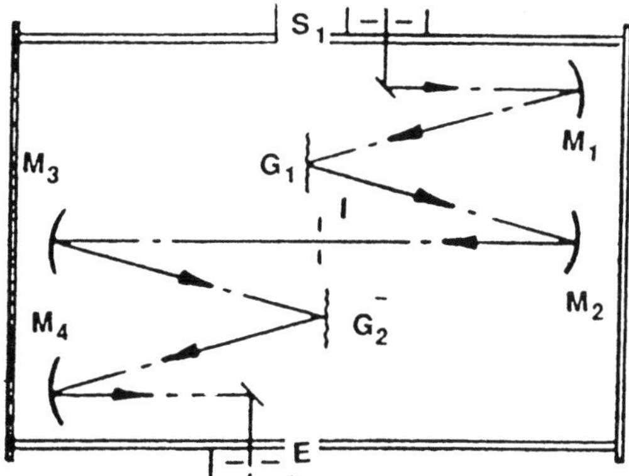

Fig.12.18 Tandem monochromator, showing two Czerny–Turner systems in series, S_1 the entrance and E the exit slit, I the intermediate slit. G_1 and G_2 are identical gratings rotating in synchronism. Collimating and focusing mirrors M_1 to M_4 must be closely matched. The 45° folding mirrors are for convenience (after [12.46]).

transmission, since bandpass is determined by the entrance and exit slits of the first monochromator. The latter serves as the intermediate slit that is the entrance to the second monochromator. The value of such a system lies not only in the obvious removal of stray light, by nearly the square of the first, but in several useful functions. One is that the light from the final exit slit is spectrally uniform, useful for fussy spectrometry such as fluorescence. If psec pulses are to be transmitted, the temporal dispersion of a single grating (i.e., the optical path difference between the two ends of the grating) will lengthen the pulses to 1 or 2 nsec (see Chapter 2.12). However, a dual monochromator will cancel this effect.

The use of multiple monochromators is not restricted to plane grating systems, and works just as well with concave gratings. An extreme example is described in [12.47] which shows that one can have as many as five successive passes of a single concave grating. The number of passes is limited only by the eventual loss of light and the skill required to adjust so many mirrors. Resolution is shown to double with the second pass. It cannot grow indefinitely, if only because the effect of residual errors gets magnified. Some simple commercial monochromators can be operated in tandem by mechanical coupling of the wavelength setting drive shafts.

12.8 Imaging Spectrometers

An imaging spectrometer is a special spectrophotometric system capable of generating a collection of images of a real scene or object, each of which is formed in one of a set of different wavelengths or wavelength bands. The first such systems were spectroheliographs. They were designed to produce monochromatic images of the sun. Developed by George Ellery Hale, long

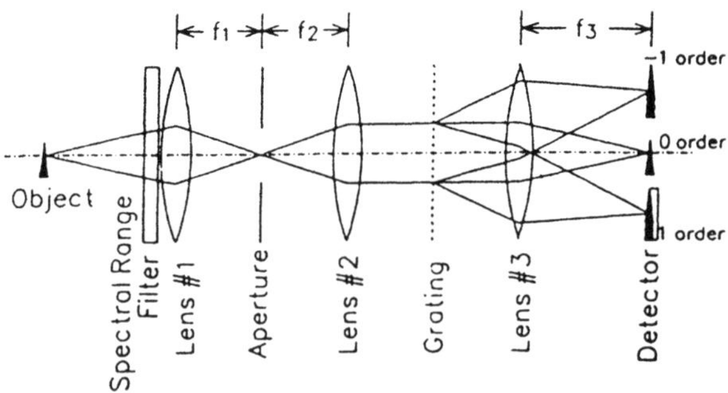

Fig.12.19 Schematic diagram of Skylab spectrograph (after [12.49]).

before he built the Mt. Palomar telescope, they were used to observe the nature of sun spots by photographing the entire solar surface sequentially at the two strong Calcium lines (396.6 and 396.86 nm) [12.48]. The concept involved is relatively simple but the instrumental complexity is severe. The exit slit image represents a slice of the solar disk as seen by the solar image projected onto the entrance slit, but as seen through a high resolution but variable filter that passes only one wavelength at a time. In order to photograph the sun a photographic plate underneath the exit slit is moved in synchronism with the entrance slit until the entire sun has been covered. It is in effect a monochromatic raster scan. The spectrograph must deliver a stigmatic image for this to work

The Skylab satellite referred to in the introduction of this chapter carried an unusually simple spectroheliograph, in the form of a 4 m concave grating Wadsworth spectrograph which produced successive images of the sun onto film over a wavelength region from 280 to 450 Å [12.49]. The images were 18 mm diameter but showed very detailed features, although the emitting

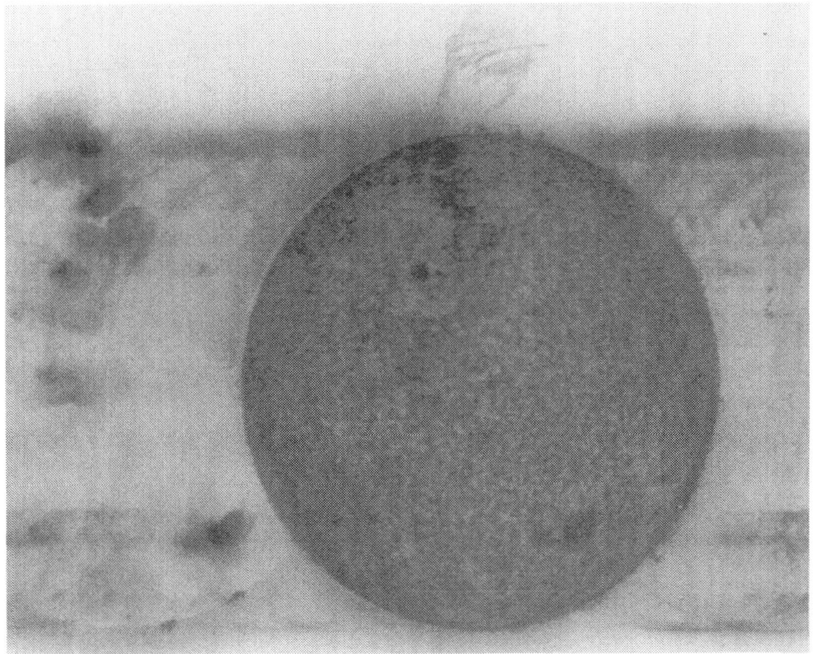

Fig.12.20 Eruption of a major He II flare (304Å) taken June 1973 with S-082 spectroheliograph on Skylab. A secondary solar image to the right is from the Fe XV line at 278 Å and has no flare (after [12.49]).

wavelengths were so close together that some of the images partially overlapped. An attempt was made to maximize dispersion through ruling the grating at 4800 gr/mm. However, that turned out to be impossible, and 3600 gr/mm was accepted as a compromise.What makes such a simple approach possible is that at these short wavelengths the sun radiates only at a few wavelengths specific to the various highly ionized elements present in the sun, especially iron, so that no complex scanning system is needed. A schematic diagram of the instrument is shown in Fig.12.19 and one of the pictures taken in the light of the Helium II line at 304 Å shows a spectacular eruption, Fig.12.20.

Even more ambitious from a data handling point of view is a project for deriving complex spectral information of earth features as seen from satellites. A telescopic image is scanned across the entrance slit of a stigmatic spectrograph. A two-dimensional detector array picks up at any instant the images arriving at the slit as observed in each of a set of successive wavelength bands. As soon as the data is transferred to a high speed computer memory the system is ready for the next strip of ground. The process makes sense only if there is provision for a large amount of data storage and processing. In effect such an imaging spectrometer acquires a three-dimensional data set, two dimensions of which are spatial (x and y) and the third is spectral. It allows many geologic and biologic surveys to be performed [12.50]. In principle it differs from a spectroheliograph only in that many wavelengths are picked up at one time, made possible by modern techniques of high speed data recording.

12.9 Multiplexing Spectrographs

If optical fibers can carry signals generated by modulated lasers, they can carry a great many more if several laser sources can be made to operate at different wavelengths. An important requirement is that there must be a means to feed the different 'colored' signals into the fiber at one end and especially to separate them at the other. One of the tools that is frequently resorted to is a miniature spectrograph, usually in the form of a modified Littrow system. Much effort has been devoted to this field because of its commercial importance, and a considerable number of patents have been issued.

Early attempts to assemble a spectrograph from separate elements served only to proove the concept. They were too expensive, especially the small high-grade collimating lens, difficult to assemble, vulnerable to environment, and far too large.

A monolithic system makes a better design, because the entire light path is through glass elements only. One of the first designs is due to Tomlinson, and is not more than one inch in size [12.8]. It uses a cylindrical gradient index lens as the collimator, which has the great advantage of flat end faces. The first face contains the input fiber and next to it is the image field, where light from the input fiber is imaged in its different 'colors', and where the output fibers must be located. The second face has a small prism cemented to it whose hypotenuse contains a suitable plane diffraction grating, typically 2x3 mm, Fig.12.21a. The principal difficulty of this type of system lies in the delicate task of locating and cementing all the fibers in exactly the right place. Since efficiency is always a critical item with such devices, and tolerances are less than 1 μm, truly elegant tooling is required, especially since there is always pressure to do this quickly. An important parameter is the number of channels to be included (for which demand seems to constantly increase), and which must have minimal cross-talk.

An important alternative approach that retains the monolithic feature replaces the GRIN lens with a glass cylinder whose end has been polished into a spherical or parabolic mirror and given a reflective coating. A plane grating is cemented to the flat face, but contains a central opening inside which the input

(a)

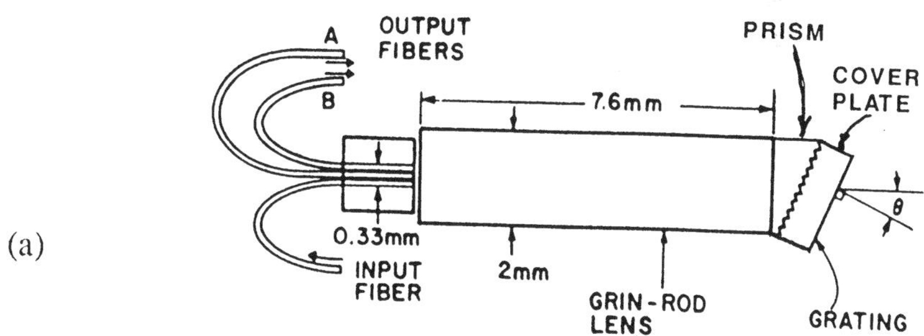

(b)

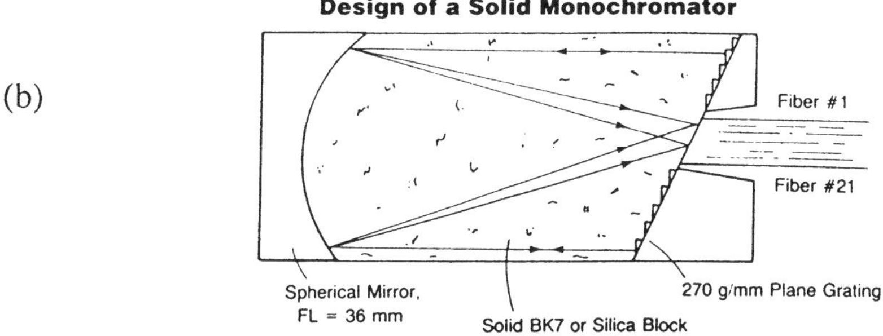

Fig.12.21 Monolithic demultiplexers for optical fibers, using miniature Littrow spectrographs (a) with GRIN rod lenses (after [12.8]) and (b) with concave surface on end of glass rod to act as collimators (after [12.51]).

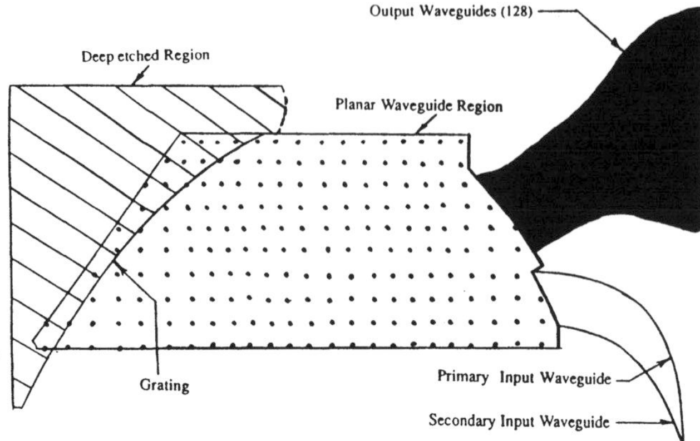

Fig.12.22 Planar waveguide grating device (top view) using short cylindrical concave grating (after [12.52]).

and output fibers are placed, Fig.12.21b [12.51]. Locating the fibers accurately is still an important and difficult job.

As the number of channels to be multiplexed keeps increasing, efforts have been made to escape the tedious problem of precision fiber location by adopting guided wave technology. Since at the same time the wavelength of interest shifted from the initial 0.8 μm region to 1.55 μm, where silica fibers have the optimal transmission characteristics, it becomes practical to operate with silicon structures [12.52]. An example is shown in Fig.12.22, where a concave grating is generated in a silicon structure by photolithographic methods. The chief advantage of such an approach is that light is no longer picked up by glass fibers but by waveguides which direct it to individual detector areas, carrying with it great potential for cost reducing mass production. Unfortunately it is difficult to generate really good grating grooves (i.e., straight and with sharp edges) by lithographic transfer from a master pattern, even if the latter is perfectly produced in an electron beam pattern generator (see Chapter 16).

An entirely different approach is to take advantage of the properties of slab waveguides, such as can be made by cementing a thin sheet of glass between two thicker pieces of lower index. Typically one end of a rectangular sandwich, about 30x50 mm, is given a cylindrical radius through grinding. This surface is converted into a concave grating by wrapping and cementing a thin sheet of a reflecting diffraction grating around it. The opposite end has to be dimensioned so as to take advantage of Rowland circle imaging, since we are now in the concave grating domain. Some authors have suggested that the grating be derived from a standard ruled plane grating, and then express surprise at the poor quality of the image. The explanation is that they have forgotten Rowland's finding that proper imaging of a concave grating requires uniform groove spacing *along the chord* of a grating, not the circumference. To act on this requires not only the difficult ruling of a non-uniformly spaced master, but locating the replica correctly around the circumference.

Several designs can be found in Figs.9.6-8.

12.10 The Role of Fiber Optics in Spectrographs

Section 12.9 (as well as Section 9.4) described how special spectrographs serve as multiplexers for fiber optics communication. In this section there is a discussion of how the design and use of spectrographic systems has been affected by the use of optical fibers as light input devices.

What makes fibers so useful is recognition that bringing light to be analyzed to the entrance slit of a spectrograph is sometimes a complex assignment that benefits from simplification. The most extreme example is found in astronomy. High resolution work used to call for coudé spectrographs of enormous cost and size, located in their own special rooms, which had light fed to them by a train of at least five mirrors. It was a great relief to replace them with echelle spectrographs small enough to be hung from the telescope at the Cassegrain focus. However, the next step is to locate the instrumentation in a fixed enclosure removed from the telescope, in which it is much easier to control temperature and vibrations and possibly interchange gratings. Light is supplied from the Nasmyth focus of the telescope by means of an optical fiber, which may be several meters long. If the spectrograph is made sufficiently stigmatic there is no need to restrict oneself to a single fiber. Instead, it is possible to have forty or many more fibers lined up to represent the entrance slit which lead to an equal number of dispersed images onto the same number of detector arrays. In the case of astronomy, the input end of the fibers are pre-aligned on the telescope image plane plate ahead of time, so that each fiber receives the image of a specific stellar object. Considering the cost of operating a large telescope and its instruments, it is a major advance to operate with so many channels in parallel.

At the other end of the scale one can attach fibers to microscope image planes in order to conveniently perform spectral analysis on a microscale.

Another application of fiber inputs is in the construction of miniature spectrophotometers, with 50 mm radius Czerny-Turner optical systems, mounted on a printed circuit board that contains the array detector and all associated electronics. Such a level of compactness would be unthinkable without the use of optical fibers [12.53].

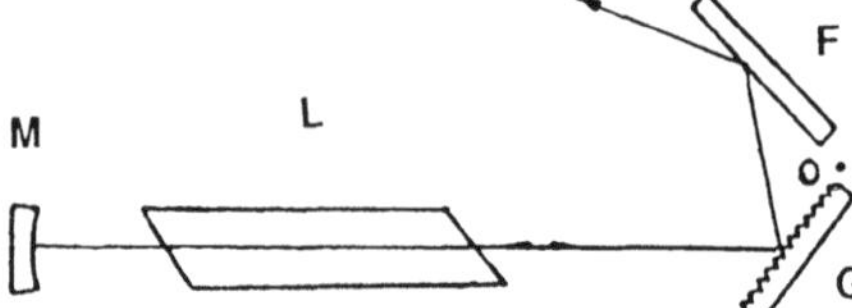

Fig.12.23 Littrow tuning of laser L via rotation of grating G. Mirror F reflects zero order output and is rigidly attached to the grating. Rotation about axis O, where planes of G and F intersect delivers fixed output beam (after [12.54]).

12.11 Laser Tuning

One of the classical methods for controlling the wavelength of a tunable laser, from UV to IR, is to insert a diffraction grating into the laser cavity. In the case of CO_2 lasers, no more is needed beyond setting up the grating in first (occasionally second order for CO lasers) in autocollimation. Such laser tuning is the only pure Littrow or autocollimation application of gratings. Besides the laser tube the only other optical element is the reflecting mirror at the other end. The latter can be made semi-transparent if output is to be drawn from that end. As a rule it is more efficient to take advantage of the fact that all gratings will diffract at least some light into the zero order. The only difficulty is that whenever the grating is rotated for tuning purposes, the zero order direction will rotate by twice the amount. The problem can be solved by the addition of two mirrors, one fixed and the other rotating with the grating as shown in Fig.12.23 [12.54]. The grating groove frequency will be chosen so that $\lambda/d > 0.5$, thus insuring that light can be diffracted only in the zero and first orders. Efficiency can be very high in the S–Plane (TM), and the groove angle or groove depth can be designed so that anywhere from 2 to 25% will be delivered into the zero order. The high energy of CO_2 lasers requires that for this application the gratings are either cut into solid metal blanks, most often aluminum, or made in the form of replicas on copper or aluminum substrates to act as heat sinks. In special cases it becomes necessary to provide the blanks with water cooling. To maximize reflection these gratings are usually gold coated. The purpose of the gratings is to separate the lasing light from the many transitions that a CO_2 laser can make.

With dye lasers the situation is quite different since the wavelength choice is continuous, rather than determined by a few molecular transitions. Wavelength purity becomes a function of the optical tuning method chosen, which must have relatively high resolution. The entire grating must be illuminated coherently, despite the usually small beam diameters, for which two

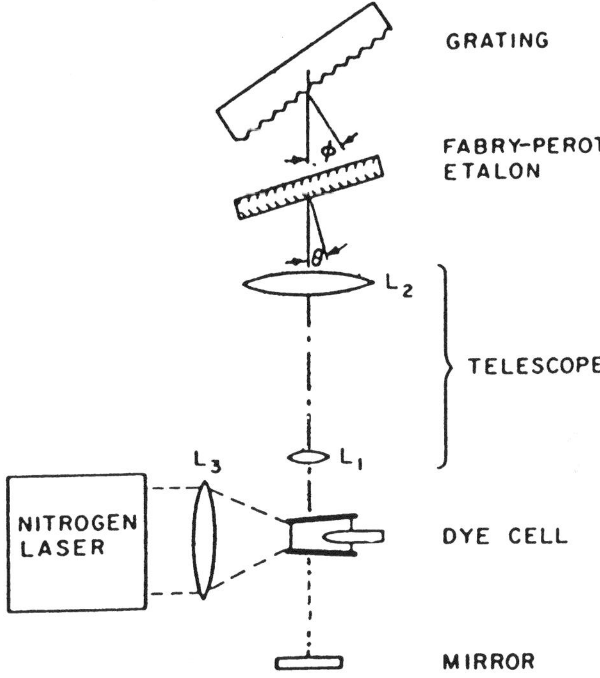

Fig.12.24 Dye laser tuning with high-angle grating G filled by inverted telescope L_1–L_2, etalon for fine tuning. Output via mirror M (after [12.55]).

equally valid schemes have emerged. One is to expand the beam and the other is to fill the grating with a narrow beam operating in grazing incidence.

The initial approach to beam expansion, due to Hänsch, is to insert an inverted diffraction limited telescope into the cavity, as shown in Fig.12.24 [12.55]. A bandwidth of 0.08 Å is attainable, but can be reduced to 0.004 Å by adding a Fabry-Perot etalon.

An alternate method of expanding the beam is with prisms, which are more compact and less expensive. The example shown in Fig.12.25 is for a KrF excimer laser, where the purpose is to maintain at all times the absolute wavelength near 248 nm within a ±2 pm FWHM band (or 8 parts per million) [12.56].

If the grating is tilted at at least 85° the laser beam will cover sufficient ruled area to provide adequate resolution, thereby delivering tuning capability with utmost simplicity. As shown in Fig.12.26 the light must be returned by a mirror, whose rotation supplies the tuning function, and which can also be in the form of a second grating [12.57, 58]. Diffraction efficiency tends to be low so that the concept is restricted to high gain lasers.

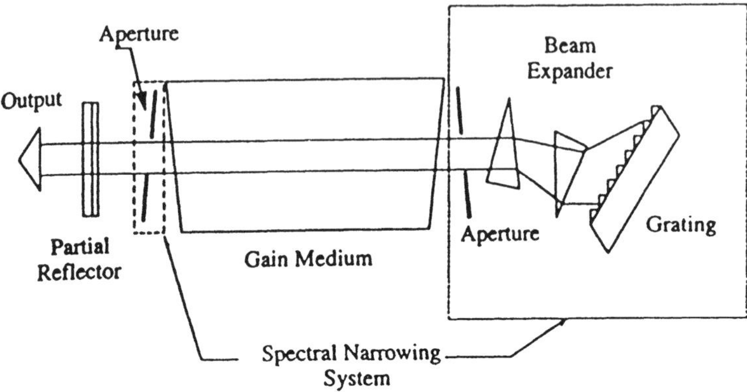

Fig.12.25 KrF excimer laser wavelength control and line narrowing system. Beam expansion by factor 15 via prisms fill high-angle grating. Apertures select spectral bandwidth (after [12.56]).

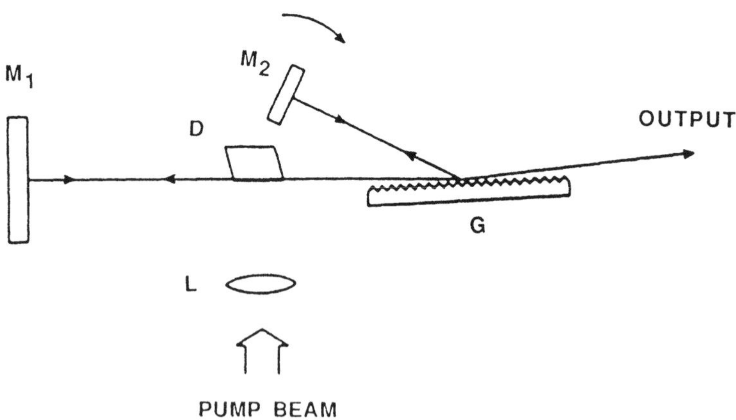

Fig.12.26 Grazing incidence tuning system for dye laser D with fine pitch grating G (1800 gr/mm in visible). Output via zero order. Tuning by rotation of mirror M_2. Mirror M_1 is cavity end mirror. Pump beam enters via lens L (after [12.57]).

12.12 On Absolute Groove Spacing

When gratings were first introduced they received particular welcome from the optics community because once groove spacing had been determined by comparison with a standard of length it became a relatively simple task to determine the wavelength of specific radiation or color in absolute terms. All it took was the grating equation and the measurement of angles of incidence and diffraction. Repeating measurements in positive and negative orders, as well as higher ones, allowed reduction of random and some systematic errors. Even given the spectrometers of the last century the weak point of the measuring chain was invariably the problem of measuring the grating spacing. The usual method was to use a line reference scale to monitor the travel of the slide that carried the grating underneath a fiducial microscope, counting lines and the distance between. Uniformity could be inferred from the quality of the spectral line images. Miscounting by even one groove would be fatal, but avoidable by repetition. Time can be saved by counting say 100 grooves, skipping to 200 or other multiple, and accepting that skipping a whole line would be unlikely. Ångström, it may be recalled, followed this scheme but neglected to assure himself that the reference scale was adequately calibrated. Although he discovered after just two years that the scale was short by 130 parts per million the secret was maintained until his assistant revealed the truth 20 years after his death.

Rowland, following a similar path, did not make such an error, and had the advantage of ruling his own gratings. Once careful calibration had established the exact pitch of his leadscrew, and maintaining a constant reference temperature, he could determine the pitch of any grating through knowledge of the change gears used.

The nature of wavelength reference was completely turned around in 1895 when Michelson succeeded with help of his recently invented interferometer to establish the wavelength of the red Cadmium line in terms of the international meter to an absolute accuracy of better than 1 PPM. The result is that from this time on grating spacing could always be determined to an accuracy set only by that of the angle measuring spectrometer. However, maximum wavelength accuracy is attained by use of Fabry-Perot etalons. Full advantage could be attained only after it became possible to prepare special lamps with isotopically pure elements in the 1940's. Cadmium represents a fortunate anomaly because it is the only natural element that delivers a line (red) that is virtually free of detectable hyperfine structure.

Instruments today invariably owe their wavelength calibrations to comparison with well established secondary wavelength references. In the case of spectrographs the most common reference spectrum is that of iron, because it has a large number lines across the entire visible and UV spectrum, all of which have had their wavelengths determined with great care. At wavelengths below 200 nm it is progressively more difficult to make interferometric calibrations so that the best technique is to make multiple order comparisons with the aid of gratings. In such work it is important that both the spectral source and the grating deliver images whose lines are completely symmetric. Modern gratings can safely be depended on in this respect, but Rowland's were not quite a good as he assumed, a fact discovered as soon as interferometric comparisons could be made. An astonishing amount of time has been devoted over the last 100 years to establishing wavelength references to ever increasing accuracy.

In most spectrometric instruments the exact groove spacing of the grating is not important because readings obtained will depend also on exact focal lengths or sine bar linkages, and are always adjusted by calibration to known spectral sources. However, if gratings are to be made interchangeable they should maintain exact numerical relationships in groove spacing, typically to 0.01%, as most gratings from a single source are likely to be. In replication, groove spacing can be expected to match to a few PPM provided that masters and replicas are made of the same material. Otherwise thermal expansion effect must be considered.

12.13 Multiple Entrance Apertures

In monochromators where throughput is a critical item, i.e., in the IR region, it would be desirable to have multiple input slits, provided the resulting multiple image field could have apertures capable of sorting out the result. This was accomplished by A. Girard, with the aid of special slit patterns of a family of hyperbolas, with a matching one at the output plane [12.59]. The results were excellent, but the instruments difficult to build because they demanded near perfect imaging. It meant, for example, that ordinary high-class replica gratings had to be replaced by specially selected ones. The effort devoted to this interesting development was effectively negated when Fourier-transform instruments were developed that could perform the same task with less effort.

References

12.1 C. F. Meyer: *The Diffraction of Light, X-rays and Material Particles*, (J. W. Edwards Co, Ann Arbor, MI, 1949).

12.2 J. F. James and R. S. Steinberg: *The Design of Optical Spectrometers*, (Chapman Hall., NY, 1969).

12.3 G. Harrison, R. Lord, and J. Loofbourow: *Practical Spectroscopy*, (Prentice Hall, NY, 1948).

12.4 R. Jarrell and R. Barnes: "Gratings and Grating Instruments" (Chapter 4) *Analytical Emission Spectroscopy*, vol. **1**, ed. E. L. Grove, (Marcel Decker, NY,

1971).

12.5 P. Bousquet: *Spectroscopy and its Instrumentation*, trans. K. M. Greenland, (Adam Hilger, London, 1971).

12.6 D. Schroeder, *Astronomical Optics*, (Academic Press, London, 1987).

12.7 W. H. Simms: "On the optical glass of the late Dr. Ritchie," Mem. Royal Astr. Soc. **12**, 165-170 (1840).

12.8 W. Tomlinson, "Wavelength multiplexing in multimode optical fibers," Appl. Opt. **16**, 2180–2185 (1977).

12.9 O. V. Littrow: "Über eine neue Einrichtung des Spectralapparats," Wien Ber. **47**, 26–32 (1863).

12.10 F. Lippich: "Vorschlag zur Construction eines neuen Spectralapparats," Z. f. Instrumentenkunde **IV**, 1–8 (1884).

12.11 H. Ebert: "Zwei Formen von Spectrographen," Wied. Ann Physik & Chemie, **38**, 489-493 (1889).

12.12 W. de W. Abney: "On the photographic method of mapping the long wavelength end of the spectrum," Phil. Trans. Royal Soc. **171**, II, 653–667 (1880).

12.13 M. Czerny, F. Turner: "Über den Astigmatismus by Spiegelspectrometern," Z. Physik, **61**, 792-797 (1930).

12.14 H. Kayser: *Handbuch der Spectroscopie*, vol.1 (S. Hirtzel, Leipzig. 1900).

12.15 W. Fastie: "A small grating monochromator," J. Opt. Soc. Am. **42**, 641-647 (1952).

12.16 G. Monk: "A mounting for the plane grating," J. Opt. Soc. Am. **17**, 358-364 (1928).

12.17 A. Gillison: "A new spectrographic diffraction grating monochromator," J. Sci. Instr. **26**, 334-339 (1949).

12.18 A. Shafer, L. Megil, and L. Droppelman: "Optimization of Czerny-Turner spectrometers," J. Opt. Soc. Am. **54**, 879-888 (1964).

12.19 J. Simon, M. Gil, and A Fantino: "Czerny-Turner monochromators: astigmatism in the classical and crossed beam disposition," Appl. Opt. **25**, 3715-3720 (1986).

12.20 J. Reader: "Optimizing Czerny-Turner spectrographs: A comparison between analytic theory and ray tracing," J. Opt. Soc. Am. **59**, 1189-1196 (1969).

12.21 V. Chupp and P. Granz: "Coma canceling monochromator with no slit mismatch," Appl. Opt. **8**, 925-929 (1969).

12.22 W. Fastie: "Image forming properties of the Ebert spectrometer," J. Opt. Soc. Am. **42**, 647-650 (1952).

12.23 W. Fastie: "Ebert spectrometer reflections," Physics Today, **44**, 37-43 (January 1991).

12.24 H. Rowland: "Preliminary notice of results accomplished on the manufacture and theory of gratings for optical purposes," Phil. Mag. Suppl. to v.**13**, 469-474 (1882).

H. Rowland: "On concave gratings for optical purposes," Phil. Mag. **16**, 197-210 (1886).

12.25 W. deW. Abney: "The solar spectrum, 7165-10,000," Phil. Trans. Royal Soc. **177**, II, 457-465 (1886).

12.26 C. Runge and F. Paschen: "Über die Strahlung des Quecksilbers im magnetischen Felde," Phys. Abh. Königl. Preussischen Acad. Wissensch., Berlin, **Abh. I**, 1-18 (1902).

12.27 M. Hasler and H. Dietert: "Direct reading instrument for spectrochemical analysis," J. Opt. Soc. Am. **34**, 751-758 (1944).

12.28 A. Eagle: "On a new mounting for a concave grating," Astroph. Jl. **31**, 120-142 (1910).

12.29 H. Straat: "Compensation of astigmatic errors in a grating spectrograph," J. Opt. Soc. Am. **43**, 593-594 (1953).

12.30 F. Wadsworth: "The modern Spectroscope," Astroph. Jl. **3**, 47–62 (1896).

12.31 R. Jarrell: "A stigmatic spectrograph for industrial laboratories," J. Opt. Soc. Am. **32**, 666-669 (1942).

12.32 J. Bartoe and G. Brueckner: "New stigmatic coma free concave grating spectrograph," J. Opt. Soc. Am. **65**, 13-21 (1975).

12.33 M. Seya: "A new mounting of concave grating suitable for a spectrometer," Science of Light **2**, 8-17 (1952).

12.34 T. Namioka: "Theory of the concave grating III: Seya-Namioka monochromator," J. Opt. Soc. Am. **49**, 951-961 (1959).

12.35 G. Hass and R. Tousey: "Reflecting coatings for the extreme ultraviolet," J. Opt. Soc. Am. **49**, 593-602 (1959).

12.36 G. Hayat, J. Flamand, and J.-P. Laude: "Designing a new generation of analytical instruments around the new types of holographic diffraction gratings," Opt. Engineer. **14**, 420-425 (1975).

12.37 G. Pieuchard, J. Flammand, and J. Cordelle: "Monochromator with a concave grating," U. S. Patent No. 3,909,134 (1975).

12.38 D. Lepere: "Monochromateur à simple rotation du réseau, et réseau holographique sur support torique pour l'ultraviolet lontain (Monochromators with single grating rotation and holographic gratings on toroidal blanks for vacuum ultraviolet)," Nouv. Rev. Optique **6**, 3, 173-178 (1975).

12.39 J. Lerner, R. Chambers, and G. Passereau: "Flat field imaging spectroscopy using aberration corrected holographic gratings," SPIE **240**, 122–128 (1981).

12.40 C. Palmer: "Absolute astigmatism correction for flat field spectrographs," Appl. Opt. **28**, 1605-1607 (1989).

12.41 M. Koike, Y. Harada, and H. Noda: "New blazed holographic grating using an aspherical recording with an ion etching method," SPIE **815**, 96-101 (1987).

12.42 C. Palmer: "Theory of second-generation holographic diffraction gratings," J. Opt. Soc. Am. A **6**, 1175-1188 (1989).

12.43 T. Kita, T. Harada, N. Nakano, and H. Kuroda: "Mechanically ruled aberration corrected concave gratings for a flat field grazing incidence spectrograph," App. Opt. **22**, 512-513 (1983).

12.44 M. Hettrick: "In focus monochromator: theory and experiment of a new grazing incidence mounting," Appl. Opt. **29**, 4531-4535 (1990).

12.45 I. Peisakkhson: "New forms of dispersive elements and their use in spectral devices," J. Opt. Techn. **62**, 203-208 (1995).

12.46 D. Murcray, F. Murcray, and W. Williams: "A balloon–born grating spectrometer," Appl. Opt. **6**, 191–196 (1967).

12.47 F. Jenkins and L. Alvarez: "Successive diffractions by a concave grating," J. Opt. Soc. Am. **42**, 699–705 (1952).

12.48 G. E. Hale and F. Ellerman: *The Rumford spectroheliograph of the Yerkes Observatory*, (U. of Chicago Press, Chicago 1903).

12.49 R. Tousey, J-D. Bartoe, G. Brueckner, and J. Purcell: "Extreme Ultraviolet spectroheliograph ATM experiment S-082A," Appl. Opt. **16**, 870–878 (1977).

12.50 A. Goetz: "Imaging spectrograph for remote sensing," SPIE **2480**, 1–8 (1995).

12.51 J.-P. Laude: "Wavelength division Multiplexing," Prentice-Hall, *Intern. Series Optoelectronics*, Hemel Hempstead, U. K. HP27EZ (1993).

12.52 K. Liu, F. Tong, and S. Bond: "Planar grating wavelength demultiplexer," SPIE **2024**, 278–283 (1993).

12.53 M. Morris: "Miniature optical fiber based spectrometer employing tandem fiber probe," SPIE Chemical, Biochemical and Environmental Sensors, **IV**, Fibers (1992).

12.54 T. M. Hard: "Laser output selection and output coupling by a grating," Appl. Opt. **9**, 1825–1830 (1970).

12.55 T. W. Hänsch: "Repetitively pulsed tunable dye laser for high resolution spectroscopy," Appl. Opt. **11**, 875-898 (1972).

12.56 U. Sengupta: "Krypton Fluoride excimer laser for advanced microlithography," Opt. Eng. **32**, 2410-2420 (1993).

12.57 M. G. Littman and H. J. Metcalf: " Spectrally narrow pulsed dye laser without beam expander," Appl. Opt. **17**, 2224-2227 (1978).

12.58 I. Shoshan, N. Danon, and U. Oppenheim: "Narrowband operation of a pulsed dye laser without intracavity beam expansion," J. Appl. Phys. **48**, 4495-4497 (1977).

12.59 A. Girard: "Spectrometre à Grilles," Appl.Opt. **2**, 79-87 (1963).

Additional Reading

W. Cash: "Far–ultraviolet spectrographs: the impact of holographic grating design," Appl. Opt. **34**, 2241–2246 (1995).

J.-C. Fontanella, A. Girard, L. Gramont, and L. Nicole: "Vertical distribution of NO, NO_2, HNO_3, as derived from stratospheric infrared absorption spectra," Appl. Opt. **14**, 825-830 (1975).

W. R. Hunter: "On-blaze scanning monochromator for the vacuum ultraviolet," Appl. Opt. 21, **9**, 1634-1642 (1982).

T. Kita and T. Harada: "Use of aberration-corrected concave gratings in optical demultiplexers," Appl. Opt. **22**, 819-825 (1983).

P. Philippe, S. Valette, O. Mata Mendez, and D. Maystre: "Wavelength demultiplexer: using echelette gratings on silicon substrate," Appl. Opt. **24**, 1006-1011 (1985).

H. A. Padmore, V. Martynov, and K. Holis: "The use of diffraction efficiency theory in the design of soft X-ray monochromators," Nucl. Instr. Meth. Phys. Res. A **347**, 206-215 (1994).

G. Passereau: "Spectrograph," U. S. patent No 4,087,183 (1978).

I. Powell and A. Bewsher: "Ultraviolet-visible spectrograph optics: ODIN project," Appl. Opt. **34**, 6446-6452 (1995).

See also the references to Chapters 7, 9, and 16.

Chapter 13

Grating Damage and Control

13.1 Introduction

Grating surfaces are inherently fine pitch delicate structures that are easily damaged. Since they are also expensive there is ample reason to treat them with care. The single greatest enemy of gratings is the human finger. There is a history of curious customs inspectors who tested with their fingernails for the presence of the grooves listed on the label. More common are fingerprints, so often the result of simple carelessness, and many spectroscopists have felt like calling the police to discover whose print is on the grating. Development of high power lasers has led to a new source of concerns: for the first time light sources can be so intense that they can lead to overheating or even ablation of portions of a grating surface.

There is an enormous difference in damage resistance between aluminum-coated replica reflection gratings that make up the bulk of spectrometric gratings and, at the opposite end of the scale, gratings whose grooves have been etched into solid fused silica. The latter are also much easier to clean, but much more difficult and expensive to produce.

During the early days of outer space exploration, i.e., the 1960's, there was widespread belief that under the extreme high vacuum conditions portions of the epoxy replica film would be sucked away, leaving a wrinkled mess in place of the grating. This led to reliance on flying master gratings only, a severe strain on their supply. Proving the safety of replicas was rather difficult with ground based facilities, but the problem was finally resolved by accident. A satellite spectrometer was flown with a standard replica grating, because its planned life was only a month or two. When the satellite and its instruments unexpectedly remained in perfect working order for over a year the difficulty vanished.

This chapter is designed to advise on the nature of grating damage, the resulting limitations under various conditions, what care should be exercised, and what techniques are available for conducting rescue operations in case of damage.

13.2 Reflection Gratings

Reflection gratings are used in the bulk of grating applications because, compared to transmission gratings, systems become more compact, there are few wavelength and diffraction angle boundaries, and efficiencies tend to be greater. In return we have to accept a thin, delicate metallic surface layer and the ease with which it can be damaged. In addition the almost universal use of Aluminum gives us a material that is easily damaged, although together with its fine reflectance properties it has excellent resistance to atmospheric tarnishing.

13.2.1 The Fingerprint Problem

In some instances fingerprints left on a grating cause no problem beyond the desire to see clean optics. However, they are bound to lead to a certain amount of stray light, and thus ought to be discouraged. In almost the same category is the damage caused by indulging in casual conversations near a grating surface, leading to annoying little spots. Like a surgeon they are easily avoided by using a face mask, or simply by being careful. The first question is always how to remove such deposits, and the one clear response is take action as soon as possible, because once there has been chemical interaction between a print and the metallic surface there is no way to remove the effect. The reactivity of skin residue with metals like aluminum and silver varies greatly between people, both acidic and basic, with red hair often associated with higher rates of attack. Noble metals such as gold and platinum by comparison are relatively free of chemical attack.

If the marks are new or small enough they can sometimes be removed by judicious application of ordinary scotch tape. Gently pressing it on and then pulling it off should do no harm, since the procedure parallels one that is used routinely to insure proper adhesion of a replica grating.

A comparable procedure is to pour collodion solution over the whole grating surface, waiting for it to dry and then pulling it off. This is also effective for removing any dust particles that could not be blown off. Before they were banned Freon sprays were sometimes used to clean gratings and other optics, but were never recommended because their lack of purity often resulted in ugly looking white streaks. Fortunately they were easily removed with a rinse of isopropyl alcohol.

Removing surface marks by a *washing process* is possible, but takes more care than most people are willing to apply. The procedure is as follows: Place the grating in a clean glass dish filled with a dilute solution of industrial detergent (free of bubbles). A two inch swab of absorbent cotton is placed on the grating, and allowed to saturate. The cotton is held with a pair of tweezers and using no pressure beyond its weight, the cotton is dragged back and forth *in the direction of the grooves.* The next step is to rinse the grating at once in another dish, with the grating inclined at a steep angle, such as 45°, with the grooves facing downhill. For rinsing fluid take a spectroscopic grade of a chlorine free solvent, such as toluene or xylene, in a squeeze or spray bottle. Follow this immediately with a similar spray of isopropyl alcohol. The aim is always to avoid leaving any dried-on residue. It is advisable to wear rubber gloves during the procedure, because the solvents are quite capable of dissolving skin oils and depositing them on the grating surface.

A new technology for cleaning optical surfaces involves the use of CO_2 snow. It works well on gratings, especially for removing fingerprints. The idea is to expand liquid CO_2 through specially designed nozzles, so that solid particles of CO_2 snow are formed that mix with the gas. Their collision with surface contamination loosens it enough so that the gas flow can remove it [13.1].

Scratches on the grating surface sometimes are a cause of great alarm, because they stand out in the eye of the observer, usually much more than the harm they might do to spectral imaging. They may be the result of careless cleaning attempts, but more likely are already present on the surface of the master. In ruled gratings they may come from small adhesions to the tip of the diamond tool. Interference, or holographic gratings have different defects, usually originating from diffraction caused by dust particles on the collimating optics.

13.2.2 Vacuum System Residues

Experience has shown that, contrary to common sense, gratings used in an ordinary vacuum environment are much more likely to be damaged than those used in open air. The explanation is simple, and has two causes. One is that oil diffusion pumps leave a small amount of oil vapor in the system, which would do no harm if it were not for the simultaneous presence of UV radiation that usually is the primary reason for putting a spectrometer under vacuum. This leads to breakdown products, principally carbon, which are deposited and baked onto the grating. The obvious solution is to use oil free pumping, such as centrifugal or cryogenic.

Once a grating has acquired the surface haze associated with this effect, and consequent loss in efficiency, there is no way to remove it by washing, since the deposits are too well baked on. Sometimes a grating can be rejuvenated by simply giving it a new overcoat of aluminum or gold, but in other cases it may turn out better to remove the deposit by plasma ashing in an oxygen atmosphere and then providing a new overcoat [13.2]. Experience has shown that visual appearance is a surprisingly poor prognosticator of which path is the better one to follow.

13.2.3 Laser Beam Damage – CW

The nature of damage caused by CW laser radiation is largely one of gratings becoming too warm from absorbing some of the incident light. If the grating is illuminated exclusively in the TE plane of polarization (the electric vector parallel to the grooves) the light absorbed will never exceed the reflectance loss and will therefore be small, no matter what the λ/d ratio or the number of possible diffracted orders. However, for laser tuning applications the polarization will usually lie in the TM plane (electric vector perpendicular to the grooves). This means that for wavelengths where λ/d is near 2/3, i.e., the Wood's anomaly region, there can be significant energy absorption beyond that predicted from reflectance measurements, and is caused by resonance effects. The obvious solution is to operate in the region $\lambda/d > 0.8$, where such effects disappear (see Chapter 4).

The unavoidable consequence of energy absorption is raising the temperature of the grating. This may cause the surface of the grating to become convex due to temperature gradients, especially if the substrate has poor thermal conductivity, such as glass. However, the main concern is that when the grating surface temperature exceeds about 65°C the standard epoxy replica resins no longer maintain their geometrical integrity, and diffraction efficiency will drop with changes in groove shape. A widely adopted solution to this problem is to replace the glass substrate with a metal, where copper, with a thermal conductivity some 100 times greater than glass is usually preferred. Also, for a given size, it has a thermal heat capacity three times that of glass or aluminum, although the latter has a thermal conductivity almost as good as copper.

The ability of metallic substrates to absorb energy input can be increased by providing means to conduct heat elsewhere. One simple means is to construct a heat radiating tail, one end of which can be screwed into a tapped hole provided in the substrate. If that is not sufficient, the blank can be equipped with hollow passages through which cooling water can be passed. This is very effective, although the additional plumbing can be a real nuisance.

Another approach is to substitute high-temperature epoxies for the standard room temperature curing versions. These have the additional advantage of allowing outbaking of gratings used in ultra-high vacuum systems. Gratings able to survive upto 200°C have been reported [13.3]. The difficulty of using high temperature resins is that they must be cured at temperatures above ambient, i.e., in ovens, so that there is always potential for uneven thermal fields which lead to less than perfect wavefronts in the final grating, especially if they are large.

The final step up the ladder for increasing temperature resistance is to abandon replicas in favor of master rulings. For use in the infrared region (λ >

10 μm) such gratings are usually made by cutting the grooves directly into solid aluminum or copper blanks. They should be able to survive surface temperatures upto 300°C. For shorter wavelengths the cutting process is incapable of delivering adequate wavefront quality, for both metallurgical and mechanical reasons. To revert to the burnishing process requires a metal surface hard enough be polished but soft enough to be rulable. The only material that fully qualifies on both counts is electroless deposited nickel, which can be put down in layers of about 0.1 mm onto most metals. It contains about 8% phosphorous and has an amorphous rather than crystalline structure. It takes careful control to achieve surfaces free of even minor defects, and in addition its reflectance is low so that it has to be overcoated with aluminum or gold. Unless the back face of the blank receives a similar coating temperature gradients may cause the blank to bend, due to the bi-metalllic effect, especially if the blanks are thin.

The most common application of these gratings is for laser wavelength tuning but there are many others such as beam steering, or beam combining. Experimentally observed limits for allowable energy densities for different types of high efficiency ruled gratings are as follows:

standard replicas on glass: 40 to 80 W/cm^2
replica gratings on copper: 100 W/cm^2
replica gratings on water cooled copper: 150 to 250 W/cm^2
master gratings on copper: 1000 W/cm^2 (ten times that for 10 seconds).

13.2.4 Laser Damage with Pulsed Lasers

Pulsed lasers occupy an important niche in technology. In metal working they are widely applied, although not usually with gratings, unless there is need to control the wavelength. In the important field of pulse compression the damage threshold sets the limit for the amount of energy, or fluence, that the grating will tolerate per unit area. Narrow linewidth grating cavity tuning systems based on broadband solid-state materials, like Alexandrite or Ti:saphire, require damage threshold for the gratings of at least 0.5 KJ/cm^2 in order to make full use of the high energy storage of these materials. The picture is complicated by the fact that there are two types of damage, melting and ablation.

Pulse power is expressed in W/cm^2, and the energy or fluence of the pulse in J/cm^2 and is the power multiplied by the pulse duration in seconds. Total energy involved is obtained by multiplying the fluence by the pulse frequency, and becomes a concern when it is high enough to heat the substrate. In some instances this effect can be greatly reduced by simply doubling the normal 1 μm thickness of the replicated metal coating.

In general the critical quantity is the energy level per pulse, but the details can be quite complex. For instance there is a small but detectable influence of wavelength. Other things being equal the shorter the wavelength the greater the damage potential. A greater role is played by pulse duration. The quality and general surface integrity of a grating will have a strong influence on the damage limit, since at the edge of otherwise minor defects the electric fields can build up.

For high quality gold coated replica gratings used in the 1 to 10 μm wavelength region, with pulse lengths from 0.1 to 100 nsec, one may expect a damage tolerance of about 10 J/cm^2, which corresponds to a power level of 1 GW/cm^2 for 10 nsec pulses. However, experimenters have found gratings with damage limits as low as 0.25 J/cm^2 as well as some as high as 400 J/cm^2 [13.4]. If pulse repetition rates are high enough to heat the grating surface then the same criteria apply that were just described for CW lasers.

The physical reason why energy limits are increased for very short pulses are traceable to a change in the damage mechanism, as dictated by the heat flow process in the surface. For metals, in which laser energy is absorbed to only a skin depth of ~ 3 nm, no more than a 10 nm layer of the material is heated. Therefore, unless the energy level was high enough to vaporize some of the metal, with obvious accumulating damage, any material that is temporarily

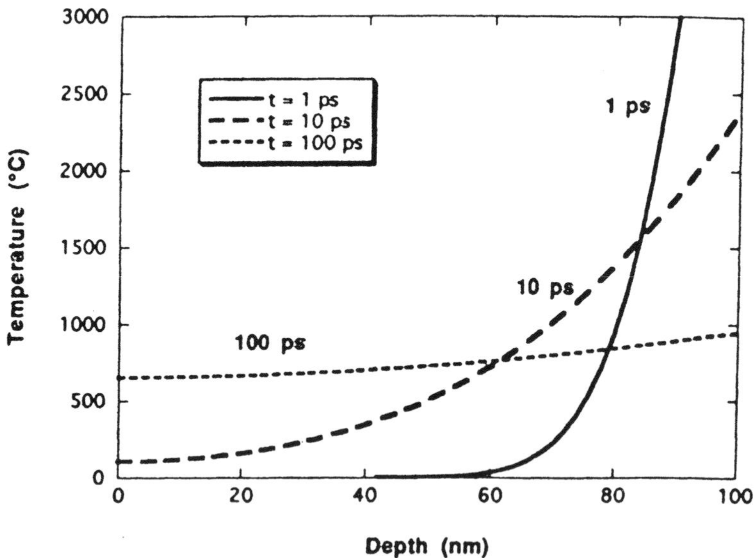

Fig.13.1 Temperature distribution T(x,t) – T_0 for an absorbed fluence of 500 mJ/cm^2, delivered to a 100 nm thick gold film, as a function of depth, for pulse durations of 1, 10, 100 psec. (after [13.4]).

melted will refreeze in place without visible damage. This is why a higher pulse power becomes acceptable when duration is short enough. The mechanism has been described by Boyd *et al.* in reference [13.4]. Fig.13.1, taken from this paper, shows the temperature distribution $T(x,t) - T_0$ for an absorbed fluence of 0.5 J/cm^2 delivered to a 100 nm thick gold film, as a function of depth for pulse durations of 1, 10, 100 psec., where T(x,t) is the temperature as a function of distance x and time t, T_0 the starting temperature. For pulses < 1 psec heat will not penetrate more than 10 nm because of the finite diffusivity of gold (143 nm^2/ps). Thus damage in this instance is not influenced by the thickness of the gold film.

For longer pulses coating thickness will influence the surface temperature, since thermal diffusivity is high enough to conduct heat into the gold, even in the short time available. For pulses > 100 psec duration conduction across the film is complete, and its temperature uniform throughout. The damage threshold will therefore be strongly influenced by the film thickness. It has been observed experimentally that in the case of replica gratings the damage threshold can be markedly increased by doubling the normal 200 nm thickness of the gold layer. In some cases there is an advantage in using a solid metal (master) grating, because there is no insulating layer like that formed by replica resin.

13.2.5 Dielectric Reflection Gratings

As described above, the damage threshold of metallic diffraction gratings is largely set by the absorption behavior of the metals, even if they have high reflectance. An obvious question is whether this can be improved by taking advantage of dielectric materials, specifically overcoatings. Experience has shown that there is no improvement to be expected from such a step, nor does theory predict it. Somewhat more promising was a suggestion to skip the metal and overcoat a standard epoxy transmission replica grating with a three layer stack of dielectric high reflection coating. The experiment gave such poor results that it was quickly abandoned, although theory had indicated some promise.

However, using an inverse approach Perry at al. [13.5] have produced all dielectric reflection gratings with high efficiency (96%) at 1 μm wavelength, in the TE plane under Littrow conditions, although limited to no more than 50% in the TM plane. Their method was to coat a dielectric substrate with several alternating layers of ZnS (n = 2.35) and ThF_4 (n = 1.52), with the higher index on top and given an optical thickness of $3\lambda/4$. The next step was to coat the surface with photoresist and expose it to an interference fringe field as normally used to make holographic gratings, and develop. This was followed by an ion etching step leading to a trapezoidal groove shape, Fig.13.2. It is

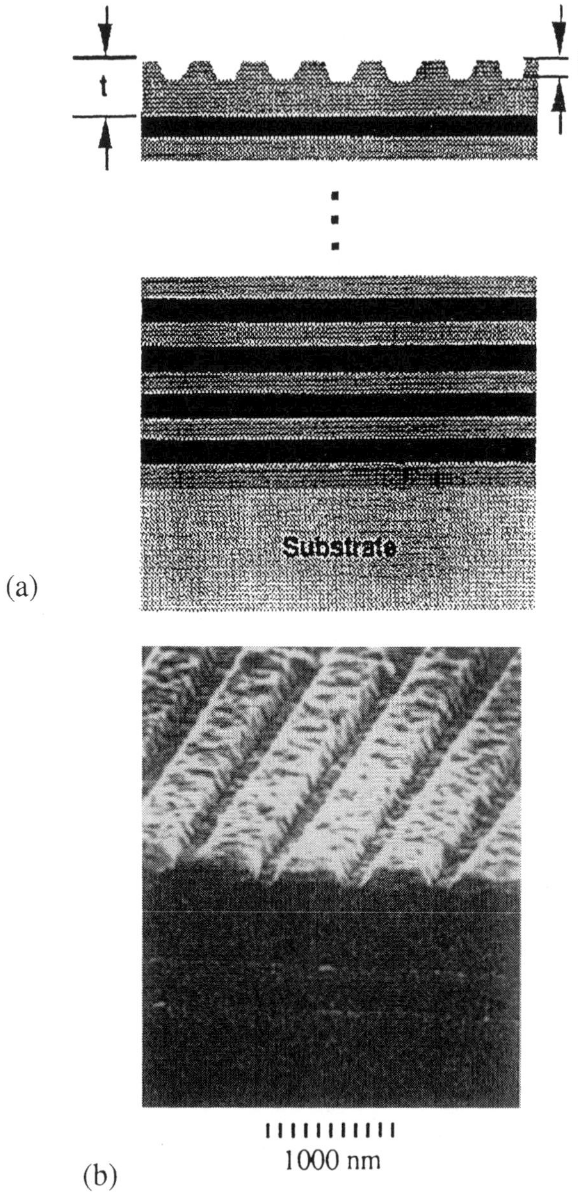

Fig. 13.2 Trapezoidal grooves generated in a dielectric multilayer stack by ion bombardment through a resist pattern. (a) Schematic, (b) SEM photograph (after [13.5]).

carried on until the groove depth reaches $\lambda/4$. High damage resistance is expected, but useful in pulse compression systems only if TE polarization is acceptable [13.6].

13.2.6 Synchrotron Grating Applications

The high power levels, short wavelengths, and ultrahigh vacuum levels typical of synchrotron monochromators present a whole set of special problems. High vacuum requirements are dictated not so much by the monochromators or experiments but by a strong need to avoid contaminating the synchrotron beam itself. The better the vacuum maintained the longer the interval between beam recharging and its accompanying personnel evacuation of the whole lab for safety reasons. To maintain a vacuum of better than 10^{-10} mm Hg requires baking of the instrument system in order to drive off any adhering water vapor or other contaminants. Standard replica gratings are sometimes the only system component unable to tolerate the 250°C baking temperature, because the resin softens too much to maintain geometry. In addition, there have been reports that standard epoxy layers being damaged by soft X-rays penetrating the gold film and destroying the integrity of the material in a matter of hours or weeks, depending on the beam intensity. In some instances this effect can be greatly reduced by simply doubling the normal 1 μm thickness of the replicated metal coating.

A few alternatives present themselves. One is to switch to special epoxy resins that can withstand 200°C temperature [13.3]. Water cooling applied to the back of the fused silica substrate has little effect here because, like all ceramics, it has low thermal conductivity. While the thermal conduction problem might be resolved by going to solid aluminum, this can lead to excessive bending (i.e., loss of figure) due to its high thermal expansion. SiC turns out to be an ideal substrate in this application.

When a grating is exposed to light from a high power undulator beamline no replica can survive, so that the only choice is a master grating. There are several alternatives to making them. One is to rule mechanically into a gold film (400 to 500 nm) vacuum deposited on the substrate onto a chrome link of 10 to 20 nm. However, if allowed to go over 300°C the surface begins to deteriorate [13.3]. A totally different approach is to generate a lamellar grating into a fused silica or SiC substrate by ion etching, as described in sec.13.3.2.

In a few instances it has proven useful to use photoresist masters whose sinusoidal grooves have been overcoated with gold. However, the resist lacquer in which the grooves are formed has a limited temperature tolerance (80°C), although greater than normal epoxy resins.

An interesting suggestion is to obtain superior temperature resistance by making replicas in nickel electroforms (see Chapter 17) [13.7]. It provides an all metal grating, with a temperature tolerance that should reach 300°C. It has shown near perfect groove shape fidelity, but residual stress problems usually limit thickness to about 0.1 mm, although there is some promise of increasing this value [13.8]. Unless very small this thickness does not provide enough rigidity or heat sinking. It therefore becomes desirable to attach them to solid nickel substrates, for which a satisfactory procedure remains to be developed. The problem is to preserve the figure (preferably curved rather than flat) with a cement that has good thermal conductivity but is free of outgassing effects.

The harmful effects of any kind of surface contamination reaches extreme levels for synchrotron gratings. In manufacture they should be handled with gloves only and great care must be taken in shipping, because so many plastic enclosures deposit invisible films that reduce reflectance and lead to outgassing. As a result such gratings are often shipped in special metal boxes.

13.3 Transmission Gratings

The care necessary in handling reflection gratings can be relaxed somewhat for the equivalent transmission gratings. There is no longer a delicate metal film subject to chemical attack from fingerprint residue, nor would they be visible to the human eye (except under UV radiation). This results from the small difference in optical properties between an epoxy surface and a fingerprint. Cleaning is simpler also because, for the same reason, residual deposits from inadequate rinsing are not so readily detected. These points are hardly enough reason to abandon the use of reflection gratings, because they have too many advantages as explained elsewhere (see Chapter 4).

This picture changes for a special application involving a high energy laser fusion experiments where unusual damage resistance becomes key to survival. In this instance there is just a single operating wavelength, since the purpose of the grating is no longer analytical spectral dispersion but temporal smoothing (SSD, or *smoothing by spectral dispersion* of the wavefront, as modified by a distributed phase plate) of a high power beam [13.9]. The objective is to present as uniform an energy input to the target surface as possible. Transmission gratings are ideal for this purpose, especially if used in the Bragg mode where high efficiency is attainable, at least for a single plane of polarization, since damage resistance will be far greater than an equivalent reflection grating. Groove frequency will be high, for example 1326 gr/mm for 1.06 μm wavelength ($\lambda/d \approx 1.4$), and modulation deep ($h/d \approx 1$).

13.3.1 Photoresist Gratings

An important question is how fine-pitch deep-modulated gratings can best be made. The standard approach is to generate them in photoresist, as described in section 15.3. This gets progressively more difficult as the gratings become larger, the pitch finer, and the demand for uniformity greater. For example, the demand for uniformity usually leads to beam truncation which wastes valuable energy. One solution suggested is not to expose the entire grating at one time, but instead take the entire Gaussian input beam in such a way that it is reduced to a relatively small but intense spot and raster it in an X–Y pattern over the stationary grating. The optical system is stationary, except that the input is displaced by motor driven tilt plates [13.10].

Deep modulated gratings are very difficult to replicate. The standard epoxy casting process fails because the large surface area in contact prevents separation. One suggestion is to derive a nickel foil electroform replica, even though the process destroys the precious original. Then a film of polyimide can be applied to a suitable base and impressed onto the electroform submaster. To maintain geometrical uniformity demands that this takes place at ambient temperature and it is not clear how one removes the electroform without damage to either master or replica.

13.3.2 Monolithic Dielectric Gratings

The most damage resistant gratings are formed into solid surfaces such as fused silica or silicon carbide, by methods already described. The same attributes that give them this property are also the ones that make them difficult to produce. Chemically assisted ion etching is probably the most powerful technique, but works only in conjunction with a resist that withstands the etching better than the substrate, but can be removed by a final etching process without attacking the substrate. Tungsten and titanium have been suggested. This generally leads to laminar groove shapes. Such gratings exhibit exceptional damage resistance. For example at 1.053 μm fused silica showed a damage limit of > 20 J/cm^2 for 1 nsec pulses and > 2 J/cm^2 for 0.4 p sec pulses [13.5]. However, if the original pattern is in the form of a triangular groove of photoresist it can be bombarded with an ion beam until completely removed. The result will be a new set of triangular grooves, whose angle will be greatly reduced in comparison to the original by the ratio of the resistance of the two materials to ion attack. This leads to a rather limited choice of angles, but is interesting in the X–ray region [13.11].

13.4 Overcoatings

Special overcoatings have long been used to give gratings enhanced efficiency in specific spectral regions. Initially it was common to overcoat aluminum replicas with silver, gold, or copper in order to enhance IR reflectance. However, this kind of after-the-fact modification makes no sense when superior results are obtainable by replicating directly into the proper metal (i.e., coating the master with the final metal, with no aluminum intermediate).

In the vacuum UV (110 nm < λ < 180 nm) it is not practical to prepare the vacuum enhanced aluminum directly for replication, because even in 10^{-5} mm vacuum there is less than 1 minute of time available to cover fast deposited Al with the 25 nm of MgF_2 necessary to protect it from oxidation. It is interesting that so thin a layer of dielectric is sufficient to protect aluminum for an indefinite time. In the 110 to 120 nm band, MgF_2 is not sufficiently transparent and is often replaced by LiF. Unfortunately the latter is hygroscopic and thus has a limited lifetime in most environments. The solution is to sacrifice performance for life by overcoating the LiF with a thin (10 nm) layer of MgF_2. At wavelengths <110 nm there is no material that reflects very well, so that the standard MgF_2 coating is often used down to 80 nm even though never originally intended. Below that wavelength heavy metals are used, especially platinum. It is highly advisable to make them in the form of direct replicas rather than overcoatings over aluminum. The reason is that overcoatings are not only rougher in a region where that is particularly undesirable, but can be subject to catastrophic destruction by one of two events. One is galvanic breakdown. For example, the surface of a platinum overcoated aluminum grating will vanish in less than a minute if the grating is submerged in water, due perhaps to a misguided cleaning attempt. The other is intermetallic diffusion. This has been observed with great anguish on gold replica gratings that have been overcoated for some reason with aluminum. After a few months atoms of gold migrate into the aluminum and vice versa, resulting in purple blotches of low reflectance. The inverse process, of overcoating an aluminum grating with gold does not give rise to such a problem because the thin layer of Al_2O_3 (5 nm) is enough to prevent atomic contact. If the gold is applied on aluminum in a vacuum coater with a short enough time interval to prevent the oxide from building up, the problem will recur.

References

13.1 R.V.Peterson, W. Krone-Schmidt, W.V.Brandt: "Jet spray cleaning of optics," SPIE **1754**, 295-305 (1992).

13.2 R. Hansen, M. Bissen, D. Wallace, J. Wolske, and T. Miller: "Ultraviolet/ozone cleaning of carbon-contaminated optics," Appl. Opt. **32**, 4114-4116 (1993).

13.3 T. Kita, T. Harada, H. Maezawa, Y. Muramatsu, and H. Namba: "High-temperature gratings for synchrotron radiation," Rev. Sci. Instr. **63**, 1424-1427 (1992).

13.4 R. Boyd, J. Britten, D. Decker, B. Shore, B. Stuart, M. Perry, and L. Li: "High-efficiency metallic diffraction gratings for laser applications," Appl. Opt. **34**,

1697-1706 (1995).

13.5 M. Perry, R. Boyd, J. Britten, D. Decker, B. Shore, C. Shannon, and E. Shults: "High-efficiency multilayer dielectric diffraction gratings," Opt. Lett., **20**, 940-942 (1995).

13.6 B. Stuart, M. Feit, B. Shore, and M. Perry: "Laser damage in dielectrics with nanosecond to subpicosecond pulses," Phys. Rev. Lett. **74**, 2248-2251 (1995).

13.7 W. McKinney and L. Bartle: "Development in replicated nickel gratings," SPIE **315**, 170-172 (1981).

13.8 S. Fawcett and D. Engelhaupt: " Development of Wolter I x–ray optics by diamond turning and electrochemical replication," Precision Engineering **17**, 290-297 (1995).

13.9 S. Skupsky, R. Short, T. Kessler, R. Craxton, S., Letzring, and J. Soures: "Improved laser-beam uniformity using the angular dispersion of frequency-modulated light," J. Appl. Phys. **66**, 3456-3462 (1989).

13.10 J. Armstrong: "Holographic generation of ultra-high efficiency large aperture transmission diffraction gratings," M. S. Thesis, University of Rochester, 1992.

13.11 P. Stuart, M. Hutley, and M. Stedman: "Photofabricated blazed x-ray diffraction gratings," Appl. Opt., **15**, 2618-2619, (1976).

Additional Reading

T. Harada, S. Yamaguchi, M. Itou, S. Mitani, H. Maezawa, A. Mikuni, W. Okamoto, and H. Yamaoka: "Ultraviolet/Ozone cleaning of a soft x-ray grating contaminated by synchrotron radiation," Appl. Opt. **30**, 1165-1168 (1991).

Chapter 14

Mechanical Ruling of Gratings

14.1 Introduction

Mechanical ruling of diffraction gratings has roots going back to Fraunhofer [14.1] and, until technological developments in other fields made interference gratings practical (see Chapter 15), it was the only technique available. Its unique attributes guarantee an important role indefinitely. While the basic concepts are simple the extraordinary demands for accuracy have made the machines required one of the ultimate challenges in precision engineering. Starting again with Fraunhofer, who never described his machine, except to boast that "its performance could never be exceeded by hand of man", there has been an aura of mystery surrounding these machines. Many of their designers and builders worked so hard and for so long that they became reluctant to disclose more than a general description, and sometimes not even that.

The object of a ruling engine is to burnish a large number of fine grooves, normally parallel and equally spaced, onto a suitable optical surface. The surface needs to be soft enough to accept local deformation and at the same time be highly polished. For nearly a century the only material that qualified on both counts was speculum metal (an alloy of tin and copper), although Fraunhofer had tried gold foils transferred to a glass substrate. In the 1930's Strong [14.2] revolutionized this aspect with a crucial update in the form of vacuum-deposited aluminum on glass. It is still the most widely used combination, although other soft metals such as gold are sometimes preferred. One obvious gain is that any less than perfect ruling can be chemically removed and the blank recoated without a repolishing cycle.

Diamond is the only tool material hard enough to generate miles of grooves and still maintain shape. Naturally produced splinters were originally the only and troublesome choice, but have long been replaced by specially shaped ones [14.3]

To generate the desired groove pattern requires two orthogonal motions. One produces the long straight groove and the other the small, but precise indexing that has always been considered the most difficult objective, since errors are measured in nanometers. With a few interesting exceptions, the mechanism of choice for the indexing motion has been the leadscrew. Since experience has shown that nature has set a limit for screw accuracy at about 200 nm, no matter how great the skill and care, much effort has been devoted to reducing the effect of this error by factors of at least ten. The major breakthrough here has been Harrison's development of interferometric feedback control, the effect of which was to transfer accuracy responsibility from the leadscrew to the wavelength of monochromatic light [14.4].

Ruling gratings one groove at a time is a slow process. Except for a few small, lightweight engines with restricted applications, the inertia effects of reciprocating motions invariably lead to vibrations as soon as the rate exceeds about 12 strokes per minute. This leads to ruling times that may vary from one day to 8 weeks. Not only must power supply and all controls operate without fault over such periods, but temperature must remain constant to 0.01°C or possibly even less.

Master gratings produced in this fashion are clearly too few and too expensive to be used for normal applications, which points to the importance of being able to make high fidelity replicas (see Chapter 17).

Compared to most other mechanisms that require ultra-precision motions, gratings ruling stands out for the relative ease with which residual errors can be detected by straightforward optical tests. This is because gratings act as their own Fourier analyzers.

A distinctive feature of ruling engines is their concentration on a single task: putting grooves where they belong, while leaving to the blank the responsibility of locating them in the correct plane.

14.2 History

The history of gratings ruling is a long and interesting one. It requires great ingenuity and skill to achieve accuracies that in ordinary engineering circles are regarded as impossible. Good reviews of early efforts have been published [14.5, 6]. Nobert was the first to make gratings on a commercial basis (1850) in Germany. His finest ruling was 500 gr/mm and 25 mm square, but his gratings served a whole generation of spectroscopists, including Ångström. Fasoldt, an Albany, N.Y. watchmaker, claimed that he achieved 4000 gr/mm in glass (Fig.14.1), although he offered them only for microscope resolution targets, rather than for spectrometry (1860).

Rutherfurd dropped his New York law practice in favor of astronomy and constructed an automatic ruling engine (1863) and later a second one, which produced scientifically useful gratings. Ghosts, resulting from periodic screw errors, were the incentive for his friend W. A. Rogers of Harvard to attack the perennial problem of making a "perfect" lead screw (1878). Together they provided the background from which Henry Rowland developed (1881)

Fig.14.1 Fasoldt microscope resolution target ruled in glass showing section with groove frequencies of 1770, 1970, 2160 lines/mm, taken with 100x oil immersion objective.

the relatively large (5 inches) and more accurate ruling engines at Johns Hopkins [14.7]. He took particular pride in this achievement, in which his technician Schneider shared, and acquired world wide fame for the advances in high resolution spectrometry made possible. His later invention (1885) of concave gratings was a great contribution in that it led to simple, but high resolution spectrometric systems. The last of his engines, dating to perhaps 1895, was in regular use for 75 years.

Albert Michelson decided in 1900 to devote some of his skills to developing a ruling engine, with the feeling that five years should be sufficient for the project at the University of Chicago. It actually occupied his attention for the next three decades, until his death. His objective was to rule much larger gratings, 8 inches on the first engine and 10 inches on a second one started in 1910. He introduced some rather important innovations. One was to make the construction much heavier than seemed necessary at the time, which turned out to have great value later. Considerable effort was devoted to the design and manufacture of the lead screw and a nut that provided a high degree of elastic averaging (Fig.14.2).

The original engine, given to M.I.T. in 1947, became the A-engine. The second engine was given to the Bausch & Lomb Co. (1947), but could not be used over more than 8 inches of its nominal 10 inches of travel because Michelson, frustrated by the slow progress of his careful technicians who were lapping the sawtooth thread, used their vacation absence to finish the job himself. Impatient, he did not not allow enough time for temperature to

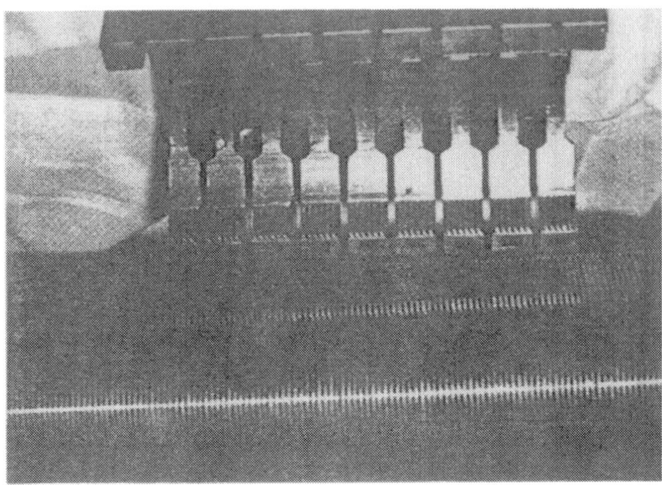

Fig.14.2 Michelson *lead screw* and *nut* from his first ruling engine. Note saw-tooth thread form and the thinned down tynes to provide elastic averaging.

stabilize after each lapping cycle and he ended up with a 3 μm accumulating error, well beyond the ability of the correction cam to handle. Not recorded are the comments of his staff. The engine was later modified several times by Wiley. It acquired its full ruling range only in 1990, when laser interferometer feedback control was added to supercede the cam.

Another incident from its early history was provided by Michelson's assistant Frank O'Donnell, who recalled that figuring of the correction cam proved impossible at first because the engine never gave the same cross ruling pattern twice [14.8]. After nearly a year of futile efforts Michelson gave permission to remove the grating carriage and it was discovered that its surface had never been scraped in for a proper fit.

Michelson provided two highly original features, rarely copied, that turned out to be crucial for success. One was to let the grating carriage rest on top of a second slave carriage, separated by balls in V-ways and driven by a second coupled lead screw. The second feature aimed to reduce to a minimum the work required by the primary screw: most of the weight of the primary carriage was supported by a piece of wood submerged in a tank of mercury located just below it. Obviously this required the secondary carriage to have an appropriate opening along its center [14.9].

Another major contributor to the art of gratings ruling was the Mt. Wilson (later Hale) Observatories, where two engines were built by H. D.

Fig.14.3 Michelson's 1910 ruling engine in David Richardson Grating Lab. Note tool carrying bridge on cylindrical guideways (courtesy Spectronic Instruments Co.).

Babcock and H. W. Babcock [14.10]. The first was so large (400 by 600 mm) that a key contribution was to teach us how severe problems become when existing concepts are scaled up by too large a ratio. The second engine (with a capacity to rule 150 x 250 mm) produced many good gratings before being decommissioned, and introduced a major new design concept in the form of a bridge that crossed the entire engine and supported the ruling tool so that its edge was located exactly on the axis of the cylindrical ways on which it moved. This improved stability of the diamond tool motion far beyond that attainable by cantilevered carriages that were used until then. As a result the design was adopted by Bausch & Lomb when their Michelson engine was modified (Fig.14.3), and by many others.

The single most important development in the recent history of gratings ruling was the introduction of interferometric feedback control by George Harrison at M. I. T., starting in 1945. The concept of using the wavelength of light as a precision reference had been clear to Michelson 30 years before, but he lacked a suitable light source, detectors, amplifiers, and servo feedback components. Thanks to wartime developments, all of them had become available, and the gift of the first Michelson engine by the University of Chicago Physics department, was the spark that started the M.I.T ruling projects that continued until Harrison's retirement in 1978. The advantage was so great that from that time on not a single new engine was built anywhere that failed to take advantage of this breakthrough.

The first interferometers used ^{198}Hg isotope lamps, but once frequency stabilized He-Ne lasers became available (1965) their advantage proved overwhelming. Electronic servo systems started out with phase analog concepts, but since Harada's first application of digital computers paved the way [14.11], the advantages of a digital approach become clear.

14.3 Generating Grooves

It is important to appreciate that, with some minor exceptions, gratings grooves are always generated by a burnishing process, that is by plastic deformation of the surface film, and that no cutting of any kind is involved. Although this sometimes makes it difficult to control the exact groove shape, there is a long list of advantages. The single most important one is the way it distributes responsibility for attaining surface accuracy. The ruling engine is responsible for locating the grooves in the right place, while the flatness of the surface is the result of properly polishing the blank. Another important attribute is that it provides maximum smoothness, as can be seen in Fig.14.4. Tools are made from gem grade diamonds, carefully brazed into steel shanks in such a way that the working faces end up close to the hardest of the crystal planes. The tightest control is exercised by X-ray crystallography [14.12]. Two shapes have proven valuable in practice. One is called chisel shape or roof edge, the other and most widely used, is boat or canoe shaped. It is generated in the form of two cones whose intersection is the ruling edge, which must be almost infinitely sharp. The tool geometry can be visualized from Fig.14.5. It is more difficult to align, but has a longer life than the chisel shape tool. When ruling on a plane surface contact is made along a single plane of intersection with an effective width of about 0.1 mm, whose location is controlled by tilting the tool around a horizontal axis. When ruling on a concave blank a whole range of intersecting planes comes into play as the angle of attack changes. While this reduces tool wear effects to a minimum, it puts a premium on a perfectly straight intersection of the cones. Otherwise the grating will display a set of visible concentric rings known as target pattern which can degrade performance.

The included angle between the two cone bases, which nominally matches the groove angle, is usually between 90° and 120°. Ruling set-up

Fig.14.4 Electron photomicrograph of the last groove of a 600 gr/mm grating ruled in aluminum. Note the burnishing action of the tool on the fine structure of the coating. Groove depth of 136 nm is made visible by slant shadow cast by 200 nm diam. asbestos fiber at an angle that magnifies vertical 3.8 times horizontal (courtesy Spectronic Instruments Co.).

involves not only adjusting tool loads to match the plastic flow properties of the coating, but optimizing angular orientation around all three axis.

As evident from Fig.14.4 the direction of ruling is towards the deep end of the groove, so that any lateral plastic flow is expressed chiefly onto the steep face, which plays a lesser role in the behavior of low groove angle gratings.

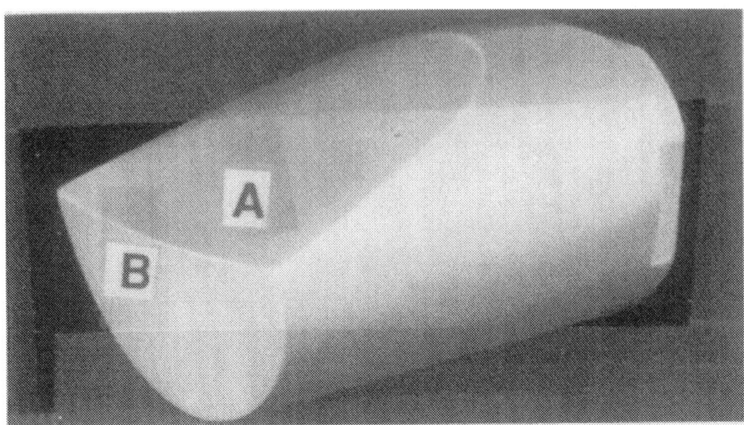

Fig.14.5 Wooden model of a diamond ruling tool. The ruling edge is intersection of the bases of two co-axial cones of different altitudes, cones (A) and (B) placed base to base. Plane of intersection, as viewed from the work, is a straight line. Ruling edge typically 1.5 mm long, and the radius of the cones about 10 mm.

Under favorable conditions one can sometimes observe good geometry on both faces of the groove, Fig.14.6.

When ruling echelle gratings the ruling direction is reversed. Here the steep face is used, with angles typically from 60° to 76°, Fig.14.7. As a result it is difficult to photograph. Because echelles are used in high diffraction orders (see Chapter 6) this face must be flat to a few nanometers.

The critical difficulty for the operator is how to tell when the tool is perfectly adjusted, especially when one considers that a typical groove width is only 1 μm, often less, and that the tool load should be just sufficient to rule full depth. Anything less will leave unruled area, and thus reduce efficiency, while an excess load will lead to rough edges that produce stray light and premature tool wear.

An obvious solution is to test rule areas large enough so that their spectral efficiency can be measured. However, even with a laser, this requires at least 1 mm per test, which can easily use up excessive space on the blank, and is also rather time consuming. As a result it is used only in special cases.

The most satisfactory approach is interference microscopy, especially when equipped with oil immersion lenses. A typical picture is shown in Fig.14.8. It is necessary to make allowance for the fact that fringe spacing is not exactly $\lambda/2$ under high N.A. conditions. Additional difficulties arise when

Fig.14.6 Electron photomicrograph of a 600 gr/mm grating, 13° groove angle, with shadow cast from an asbestos fiber to show exact shape (courtesy Spectronic Instruments Co.).

groove angles are too low to resolve step height with sufficient accuracy, or when the groove width is too small to be properly resolved, and with echelles the active face cannot be seen at all. Experience, gained from many less than perfect attempts is the classical solution. It should be noted that it is usually necessary to remove the blank with its test rulings from the engine in order to examine them, which requires carefully designed kinematic mounting in order to return it to exactly the same position.

Transmission electron microscopy is excellent for observing the shape of very shallow grooves, as some of the figures indicate, but sample preparation

Fig.14.7 Electron photomicrograph of carbon replica of 79 gr/mm r-4 echelle. Smoothness of groove face verified by white shadow of asbestos fiber (courtesy Spectronic Instruments Co.).

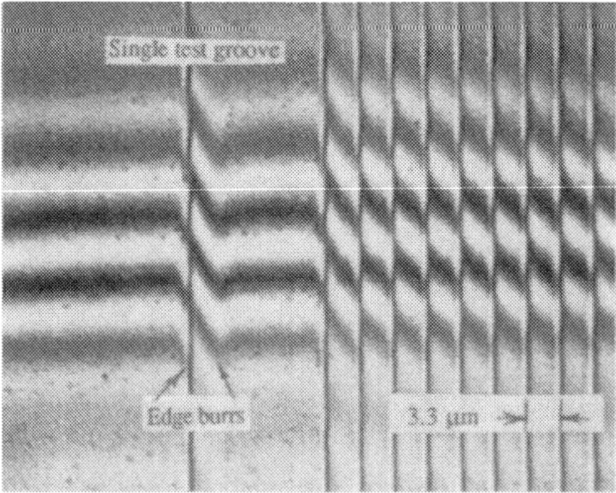

Fig.14.8 White-light interference photomicrograph with 100x oil immersion lens of a 300 gr/mm 300 nm blaze grating with a groove angle of 2.6° and 150 nm groove depth. Single test groove shows edge burrs due to plastic flow of aluminum coating. Sharp dividing line is the groove bottom. First nine ruled grooves on the right (courtesy Spectronic Instruments Co.).

makes it too slow for routine use. SEM instruments are faster, but not sufficiently quantitative. High resolution stylus instruments have been used, but suffer from constraints imposed by the solid angle of the diamond stylus, and thus seem to offer no great advantage, except that they may be used without removing the blank from the engine.

The development of scanning tunneling microscopes (STM) has made available a fast surface analyzer with very high resolution, both horizontal and vertical. This makes it ideal for examining shallow grooves, especially since such gratings are likely to be ruled in gold. They do not work with aluminum because the thin surface coating of oxide acts as an effective insulator. A derivative instrument, the atomic force microscope (AFM), does not have this restriction, but shares the need for effective calibration. Specially shaped tips are needed to obtain a trace of an entire groove (see also section 11.3.4).

14.3.1 Metallic Ruling Coatings

In order to achieve good groove geometry, a primary aim of mechanical ruling, the metallic coating in which the grooves are to be embossed plays a key role. The coating must be soft enough to readily deform plastically, must adhere strongly to the surface of the blank material, and it is important that a thick but extremely flat layer be available for coarse ruling. The material must also allow itself to repeatedly undergo a replication procedure. Aluminum is outstanding on all these counts, but has limitations for ruling short wavelength gratings, because of its deformation behavior, and especially its crystal structure that shows up as surface micro-roughness. The most popular alternative is pure gold, as pictured in Fig.14.9.

A generally accepted rule is that a high degree of vacuum (i.e., $< 10^{-5}$ mm Hg) leads to the best ruling films. To achieve the necessary flatness with thick films requires deposition uniformities of better than 1%, attained by either long distance from the evaporation source, shadow masks or planetary motions. A rule of thumb is to make the coating about 25% thicker than the maximum depth of the groove in the case of steep angle grooves, but for lower angles 50 to 100% of groove depth may be adequate.

Although the hardness of diamond is without equal, its shape is still subject to wear, especially against aluminum. While at first that may seem surprising, given the softness of the metal, one must consider that accumulated groove length produced without interruption can sometimes exceed 50 km. The ability to tolerate tool wear depends on the type of grating, and is most critical at either end of the groove angle range, where changes in shape cause the most harm. The layer of oxide that always accompanies aluminum is hard, but it is believed that the wear mechanism is principally a physio-chemical one. A layer of oil normally covers a blank during ruling, where it may decrease contact

Fig.14.9 Transmission electron photomicrograph of a 300 gr/mm, 2 m radius concave grating ruled in vacuum deposited gold. Groove depth 116 nm (courtesy Spectronic Instruments Co.).

friction and supply a small amount of mechanical damping, but certainly serves to keep dirt particles from interfering.

14.3.2 Master Blanks for Gratings Ruling

The first requirement of master blanks is that their surface geometry be such as to contribute negligibly to any errors in the diffracted wavefront. Typically this means a flatness of $\lambda/10$ for plane blanks, not difficult until size exceeds 200 x 300 mm. Since the sensitivity of blank geometry to temperature gradients varies with the square of its diagonal it has become standard practice to make all larger ones (>250 mm) from low expansion material such as fused silica, ULE™ glass, or glass-ceramic. Solid metal blanks are used only in special cases, and require a nickel phosphide coating to make them polishable. Concave blanks are treated to similar thoughts, but long radius surfaces are made to even lower figure tolerances in order not to compromise imaging behavior.

For low diffraction angle gratings, errors in blank figure carry over into the diffracted wavefront almost directly, since they are multiplied by cos θ_d, where θ_d is the angle of diffraction. At high angles of diffraction, where other requirements become fussier, this aspect is relaxed slightly. As is the case for many optical elements, spherical departure from ideal geometry can sometimes be accepted, since it leads merely to refocusing. The same is true for cylindrical departures, provided their axis are parallel or perpendicular to the grooves. All other defects must be kept below $(\lambda/4)\cos\theta_d$.

It is clear that in the mounting of blanks on the engine table, kinematic principles must prevail to avoid bending constraints. The supports should be adjustable for height so that the blank surface always lies in the same plane. Blank thickness is usually specified according to the well known 1:6 rule for the ratio of thickness to diagonal of precision mirrors, which assures that gravitational deflection will be negligible when optimal 3-point supports are used.

14.4 Accuracy Requirements

The degree of perfection of a diffracted wavefront determines the quality of its imaging and the degree to which theoretical resolution will be attained. Disregarding imperfections of the blank surface, as described above, or that of the incident wavefront (i.e., quality of the collimation), it will be determined solely by the perfection of the groove pattern.

How various types of defects influence imaging behavior is best discussed by taking them one at a time. Following Hutley [14.13] the picture can safely be simplified by using the grating equation for Littrow conditions ($\theta_i = \theta_d$), in which case

$$m\lambda/d = 2 \sin \theta_d \tag{14.1}$$

Errors that need to be considered are those of groove spacing, parallelism, and curvature. Since it is reasonable to expect contributions from all these sources and since the Rayleigh criterion points to an accumulating limit of $< \lambda/4$, it seems appropriate to limit the effect of any one of them to $\lambda/10$, peak to peak.

14.4.1 Constancy of Spacing

If a groove is displaced from its correct position with respect to the first groove by an amount ε_1, the corresponding error of the wavefront will be $2\varepsilon_1 \sin\theta_d$, where the subscript 1 refers to this particular error. If it is to be less than $\lambda/10$, one can write

$$2\, \varepsilon_1 \sin \theta_d < \lambda/10 \tag{14.2}$$

Thus the peak to peak groove spacing error allowed would be given by

$$\varepsilon_1 < \lambda / (20 \sin \theta_d) \quad . \tag{14.3}$$

From this one can conclude that there are just two criteria which determine the accuracy required for a specific application in first order, namely the diffracted wavelength λ and the angle of diffraction θ_d. However, since the two are related via the grating equation it is equally correct to write

$$\varepsilon_1 < d / (10 \text{ m}) \quad , \tag{14.4}$$

which relates the allowable error to the groove spacing d and the order number m.

To obtain an insight into typical tolerances one can take a grating diffracting at 17.5° in the visible spectrum (λ = 500 nm, 1200 gr/mm) where both the above expressions lead to a 42 nm spacing error limit in the first order. Note that as long as the grating is used in first order the spacing tolerance will be a fixed fraction of the groove spacing, because as the wavelength changes so does the angle of diffraction. For a steep angle (63.5°) echelle grating eq.(14.3) is more appropriate because here the angle of use always remains near the groove angle, while the orders change with wavelength. Thus an echelle at the same wavelength above leads to an error limit $\varepsilon_1 <$ 28 nm, but this time the error limit decreases with wavelength.

The most common origin of spacing errors is deficiencies in the lead screw itself, its mounting, and the gear and indexing mechanisms that drive it. Temperature changes always have an effect on groove position, and this holds even with interferometer feedback. A major concern, and one not always sufficiently appreciated, is the guiding system that determines how precisely the diamond tip repeats its location during each of its many strokes. In addition to these major effects there may be a series of smaller ones, and the challenge lies in the fact that it is the total sum that should remain within the limits given above.

Fanning error is the term applied to grooves whose parallelism undergoes a progressive change, normally caused by grating carriage ways that are not sufficiently straight. For a given curvature the effect will be proportional to groove length, and thus is of most concern with larger gratings. Since the result is to change groove spacing from a minimum at one end of a groove to a maximum at the other, its influence is covered by the same ε_1 limit given above. How the error budget is apportioned becomes a major concern to engine builders.

14.4.2 Groove Straightness

Variation in groove straightness can be considered separately, as long as it repeats for each groove. For example a crank drive mechanism on a continuous motion machine will make the grooves slightly S-shaped, leading to a comatic flare along a spectral line image, which may be considered negligible as long as the peak to peak value, ε_2, is limited by

$$\varepsilon_2 < \lambda / (6 \sin \theta_d) \quad \text{or} \quad \varepsilon_2 < d / (3\,m) \quad . \tag{14.5}$$

14.4.3 Random Spacing Errors

Of special concern with all ruled gratings is low intensity scatter over the whole spectrum, caused by random errors, often termed grass, because when Foucault testing with the Hg green line it has the appearance of a strip of lawn. According to Maréchal [14.14] the fraction of total grass energy I_g to total energy I_p is given by

$$\frac{I_g}{I_p} = \left(\frac{4\pi}{\lambda} \varepsilon_3 \sin \theta_d \right)^2 \quad , \tag{14.6}$$

where ε_3 is the RMS random error in groove spacing. If the grass level is to be held to a 1% limit then

$$\varepsilon_3 < \lambda / (125 \sin \theta_d) \quad \text{or} \quad \varepsilon_3 < d / (62\,m) \quad , \tag{14.7}$$

from which it is evident that this tolerance is at least 6 times more stringent than that of ε_1, and 20 times more if the level is kept to 0.1%. This explains why total elimination of grass is so very difficult.

14.4.4 Periodic Errors

When Rowland ghost intensity with respect to the parent line is to be controlled to a low level, a frequent requirement, this becomes another rather critical tolerance. The principal ghost intensity I_{Gh} compared to the parent line I_p is given approximately by (see eq.11.1):

$$\frac{I_{Gh}}{I_p} = \left(\frac{2\pi}{\lambda} \varepsilon_4 \sin \theta_d \right)^2 \quad , \tag{14.8}$$

where ε_4 is the peak to peak error amplitude. Solving for ε_4 gives

$$\varepsilon_4 = \frac{\lambda}{2\pi \sin\theta_d}\sqrt{\frac{I_{Gh}}{I_p}} \quad \text{or} \quad \frac{d}{\pi\, m}\sqrt{\frac{I_{Gh}}{I_p}} \,. \tag{14.9}$$

If the ghost intensity is required to be less than 0.1% for a 1200 gr/mm grating at 500 nm, i.e., for a diffraction angle of 17.5°, the periodic error amplitude ε_4 < 8 nm. For high angle echelle gratings the value ε_4 should be held to less than 0.25 nm if ghosts are to be invisible, i.e., $I_{Gh} / I_p < 10^{-6}$.

There is no advantage to combining the various values of ε since they have such different origins and limits.

14.5 Ruling Engine Design Concepts

If one considers the extraordinary tolerances that are routinely required for mechanically ruled gratings it is perhaps surprising that so many differing design approaches have been successfully implemented. However, as engine size increases so do performance requirements, and design criteria become more restrictive as well as more demanding.

Among the critical decisions to be made are the choice of continuous motion vs. start-stop indexing, screw drive vs. alternative indexing methods, whether to index the grating carriage or the diamond carriage, the choice of methods to control the effects of friction, and always how to reduce the influence of geometric deficiencies. While interferometric feedback is now considered a necessity, there are a number of choices for its implementation, both regarding optics and electronics. Methods used for controlling the environment, vibration and temperature are often critical for success, again more emphatically for larger engines.

Disregard for any of the key requirements has proven fatal to more than one attempt, as performance is always the sum of its elements. Interesting is that there have been several projects where the objective was not primarily to rule gratings, but to study precision engineering design concepts which in the nanometer domain are more easily analyzed through ruling a grating than by any other means.

14.5.1 The Mechanical Motions

All mechanical ruling engines are based on two orthogonal motions, one of which provides the slow indexing between grooves, the other a relatively fast motion along the grooves. The classical design, which has stood the test of time, is the "shaper" concept, in which the grating carriage indexes while the

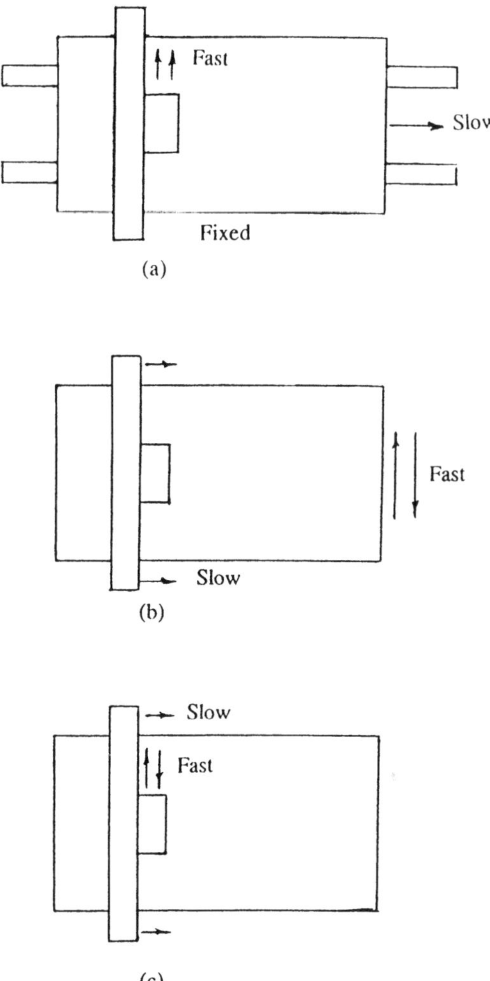

Fig.14.10 The three basic design configurations of ruling engines. (a) The "shaper" design, (b) the "planer" design, and (c) moving bridge design.

diamond is made to reciprocate along a line fixed with respect to the machine base, Fig.14.10a. Probably 90% of the engines built, from the small ones of the last century [14.7] to the largest of modern machines [14.15], follow this principle. One reason for its dominance is that it assigns the highest speed motions to the part with the lowest mass.

The "planer" concept is due to Strong [14.16]. At least three laboratories adopted it. Although some good gratings were produced, none of the engines remain operational at this writing. The idea was to index a bridge structure that carried the diamond tool, while the grating carriage reciprocated in place to generate grooves, Fig.14.10b. Its clear advantage was that there would be no fanning errors, although that was a concern mainly for larger gratings (> 100 mm). The penalty was that the reciprocating carriage had considerable mass, which led to vibrations unless operated slowly.

The only remaining alternative was based on a misconception: when it came to the ruling of a very large grating (> 400 mm), weighing more than 50 kg and necessarily supported on a carriage weighing several times that, it was presumed impossible to control indexing to nanometer precision. Once believed, there was no choice but to leave the blank stationary underneath a bridge that was given slow precision indexing and incorporated a relatively fast moving diamond tool carriage, Fig.14.10c. The chief problem, as it eventually turned out, was that what was really impossible was to generate the desired rapid diamond motion on a moving bridge without generating excessive vibrations. Too late came the discovery that nm control is entirely feasible for massive tables, using double V-ways with precision rollers for the grating support carriage, a finding that led to the construction of the MIT C-engine, with a capacity of 400 x 600 mm [14.17].

All early ruling engines operated in a start-stop mode, in which index motion of the grating carriage by one groove space occured during the return stroke of the raised tool. During the subsequent ruling stroke, the tool was lowered and moved across the stationary blank. This had the advantage of simplicity, involving no accurate timing between the movements of the gratings and the diamond carriages, but in return demanded highly consistent behavior of all oil films, both during the build-up while indexing and subsequent relaxation while stationary. Included here were not only carriage ways but also lead-screw nut and thrust bearing systems. Despite these caveats the possibilities of such an approach may be appreciated by noting the ability of Michelson's engine (Fig.14.3) to rule uniform intervals as small as 100 nm [14.12].

Interferometric feedback control gave designers the choice of either continuing in the classical start-stop mode, or switching to an entirely new one based on continuous indexing motion of the grating carriage [14.17b]. This naturally leads to a small angle between the grooves and the ruling direction. The angle is d/2b where b is the length of the groove, with 2 arc seconds a typical value. In the unlikely case that this matters one needs only to rotate the blank by such an angle. For continuous indexing to work the diamond carriage must maintain a corresponding uniformity with respect to time to maintain a full synchronization with the grating carriage motion.

There are advantages to both. The start-stop approach avoids timing problems, and the long stationary interval leads to simple integration of position information. The constant concern is introduction of mechanical noise due to friction loading that is variable because of oil films that partially collapse during still periods.

Continuous motion feedback has a truly critical advantage in that oil films are never interrupted, so that for sub nanometer control it is usually the preferred approach. Real time control, without a delay, is possible in principle, but in the interest of noise reduction, signal integration is usually included despite the time lag that it introduces. An engine would be of little value if error frequencies were so high that such delay could not be accepted.

Another concern is that crank drives do not deliver constant speed to the diamond travel, resulting in slightly S-shaped grooves. Attempts to linearize motion with complex cams [14.17] turned out later to be unnecessary because the problem can be reduced to any degree desired by a simple expedient. The engine is set up to index in increments fine enough to make the effect negligible. A frequency is chosen that is an integral multiple n of the one desired, to which one adds a mechanism that allows the tool to drop only every n-th stroke. Although ruling time increases by the same factor, that is a minor concession to increased quality.

14.5.2 Grating Carriage Drives

While the overwhelming majority of ruling engines have relied on lead screws to control the precision index motion, two alternatives have proven quite successful, at least for smaller machines. Unique is a hydraulic cylinder drive, due to Horsefield [14.18]. Normally such devices have far too much friction to be seriously considered, but Horsefield solved that problem by adopting a unique Teflon® cylinder with a mercury ring seal that behaved like a frictionless O-ring. An ingenuous micropump supplied the oil for interferometrically controlled indexing.

The other off-beat design involves piezo crystal driven inch-worms, also feasible only in a start-stop mode and with interferometric control, as conceived by Gee [14.19] and Bartlett and Wildy [14.20]. Both of these achieved some measure of success, traceable in part to a geometric advantage derived from locating the line of action of the drive very close to the plane of ruling, something not practical with lead screws.

Lead screws have an advantage that they serve not only to generate motion but do so inherently with a certain accuracy. Unfortunately, a century of experience has shown that despite great skill in not only making screws, but in designing nuts that perform feats of elastic and geometrical averaging, and elegant mechanisms for connecting the nut to the carriage, there is simply no possibility for achieving unaided the near nanometer uniformity that modern gratings have come to demand.

14.5.3 Concepts for Error Reduction

There are several factors that limit the performance of lead screws. The thread, regardless of whether it is a standard 60 degree shape, or the 29 degree Acme shape preferred by machine tool builders, or the saw-tooth shape adopted by Michelson, has a complex helical geometry. It is generated by a special lathe or grinder, which inevitably leaves some geometrical errors, especially periodic ones. The only technology known to improve it is lapping, where a series of well-fitted nuts of different lengths, made of soft metals and sometimes in two segments, is run back and forth with fine abrasive. High spots will be worn down naturally by an averaging process. Elasticity of the metals and compliance of the oil-abrasive films, leads to a finite correction limit, found to be about 0.2 μm. It is also a tedious task, because lapping generates heat that deforms both screw and nut, and can therefore be performed safely for just a few minutes at a time.

The axis around which a screw rotates must be made concentric with the centroid of the helical thread, i.e., the pitch diameter. However, this is a virtual diameter and not necessarily concentric with the journal surfaces that were turned on the lathe prior to lapping, pointing to another high skill effort.

Also important is that the thrust bearing that locates the screw axially with respect to the machine structure does not introduce any longitudinal motion. One solution is a hard sphere at the end of the screw, made to bear against an optical flat made of diamond.

Since the most crucial defects are always associated with rotation of the lead screw, one can conceive of a correction cam that rotates with the screw, acting on a lever system that modifies the input drive, most often by shifting along its axis the worm that drives the wormwheel mounted on the input end of the screw. In practice the errors will not be constant along the screw, so that the cam needs to have a complex barrel shape. It is a tribute to the skill and patience of the early craftsmen that such systems were made to work, at least adequately for the needs of the day.

To perform these feats called for an equivalent feat in determining the amount of correction required over the entire travel, initially without the help of electronics. The most sensitive method available was to lightly rule a reasonably fine pitch grating, reset the engine, and rule for a second time after rotating the blank through a small angle. This leads to a moiré pattern with a visible zig-zag representing periodic errors greatly magnified.

The modern solution is to abandon the futile quest for a perfect screw and adopt the concept that Michelson put aside long ago, i.e., direct error control via interferometry.

An interesting alternative was developed by Gerasimov [14.21], who used a previously ruled grating as the indexing reference. It was illuminated via a transmission grating of twice the groove frequency which generated moiré fringes from a broad band source. By averaging information from a relatively large area of the reference grating, periodic errors were almost completely eliminated. Beyond that the servo feedback control system was similar to ordinary interferometers, but required no attention to atmospheric effects. There was one built in restriction in that rulings were limited to integral multiples of the groove spacing of the reference grating.

14.5.4 Interferometer Feedback Control

The wavelengths of several monochromatic sources are defined to 1 part in 10^8, 10 times more than needed even in this application, but it is not a completely straightforward process to take advantage of this for a ruling engine control system. While ^{198}Hg was used successfully for ruled widths to 250 mm, the He-Ne laser has completely replaced it, because of its greater coherence, which offers no length restrictions in this context, its intensity, which improves the signal to noise ratio, and finally its small beam diameter, which greatly reduces the size of all optical components.

A basic concern is that wavelengths, even of frequency-stabilized lasers, are constant only in a vacuum, so that a correction must be applied for changes in the refractive index of air. For all practical purposes this represents variations in barometric pressure (temperature must be stable for other reasons, and effects of humidity changes are usually too small). Typical pressure extremes of ±25 mm Hg correspond to ± 10 PPM wavelength shift, and thus need to be followed to 1% accuracy. While hardly new to length interferometry, it plays a more prominent role here because ruling times are so long, and because the effect must be convoluted with the actual distance from the position of zero path difference, where the effect vanishes, to a maximum at the extremes. On early systems zero path position was carefully located near the center of ruling in order to maximize the ruled width, but with lasers there is no such restriction. The zero path position with respect to the blank must be known, and accounted for by appropriate software.

Since the objective is to control in the nanometer domain, interference fringe location must be detected to less than 1/100 fringe. This has been accomplished successfully by either setting exactly onto the center of a fringe, by continuous phase matching, or by digital subdivision. Most interferometers have been d.c. systems, some of them polarizing, but modern double-frequency lasers lead directly to operating in the digital domain.

Fringe detection can be configured for start-stop ruling control in one of two ways. In the first, indexing proceeds until the detection system senses that it is exactly centered on a fringe, at which point the position is locked by carefully designed magnetic clamps. Implied here is piezo type stepping, rather than leadscrew drives. Such an approach is described in 14.5.5.3. The more versatile alternative is to perform high resolution digital subdivision, and compare accumulating counts with the nominal values as continually generated by a simple microprocessor. An advantage of digital control is that any spacing can be obtained from simple keyboard input. Furthermore, any mathematically definable departure from uniform spacing can readily be programmed.

In both systems it is important to integrate position information acquired during the stationary period. In the second case correction is necessarily based on the error of the previous groove ruled [14.3], rather than the one actually being ruled. The penalty of being 'one groove late' is usually acceptable.

For engines where the grating carriage moves continually the control concepts differ somewhat. In particular we need tight timing control between the two motions. For a medium groove frequency of 300 gr/mm, and a ruling rate of 10 strokes per min., the indexing rate is 48 mm/day, which seems slow, but corresponds to 550 nm/sec., which means that timing cannot be allowed to drift periodically more than 1 msec. However, it is also true that feedback loops should never aim for msec response. All successful systems have used signal integration over distances of typically one fringe because it greatly increases the signal to noise ratio. It does require mechanical drive smoothness good enough so that this delay does not matter, but that should be part of a proper design. In general, it is a great mistake to think that sloppy mechanical behavior becomes acceptable the moment computers are added to a control system.

A system design choice is whether the diamond carriage drive or the leadscrew of the grating carriage is designated as the dependent variable. Both have been successfully applied. There seems a slight advantage to making the diamond carriage drive the dependent one because operating at a higher speed digital damping is more readily introduced.

Optical Systems

Interferometers for ruling engine controls are all derivatives of the basic Michelson configuration, Fig.14.11a. With laser sources attention must be paid to avoiding light reflecting back into the laser, which would give rise to frequency pulling. This can be accomplished either with polarization isolators, or with beam offsetting optics. Early designs used the classical Michelson system, in which all the optics were on a single platform from which just a measuring beam emerged towards a mirror that was mounted on the grating carriage. Effective for small engines, it puts a premium on structural rigidity between the platform and the diamond carriage system.

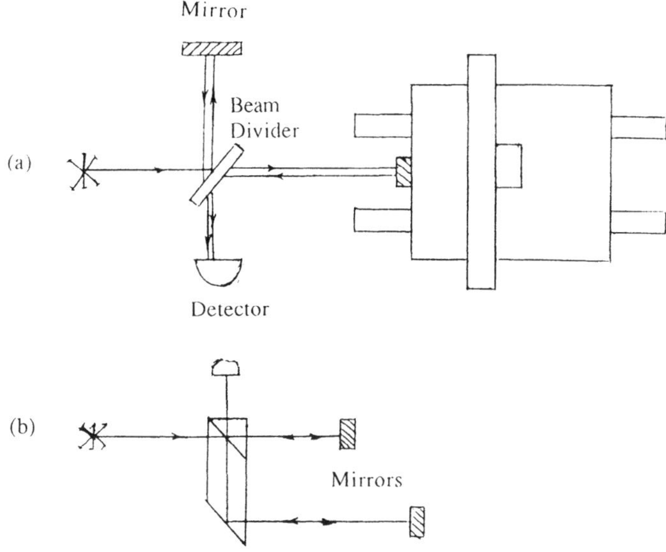

Fig.14.11 Schematic of Michelson interferometers for ruling engine control. (a) Standard and (b) parallel beam modification for differential measurement.

For larger engines it is much safer to adopt the parallel beam modification, Fig.14.11b. It has two advantages. The first is that optics platform stability is not quite so critical because the measurement becomes differential between the two mirrors, one on the grating carriage and the other on the engine structure, but tied in an optimal fashion to the diamond carriage. Since, for mechanical reasons, the two beams must lie above one another they cannot both be made to lie in the ruling plane, so that a slight violation of the Abbé comparator principle must be accepted. Located close to each other, and along the centerline of the blank, will reduce differential air path effects to a minimum, but sensitivity to pitch errors always remains and can cause mischief.

Ideally the reference beam would be tied directly to the diamond carriage itself, but its rapid reciprocating motion makes that impractical for nm control, so that we must be satisfied with a mechanical equivalent. To make this tie with the uniformity and reliability required is one of the most severe mechanical design challenges to the engine builder. Since by definition this mechanical tie is always outside the feedback control loop, it is the single most important reason why a ruling engine must operate in a highly temperature stable environment, 0.01°C for small engines and 0.005°C for large ones.

14.5.5 Examples of Ruling Engines

If one sums up all diffraction gratings ruling engines built since Fraunhofer (1823) for which there is some written record, the total is no more than 80. Approximately 20 of these are considered operational at this writing, of which only 8 have a capacity to rule 250 mm width or greater. To go into detail on the designs for which published information exists, even though a small fraction, would easily constitute an entire book, so that only four examples will be described here. Two are located in the David Richardson grating laboratory[1] in Rochester N.Y., the third is at the CSIRO lab in Melbourne, Australia, and the fourth at the Hitachi Corp. in Japan.

The Michelson Engine

The second of the two ruling engines built by Michelson, to which reference has been made in preceding sections, is an exceptionally versatile machine. It covers the widest range of groove spacings of any engine known, from 20 to 10,000 gr/mm. It is relatively large, with a capacity up to 200x250 mm ruled area, and handles both plane and concave gratings. Like any older machine it has been modified a number of times since it first produced gratings in the 1920's, so that it now routinely produces much better gratings than those Michelson described as "perfect".

Already described in section 14.2 is the unique double carriage system sketched in Fig.14.12, with its mercury pool concept for supporting most of the mass of the grating carriage. The important addition of the diamond tool bridge was shown in Fig.14.3, driven by a relatively simple crank mechanism connected with a segmented belt to a drive motor located outside the room. It is guided by three precision cylindrical guide bars, two of which are visible, and which are straight to 0.05 μm over most of the 200 mm travel. Contact between the bridge and the guide bar is with carefully fitted pads of inverted V's, made of Graphitar, a constant friction material. Wear effects thus tend to be self-canceling.

The lead screw has a 2 mm pitch and is unusual not only in its high accuracy but in having a buttress-shaped thread, whose leading face is nominally square to the axis, and therefore not too sensitive to misalignment of the nut. It can be seen in Fig.14.13.

Indexing originates with a pawl that engages an accurate indexing wheel. A correction cam was the original tool for reducing the periodic error of the screw from 200 to 20 nm, a ratio typical of such mechanisms. A major concern of all start-stop drives is how to limit coasting after the index motion is

---

[1] Of Spectronic Instruments Co, Inc., formerly Milton Roy Co., formerly Bausch & Lomb Co.

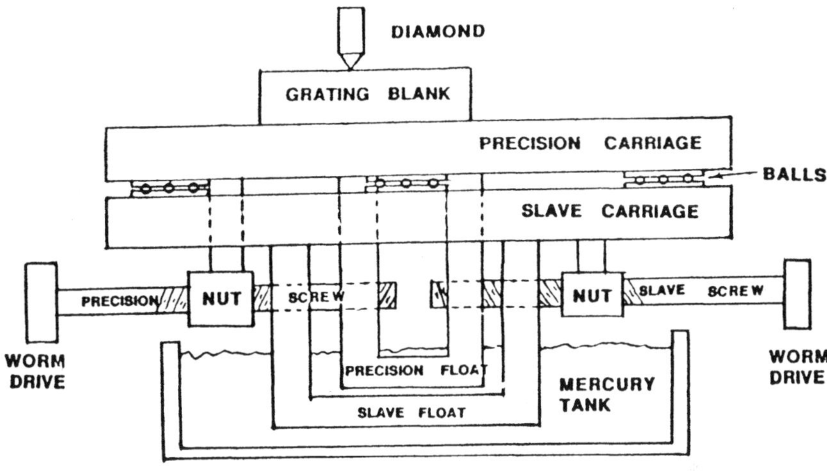

Fig.14.12 Schematic of Michelson *double carriage* system. Note both precision and slave carriages have their mass supported by floats submerged in mercury.

complete, a natural result of stored energy in the drive mechanism. Due to the design and great care in all the precision fits, it takes only 1 in-oz (0.007 Nm) of torque to drive the screw and the result is that by the time the ruling diamond comes down on the blank coasting, if any, is so small that it is not detectable with an interferometer of 1.2 nm resolution.

The principal purpose of adding an active interferometer feedback system is to reduce periodic as well as random errors to an order of magnitude less than attainable with a cam. It is based on a frequency stabilized He-Ne laser fringe counting interferometer, along the lines explained in sec.14.5.4. Accumulating errors are eliminated at the same time.

An important aspect, not much discussed in the literature, is the mechanism for raising and lowering the diamond tool. One requirement is that this take place at exactly the same place along the stroke, but even more, that it be done with a minimum of mechanical commotion. Otherwise the tool will start by bouncing along the groove, producing an unsightly surface that has to be masked to avoid stray light.

It is perhaps obvious that in the lowered position the tool must repeat its location with respect to the diamond carriage guideways to within a few nm every time. Otherwise random errors are produced that lead to diffracted stray

Fig.14.13 Closeup of Michelson ruling engine. Precision screw is visible between ways, driving the grating carriage in right foreground. Visible in extreme left center is the correction cam. Note one of the mercury troughs seen behind the screw (courtesy Spectronic Instruments Co.).

light. Flexure pivots are normally used, since for short travel their properties are ideal. They may be single or double cantilever design, or use cross-spring pivots. The load imposed on the diamond can be set with an adjustable spring or dead weights.

When ruling on a concave surface the mechanics become much more difficult, because flexure pivots do not have enough range of constant force to follow blank curvatures that can reach 1.5 mm sag. The classical solution is to use a pair of V-type pivots similar to those used in fine watches, but they must operate with near zero friction in order to rule with the low pressures needed by fine pitch rulings. At the same time they must also be completely free of play if grooves are to be properly located. One solution is to give the pivot supports some lateral flexural compliance.

The B - Engine

The B-engine, located at the David Richardson Grating Lab of the Spectronic Instrument Co., is large enough to rule 400 mm width, rigid enough to rule from 20 to 1500 gr/mm, accurate enough to provide near theoretical resolution at high angles over the full width, precise enough to eliminate Rowland ghosts and keep stray light to a bare minimum. Such performance is attainable only by paying great attention to every detail.

The base of the machine is a standard Moore No. 3 measuring machine, whose double V table way concept appears at first to violate basic kinematic design principles against overconstraint. However with sufficiently high geometric accuracy overconstraint leads to elastic averaging that not only increases effective rigidity but at the same time averages geometric errors. The V-ways operate with self-generated oil films, and are thus not usable with start stop indexing, but are entirely compatible with continuous indexing motion. The table is driven by a slightly modified version of the standard Moore lead screw, which is made of highly stable Nitralloy, and has a standard 10 threads per inch Acme thread, ground and lapped to fit a solid bronze nut that is rigidly clamped to the table. The screw axis is aligned to the table motion in both directions to within 1 μm over its full length, and has a special thrust bearing. The attention paid to all aspects of accuracy, which also includes the use of a high accuracy worm gear to drive the screw (derived from a standard precision indexing table), leads to a drive smoothness of better than 1 nm.

One aspect of table motion geometry that has defied even the best craftsmanship is the need for 0.05 arc sec straightness (yaw motion) for a grating carriage, required in order to avoid fanning errors for large high-angle echelles. Because of inevitable temperature effects in manufacturing this exceeds even the best practice by an order of magnitude, and calls for a special compensation system. This takes the form of a second table, mounted above the first, and designed to rotate about a central axis by a small angle. This rotation is controlled by a set of compound reduction levers, driven by a servomotor operated micrometer screw or a piezoelectric driver. Error information to control it is derived from a second interferometer system, whose differential output with respect to the translation interferometer is a function of yaw error only.

Most critical of mechanical elements of a ruling engine is the one that determines to a few nm the constancy of the plane of motion of the diamond, for a period that may last several weeks. No servo correction schemes are available, because none can operate at such a high sensitivity and high rate without excessive noise. The elegant design solution described in the previous section is not appropriate here, because a bridge long enough to support 300 mm of ruled travel would be far too heavy and bulky to fulfill its assignment.

As a result it was again necessary to violate kinematic ideals. The alternative adopted was to use a fused silica straightedge to guide the motion of the diamond carriage, which was hung from a cylindrical monorail overhead, and used gravity to urge the carriage against the straightedge, Fig.14.14. Obviously much depended on the quality of the straightedge, made straight to 0.05μm over 300 mm, but actually slightly convex to make up for the load applied to it. Separating the carriage and the fused silica surface is a thin button of a stable but low friction material. Its own wear must be low (or at least highly uniform) and it must never attack the fused silica, even after millions of traverses.

The overriding concern of the crank mechanism that drives the diamond carriage is that of minimum noise. This starts with the drive motor, the subsequent reduction gears, and then the crank mechanism itself and the guideways for the linkage. There is little opportunity for introducing damping, so that any vibrations are easily transmitted to the diamond, leading to a visible disturbance on the ruled surface.

The control system installed in 1987 is a second generation digital computer version of the original design that depended on a complex gear

Fig.14.14 The B-engine: On the left is the fixed reference mirror and in the center is the fused silica straightedge that guides the diamond carriage hanging from the cylindrical monorail. Interferometer optics and detector preamps upper extreme left (courtesy Spectronic Instruments Co.).

system to feed in corrections [14.15]. The interferometer is a Michelson parallel beam type, using polarization optics to obtain a sin-cos pair of output signals that represent the carriage motion. Signal rate would be constant if the drive screw system were free of error. The phase signals are digitized into 0.088 nm steps, which corresponds to 0.1 phase degrees. They are averaged over a complete fringe (360 phase degrees) in order to cancel the effects of any polarization imperfections of the optical components, as well as other noise. The counting rate is continually compared to a reference, derived in the computer in accordance to the spacing desired. Any difference constitutes the error signal, which then modifies the speed of the diamond carriage drive servo motor to cancel the error.

The result is to reduce periodic errors to a level where even under the most severe testing they are only just spectroscopically detectable, implying a peak to peak error amplitude of ~ 0.5 nm. Random errors are estimated at 5 nm. There is also a second interferometer system, identical to the first, except that instead of being located over the table center it is displaced sideways by 100

Fig.14.15 The B-engine: interferometer optics and detectors in rear, and, diamond carriage in the center, showing monorail support. Grating blank and support carriage are in front (courtesy Spectronic Instruments Co.).

mm. Any difference between the two represents the yaw error.

The lower arm of the interferometer is located close to the plane of ruling, reflected by a plane mirror mounted on the carriage just behind the blank, and adjusted square to the direction of motion within 1 arc sec. The same tolerance holds for the fixed reference mirror located on the same casting that holds the reference straightedge, except this one is necessarily about 50 mm above the first, Fig.14.15. While the system remains potentially sensitive to pitch type deformations, that remains negligible because of the kinematic support given to the base structure.

A larger version of the B-engine was built at MIT. Named C-engine it used a Moore No.4 measuring machine as its base and shared most of the basic construction and control features. In an attempt to make it less sensitive to temperature the return mirrors of the interferometers were attached respectively to the back of the silica straightedge and the front edge of the blank. It ruled only one grating at its maximum capacity of 400x600 mm, but while located at the Kitt Peak observatory ruled a number of smaller ones.

Fig.14.16 The Bartlett-Wildy engine: shown are the inclined mirror in lower end of grating carriage and the push rod at upper end with its two magnetic clamps on the lower right. Light weight diamond carriage is visible in center.

The Bartlett – Wildy Engine

This is an engine with several unusual aspects. One is that unlike other engines that took years to build, this one produced acceptable gratings just two months after its assembly began [14.20]. Also, it operates five times faster than any currently operating machine, a feat traceable to the low mass of its diamond carriage, which in turn becomes feasible by limiting the stroke to 75 mm and restricting groove frequencies to a narrow range (600 and 1200 per mm), Fig.14.16.

Indexing is controlled by a ball-ended push rod against a small spring-loaded air bearing carriage, locating it near its center of action. The rod is controlled by a piezo-electric crystal pusher, with the aid of two carefully designed drift free magnetic clamps on either side that are activated alternately to press the rod against its V supports. The amount of advance is programmed to be an integral number of laser fringes, which leads to simple photo-electric symmetry sensing. To accommodate the 316 nm fringe spacing to the usual 600 or 1200 per mm groove frequency the interferometer axis is tilted through an appropriate angle with a steering mirror. The return mirror on the carriage must be tilted the same angle.

The Hitachi Ruling Engine

This modern engine was designed to rule relatively large (200x300 mm) gratings with groove frequencies as high as 10,000 gr/mm, as well as to provide capability for programming variable spacings [14.22]. Such fine pitches are not compatible with the usual continuous motion control program, because the indexing motion (1 μm per min.) is too slow. On the other hand, start-stop motion presents severe difficulties in achieving position lock-ins to at least 5 nm precision over long periods of time (8 weeks). The elegant solution developed by Kita and Harada is a hybrid system. The indexing table is constructed in dual form, with a lower table continuously driven by a lead screw at the appropriate rate and the upper one, which supports the grating blank, moves separately with respect to the lower.

Unlike the Michelson approach described above, the upper carriage is supported by frictionless leaf springs in place of balls in V's, and instead of a special lead screw the small differential motion required is supplied by piezo-electric pushers. Not only does this lead to greater compactness, but it allows interferometer control in real time rather than one fringe late. The system schematic is shown in Fig.14.17, and its overall appearance in Fig.14.18.

It is unavoidable that the application of a piezo force to a mass-spring system like the table sets up vibrations, but they are damped in less than 0.1 sec. and do not affect the ruling. This is an important point, because the low tool forces that are inherent with fine pitch rulings make the engine especially

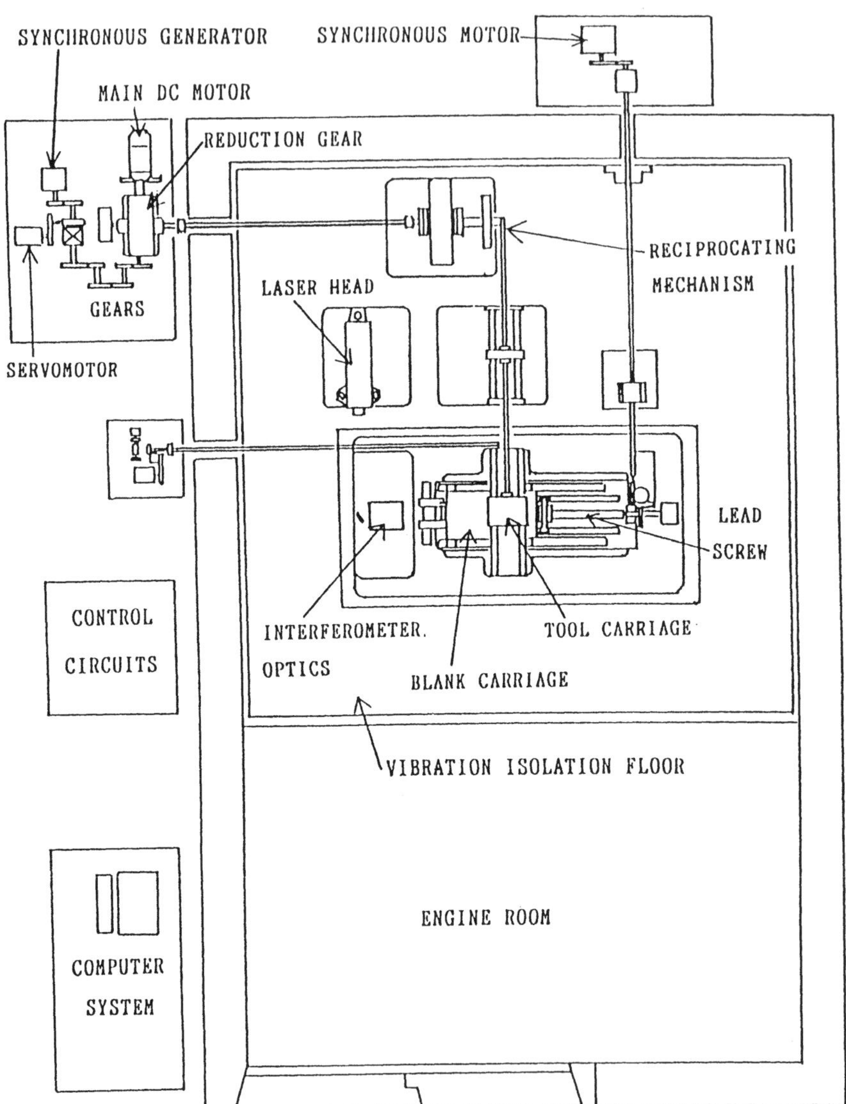

Fig.14.17 Schematic layout of the Hitachi hybrid ruling engine (after [14.22]).

Fig.14.18 Photograph of the Hitachi ruling engine (after [14.22]).

sensitive to vibrations.

An obvious advantage of start-stop ruling is that since the timing of the diamond carriage motion is no longer critical a greater portion of the crank stroke can be used. The effect is again to reduce the size of the system, and with it the mass.

14.5.6 Environmental Factors

It has long been known that nanometer tolerances are attainable only when structures are well isolated from changes in temperature and from effects of external vibrations. Both influences can be taken care of with modern technology, albeit at significant expense, but special attention needs to be paid to factors generated by the engine itself.

Temperature Control

The key to controlling engine temperatures lies in adopting a cascade approach. The engine will have its own enclosure, typically made of aluminum, although wood and Plexiglass™ have also been used. Sometimes this enclosure is supplied with carefully controlled air, but especially for larger machines it is better to keep air paths quiet. This engine cell will be located in a room whose air supply is controlled to a level that varies with size, from 0.01°C for a small engine to 0.005°C or less for a large one. This by itself is never sufficient, because unless the walls are kept at a constant temperature their radiative heat exchange will have a major effect, for example the day-night difference can be quite disturbing.

As a result it is essential to surround the ruling room by an envelope area whose *average* temperature is also held constant, although an order less stringent. Temperature swings are of no importance provided their frequency is high enough for walls to smooth them out. In some cases even the envelope area needs a third temperature controlled area around it.

Thermal sources inside the ruling lab must not be ignored. Most likely to cause trouble is the laser light source, which can dissipate as much as 30W. Far too much to tolerate, it can be readily taken care of with appropriate shielding, and exhausting heated air to the outside with a small blower fan. Modern d.c. servo motors with permanent magnets run cool enough to eliminate an old concern.

Vibration Isolation

Avoiding vibrations from the outside world is no longer considered a difficult problem. It merely takes a large support mass, typically concrete, suspended on air springs soft enough to give a natural frequency of 2 to 3 Hz, as the base of the ruling engine. The low frequencies that pass through are of no concern, while the high ones (> 60 Hz) are usually too low in amplitude. The base should not be supported from the bottom, but somewhere near its center of mass (the engine included) in order to make the system more stable against the effect of the moving masses or slides.

Vibrations generated by the engine itself are inherent with any reciprocating mechanism. Other than care in balancing any rotary devices, experience has shown that the most important guard is to limit the ruling rate to no more than 12 strokes per minute. As mentioned above this rule can be successfully violated for very small machines using exceptionally light diamond carriages, which, as a result, can only rule a limited range of grating spacings.

References

14.1 J. Fraunhofer: "Kurtzer Bericht von den Resultaten neuerer Versuche über die Gesetze des Lichtes und die Theory desselben," Ann. Phys. **74**, 337-338 (1823).

14.2 J. Strong: "The evaporation process and its application to the aluminizing of large telescope mirrors," Astrophys. Jl. **83**, 401-423 (1936).

14.3 H. W. Babcock: "Ruling of diffraction gratings at the Mt. Wilson Observatory,"

J. Opt. Soc. Am. **41**, 776-786 (1951).

14.4 G. R. Harrison: "Production of diffraction gratings: I. Development of the ruling art," J. Opt. Soc. Am. **39**, 413-426 (1949).

14.5 C. Evans: *Precision Engineering, and Evolutionary View*, (Cranfield Press, Cranfield Institute of Technology, Bedford UK 1989).

14.6 G. W. Stroke: "Diffraction gratings" in *Handbook of Physics*, v.**29**, *Optical Instruments*, ed. S. Flugge (Springer, Berlin, 1967).

14.7 H. Rowland: "Preliminary notice of results accomplished on the manufacture and theory of gratings for optical purposes," Phil. Mag. Suppl. to v.**13**, 469-474 (1882).

14.8 F. O. O'Donnell: Personal communication (1965).

14.9 A. A. Michelson: "The ruling and performance of a ten inch diffraction grating," Proc. Am. Phil. Soc. **54**, 137-142 (1915).

14.10 H. W. Babcock: "Diffraction gratings at the Mount Wilson Observatory," Physics Today, **39**, no. 7, 34-42 (July 1986).

14.11 T. Harada, S. Moriyama, and T. Kita: "Mechanically ruled stigmatic concave gratings," Jpn J. Appl. Phys. **14**, suppl. 14-1, 175-179 (1974).

14.12 R. S. Wiley: "Diamond's role in ruling diffraction gratings," Proc. Intern. Diamond Conf., Columbus, Ohio, 249-256 (1967).

14.13 M. C. Hutley: *Diffraction Gratings*, (Academic Press, London 1982).

14.14 A. Marechal, "La diffusion résiduelle des surfaces polies et des réseaux," Opt. Acta **5**, 70-74 (1958).

14.15 E. G. Loewen, R. S. Wiley, and G. Portas: "Large diffraction gratings ruling engine with nanometer digital control system," SPIE **815**, 88-95 (1987).

14.16 J. Strong: "New Johns Hopkins ruling engine," J. Opt. Soc. Am. **41**, 3-15 (1951).

14.17a G. R. Harrison: "The diffraction gratings - an opinionated appraisal," Appl. Opt. **12**, 2039-2049 (1973).

14.17b G. R. Harrison and S. W. Thompson: "Large diffraction gratings ruled on a commercial measuring machine controlled interferometrically," J. Opt. Soc. Am. **60**, 591-595 (1970).

14.18 W. R. Horsefield: "Ruling engine with hydraulic drive," J. Appl. Opt. **4**, 189-193 (1965).

14.19 A. E. Gee: "A piezo-stepping diffraction grating ruling engine with continuous grating-blank position control," Jap. J. Appl. Phys., **14**, Suppl.14-1, 169-174 (1975).

14.20 I. R. Bartlett and P. C. Wildy: "Diffraction grating ruling engine with piezo-electric drive," J. Appl. Opt. **14**, 1-3 (1975).

14.21 F. M. Gerasimov: "Use of diffraction gratings for controlling a ruling engine," J. Appl. Opt. **6**, 1861-1865 (1967).

14.22 T. Kita and T. Harada: "Ruling engine using piezo electric device for large and high groove density gratings," J. Appl. Opt. **21**,1399-1406 (1992).

Additional Reading

G. R. Harrison: The production of diffraction gratings: II. The design of echelle gratings and spectrographs," J. Opt. Soc. Am., **39**, 522-528 (1949).

E. G. Loewen: *Diffraction Grating Handbook* (Bausch & Lomb, Rochester, 1970).

E. G. Loewen, R. H. Burns, and K. H. Kreckel: "Numerically controlled ruling engine with 1/10 fringe interferometric feedback for grids 600 mm square and scales 760 mm long," J. Opt. Soc. Am. **60**, 726A (1970).

J. F. Verrill: "A study of blazed diffraction grating groove profiles using an improved Talystep stylus," Opt. Acta **23**, 425-932 (1976).

Chapter 15

Holographic Gratings Recording

15.1 Introduction

The traditional method of generating master gratings is by burnishing one groove at a time with a diamond tool against a thin evaporated metal coating deposited on the surface of a plane or curved substrate. Quality of the diffracted wavefront depends on the ability of a mechanical ruling engine to operate at levels of precision that are measured in nanometers. Thanks to interferometric feedback control such results are routinely achieved. However, there are spectrometric problems, such as Raman spectrometry, where even the smallest residual mechanical noise of the best ruled gratings shows up in measurements. These difficulties alone are enough to provoke a search for alternative methods of grating manufacturing. There is an old and simple idea: a textbook picture of interference between two monochromatic collimated beams looks like an ideal diffraction grating. Technical tools for transfering the picture into phase or relief index (surface) change are available since invention of the photographic process. Cotton [15.1] was the first to use the simple idea of interference pattern recording for producing a grating. Wiener's classical experiment [15.2] to prove the electrical field vector to be the photographically active one also produced an interferometric grating as a side product. Michelson [15.3] regarded this approach as obvious in 1927, but made no attempt to put the idea into practice. Genius of the details, he recognized that he lacked both an intense, highly monochromatic, coherent source of light and an adequate grainless photosensitive material. Both became available in the mid-1960s in the form of ion lasers (Ar^+ and Kr^+) and photoresist, especially the diazo type, which opened up the possibility to produce high grade gratings. One can only imagine what Michelson would have been able to contribute to physics with recent tools!

Another non-conventional optical method for grating production consists of photographic reduction of larger and coarser gratings. It was proposed in 1872 by Lord Rayleigh [15.4] and is today one of the basic tools for manufacturing metrological gratings using step and repeat cameras. Interested readers can see some details in chapter 9 of ref. [15.5].

Since experimental tools and techniques involved in the recording process of such gratings are similar to those developed for holography, they acquired the name *holographic gratings*, although terms like "photoresist" or "interference" are more descriptive. In order to distinguish them from the gratings recorded in standard holographic materials (dichromated gelatine, for example), they are called in the specialized literature "holographic surface relief gratings."

The principal advantage of such gratings is that, *if properly made*, they are completely free of both the small periodic and the random groove placement errors that are found on even the best mechanically ruled gratings (see Chapters 2, 11, and 14). Although they still share or even slightly exceed the surface roughness that leads to three dimensional scatter, they offer significant advantages in spectrometric systems where stray light is performance limiting, such as near-UV spectrometry and Raman spectroscopy of solid samples. A typical example of a direct comparison between stray light from a holographic grating and a ruled one can be found in Fig.15.1. However, unless the recording optical system components are of diffraction limited quality and unless the mirrors and pinholes are adjusted perfectly, the resulting gratings will not have the near-theoretical resolution that is expected from modern ruled gratings. Any object in the optical system receiving laser illumination will be holographically recorded, will contribute to stray light, and can be observed in simple holographic reconstruction. Ghosts can also appear due to a second wavelength coming from the laser if not properly tuned, or from secondary reflection from such sources as bevels of the blank edges.

Another important advantage of holographic recording, in principle at least, is that it is faster than ruling. The grooves over the entire area are formed simultaneously and there is no diamond wear, so that one can expect the groove profile to be more uniform. Unfortunately, there are two significant disadvantages of photoresist recording compared to ruling. First, the optical recording and the following development of the photoresist does not lead to precise control of the groove form. Second, in order to have a homogeneous exposure and thus uniform groove depth over the surface of large gratings, it is necessary to use high-power lasers with large expansion of the beams, large mirrors, collimators, etc. Thus the key advantage of photoresist gratings compared to ruled ones is often under question. It is valid in two cases: when the lack of precision is not too important, and when the grating parameters fall out of the usual standards. The most typical case are gratings for integrated optics. Laboratory investigations require only small corrugated surfaces, usually over optical waveguides, so that classical replication technology is not applicable. Maintaining wavefront is usually not a requirement, so that holographic recording without high-precision optical and mechanical equipment works well. In fact, it is possible to record small and mediocre gratings without stable and powerful lasers and do without expensive antivibration equipment, whereas it is impossible to build a useful ruling engine

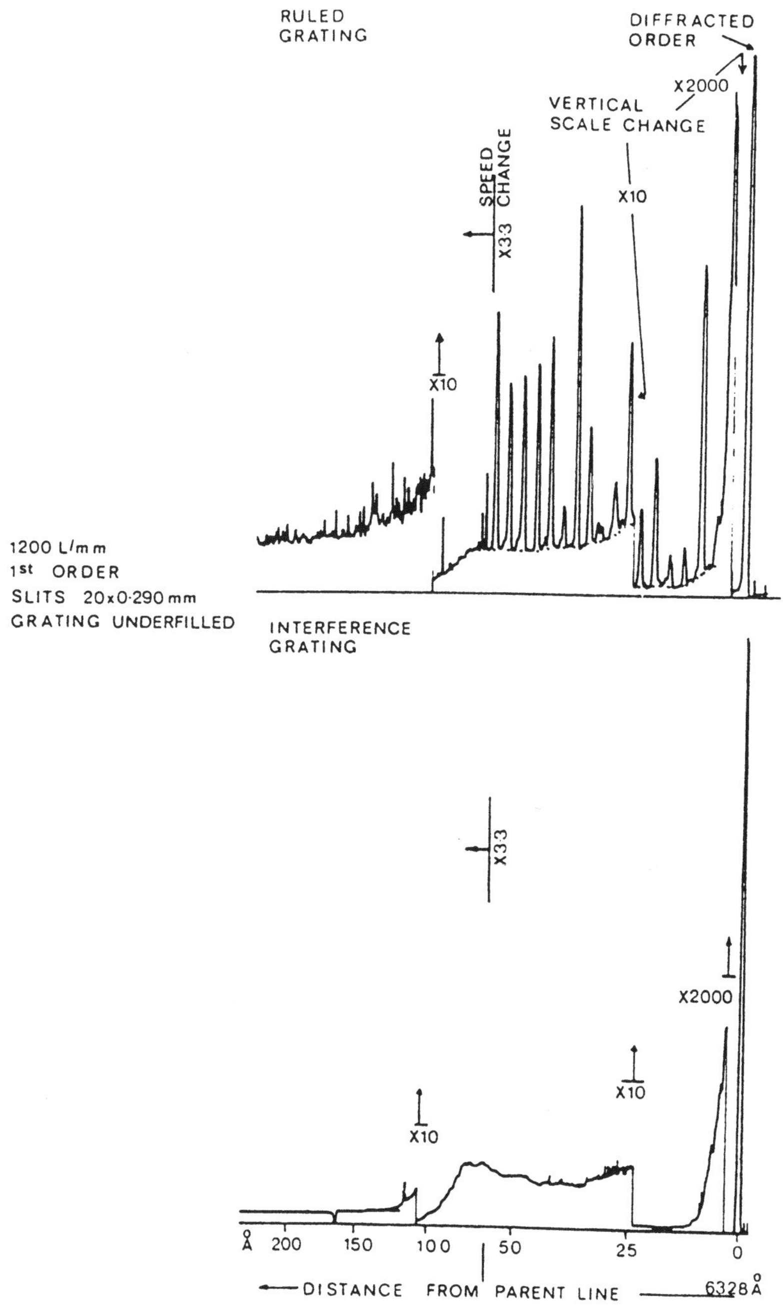

Fig.15.1 Stray light comparison between ruled and holographic grating (after [15.6]).

with simple components.

Otherwise, with the recent replication techniques, high-precision large photoresist gratings are just as expensive and difficult to make as ruled ones.

15.2 Photoresist Layer and Groove Formation

Let us imagine for simplicity that somehow we are able to produce two absolutely coherent perfect plane waves. How to approach this goal and what difficulties are encountered is discussed in detail in the following subsection. The interference pattern of the two waves consists of parallel planes, with intensity varying in the transverse direction according to $\sin^2$ (Fig.15.2). The period of intensity variation is given simply by

$$d = \lambda_0/2\sin\theta \quad , \tag{15.1}$$

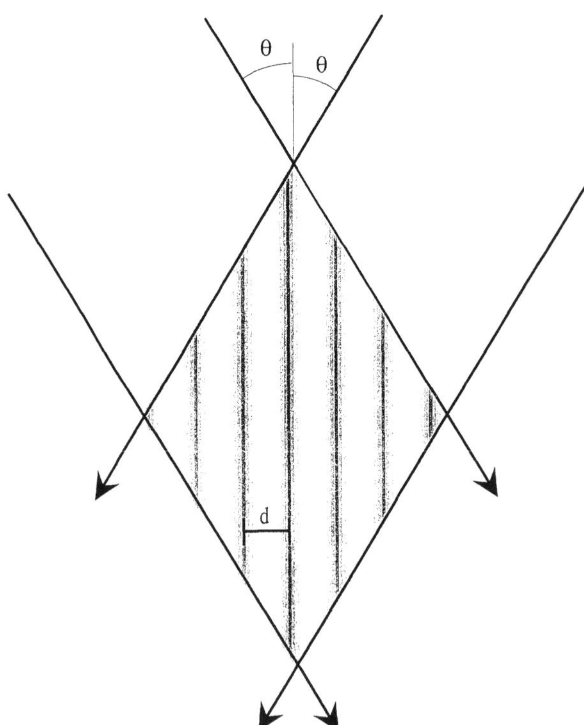

Fig.15.2 The generation of interference fringes of two coherent beams of light (after [15.5]).

where λ_0 is the recording wavelength and θ is half the angular separation of the waves. When this interference pattern illuminates the photoresist layer, some of the polymer connections are broken (provided the wavelength is shorter than some critical value). A developer removes the "broken" regions at a rate much higher than the unexposed surface so that a surface relief pattern is formed with a depth proportional to the exposure (intensity multiplied by time), to the developer concentration and to the development time. The proportionality is not linear [15.7], so that special attention has to be paid to control overexposure of the photoresist. Often recommended is to make a weak pre-exposure with a single uniform beam so that one can work in the linear part of Fig.15.3. Otherwise the groove bottom is slightly sharper than the groove top (Fig.15.4), although that is not a serious disadvantage. More important is to achieve a predictable groove depth.

When the blank is inclined with respect to the incident wave bisector, the groove period is slightly modified acording to eq.15.1, and its form departs from Fig.15.4. The planes of equal intensity of the interference pattern inside the photoresist (they determine the direction of fastest development) are separated by a distance (d) and the resulting groove form is asymmetrical, resembling blazed gratings. From simple geometrical considerations it can be shown [15.5, 9-11] that the effective blaze angle is half the angle of blank inclination. The possibilities of recording holographic gratings in this fashion with asymmetrical profile are discussed in section 15.4.

In order to obtain holographic gratings of any reasonable quality, it is necessary to have a homogeneous photoresist layer with a plane surface

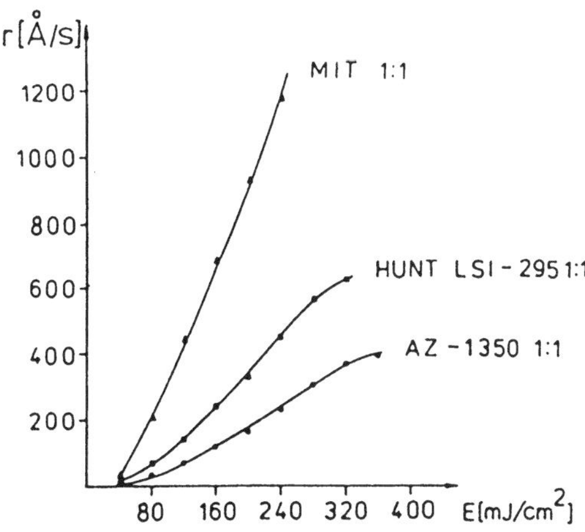

Fig.15.3 Development rate as a function of exposure for different types of photoresists with initial thickness 470 nm (after [15.8]).

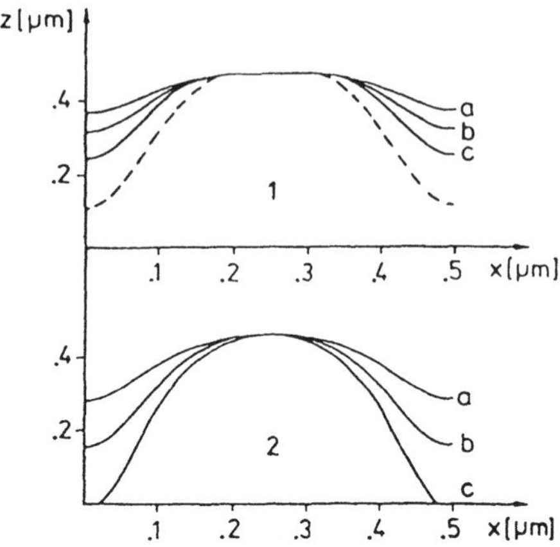

Fig.15.4 Groove shape as a function of (1) exposure energy: (a) 50 mJ/cm^2, (b) 60 mJ/cm^2, (c) 70 mJ/cm^2; and (2) developing time - (a) 20 sec, (b) 30 sec, (c) 40 sec. (after [15.8]).

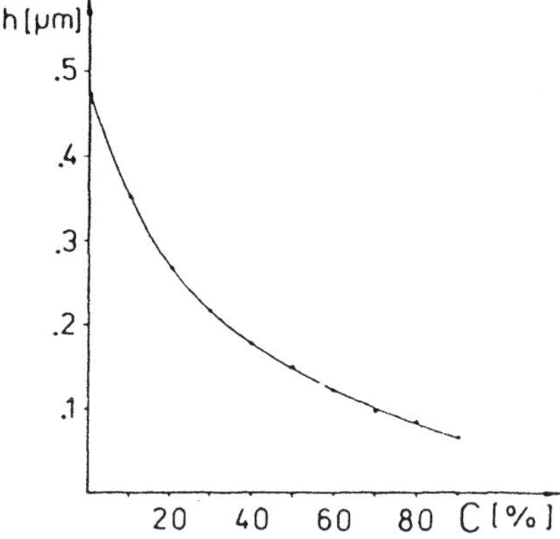

Fig.15.5 Photoresist thickness as a function of a AZ-thinner concentration. Centrifuging at 3000 rpm (after [15.8]).

virtually defect free. There are two main methods to deposit the photoresist, usually sold as a dense liquid: by spinning or by pulling. Smaller blanks are spin-coated by centrifuging. The photoresist is diluted to ensure better

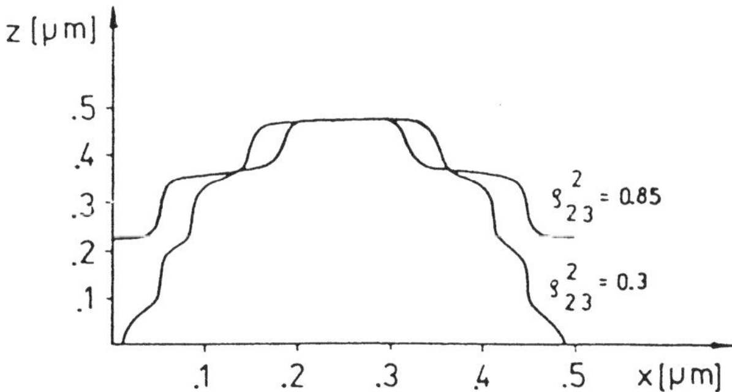

Fig.15.6 Groove shape when the substrate has significant reflectivity r: silver (r=0.85) and GaAS (r=0.30) (after [15.8]).

homogeneity of the layer and control of thickness. The centrifuge must ensure a sharp starting up of rotation. Within 30 - 60 seconds the layer is thin enough and its diluter partially evaporated, so that further spinning no longer affects its thickness, which depends on initial viscosity and on the speed of rotation (Fig.15.5). Any defect of the blank surface or inhomogeneity in the photoresist or dust particles causes comet-like defects of the surface several orders of magnitude larger than the defect, so that the process demands blanks of good quality, photoresist filtering and laminar flow enviroment to control dust. Humidity control is necessary because photoresist properties are significantly affected by humidity (above 50% is necessary, but higher values have harmful effect on auxillary equipment).

Large and heavy blanks cannot be spin-coated due to limitations of the centrifuge or even safety concerns. They are put in tubs full of photoresist and pulled out slowly and uniformly so that when the photoresist flows from their surface, it forms a layer whose thickness depends on its dilution, on the speed of pull-out and on the inclination angle of the blank. An inverse procedure is to drain the tank at a uniform rate while maintaining the blank stationary. Simple as these procedures appear to be, it takes a considerable amount of skill and process control to achieve the desired goals.

Unless there is good index matching between the photoresist layer and the blank, incident light is partially reflected at the blank front surface so that an additional interference pattern is formed due to standing wave formation inside the photoresist layer. The result can be a significant profile deformation, as shown in Fig.15.6. Similar sources of interference appear when light transmitted through the substrate is partially back reflected.

15.3 Two-Beam Symmetrical Recording

The more common of the two basic concepts for plane grating optical recording is to expose a photoresist-coated blank to a fringe field nearly normal to it, generated by the intersection of two collimated monochromatic coherent beams of light. Several typical arrangements of the optical system are shown in Fig.15.7.

The system, presented in Fig.15.7a is the most commonly used, 15.7c reserved for large gratings with central aperture (e.g., for coudé spectrographs of large telescopes). The arrangements seem simple, but call for high-quality requirements in all the elements. First, due to the low sensitivity of photoresist, exposure times are from minutes to hours, so that the laser must work in a single mode regime maintained for a long time. Optically stabilized ion lasers are the most suitable source and are available with light power exceeding 100 mW in the blue, where photoresists are sensitive. Second, the optical elements must not introduce wavefront deformations larger than the aberrations of the optical elements of the device for which the grating is intended. This can easily be understood by assuming that one of the recording beams wavefront is perfect. The aberrations of the second beam are recorded and then reconstructed during usage in the +1st order. Working at shorter wavelengths and higher orders magnifies the aberrations by a factor of

$$\Delta_U(\lambda, N) = \Delta_R(\lambda_0)\frac{N\lambda}{\lambda_0} , \qquad (15.2)$$

where Δ_U and Δ_R denote the usage and recording aberrations, correspondingly, N is the order in which the gratings is used at wavelength λ, and λ_0 is the recording wavelength.

Of course, these simple considerations do not take into account the deviations of the blank from a plane and the defects of the photoresist layer. Let us consider one at a time the optical elements involved in the scheme. They must be firmly mounted on a vibration isolated table heavy enough so that the natural frequency is no more than a few Hertz. One must take into account that a change in the position of any element to more than, say, $\lambda/10$ during the recording lowers the contrast of the grating and introduces moiré patterns. Even a telephone bell ringing in the recording room may be enough to spoil an exposure. Standard holographic tables are the best solution, but must be used in a proper sound-proof room with dust, temperature, and humidity control, fortunately not quite so precise as for the ruling engines, because optical recording is much faster than mechanical ruling. Unfortunately standard optical tables are narrower and shorter than required by such an optical system, so that

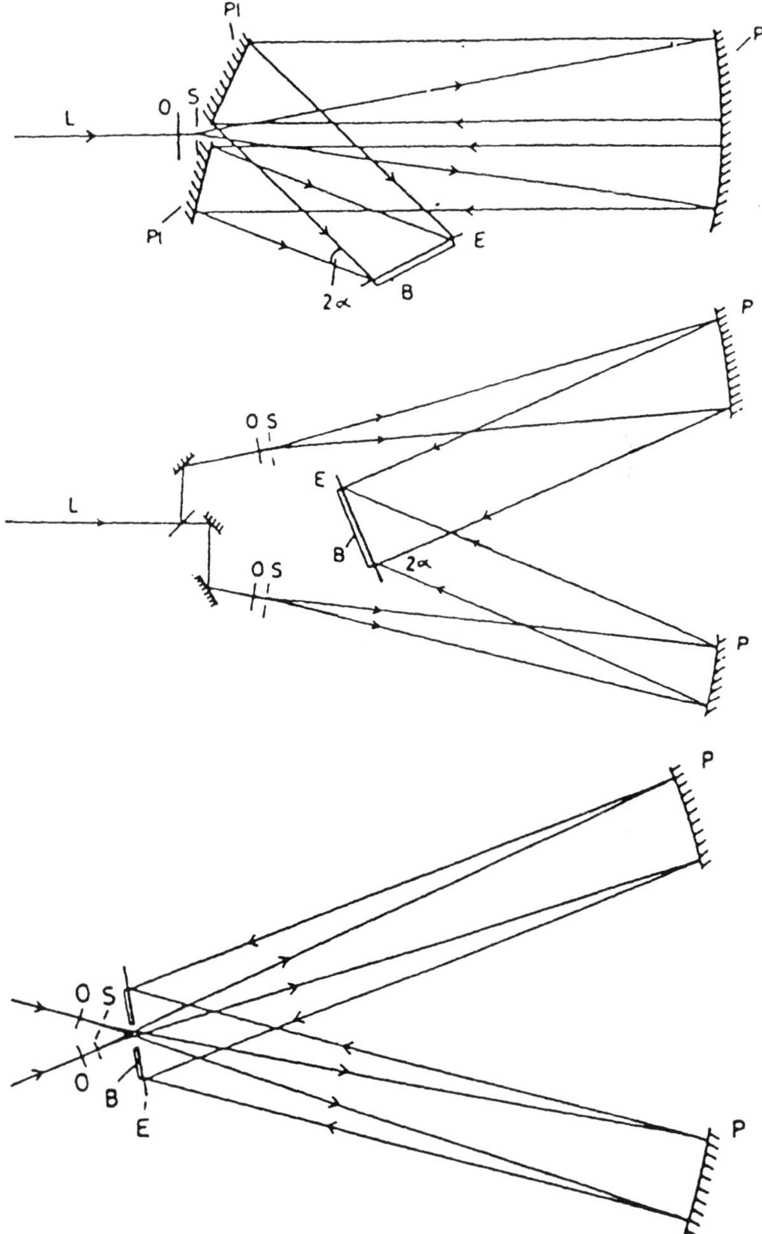

Fig.15.7 Optical arrangements (after [15.12]).

in reality the arrangement is more complicated than shown in Fig.15.7. A summary of different optical schemes and a brief discussion of their advantages and disadvantages can be found in ref. [15.5] (pp.100-102). In brief, transmission optical elements can be used only for small aperture gratings, due to bulk defects in large prisms and lenses.

Involving additional mirrors reduces the recording intensity, but much worse can be their contribution to stray light. Even a blackened metallic holder can reflect a lot of light under grazing incidence, so that the blank holder in Fig.15.7 can act as an additional source by reflecting small amounts of light coming from the pinholes. As a result one can sometimes observe ghosts in holographic gratings due to these parasitic light sources, coherent to the main ones. In addition, holders of the blank, the pinholes and the objectives require micropositioning adjusting screws that enlarge the actual dimensions, so that proper masking and baffling are critical for success. As a consequence, the pinholes must not be too close to the blank and collimating is preferably done with off-axis parabolic mirrors. Spherical ones can serve only when used in on-axis mount or at high f numbers, otherwise their aberrations are too high. Unfortunately, on-axis recording is rare (e.g., Fig.15.7c) and often grating applications require low spherical aberrations, so that again parabolic mirrors are prefered. Unfortunately, they are the last optical element between the recording system and the recorded blank, so that any defect on their surface is holographically recorded and will contribute to stray light level, as discussed in Chapter 11. It is very difficult to achieve in such a mirror simultaneously a near perfect figure and freedom from any polishing marks. Cleaning them also requires great care.

It is not enough to have high quality mirrors. They must also be adjusted to provide the required grating period and to reduce the aberration error due to divergence or convergence of the beams. The angle between the beams can be easily controlled provided a ruled grating with the required period is already available. It is positioned in place of the blank which leads to easily observed moiré fringes. Perfect adjustment corresponds to fluffed out moiré fringes. Much more difficult is to ensure the necessary collimation of the beams. Fortunately, similar aberration of the two beams can serve to mutually compensate. The superposition of two divergent or two convergent beams yields an interference pattern described by two-sheet hyperboloids of revolution, whereas the superposition of one convergent beam and one divergent beam can be described by a family of two-sheet ellipsoids of revolution. Assuming the same angle of aperture (divergence or convergence angle) for the two beams, the hyperbolic case gives imaging errors several orders of magnitudes smaller than the elliptic case. If the aberrations of the recording wavefronts are equal, the angle between their normals is equal along their intersection plane, and the difference of the period of the recorded grating comes from the different inclination of the blank to the interference fringes. However, since the period varies as the cosine of this angle, deviations from the desired period are small. In order to obtain the necessary collimation of the recording beams, several tricks can be applied. The simplest is to rotate the blank successively in a direction normal to each of the beams so that it will focus the reflected light back next to the pinhole that serves as a point source. This ensures that the pinhole is positioned at the focus of the parabolic mirror. The main problem is how to preserve the positions of the objectives, mirrors and the blank during the adjustment of spatial filters and interchanging of blanks. Several simple optical elements can serve as a source of trouble. First, the beamsplitter must be thick enough so that the beams inside it that are due to multireflection can be separated from each other. Second, recording beam intensity must be equalized *after* the spatial filtering in order to have maximum contrast of the fringes, which requires an additional optical element with varying reflectivity to be introduced in one of the beams between the laser and the objectives. Third, there may be several intermediate mirrors. They all shift the beam axis when rotated, so that the position of the beam at the entrance of the objective can be altered. It is common practice to use piezo driven active feedback to maintain the fringe field constant in the face of mechanical drift or even barometric pressure changes. The error signal is derived from photoelectric observation of a moiré fringe reference pattern just outside the exposure area of the blank.

Spatial filtering is necessary to remove from the beam all the rays that are due to scattering before and especially inside the objective. It usually consists of a simple pinhole placed at the objective focus. Objectives with different magnification or aperture require different pinhole sizes. Problems arise when one tries to locate the pinhole and the objective so that the axis of the filtered beam coincides with the center of the mirror, a time consuming operation that requires more patience than skill. Unfortunately, it has to be repeated each time something between the pinholes and laser source is changed. Routine hitting of the table is enough, but even the slightest adjustment of the laser mirror (usually to increase the intensity) moves the beam axis.

Finally, the grating blank holder must be firmly held, but not so strongly as to deform it. Special attention has to be paid to the temperature difference between the holder (and the environment) and the blank, which is usually baked after the deposition of the photoresist layer and kept in a separate location. Otherwise temperature variations during the recording will lead to deformations of the groove period. It is evident that resist exposure time is but a small fraction of this initial set-up time.

15.4 Blazing of Holographic Gratings

One of the main disadvantages of holographic gratings is that their groove shape cannot easily be controlled or modified, unlike ruled ones. Provided the grating works in the -1st diffraction order, and no other can propagate except the specular one, this does no harm, as explained by the equivalence rule (see Chapter 4). Quasisinusoidal grooves (Fig.15.4) are not desirable when several diffraction orders can propagate and when comparatively large efficiency is desired. Several possible approaches are available for modifying the groove shape. Moving from a quasi-sinusoidal to a quasi-trapezoidal form is easy: change the pre-exposure and development process. Asymmetrical (quasi-triangular) profiles can be obtained by three different methods: using a symmetrical recording and asymmetrical ion-beam milling; using asymmetrical 2-beam recording; and finally by a symmetrical multiple (usually 4) beam recording Fourier synthesis.

15.4.1 Asymmetrical 2-Beam Recording

The oldest and most direct method of blazing holographic gratings is to incline the blank with respect to the bisector of the recording beams. Then the planes of maximum intensity along which the photoresist is developed are inclined with respect to the photoresist plane and the resulting profile is quasi-triangular with a blaze angle equal to half the inclination angle of the fringes *inside* the photoresist. However, the index of the photoresist is larger than 1 so that the angle of inclination of the fringes is always smaller than the inclination of the blank. Due to this limitation the effective blaze angle achievable is very small, when the classical 2-beam scheme is used with the beams hitting the photoresist surface from air. Using a prism, index-matched to the photoresist, provides a larger choice of angles of incidence (and thus blaze angles), but not nearly as great as desirable. Michelson [15.13] proposed a recording 1-beam method for obtaining blazed gratings (Fig.15.8), a scheme that greatly resembles the classical experiment of Wiener [15.2]. Much later Sheridon [15.14] proposed to use the classical 2-beam method, with the fringes inclined to the photoresist layer so that the grooves take an asymmetrical form.

Fig.15.9 summarizes all the possibilities, including the Sheridon scheme as a particular case of beams incident along directions 0 and 4. Geometrical considerations determine the range of available blazing wavelength-to-period ratios attained by the different schemes of Fig.15.9 and the results are summarized in Fig.15.10, where the groove frequency ν is simply equal to the inverse period 1/d. Regions of the λ-ν plane, where different numbers of diffraction orders can diffract in Littrow mount, are separated by dashed curves. Above $\lambda/d = 2$ only the specular (zero) order can exist. The limits $2/3 < \lambda/d < 2$

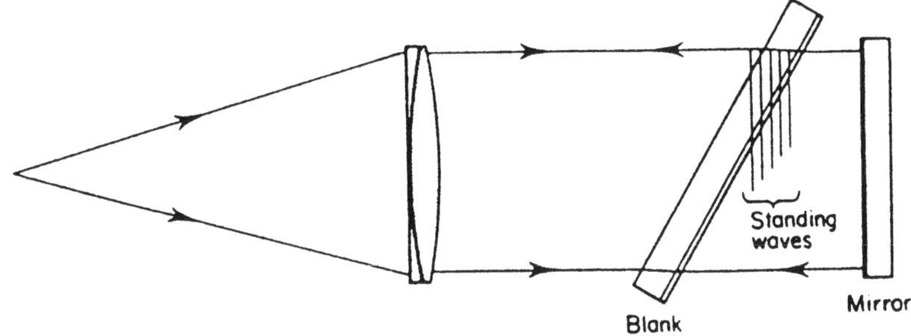

Fig.15.8 A simple apparatus for the production of blazed gratings in photoresist after [15.5]).

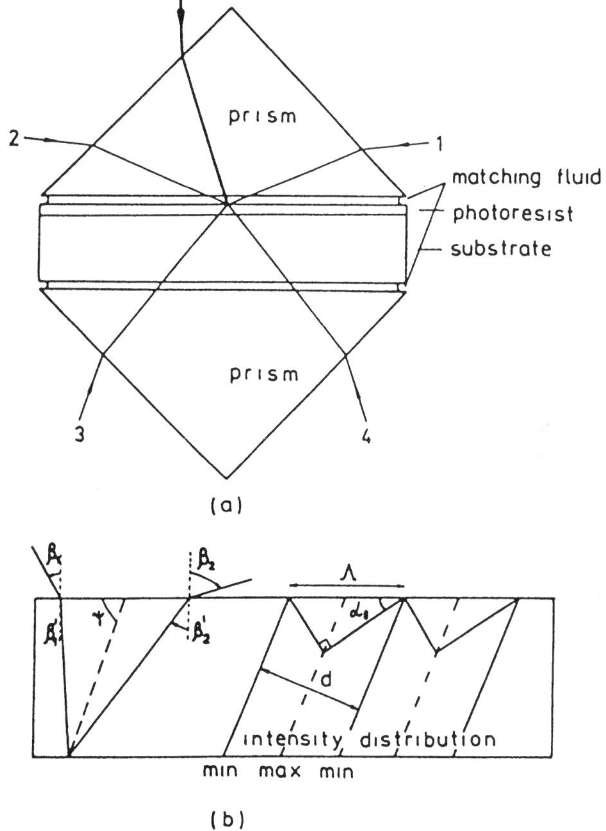

Fig.15.9 Different schemes for 2-beam recording of blazed gratings. The first beam is incident from direction 0 (after [15.9]).

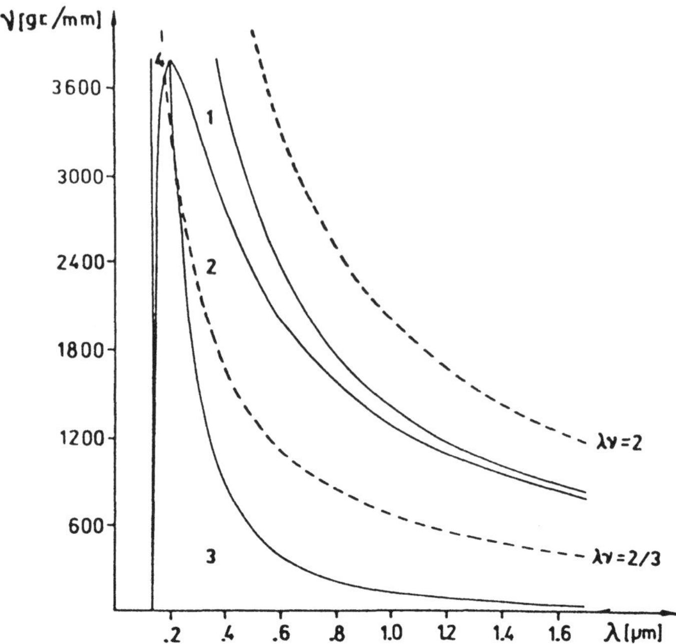

Fig.15.10 Regions in the wavelength-groove frequency plane in which the corresponding schemes from Fig.15.9 can result in blaze profile (after [15.9]).

bound the region with -1st order propagating, so that it is obvious that scheme 1 (recording with two beams from one side of the blank with the blank normal between the beams) does not contribute to groove blazing. In Scheme 4, utilized by Sheridon, the beams propagate almost in the contrary direction and the blazing is obtained in the UV part of the spectrum. Most promising are the schemes denoted by 2 and 3 as they can be used to record gratings with high blaze angles (Fig.15.11). The key limitation is that they require large prisms of high optical quality, because any surface or volume optical defect is recorded holographically in the photoresist and gives rise to stray light.

15.4.2 Fourier Synthesis (Multiple-Beam Recording)

Going back to the equivalence rule (Ch.4), it is obvious that a grating with the desired performance (blazing) does not necessarily require a perfectly triangular profile. It is enough to ensure that the Fourier component of the profile function, which is mainly responsible for the diffraction order under consideration, is equal to the corresponding Fourier harmonic of the desired

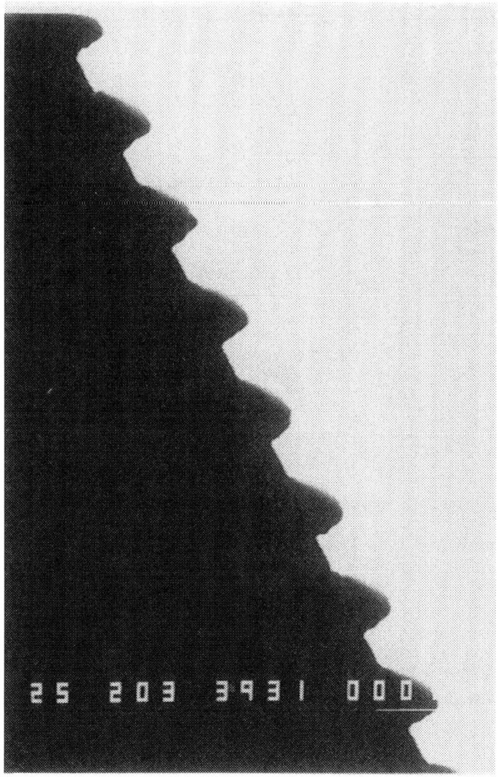

Fig.15.11 SEM photograph of a blazed holographic grating obtained according to arrangement 3 of Fig.15.9 (courtesy S. Tonchev of the Institute of Solid State Physics, Sofia).

triangular profile. For that reason it is necessary to separately record the principal Fourier harmonic as well as the one required for blazing. As the phase difference between these two Fourier harmonics is critical, special attention has to be devoted to obtain and then maintain not only the directions but the phase differences between the beams. All possible solutions have been tried. The first logical possibility is to divide the laser beam into several (at least 4) parts, to make them interfere at the required angles and to control their phases with optical plates. However, this requires a real-time control of the two interference patterns together with their mutual phase difference and the possibility to adjust the phases during recording: a good excercise for an advanced electro-optical laboratory but hardly practical in production.

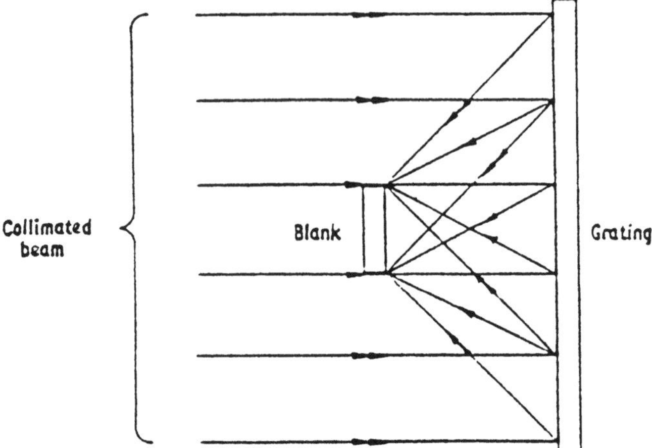

Fig.15.12 Scheme for producing asymmetrical groove profiles by Fourier synthesis from fringe patterns with different spacial frequencies (after [15.15]).

More realistic is to use another grating as a source of multiple beams. The required phase and intensity correlations between the diffraction orders are completely assured by this grating so that the recording scheme becomes quite simple (Fig.15.12) and does not need sophisticated and capricious control elements. The only disadvantage is that both the initial beam and the recording grating must be at least twice as large as the final grating [15.15], a severe limitation.

The third and probably the most promising possibility is to use two-step recording, each one with only 2 beams. At first only a part (for example, an outermost ring of the grating area) is exposed to the two-beam interference pattern corresponding to the fundamental Fourier harmonic of the desired groove profile. Development of the grating at this point does not affect the central part and forms a transmission diffraction grating in the outer ring (Fig.15.13). The Moiré pattern of this grating when put in the same scheme is used to adjust the phases of the two recordings so that the central part is recorded in phase with the outer part. Covering the central part again, an almost exact doubling of the groove frequency can be made with the necessary phase matching the first recording, again using the moiré pattern that covers the reference transmission grating. The two Fourier harmonics that had been recorded in the central region are developed together so that the desired profile is obtained (Fig.15.14). The outermost grating is overexposed during the

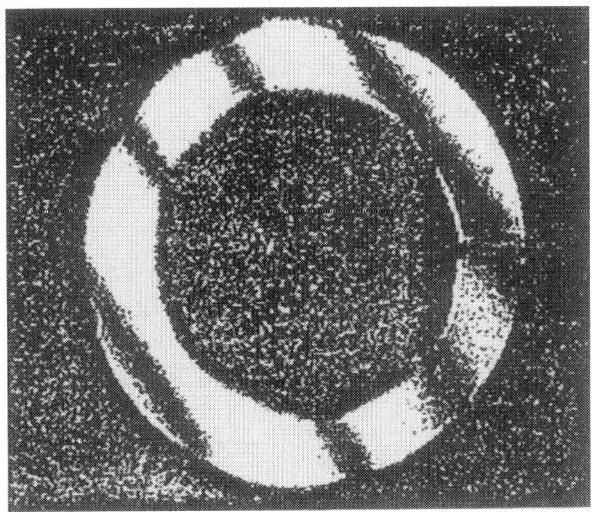

Fig.15.13 Photograph of the interferogram obtained at the reference grating, demonstrating a non-perfect adjustment of both spatial frequency and angular orientation of the irradiance distribution (after [15.16]).

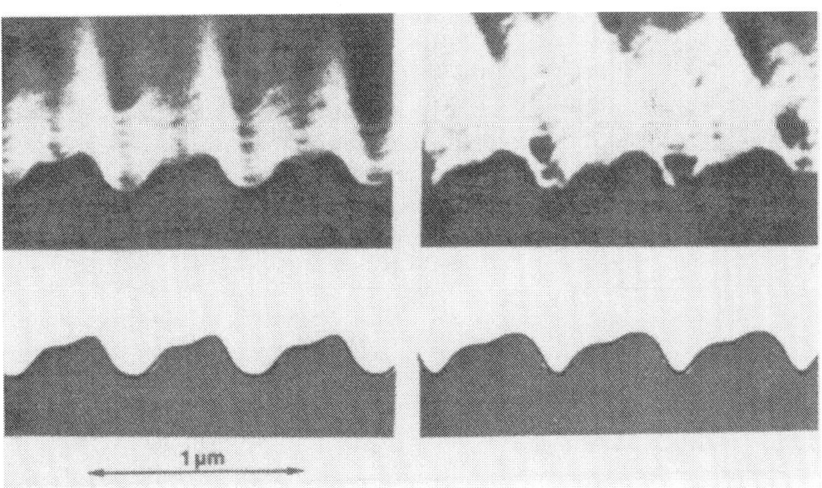

Fig.15.14 Pictures of two different parts of the same Fourier grating: without coating and coated with aluminum by sputtering (after [15.18]).

adjustment so that the photoresist in that area is completely removed during the second development [15.16-18]. This method is limited mainly by the requirement to obtain and preserve almost perfect wavefronts during the adjustment and recording. It is not so easy to take into account that the second exposure must be made at exactly double the fundamental frequency. Difficulties always increase with the depth of the desired groove and are hightened by non-linear response of the resist.

15.4.3 Blazing Through Ion Etching

Two quite different approaches have been used to transform a resist grating into a different shape with the aid of ion bombardment. The first is intended for producing very shallow triangular grooves, blazed for the vacuum UV. One starts with a resist grating made by one of the methods described in 15.4.2. It is then subjected to bombardment by a uniform beam of Argon ions until all the resist has been removed. What is left behind is a blazed grating etched into the substrate, whose groove angle is reduced by the ratio of the rates with which an ion beam removes resist vs. the substrate material. A typical choice for the latter is fused silica. An obvious problem is that there is not a continuous choice of angles and materials, so that only limited blaze angles are attainable.

The second approach is to generate the master grating in resist and give it enough exposure so that after development there is left a series of narrow strips (bumps). For these to be sufficiently uniform over the entire blank is not a simple matter, given the naturally Gaussian intensity distribution of laser beams. This pattern is next exposed to a uniform argon ion beam bombardment at angles fairly steep with respect to the normal, again until the resist is substantialy removed. Provided that the substrate removal rate is large enough compared to the resist, this leaves a grating with a good choice of blaze angles. It turns out that this ends up with a rather restricted. This works with only a limited list of substrate surfaces, principally PMMA (polymethylmethacrylate) and CdS.

15.4.4 The Practical Result of Blazing

As often in life, one gains once and looses twice. Simpler (or cheaper) technology rarely gives better results. Several different approximations to the ideal triangular profile are presented in Fig.15.15. Aiming at No.1, the holographic blazed gratings usually finish with profile No.0. Whereas, due to the equivalence rule (see Chapter 4) the difference is not fatal in the region where only two orders propagate (Fig.15.16), this is not the most interesting case for blazed gratings. Given a greater number of diffraction orders, the deformation of the groove top, which seems inevitable, is enough to drastically

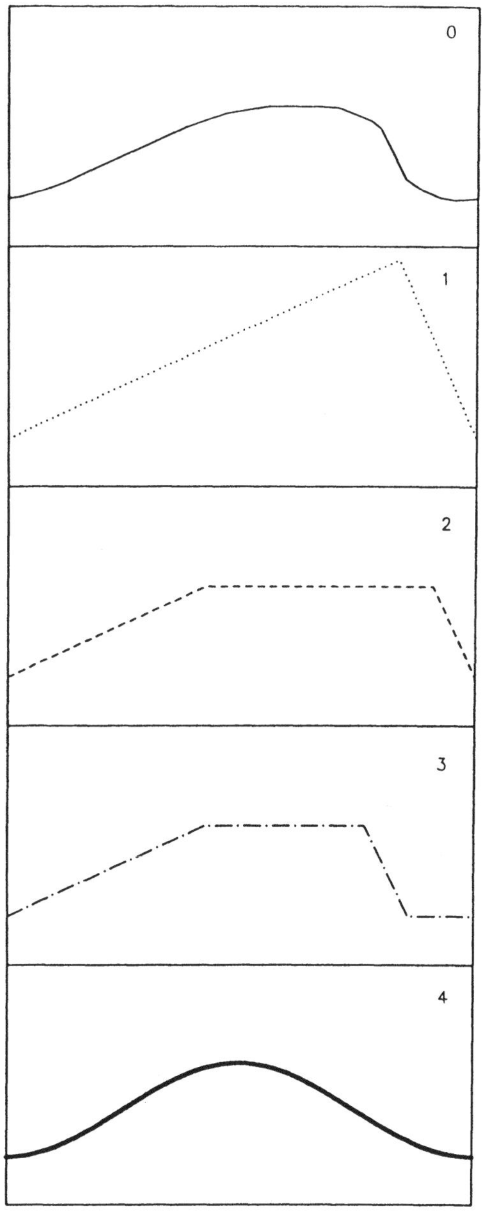

Fig.15.15 Several profiles of diffraction gratings with efficiencies given in Figs.15.16 and 15.17. No. 0 profile of a real blazed holographic grating, No.1 of the ideal blazed one, No.2 and 3 give two intermediate forms, and No.4 the sinusoidal one (after [15.19]).

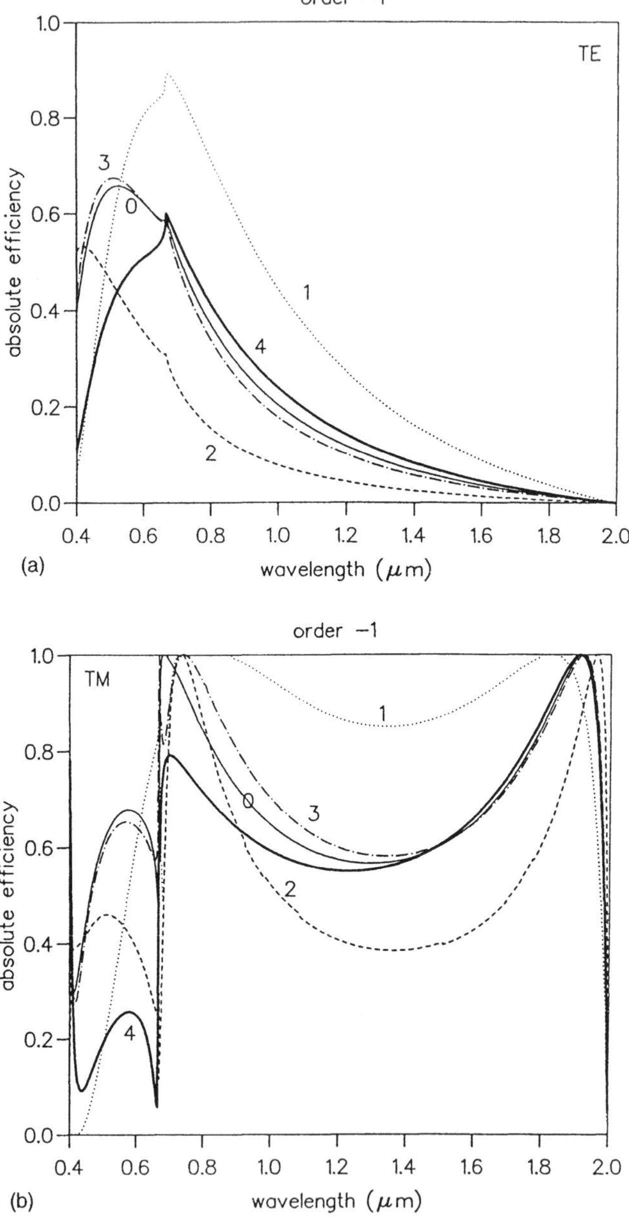

Fig.15.16 Spectral dependence of -1st order in Littrow mount of perfectly conducting gratings with profiles given in Fig.15.15 with period d = 1 μm. (a) TE case, (b) TM case (after [15.19]).

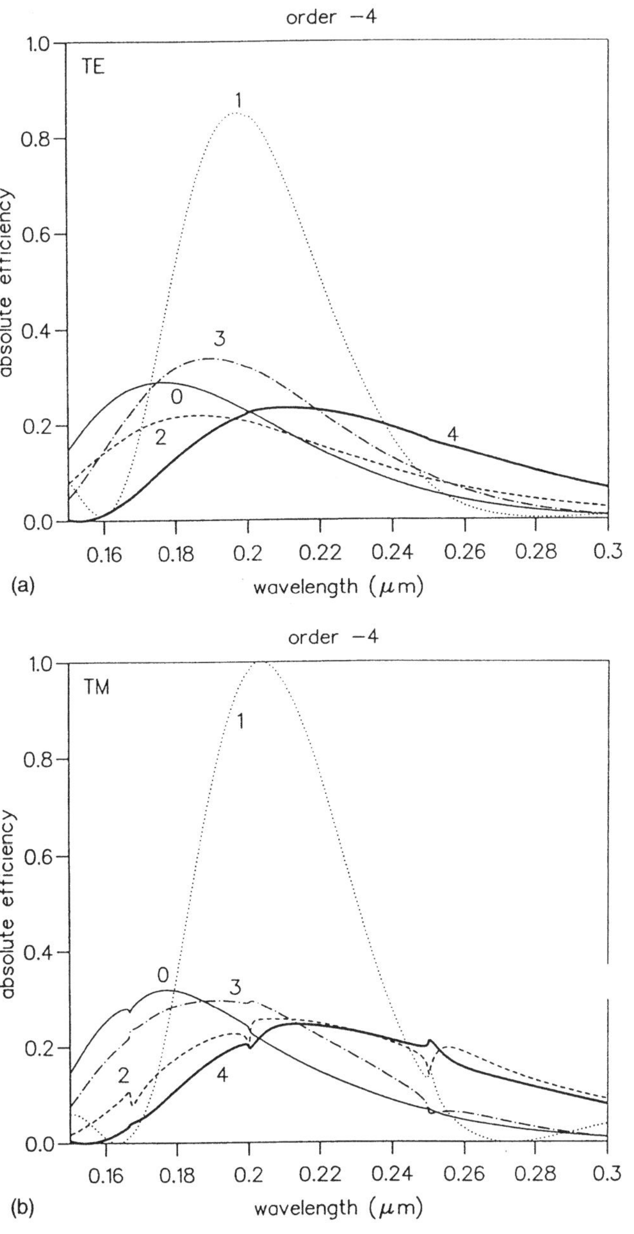

Fig.15.17 As in Fig.15.16 for shorter wavelength in the -4th order Littrow mount (after [15.19]).

disappoint expectations (compare the performance of different profiles from Fig.15.15 in the -4th order Littrow mount as given in Fig.15.17), so that the simplest sinusoidal profile (No.4) has similar performance. Unfortunately, efforts to make the profile sharper or the large facet straighter take away the simplicity of holographic technology. In some important cases, like abberation-reduced concave gratings, there is no reasonable alternative.

References

15.1 A. Cotton: "Gratings obtained by photographing stationary waves," Bulletin des Sciences de la Societe Francoise de Physique, June 7 1901, pp. 71-73.

15.2 O. Wiener: "Stationary light waves and the vibration direction of polarized light (transl.)," Annalen der Physik, **40**, 203-243 (1890).

15.3 A. A. Michelson: *Studies in Optics*, (U. P., Chicago, 1927).

15.4 Lord Rayleigh: "Preliminary note on the reproduction of diffraction gratings by means of photography," Proc. Royal Soc. London, **20**, 414-417 (1872).

15.5 M. C. Hutley: *Diffraction Gratings*, (Academic Press, London, 1982).

15.6 M. C. Hutley: "The spectroscopic properties of interference diffraction gratings," NPL Report MOM **1**, 5-25 (1973).

15.7 F. H. Dill, W. P. Hornberger, P. S. Hauge, and J. M. Shaw: "Characterization of positive photoresist," IEEE Trans. electron. Dev., ED-**22**, 445-452 (1975).

15.8 L. Mashev and S. Tonchev: "Formation of holographic diffraction gratings in photoresist," Appl. Phys. A **26**, 143-149 (1981).

15.9 L. Mashev and S. Tonchev: "Formation of blazed holographic gratings," Appl. Phys. B **28**, 349-353 (1982).

15.10 B. J. Brown and I. J. Wilson: "A numerical study of blazed holographic gratings," Opt. Commun. **20**, 418-421 (1977).

15.11 L. Mashev: "Holographic diffraction gratings," PhD Thesis, Institute of Solid State Physics, Sofia, 1986.

15.12 G. Schmahl: "Holographically made diffraction gratings for the visible, UV and soft X-ray region," J. Spectr. Soc. Jpn **23**, suppl. 1, 3-11 (1974).

15.13 A. A. Michelson: "The ruling and performance of a ten inch diffraction grating," Proc. Am. Phil. Soc., **54**, 137-143 (1915).

15.14 N. K. Sheridon: "Production of blazed holograms," Appl. Phys. Lett. **12**, 316-318 (1968).

15.15 E. W. Palmer, M. C. Hutley, A. Franc, J. F. Verril, and B. Gale: "Diffraction gratings," Rep. prog. Phys. **38**, 975-1048 (1975).

15.16 S. Johansson, L.-E. Nilsson, K. Beidermann, and K. Kleveby: "Holographic diffraction gratings with asymmetrical groove profiles," Proc. Conf. Appl. of Holography and Opt. Data Processing, Jerusalem 1976 (Pergamon, 1977).

15.17 S. Lindau: "Holographic techniques for manufacturing high efficiency spectroscopic gratings," Thesis, Royal Institute of Technology, Stockholm, 1986.

15.18 S. Lindau: "The groove profile formation of holographic gratings," Opt. Acta **29**, 1371-1381 (1982).

15.19 E. Popov, B. Bozkov, M. Sabeva, and D. Maystre: "Blazed holographic grating efficiency - numerical comparison with diffrent profiles," Opt. Commin. **117**, 413-416 (1995).

Additional Reading

Y. Aoyagi, K. Sano, and S. Namba:" High spectroscopic qualities in blazed ion-etched holographic gratings," Opt. Commun. **3**, 253-255 (1979).

R. A. Bartolini: "Characteristics of relief phase holograms recorded in photoresist," Appl. Opt. **13**, 129-139 (1974).

M. J. Beesley and J. G. Castledine: "The use of photoresist as holographic recording medium," Appl. Opt. **9**, 2720-2724 (1970).

M. Breidne, S. Johansson, L-E. Nilsson, and H. Ahlèn: "Blazed holographic gratings," Opt. Acta **26**, 1427-1441 (1979).

O. Bryngdahl: "Evanescent waves in optical imaging," in *Progress in Optics*, E. Wolf ed. (North Holland, Amsterdam, 1973), v. **XXI**, pp.167-221.

L. Cescato, G. F. Mendes, and J. Frejlich: "Fourier synthesis for fabricating blazed gratings using real-time recording effects in a positive photoresist," Appl. Opt. **27**, 1988-1991 (1988).

J. Cowan: "Blazed holographic gratings - formation by surface waves and replication by metal electroforming," in *Periodic Structures, Gratings, Moire Patterns, and Diffraction Phenomena I*, C. H. Chi, E. G. Loewen, and C. L. O'Bryen III, eds., SPIE **240**, 5-12 (1980).

D. A. Darbyshire, A. P. Overbury, and C. W. Pitt: "Ion and plasma assited etching of holographic gratings," Vacuum **36**, 55-60 (1986).

P. Ding and B. Zheng: "Proposed new method for producing blazed holographic gratings," J. Opt. Soc. Am. A **6**, 1228-1232 (1989).

J. Frejlich, L. Cescato, and G. F. Mendes: "Analysis of an active stabilization system for a holographic setup," Appl. Opt. **27**, 1967-1976 (1988).

R. W. Gruhlke and M. F. Becker: "Recording ultrafine interference patterns of evanescent waves at a silver-photoresist interface," J. Opt. Soc. Am. A **9**, 1280-1284 (1992).

M. C. Hutley: "Coherent Photofabrication," Opt. Engineer. **15**, 190-196 (1976).

L. F. Johnson, G. W. Kammlott, and K. A. Ingersoll: "Generation of periodic surface corrugations," Appl. Opt. **17**, 1165-1181 (1978).

J. P. Laude, J. Flamand, A. Thévenon, and D. Lepére: "Classical and holographic grating design and manufacture," ESO Confr. on Very Large Telescopes and Their Instrumentation, v. **II**, 967-989 (1988).

A. Labeyrie: Proc. Conf. Optics, Marseille Centre Nat. d'Etudes Spaciales, Rep. no 00015/PR/ED (1967).

A. Labeyrie and J. Flamand: "Spectrographic performance of holographically made diffraction gratings," Opt. Comm. **1**, 5-8 (1969).

P. Lehmann: "Theory of blazed holographic gratings, J. Mod. Opt. **36**, 1471-1487 (1989).

L. Mashev: "Diffraction efficiency of sinusoidal holographic gratings," Bulg. J. Phys. **11**, 297-304 (1984).

L. Mashev and S. Tonchev: "Diffraction efficiency of blazed holographic gratings," Opt. Commun. **47**, 5-7 (1983).

D. Rudolph and G. Schmahl: "Verfahren zur Herstellung von Röngtenlinsen und Beugungsgittern (Method for producing X-ray lenses and diffraction gratings)," Umschau im Wissenschaft und Technik **67**, 225 (1967).

U. Unrau and R. Nietz: "Quick precision alignment of interferometric equipment," J. Phys. E: Sci. Instrum. **13**, 608-610 (1980).

L. Wosinski and M. Breidne: "Large holographic diffraction gratings made by a multiple exposure technique," Research in Optics, Techn. Report The Royal Inst. of Technolog. v. **210**, 10-11 (Stockholm, 1988).

Chapter 16

Alternative Methods of Gratings Manufacture

16.1 Introduction

There are two classical methods of making the large, high quality gratings used in spectroscopy: mechanical ruling and photographically recording of interference fringe fields. They are likely to retain this dominance, simply because they effectively deliver an accuracy over large areas that other approaches do not match. They are described elsewhere in this book. However, there are several alternative approaches with useful fields of application over small areas which will be discussed in this chapter.

Nature offers one such example, although too slow and restricted in size to be of interest outside the jewelry trade. These are opals, a rare stone that is distinguished by small glittering multicolored internal facets that change color with angle of viewing. This is a characteristic of gratings, which in this instance are formed by hydrated SiO_2 crystals deposited slowly and under special conditions that provide the uniform spacing needed over areas large enough to be observed with a naked eye. Just how special the conditions are can be judged by the fact that opals are found only in two remote areas of Australia and Brazil. There are also a few tropical beetles as well as butterflies which owe glittering colors to a fine pitch (~1 μm) regular surface structures that give them grating type diffraction. Again they can be recognized by color change with angle of viewing.

Two separate developments have set the stage that gives some of the alternative methods important fields of application. One is a desire to generate transmission grating patterns on plane or curved surfaces in order to give them new imaging properties, or at least to modify the refractive properties already present. It represents a new field of optics termed *diffractive optics*. The other base for success is that technology for generating such patterns did not have to be developed from scratch, but could be borrowed from the well developed tools of microlithography. These tools are so effective that it is easy to appreciate why the inverse assumption is sometimes made that all gratings should be more easily generated by their use. While this is hardly the case, there are many useful applications whose methods and limitations are the subject of this chapter.

16.2 Tools of Alternative Methods for Generating Gratings

The distinguishing feature of most of the alternative methods considered here is that individual grooves are still generated one at a time, but by means of energetic beams instead of the diamond tools of classical ruling. This is a critical issue because simple beams do not naturally lead to the triangular blazed groove shapes desired for maximum efficiency. Beams may be in the form of charged particles such as electrons or ions, or alternately take the form of intense narrowly focussed light beams, typically originating from laser sources.

The material on which the beams impinge must have properties that respond adequately (i.e., differentially, to the beams). For example electron beams require layers of electron sensitive materials such as PMMA, while light beams are associated with materials whose development properties are modified by photons (i.e., photoresists). Left open at this point is whether the interaction of beams with their appropriate films leads directly to the desired grating structures, or whether additional lithographic steps are required to achieve the final result. Also left open is the distinction between patterns of straight parallel grooves that resemble diffraction gratings and those that are circular and axi-symmetric in order to provide focussing properties. Another important difference is between two–dimensional patterns, such as Ronchi rulings, and those with the three–dimensional properties needed to get the high efficiency associated with blazing. Most diffractive optic devices are based on transmission optics, but reflection has also been used.

Finally we must consider the geometric accuracies that are needed for the desired goals and what we can expect from modern pattern generators. An important advantage of pattern generators is that they are always computer controlled, which means that patterns can be generated that not only have non–uniform spacing, but can have radii that are also not necessarily uniform. The last step in making a diffraction optic usually involves replication, so that this aspect must also be considered. For example, can it be photographic or must it be by some type of molding?

16.3 The Problem of Blazing

A fundamental problem with diffraction optics is how to achieve maximum efficiency. It is well known (see Chapters 2 and 4) that when the grooves are small enough compared to wavelength, the grating can support only two diffraction orders. In that case high efficiency is attainable with any groove shape, sinusoidal, triangular, or lamellar. Unfortunately diffraction optics

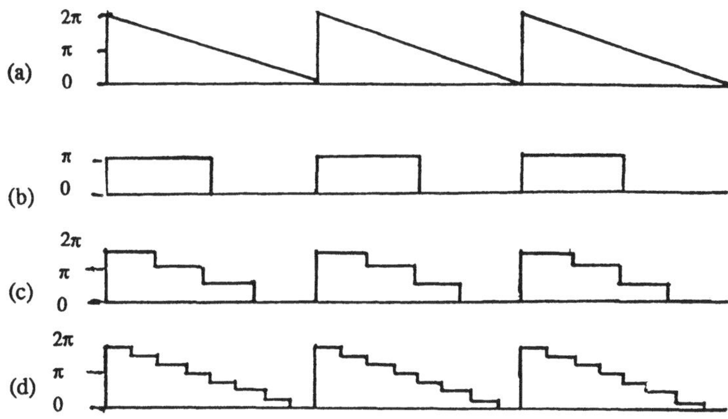

Fig.16.1 Binary transmission grating approximations (b) to (d) of triangular groove (a). Even with multiple diffraction orders (a) can have 100% transmission if reflectance is excluded. (b) has 40.5%, (c) 81% efficiency, and (d) 95% theoretical efficiency.

applications of this chapter never call for such small groove widths and therefore need special control of their shape. Although the goal is a triangular profile, it is generated by a basic *tool* that by its nature tends to generate rectangular grooves. The original solution is to approximate a triangle with a series of rectangular steps, as demonstrated in Fig.16.1. The degree to which a triangular shape is attained can be seen in the figure, and the effect of going from two to four and then 8 steps is quite evident to the eye. As can be noted in Fig.16.1 the diffraction efficiency theoretically attainable increases rapidly with the number of phase steps. The single zone with just two phase levels leads to a usually unacceptable 40.5% efficiency, where it should be noted that reflection losses are not included. Adding a second zone with an appropriate masks leads to four phase levels and doubles the efficiency to 81%. Eight levels provide 95% efficiency. Since the difficulty and cost of preparing such patterns goes up sharply with the number of zones (and masks) they rarely exceed three. In fact, unless the total groove width is large, i.e., intended for long wavelengths, a fairly common practice is to stop at two.

16.3.1 Blazing With Multiple Mask Lithography

Lithographic technology is based on covering the substrate, usually flat, with a layer of photoresist in which is generated a zone plate pattern, either by

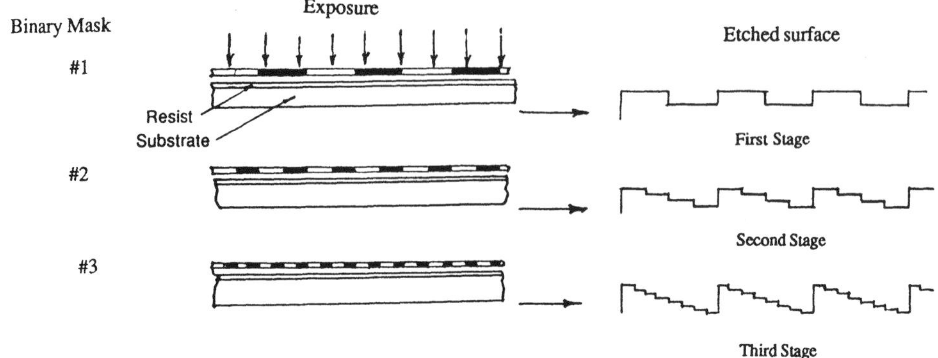

Fig.16.2 Lithographic sequence for 4–step pattern.

contact printing or by projection from an appropriate master pattern. Subsequent development lays bare the areas that are to be etched into the substrate to the required phase depth, either by purely chemical means or chemically assisted ion etching, Fig.16.2. If the resist is not capable by itself to stand up to the somewhat violent action, the process can be modified by adding an extra step. This can be a metallic layer, such as titanium or tungsten between the resist and the substrate, because although readily etched chemically, they strongly resist the impact of ion beams. The substrate can be glass or silica for use in the visible and near IR, or germanium, silicon, or zinc selenide for use in the IR where they are transparent [16.1, 2].

A second masking step is nearly always used, with a line frequency twice that of the first, and serves to double the transmitted efficiency, with grooves etched to half the depth. The accuracy required of both zone patterns is doubled, or perhaps more, in order to allow a small leeway in centering between the two exposures (for circular patterns). Typical tolerances are about 1/4 of the width of the finest zone. Techniques for accomplishing this task are well developed in the microelectronic industry.

The frequent use of such binary patterns has given rise to the term of binary optics for this class of devices. In the case of IR optics that are large and not made in large volume it is common to generate such patterns directly into the final product. For optics used at shorter wavelengths at which epoxy resins are transparent, it becomes economical to replicate the pattern by casting from a master surface, where for axisymmetric patterns it is important to use tooling that maintains the pattern concentric to the optical axis.

16.3.2 Blazing by Direct Methods

An alternative approach, especially using germanium, is to generate the patterns directly into the surface by means of diamond turning on a high precision lathe. One advantage is that in this case the modified surface can be concave, convex, or even aspheric if the lathe is so configured. Whether it makes sense to improve on the behavior of an aspheric by additional diffraction is a matter for the designer to judge. A constant problem is that success (i.e., high efficiency) is contingent on not only making the diamond tools with extremely sharp edges, but also with sharp points that are necessarily delicate. The sharper the point the more it becomes possible to generate a fine pitch pattern, but the finer the pitch the longer the total tool travel and with it the hazards of tool wear. We can think of such surfaces as circular diffraction gratings, except generated by cutting instead of burnishing. Techniques have also been developed for generating grooves by diamond turning of electroless deposited nickel, which then serve as tools for generating replicas, either by injection molding or the other methods, discussed in Chapter 17.

Worth considering is to what extent blazed groove shapes can be generated by direct ablation (i.e., without a transfer from a set of intermediate masks). Fine, but relatively high energy beams impinge directly onto the substrate or onto a specially deposited coating. The choice is between beams of ions, electrons or photons. Blazing of the profile can be done either by using several scans of constant intensity or by varying the intensity during the scan motion within one groove.

The Use of Charged Beams

In principle focused ion beams work well with dielectric materials, but in this application there are two severe drawbacks which restrict their use. One is that it is always awkward to conduct a sizable high precision operation inside a high vacuum enclosure. The other is that in a vacuum there is no air whose convection helps maintain the constant temperature that is essential for achieving structural dimensional stability for adequate periods of time. An example given in reference [16.3] involves writing a 0.72 μm grating over a 1x1 mm square area in silica. Despite a scanning rate of 6 lines per sec, this took 40 minutes. By moving the stage 10 times ('stitching'), a 9.6 mm wide pattern could be generated, but it takes special efforts to get sub-micrometer stitching accuracy. What makes ion beams attractive is that they have no latteral scatter and in addition they need no resist, because the exposed surface in silica has increased chemical reactivity wherever it has been exposed. Etching the surface with dilute HF leads to relatively deep modulation. However, there is no denying that it is an expensive and slow process.

There has been much more activity with electron beams. If necessary

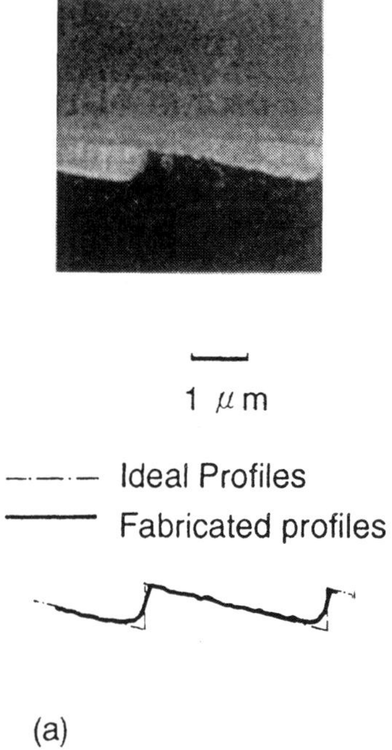

Fig.16.3 SEM cross sectional photographs of blazed grating grooves of an e-beam fabricated micro-Fresnel lens, with grating groove width of (a) 6 μm, (b) 4 μm, (c) 2.5 μm. Profile outlines are compared with ideal shape (dotted) (after [16.4]).

(b)

(c)

they can be focused to spots as small as 0.05 μm, but in the interest of surface smoothness it is usually better to spread the beam by at least a factor of 10. Another reason why excessively narrow beams are of little use is that there is no way to eliminate backward scattering, which effectively spreads out the beam by about one μm inside the PMMA resist. It is a basic handicap, despite valiant efforts to write control programs that attempt to compensate. A 2.5 μm wide groove is difficult to control adequately, as can be observed in Fig.16.3, and even the 4 μm wide groove shows some deficiencies, and only the 6 μm wide groove has a correct appearance. Micro-Fresnel lenses of 4.5 mm diameter are produced by this method, with a numerical aperture (N.A.) of 0.45, which is required for use in reading optical disks [16.4]. The electron beam operates with an accelerating voltage of 20 to 30 keV and a current of about 0.1 nA (i.e., about 2 μW). Note that for a writing spot of 0.1 μm diameter this corresponds to an energy density of 4 MW/cm^2. A modification of this approach uses a negative resist (e.g. chloromethylated polystyrene) that requires a thin conducting overcoat to prevent charge build-up. The pattern aimed for is a Fresnel lens made elliptical in order to use it in reflection at an oblique angle. A 1 mm diameter lens pattern with 0.1 N.A. could be written in about 10 minutes, with an efficiency of 78%, this time in reflection [16.5]. A basic problem of micro Fresnel lenses is that transmission efficiency drops rather sharply as the N.A. increases to the 0.45 level. This is because the larger the N.A. the smaller the width of the outer zone rings. Even if it were possible to achieve good geometry for grooves less than 2 μm wide, i.e., with steep blaze angles, the diffraction efficiency cannot exceed 10% for grooves whose width approximates the transmitted wavelength (see Chapter 5). This is readily appreciated by noting that corresponding blaze angles exceed the angle of total internal reflection.

The Use of Light Beams

To sculpt a triangular groove with a focused laser beam requires a high resolution photoresist, typically the positive type commonly used with microelectronic masks, usually applied by spin coating. The main difference between this process and normal lithography is that resist thickness must be almost an order of magnitude greater than the usual 0.5 μm, which may complicate obtaining the desired uniformity. The need for greater thickness derives from the fact that in order to obtain the one wavelength phase retardation 'om successive grooves in transmission the groove depth h must equal $\lambda/(n-1)$, where n is the index of the final material, usually a replica resin. For $\lambda = 0.9$ μm and $n = 1.6$ the value of h is 1.5 μm. It is clear that groove widths < 2 μm are hardly attainable, no matter how small the light spot. Beam diameter D is determined by the N.A. of the focusing objective from the simple relationship

$$D = 1.2\,\lambda\,/\,\text{N.A.} \tag{16.1}$$

Time to write such patterns is about 4 minutes per mm^2, about twice as fast as with equivalent electron beams [16.6]. Others have succeeded with ion lasers to increase this rate to 1 mm^2 per minute [16.7]. It is clear that at such low rates this is a method for making master gratings only.

16.4 Pattern-Generating Equipment

Classical pattern generators are designed with pairs of slides that move in an X–Y configuration, well suited for the usual microelectronic layouts. The normal approach is to operate them in a modulated raster mode, in which the X–slide moves at a constant rate while the pattern generating beam is turned on and off at high speed, producing either lines or dots. If necessary the intensity can also be modulated by the control program. Before the slide reverses at the end of its travel the Y–axis is indexed by about 1/2 the effective beam diameter. The Y–slide may be configured to lie below the X–axis, or it could take in the form of a bridge to which the beam generator is attached. The latter concept will work only if the load is not too great, which may well exclude electron beam columns.

Slides must move in lines sufficiently straight to achieve the accuracy desired. With air bearing slides 0.1 μm is attainable over a 200 mm travel. Although it is quite possible to incorporate beam steering devices to compensate for known residual slide errors, this adds appreciable complication to the control system. Orthogonality of the two slides must also be held to tight tolerances, where 2 arc-seconds may be considered standard, and 0.5 arc–second exceptional. If that proves inadequate, error compensation is no longer a luxury, but becomes essential.

The slides are usually driven by precision lead screws, whose working accuracy is seldom better than 0.5 μm. To improve on this to the often desired 0.1 μm level use is made of precision scales combined with feedback control loops to impose their accuracy on the drive mechanism. Scales may be replaced by laser fringe counting interferometers, which can be considered a special kind of scale. They have the potential of working at resolutions of 0.05 μm, even less if necessary. A basic problem is that for normal slide speeds (4 mm/sec) the control bandwidth must be quite high (100 kHz) which contributes enough noise so that accuracy may not quite match resolution. An elegant solution that has also been used for grating ruling engines (see Chapter 14) is to adopt a cascade approach [16.6]. The slide itself is driven directly by the lead screw without any special control. However, mounted on top of the slide is a second

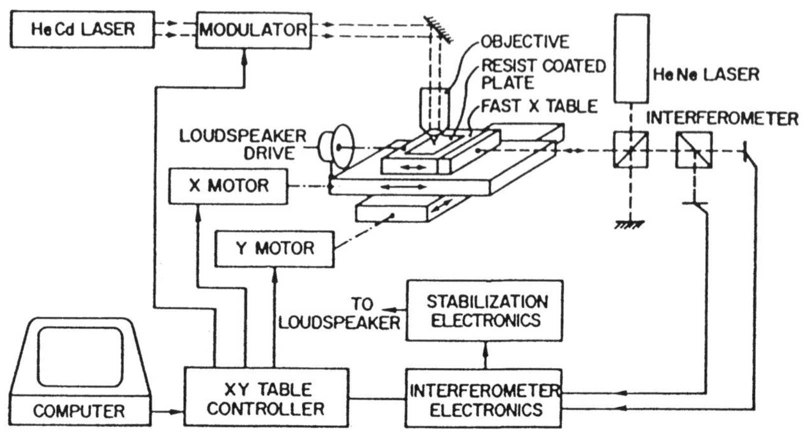

(a)

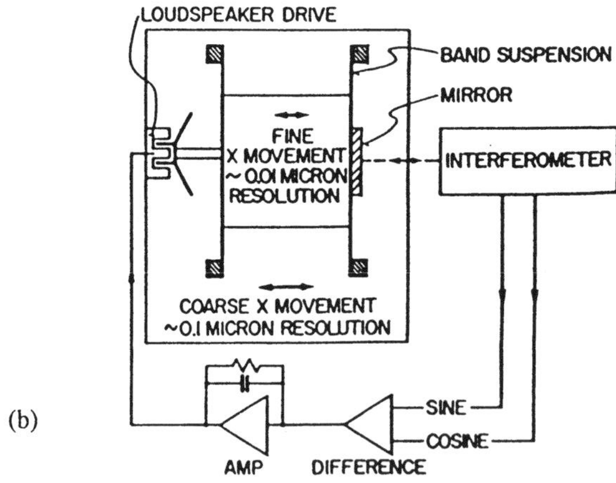

(b)

Fig.16.4 a) System sketch of X–Y laser pattern generator with motor driven slides; b) high speed X–translation stage detail (after [16.6]).

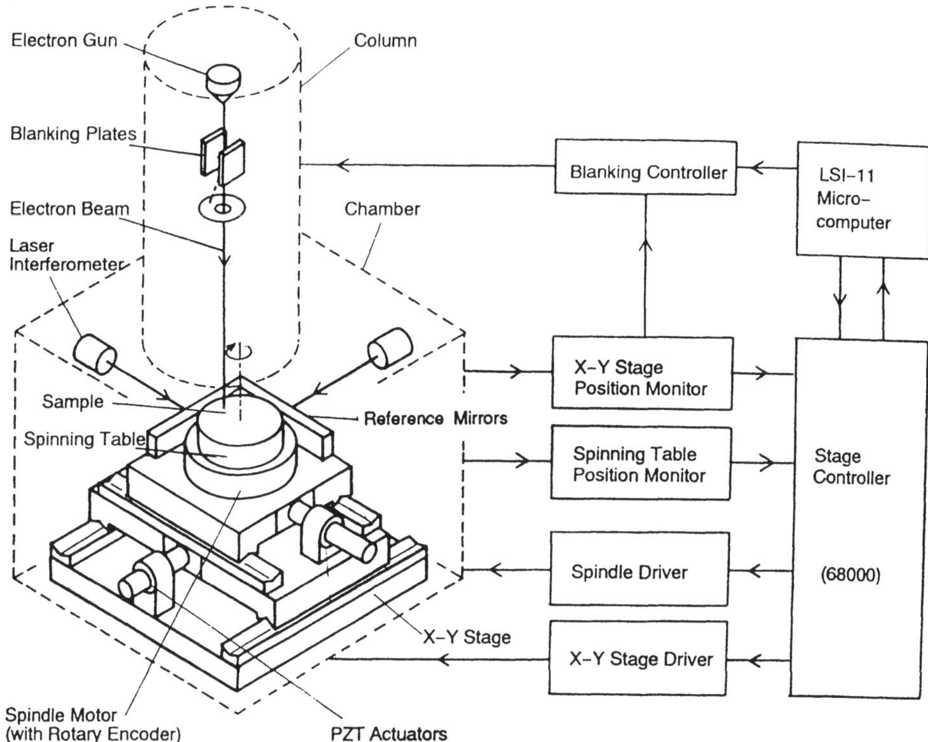

Fig.16.5 Sketch of a proposed e–beam pattern writing system. The X–Y stage is driven by PZT inch worm devices according to position information derived from laser interferometer sighting reference mirrors on the stage that supports the air bearing spin table (after [16.4]).

plate that supports the workpiece and is in turn supported by a set of flexure springs made stiff in all directions except the X. This makes it easy for a high speed actuator, such as a loudspeaker coil or piezo driver, to move the small distances required to make up for the screw errors and do so at high speed, Fig.16.4. In the instance reported in [16.6], dynamic errors are reduced from 80 to 10 nm at a writing speed of 4 mm/sec. It is clear that once the pattern writing process begins the writing axis must remain constant with respect to the mechanical zero of both slide motions, which demands a high degree of temperature control. If circular patterns are to be produced on an X–Y generator there will always be digital discontinuities. The smaller they are the longer the writing process and the greater the demand that is placed on the control computer and its memory. If left too large the microsteps lead to stray light.

A good solution to this problem is to build pattern generators designed specifically for this assignment [16.7, 8]. This means that circular motion instead of being derived by rastering is supplied directly by a precision air-bearing spindle, available with radial run-outs as small as 0.03 μm. The writing head is located on the single remaining slide, with the important provision that no work can begin until the axis of the beam is made to coincide with the spindle axis within a fraction of the smallest zone width. This usually requires several cycles of trial and error. A sketch of such equipment is shown in Fig.16.5. An interesting and useful feature of this system is the use of straight reference mirrors on the edge of the work support table normal to their directions of motion, and 90° to each other. Two interferometer axes sight against these mirrors for position information, their axes intersecting the e-beam axis in the working plane. This approach completely eliminates any Abbé offset errors that so often degrade the accuracy of displacement measurement, but still depends for its working accuracy on the X–Y invariance of the platforms that support the interferometer optics with respect to the beam axis. More easily maintained is the dimensional stability of the workpiece with respect to the reference mirrors.

In some cases it may be desirable to generate small intricate patterns by moving the electron beam with the electrostatic deflection plates. While their resolution is nearly infinite the range over which one can expect linear behavior is finite. Thus if the pattern area is more than 1 mm or so the table must be indexed in steps. Unless the e-beam traverse control is perfectly calibrated there will be 'stitching' errors as this process proceeds. They are very difficult to avoid at the sub-micron level.

16.5 Single Beam Writing with Surface Waves

A process that has more academic than practical interest is to form grating patterns with a single high intensity laser beam impinging onto a flat metallic surface [16.9]. The idea is to take advantage of the presence of micro-roughness which gives rise to a surface plasmon wave capable of automatic phase matching with the incident beam. Given enough power, material can be partially melted in regions of maximum intensity. This leads to formation of a surface relief grating which in turn increases the coupling to the incident beam, and thus again the amplitude of the grating. The incident beam must be intense enough to start the process, but not so strong as to melt the entire surface.

What kind of grating is obtained depends on the material properties (melting point, viscosity near the melting point, thermal conductivity) as well as roughness and beam intensity. To control such a process for useful and repeatable results seems very difficult.

16.6 Photomask Interference Method

In the field of integrated optics there are many applications for small fine pitch grating structures (d = 0.1 to 1.0 μm). It was clear from the beginning that these could not be made by simple contact printing from fine pitch masks (Ronchi rulings). Diffraction effects from the line edges are too disturbing to allow a 1:1 transfer into photoresist. The obvious alternative, which was used in the early stages, was to use the same laser interferometers described in Chapter 15 for making holographic gratings. While adequate for the purpose, they are inefficient and bulky, and as a result have been replaced by what has been termed phase masks, which can be thought of as particular versions of lamellar gratings that were described in Chapter 5.

Special impetus to this concept was given by the important discovery that if the normally pure silica in optical fibers is doped with a small amount of germania it becomes photosensitive in the 240 nm region. This means that Bragg gratings can be generated inside the fiber, which leads to a host of important applications [16.10]. These gratings can act as highly efficient mirrors at a single wavelength, which allows them to act as filters and even as passive devices for correction of wavelength dispersion that spreads pulses in long fiber cables. Other applications are in fiber lasers as well as in

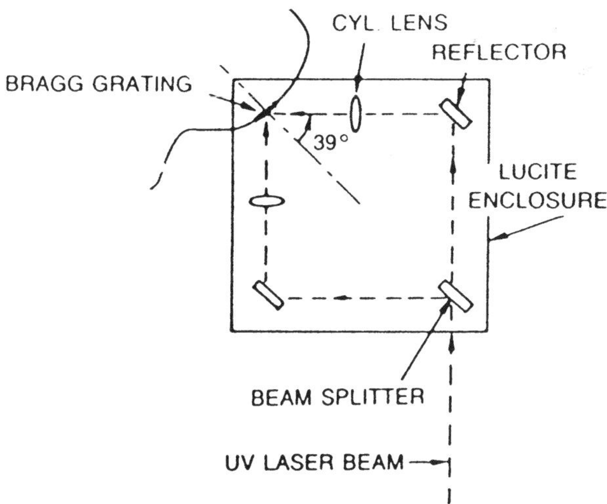

Fig.16.6 Optical schematic of interferometer for generating Bragg fiber gratings by the transverse holographic method (after [16.11]).

multiplexing, as given in Chapter 9.

Reference [16.11] is the first publication to describe the transverse method for generating Bragg gratings inside fibers, which is essential for design freedom for such gratings. It makes use of the standard laser interferometer arrangement although modified by operating in the UV (frequency doubled Ar^+ laser at 488 nm) and by the addition of cylindrical lenses to concentrate light onto the narrow fiber, Fig.16.6.

A glance at Fig.16.7 is sufficient to show how much more elegant the exposure system becomes when the interferometer is replaced by a lamellar grating, generally termed a *phase mask* [16.12]. While many of the gratings used in integrated optics are only 1 mm in width, or even less, fiber gratings

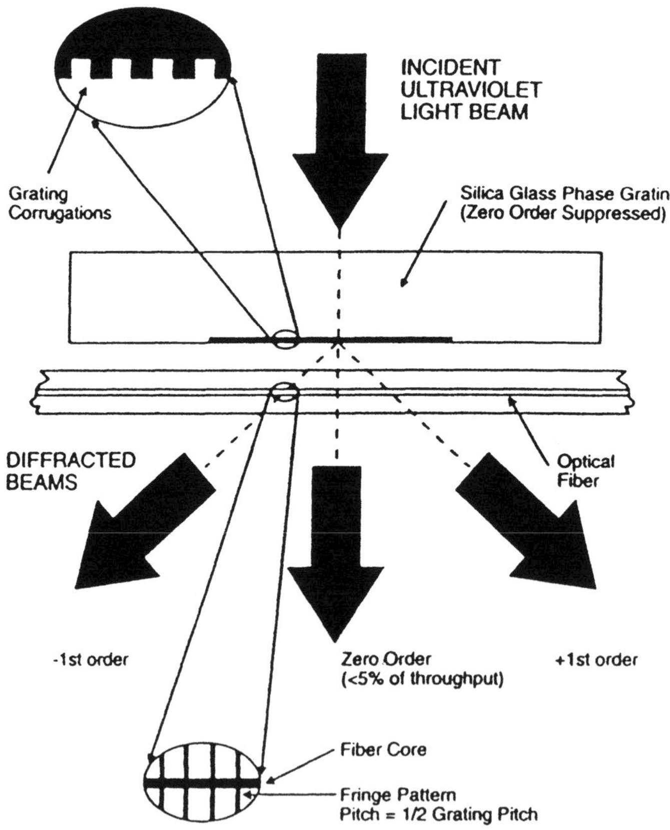

Fig.16.7 Phase mask (lamellar) grating for exposing photosensitive fibers to generate Bragg gratings (after [16.12]).

require much wider gratings for high efficiency, 100 mm or even more. To etch the pattern into the fused silica blank necessary for UV transmission, demands a resist pattern resistant to ion beams or reactive ion etching. Original patterns are usually written with electron beams, as described in sections 16.3 and 16.4. Their control versatility allows writing chirped patterns when desired, but it is difficult to suppress stitching errors for long widths. The patterns can also be produced by standard holographic methods into photoresist.

A required characteristic of such a phase mask is suppression of the zero order. Since for exposures in doped silica UV light is required (n = 1.5) the depth of the grooves h must always be close to $h = 2(n-1)\lambda$, or 0.24 μm. The period of the Bragg grating generated in the fiber will be 1/2 that of the phase mask. Given a mask period of 1.06 μm that of the Bragg grating will be 0.53 μm. The corresponding resonance wavelength will be 0.53 x 2 x 1.46 = 1.547 μm, ideal for long distance silica fiber (where n = 1.46). As long as it remains below 2% the zero order will have minimal effect, and for a 50:50 duty cycle the second order will be negligible, as will be any of the higher orders allowed.

It has been suggested that lamellar grooves in phase masks could be replaced by sinusoidal ones, but this does not seem advisable. Such grooves, which require deeper modulations, will send 10 to 12% of incident light into the second order (see Chapter 5), which is wasteful and can generate higher order harmonic patterns.

An interesting application is the use of a high intensity pulsed ArF excimer laser with a pulse so short (1 psec) that the Bragg grating can be written into the fiber while it is drawn at a sufficiently slow rate (1 m/min) past an illuminated phase mask [16.13].

The lamellar grating (or phase mask) is not the only simple approach to generating a pair of intersecting laser beams. A Ronchi mask can accomplish the same thing, and requires no reactive ion process for generating 0.24 μm deep grooves. However, some special concessions must be accepted. A minor one is that the incident light has to be inclined so that one of the exiting beams can be the zero order, the other the first order, Fig.16.8b. There is no longer a central order to suppress, but the laser has to be polarized with the electric vector parallel to the slits. There is one additional restriction, which is that the spacing between the slits has to be between 1/2 and 3/2 of the illuminating wavelength [16.14]. This approach is hardly suitable in the UV, because the required patterns become so fine that they are very difficult to produce, but at the 0.48 μm wavelength at which photoresist is useful, spacings lie between 0.24 and 0.72 μm, highly suitable for such devices as DFB lasers.

A similar approach to generating DFB laser gratings was developed by Okai [16.15]. It shares with the one-dimensional Ronchi mask just described the concept of an inclined zero order transmission symmetrical with the first order, to generate gratings that have the same groove frequency as the mask. It

(a)

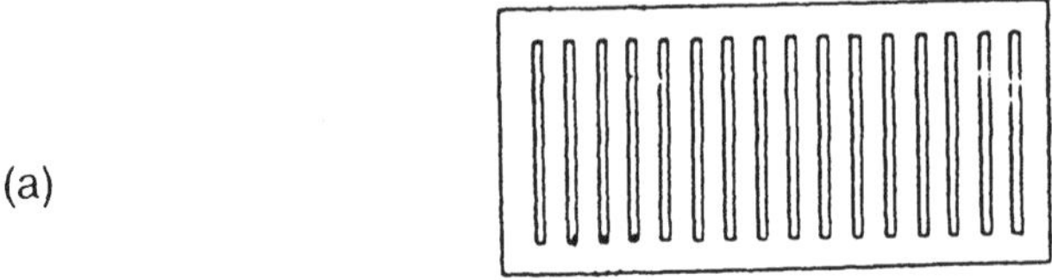

(b)

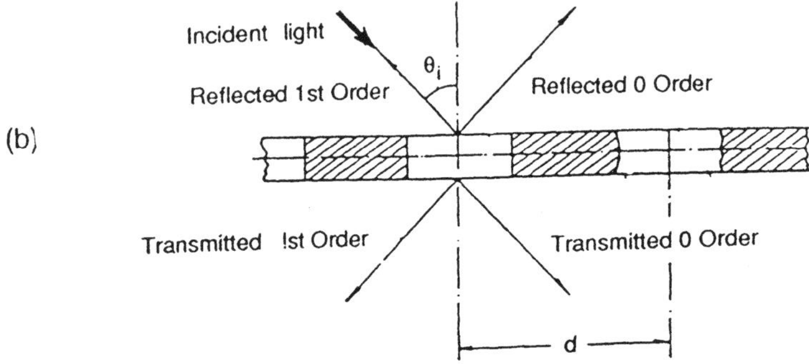

(c)

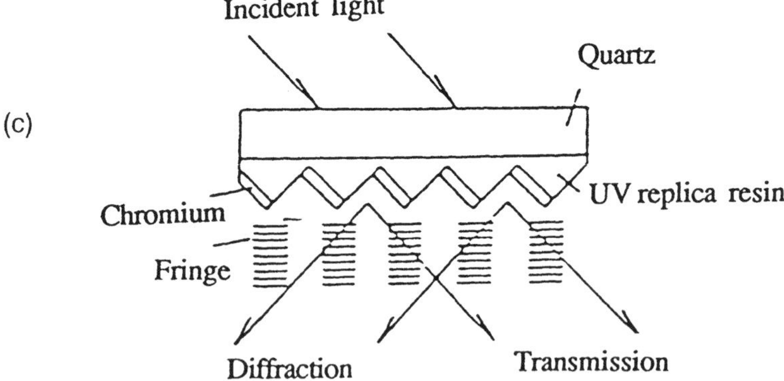

Fig.16.8 Slant illumination phase masks for generating fine pitch gratings: a) schematic of Ronchi mask; b) ray paths (after [16.14]); c) near-field pattern derived from fine pitch ruled grating with chrome applied to one face to equalize the intensities of the zero and first orders (after [16.15]).

differs by using three-dimensional V-grooves to produce the phase difference between successive grooves. It is much easier to produce in the fine pitches required by some DFB lasers. In addition they can be replicated by standard casting methods, provided the resin can transmit in the near UV (325 nm). However, one additional concern needs to be addressed: the two interfering beams must be of roughly equal intensity in order to have maximum fringe contrast. This is achieved by depositing a thin layer of chrome on one face of the groove by slant evaporation, as indicated in Fig.16.8c.

16.7 Single Beam Writing of Fiber Gratings

Special applications often require extremely long fiber gratings with a relatively large period. One example is the case of co–directional mode coupling when the grating vector, inversely proportional to the period, must be equal to the *difference* between the propagation constants of the modes. These applications are found in long range fiber communications in order to compensate for the dispersion [16.17]. Long periods are just as difficult to produce by interference methods as short ones. An alternative approach is to write the grating directly with a focused laser beam properly chopped, Fig.16.9, [16.18]. The chopper frequency and the fiber drawing speed determine the grating period, with no practical limitation as to its length. Chirping of the period can easily be added if necessary.

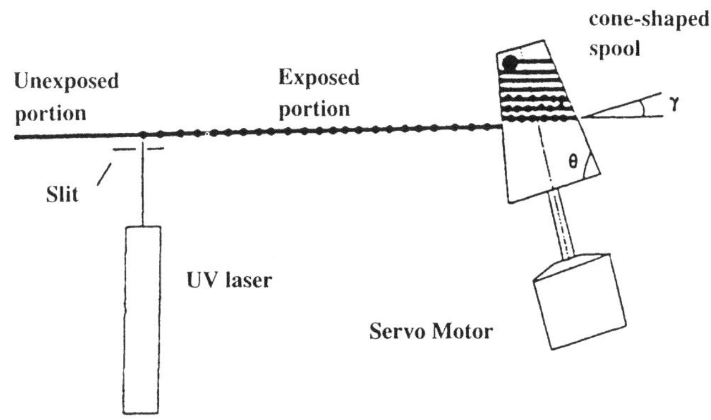

Fig.16.9 Fabrication of a fiber grating with a chirped period using external exposure method (after [16.18]).

16.8 Grating Etched Inside Planar Waveguide

A grating system quite different from any described in this chapter is required for the dispersion element of a monolithic waveguide grating, discussed in Section 12.9. Light diverging from a single mode input fiber is guided to a cylindrical concave grating only 20 μm thick (see Fig.12.22). The diffracted light is directed with the help of suitable waveguides to an array of photodetectors, one for each of the multiplexed channels. Working in the 1.5 μm region they are separated in wavelength by only 1 or 2 nm. The grating is fabricated in a thin layer of SiO_2 grown on a 1 mm thick Si wafer. The grating grooves themselves are the result of reactive ion etching controlled by a photoresist mask on the surface. The mask, in turn, is generated by contact printing from a master pattern made on an e–beam pattern generator. Since the grating surface is about 20 mm long the e–beam process is subject to many stitching steps. Fidelity of groove geometry (groove pitch 8.7 μm) depends on the digitizing steps as well as stitching accuracies, both required to generate a complex geometry from X–Y motions in the pattern generator. Of even greater concern in this instance is the optical fidelity with which the master pattern can be transferred in the contact printing process, Fig.16.10 [16.19]. Inevitable diffraction effects serve to degrade the pattern, especially at the groove tips, Fig.16.10. In principle at least, the transfer problems are minimized if the transfer is made with soft X–ray illumination.

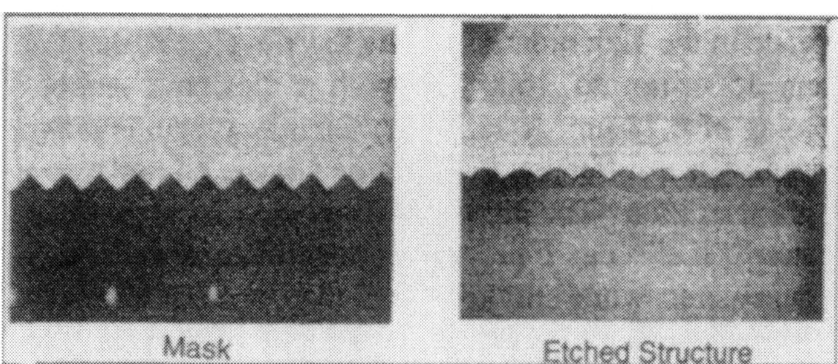

Fig.16.10 Comparison of photographs of the photomask and the resulting etched grating (after [16.19]).

16.9 Conclusions

Alternative methods have become important tools in two entirely different fields of application. One is diffractive optics, which uses relatively coarse groove spacings, but still needs high accuracy in order to maximize efficiency and minimize scatter. Replication is an important facet here in order to produce low cost product, just as with standard gratings.

The other method is phase mask technology, which has carved an important field for itself in the area of integrated optics in general and Bragg gratings inside optical fibers in particular. Here grating replication plays a relatively minor role, in that the purpose of the masks is to generate product directly with the help of lasers.

References

16.1 G. J. Swanson and W. B. Veldkamp: "Infrared applications of diffractive optical elements," SPIE, **883**, 155-162 (1988).

16.2 T. Fujita, H. Nishihara and J. Kozoma: "Blazed grating Fresnel lenses fabricated by e–beam lithography," Optics Letters, **7**, 578–580 (1982).

16.3 J. Albert, B. Malo, F. Bilodeau, D. Johnson, K. Hill, I. Templeton, and J. Brebner: "Fabrication and characterization of submicron gratings written in planar silica glass with a focussed Ion beam," SPIE., **2213**, 78-87 (1994).

16.4 S. Ogata, M. Tada, and M. Yoneda: "Electron–beam writing system and its application to large and high density diffractive optical elements," Appl. Optics, **33**, 2032–2038 (1994).

16.5 T. Shiono and H. Ogawa: "Diffraction–limited blazed reflection diffractive microlenses for oblique incidence fabricated by electron-beam lithography," Appl. Optics, **30**, 3643–3649 (1991).

16.6 M. T. Gale and K. Knop: "The fabrication of fine lens arrays by laser beam writing," SPIE, **398**, 347–353 (1983).

16.7 T. Nomura, K. Kamiya, H. Miyashiro, K. Yoshikawa, H. Tashino, M. Suzuki, S. Ozono, F. Kobayashi, and M. Usuki: "An instrument for manufacturing zone plates by using a lathe," Precision Engineering, **16**, 290–295 (1994).

16.8 V. P. Koronkevich: "Computer synthesis of diffraction optical elements," Chapter 9 in *Optical Processing and Computing*, eds. H. H. Arsenault, T. Szoplik, and B. Macukow, English Transl., pp. 277-313 (Academic Press, N. Y., 1989).

16.9 A. E. Siegman, and P. M. Fauchet: "Stimulated Wood's anomalies on laser illuminated surfaces," IEEE J. Quantum Electron., **QE–22**, 1384–1402 (1986).

16.10 R. Campbell and R. Kashyap: "The properties and applications of photosensitive germanosilicate fibre," Intern. Jl. of Optoelectronics, **9**, 33-57 (1994).

16.11 G. Meltz, W. Morey, and W. Glenn: "Formation of Bragg gratings in optical fibers by a transverse holographic method," Opt. Lett., **14**, 823-825 (1989).

16.12 K. Hill, B. Malo, F. Bilodeau, D. Johnson, and J. Albert: "Bragg gratings in monomode photosensitive optical fiber by UV exposure through a phase mask," Appl. Phys. Let., **62**, 1035-1037 (1993).

16.13 L Dong, J-L. Archambault, L. Reckie, P. Russell, and D. Payne: "Single pulse Bragg gratings written during fiber drawing," Electron. Lett., **29**, 1577-1578 (1993).

16.14 J.-L. Roumigueres and M. Nevière: "Process for casting on a support the faithful reproduction of a mask with periodically distributed slits," U. S. Patent No.4,389,094, (June 21 1983).

16.15 M. Okai, S. Tsuji, N. Chinone, and T. Harada: "Novel method to fabricate corrugation for a λ/4 shifted distributed feedback laser using a grating photomask," Appl. Phys. Lett, **55**, 415–417 (1989).

16.16 D. M. Tennant, T. L. Kocj, P. P. Mulgrew, R. P. Gnall, F. Ostermeyer, and J.–M. Verdiell: "Characterization of near–field holography grating masks for optoelectronics fabricated by electron–beam lithography," J. Vac. Sci. Techn., **B10**, 2530– 2535 (1992).

16.17 O. E. Martinez: "3000 times grating compression with positive group velocity dispersion: application to fiber compensation in 1.3 - 1.6 μm region," IEEE J. Quant. Electron., **QE-23,** 59-64 (1987).

F. Ouellette: "Dispersion cancellation using linearly chirped Bragg grating filters in optical waveguide," Opt. Lett., **12**, 847-849 (1987).

F. Ouellette: "All–fiber filter for efficient dispersion compensation," Opt. Lett., **16**, 303-305 (1991).

16.18 L. S. Tamil, Y. Li, J. M. Dugan, and K. A. Prabhu: "Dispersion compensation for high bit rate fiber-optic communication using a dynamically tunable optical filter," Appl. Opt. **33**, 1697-1706 (1994).

16.19 K. Liu, F. Tong, and S. W. Bond: "Planar grating wavelength demultiplexer," SPIE, **2024**, 278–283 (1993).

Additional Reading

L. d'Auria, J. P. Huignard, A. M. Roy, and E. Spitz: "Photolithographic fabrication of thin film lenses," Opt. Commun. **5**, 232-235 (1972).

J. Albert, K. Hill, B. Malo, D. Johnson, F. Bilodeau, I. Templeton, and J. Brebner: "Maskless writing of sub-micrometer gratings in fused silica by focussed ion beam implantation and differential wet etching," Appl. Phys. Lett. **63**, 2309-2311 (1993).

D. Anderson, V. Mizrahi, T. Erdogan, and A. White: "Production of in-fibre grating using a diffractive element," Electron. Lett. **29**, 566-568 (1993).

J. Cowan: "Blazed holographic gratings - formation by surface waves and replication by metal electroforming," in *Periodic Structures, Gratings, Moire Patterns, and Diffraction Phenomena I*, C. H. Chi, E. G. Loewen, and C. L. O'Bryen III, eds., SPIE **240**, 5-12 (1980).

D. Daly, S. M. Hodson, and M. C. Hutley: "Fan-out gratings with a continuous profile," Opt. Commun. **82**, 183-187 (1991).

M. Duignan: "Micromachining of diffractive optics with eximer lasers," Tech. Digest Series Opt. Soc. Am. v.**11**: *Diffractive Optics: Design Fabrication, and Applications*, Rochester, 129-132 (1994).

T. Erdogan and V. Mizrahi: "Fiber phase gratings reflect advances in lightwave technology," Laser Focus World, (Feb. 1994).

M. T. Gale, M. Rossi, R. F. Kunz, and G. L. Bona: "Laser writing and replication of continuous-relief Fresnel microlenses," Tech. Digest Series Opt. Soc. Am. v.**11**: *Diffractive Optics: Design Fabrication, and Applications*, Rochester, 306-309 (1994).

H. J. Gerritsen and M. E. Heller: "Thermally engraved gratings using a giant-pulse laser," J. Appl. Phys. **38**, 2054-2057 (1967).

E. N. Glytsis and T. K. Gaylord: "High-spatial-frequency binary and multilevel stairstep gratings: polarization selective mirrors and broadband antireflection surfaces," Appl. Opt. **31**, 4459-4470 (1992).

R. W. Gruhlke and M. F. Becker: "Recording ultrafine interference patterns of evanescent waves at a silver-photoresist interface," J. Opt. Soc. Am. A **9**, 1280-1284 (1992).

Y. Handa, T. Suhara, H. Nishihara, and J. Koyama: "Microgratings for high-efficiency guided-beam deflection fabricated by electron-beam direct-writing techniques," Appl. Opt. **19**, 2842-2847 (1980).

M. Haruna, T. Kato, K. Yasuda, and H. Nishihara: "Laser beam periodic-dot writing for fabrication of Ti:$LiNbO_3$ waveguide wavelength filters," Appl. Opt. **33**, 2317-2322 (1994).

K. Hill, B. Malo, F. Bilodeau, and D. Johnson: "Photosensitivity in optical fibers," Ann. Rev. Mater. Sci. **23**, 125-157 (1993).

K. Hill, B. Malo, F. Bilodeau, D. Johnson, and J. Albert: "Bragg gratings fabricated in monomode photosensitive optical fiber by UV exposure through a phase mask," Appl. Phys. Lett. **62**, 1035-1037 (1993).

K. J. Ilcisin and R. Fedosejev: "Direct production of gratings on plastic substrates using 248-nm KrF laser radiation," Appl. Opt. **26**, 396-400 (1987).

R. Kashyap, J. Armitage, R. Campbell, D. Williams, G. Maxwell, B. Ainslie, and C. Millar: "Light–sensitive optical fibers and planar waveguides," Brit. Techn. Jl., **11**, 150-160 (1993).

O. H. Kenneth, B. Y. Malo, F. Bilodeau, and D. C. Johnson: "Method of creating an index grating in an optical fiber and a mode of converter using the index grating," U. S. Patent No. 5,216,735 (1993).

J. Martin and F. Ouellette: " Novel writing technique of long and highly reflective in-fibre gratings, Electron. Lett., **30**, 811-812 (1993).

W. W. Morey, G. A. Ball, and G. Meltz: "Photoinduced Bragg gratings in optical fibers," Opt. Photon. News, pp.8-14 (February 1994).

H. Nishihara, T. Suhara, L. Rothberg, P. Hariharan, K. E. Oughstun, and I. Glaser: "Micro Fresnel lenses," in *Progress in Optics*, E. Wolf, ed. (Elsevier, North-Holland, Amsterdam, 1987), v. **XXIV**, ch. I, pp.1-37.

G. Pakulski, R. Moore, C. Maritan, and F. Shepherd: "Fused silica masks for printing uniform and phase adjusted gratings for distributed feedback lasers," Appl. Phys. Lett., **62**, 222-224 (1993).

J. D. Prohaska, E. Snitzer, S. Rishton, and V. Boegli: "Magnification of mask fabricated fibre Bragg gratings," Electron. Lett. **29**, 1614-1615 (1993).

Lord Rayleigh: "On the application of photography to copy diffraction gratings," British Assoc. Report, ch. 18, p.39 (1872).

P. Russell, J-L Archambault, L. Reekie, "Fibre gratings," Physics World, pp.41-46 (Oct. 1993).

S. M. Shank, M. Skvarla, F. T. Chen, H. G. Craighead, P. Cook, R. Bussjager, F. Haas, and D. A. Hone: "Fabrication of multi-level phase gratings using focused ion beam milling and electron beam lithography," Tech. Digest Series Opt. Soc. Am. v.**11**: *Diffractive Optics: Design Fabrication, and Applications*, Rochester, 302-305 (1994).

J. Söchtig, H. Schütz, R. Widmer, and H. Lehmann: "Grating reflectors for erbium-doped lithium niobate waveguide lasers," SPIE., **2213**, 98-107 (1993).

J.-M. Verdiell, T. L. Koch, D. M. Tennant, R. P. Gnall, K. Feder, M. G. Young, B. I. Miller, U. KOREN, M. A. Newkirk, and B. Tell: "Single step contact printing of Bragg gratings using a conventional incoherent source and a phase mask: Application to a multi-wavelenght BBR lase array," Proc. 6th Europ. Conf. Integr. Opt., P. Roth, ed., Neuchatel (1993) p. 4.8-4.9.

R. Waldhausl, P. Dannberg, E. B. Kley, A. Brauer, and W. Karthe: "Assymetric triangular (blazed) grating couplers in planar polymer waveguides," Proc. 6th Europ. Conf. Integr. Opt., P. Roth, ed., Neuchatel (1993) p.4.14-4.15.

See also the refences to Chapter 9.

Chapter 17

Replication of Gratings

17.1 Introduction

For their first century of use, diffraction gratings represented useful but somewhat academic curiosities. Not until methods were developed in the 1950s that were capable of producing sizable quantities of optically identical replica gratings, could they become the dispersion element of choice in the majority of spectrometric applications: a major revolution in the field. Earlier attempts had no difficulty in replicating local groove geometry, with cast collodion films for example, but even their most careful transfer to a new substrate failed to reproduce the surface figure well enough to qualify as a true copy of the master. Missing was the insight to make a direct cementing transfer from a rigid master to a rigid replica blank, skipping the flimsy intermediate film, first described in the White-Frazer patent [17.1]. It is important to recall here that the properties of a grating are three-dimensional, which is why they must be replicated by a casting process; a photographic approach is of only marginal value.

With a need so great it is hardly surprising that gratings were the first significant application of precision optical replication that was later extended to special mirrors as well as lenses. There seems to be no basic limit to the size that can be replicated, as long as there is appropriate tooling, nor is there a serious boundary to attainable groove frequencies. If sufficient care is taken, a diffraction limited master will produce a diffraction limited replica. Of crucial importance commercially is the fact that replicas are good enough to in turn produce additional replicas. The result is a *replication tree* that can lead to a considerable number of replicas from a single master.

More recently methods were developed to produce gratings by photographic recording in photoresist (see Chapter 15) with which it became possible to produce certain types at a cost low enough to supply them routinely as masters rather than replicas. However, even here replication retains an important role in being able to produce large families of identical gratings.

17.2 The Basic Grating Replication Process

The basic process consists of cementing a replica blank to the master with the aid of a thin layer of a suitable resin, followed by a separation. For this to be successful requires easy separation from the master, in order not to damage it, and at the same time high adhesion to the replica blank. There are several approaches to reaching this goal. The original method, still widely used, overcoats the metallic surface of the master with an extremely thin film of a separating compound, followed by vacuum deposition of the metal surface required on the replica (which may or may not be the same as the metal on the master). The master is then cemented with the aid of a thin film of a low viscosity resin to a carefully cleaned replica blank, allowing the resin to polymerize at a constant temperature, generally a slow process. The final step is the separation, as indicated in the sketch (forgetting the negligibly thin separation compound), except that the groove will be inverted (i.e., what was at the bottom of the master groove becomes the top on the replica), often a useful feature, Fig.17.1. Also, if the master was concave the replica must be convex and vice versa. The number of generations in a replica tree that can be safely used will depend on many factors; some degradation will eventually lead to unacceptable product.

If all has gone well the surface of the replica will be a direct copy of the

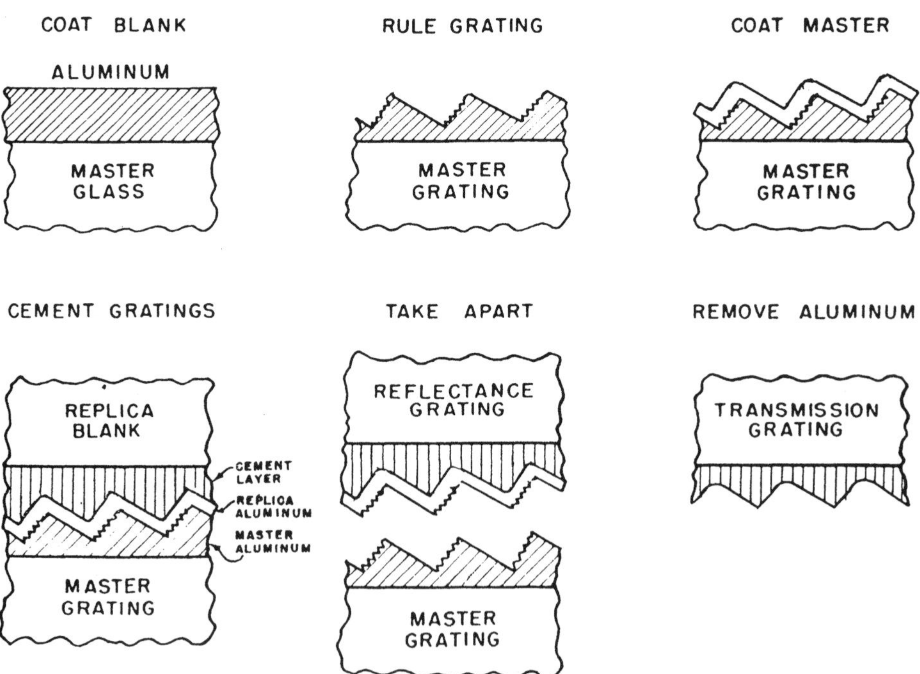

Fig.17.1 The basic steps of the replication process.

original where typically the number lies between three and seven. If that number is taken as n, and if the number of successful replications from one grating is assumed a constant R, the total number of the highest generation gratings that can be replicated from a single master will be R^n. If R and n are assumed to be 8 and 5 respectively the nominal number of fifth generation replicas that can be made from a single master would be 32,768. Restriction to third generation leads to a limit of just 512, showing the enormous influence of this factor.

The alignment of grating grooves with respect to the edge of a blank will be a function of the straightness of the reference edge, the care with which the master is set up with respect to that edge, and the methods used to maintain parallelism of the edges during replication. For ordinary applications a 5 arc min. tolerance is routine and adequate, but for special purposes it is possible to maintain it to a few arcseconds with the help of suitable alignment tooling.

17.2.1 The Substrate

Since replication permanently fixes a grating to a substrate its physical properties such as dimensional stability and thermal conductivity will obviously be important.

Choice of Materials

Most gratings are replicated onto glass substrates simply because glass is a relatively inexpensive and stable material as well as readily producable. Being able to see through the blank aids in monitoring the process. Standard optical glass, such as properly annealed BK-7 is a common choice. If thermal expansion has to be considered, as often the case for larger gratings, glass can be replaced by either Pyrex or, especially for the largest gratings, by ZeroDur™, ULE® or fused silica. For low cost instrument gratings 6 mm plate glass is entirely adequate, except, that to limit mechanical deflection size is restricted to around 50x50 mm. For smaller gratings 3 mm thickness is often used, and for miniature light weight gratings even 1 mm is adequate.

Metallic substrates are sometimes necessary for their special advantages. For example, copper is a far superior thermal sink for gratings subject to high thermal stress, such as generated by high power lasers. Aluminum can serve a similar purpose, and is also appreciated for its stability at very low temperature. Careful surface treatment of aluminum is important to achieve the high adhesion needed to survive the severe effects of thermal cycling involved. Beryllium has by far the highest stiffness to weight ratio of any solid material. This makes it attractive for space spectrometers where mass can be a serious problem and cost is not a factor. Another alternative is to use light-weight composite structures.

Surface Properties

Two substrate properties are of prime importance. One is surface figure (the conformance of the surface shape to that of the next higher generation) the other is surface roughness.

The degree of figure conformance depends largely on the amount of shrinkage that the replica resin undergoes when it polymerizes (see Section 17.2.2). For example 5% shrinkage implies that any mismatch will be reduced by a factor of 20, so that if flatness of $\lambda/4$ is desired the blank must be held to 10 fringes or better.

Surface finish must be free of any scratches deep enough to show through the replica resin. Beyond that a standard optical polish is useful since it simplifies figure testing. However, in some cases the surface must be *grayed* to increase the resin adhesion.

It is easy to appreciate that for ease in controlling separation from a master the replica blank should have the same outside geometry, as then there will be no edge effects at the interface. However, gratings are often needed in many more sizes than it is feasible to rule individual masters or sometimes they need to be made in odd shapes (corners removed, central holes, round, etc.) or perhaps with grooves going right to the edges. This can only be done via replication, and the question becomes whether damage on the submaster surface, resulting from a rim necessarily left over along the edges, can be controlled well enough to allow more than one such replication. Otherwise the submaster becomes restricted to making ever smaller products.

Aspheric Replication

A special case presents itself when the grating shape is aspheric (i.e., when ruled on an aspheric master blank). This is sometimes necessary to achieve special imaging behavior such as can be derived from toroidal, ellipsoidal, or parabolic forms. Since such geometries are always expensive to produce, it becomes highly desirable to replicate them onto the nearest "best fit" sphere. The hope is that resin shrinkage will be low enough to bridge the mismatch in surface topography. As an example we can take a 500 mm focal length 100 mm diameter parabola. Here the optimum spherical radius turns out to be 1000.6 mm and its greatest mismatch is a trivial 0.2 μm. If we double the diameter to 200 mm, the nearest spherical radius becomes 1002.5 mm with a maximum mismatch of 3.12 μm, which is much more difficult to handle.

For larger amounts of mismatch there appear to be just two solutions. One is to polish the blank nearer to its aspheric shape; the other to simply repeat the entire replication process on the same aspheric blank. The choice depends on relative costs.

17.2.2 Replication Resins

A good replica resin must combine several properties, of which low shrinkage is merely the first, since it controls not only surface figure, as mentioned above, but is also necessary to preserve groove geometry through the series of generations. For example, 5% shrinkage would reduce the groove depth of a fifth generation product to $(1 - 0.05)^5$, or 77% that of the master, obviously not acccptable. With 2% shrinkage the blaze would be reduced by 10%, the maximum that could be considered useful. A high degree of adhesion to the replica blank surface is essential both for proper separation and long life. It also assures that thermal expansion behavior of the replica grating is totally dominated by the substrate. The resin should be easy to mix and kept free of air bubbles, have a reasonable shelf life, and be capable of completely polymerizing at room temperature at an acceptable rate. It should survive unchanged at temperatures up to 60°C as well as down to -260°C, and remain unaffected for decades by the vacuum of outer space.

Both polyester and epoxy based resins have been found to meet these requirements well enough. Seya and Goto [17.2] used Epon 828 epoxy resin with diethylene triamine as hardener, or methyl phenylenediamine, according to Hass and Erbe [17.3].

Epoxy cements tend to have lower shrinkage (2 to 6%), as measured from the gel point, compared to polyesters (6-10%), but are more difficult to mix free of bubbles. An important new class of resins has shown promise in optics replication: UV activated epoxies. Requiring no mixing they have a long shelf life, and, once activated by UV radiation, cure at a much higher rate. Unfortunately, shrinkage rate tends to be somewhat higher and the process can work only when at least one of the blanks is transparent to UV radiation.

Adhesion of polyester to glass is relatively low, so that the surface may have to be treated with agents that improve this property. Dew suggests vacuum coated layers of SiO or SnO_2 [17.4]. Alternately the surface my be chemically treated with methacryl hydroxymethyl triethoxysilane. If there is insufficient adhesion to the aluminum layer coated onto the master blank, this may be remedied with the addition of a thin overcoat of SiO. Fortunately epoxies have adhesion strengths to glass and aluminum of about 15 and 30 N/cm^2 respectively, so such help may not be needed.

Filling resins with finely dispersed solids would seem a logical way to reduce shrinkage, but unfortunately it leads to surface roughness that invariably leads to unacceptable stray light. If such particles were metallic they might also serve to increase the otherwise low thermal conductivity of plastics, thereby increasing damage resistance to high light intensities, but again this would increase scatter. Perhaps colloidal particles might avoid such restrictions.

Thickness of Replica Films

The thickness of replica resin films is determined by the viscosity and surface tension of the resin-hardener mixture, but also depends on the blank area. Typical values are 10 μm for small gratings (< 20 mm), increasing to 40 μm for large gratings (300 x 400 mm).

For a given grating the film thickness can be expected to remain constant within 2 μm, provided the blanks are kept level.

17.2.3 High Temperature Resistance

There are applications where gratings must survive at temperatures above 100°C, for example when high vacuum requirements call for outbaking of an entire spectrometer, or when high intensity beams must be handled.

The choice of resins now becomes more difficult, because the higher the glass point, above which dimensional integrity starts to deteriorate, the higher will be the resin curing temperature. Sometimes the solution lies in a partial cure at room temperature, adequate to allow separation, followed by a higher temperature final cure. If that is not satisfactory, the entire replication procedure must be carried out at a higher temperature, taking special care to maintain uniformity.

If the incident light is in the form of very short pulses, the average energy may be relatively modest, but instantaneous values very high. In such cases it may help to make the metallic replica film extra thick (> 2μm) to allow the thermal wave to dissipate without damage to the resin film.

For higher thermal performance, it would be better still if replicas could be made without any resin at all. One possibility is nickel electroforming. Starting with a gold replica in a nickel sulfamate bath, it is not too difficult to produce a gold surface replica on a nickel substrate with excellent reproduction of the groove geometry. Unfortunately the result has limited practical value, because when left in thin plate form (< 0.1 mm thick) it is too flimsy to use if more than a few mm in size; if built up to more rigid sections the result is equally useless, because residual stress invariably leads to excessive deformations.

17.2.4 Environmental Resistance

The principal environmental enemy of resins, especially epoxies, is water vapor. However, if fully cured, the normal metal overcoat of reflection gratings is adequate to protect the resin almost indefinitely. Transmission gratings, by definition, do not have this protection. If the environment is so severe as to be a problem, which is rarely the case, one needs merely to overcoat the grating with a dielectric film, choosing one that can be deposited without heating the substrate, and thin enough ($< \lambda/4$) to cause no optical problems. SiO is the most likely candidate.

Under ultra-high vacuum conditions (10^{-8} mm Hg), as necessary in Synchrotron beam lines, even high temperature resins may outgass more than an acceptable degree, so that, for once, instruments must be equipped with original rulings, preferably with grooves ion etched into the substrate, either metallic or ceramic. The latter is preferred, because if damaged by long time use, it can be chemically stripped of its coating and metallized again.

17.2.5 Transmission Grating Replication

A standard method for making transmission gratings is to prepare them as reflection gratings and use chemical methods to remove the metal layer. Aluminum can be dissolved in both alkaline and activated acid solution, and silver even more easily with potassium ferrocyanide solution. The grating groove surface may be somewhat roughened in the process, but usually with little influence on optical behavior because in transmission geometrical effects on wavefront are a function of the refractive index difference (n - 1), rather than being doubled, as they are in reflection gratings.

For special applications where smooth surfaces are essential, such as certain laser beam splitters, it becomes necessary to make 'direct' replicas, without the presence of an intermediate metal film. Special methods have been developed that can accomplish this task.

One approach is to coat the master first with a thin layer of evaporateable glass or SiO. This can then be given hydrophobic properties through deposition of a partially hydrolized mixture of mono- and dichlorosilane [17.5]. Replication from such a master may be modified by using a UV activated resin [17.6]. It can be converted back to a reflection grating by giving it a final metal overcoat.

17.2.6 Overcoatings

Both metallic and dielectric overcoatings can be applied to replica gratings to obtain special properties. In some cases they are applied after the replication process is completed, in others they go onto the master prior to cementing.

Most frequently used is one that enhances the reflectance of aluminum in the 110 to 180 nm domain, in which the natural 5 nm thick layer of Al_2O_3 becomes more and more opaque. It consists of a rapid deposition of 0.5 µm aluminum, followed immediately by a 25 nm thick layer of MgF_2, which prevents the Al from oxidizing. A small optical enhancement can be noted near 120 nm. If the efficiency is to be maximized near 160 nm, the 25 nm thickness is increased to 40 nm.

The standard λ/4 dielectric layers used for mirrors are rarely effective on reflection gratings, because they give rise to complicated guided wave effects that are seldom of practical value. The coatings lead to strong anomalies in the P-plane where there were none before and if they enhance the efficiency behavior in the S-plane, it is over a limited region and at the expense of dips elsewhere (see Chapter 8).

Overcoatings of heavy metals are sometimes used to enhance reflectance at very short wavelengths, such as 20 to 100 nm. Typical are gold, platinum, iridium, and osmium (if available). Their reflectance values over the wavelengths of interest are discussed in section 4.2.1. In most cases they will serve as the replica metals, rather than overcoatings. Not only does this save a step in the process, but it leads to a smoother groove surface, since, except for the thin separation layer, the deposition is directly on the master surface. An important restriction is that the metals must be deposited at ambient temperature to avoid damage to the replica resin. In no case should a gold replica be overcoated with aluminum: over a matter of months intermetallic diffusion will destroy the grating. The inverse is no problem because of the layer of oxide provides the needed protection.

17.3 Separation of Master and Replica

There are only a limited number of basic techniques available for the critical step of separating master from replica. All have been used successfully, but much depends on the details of techniques that are rarely published. Skill and experience determine how many replications can be derived from a single master or sub-master.

The first is based on wedging the two apart, i.e., with a knife or razor blade, applied perpendicular to the grating grooves at the dividing line. This task is aided by giving both master and replica matching bevels. It is difficult to avoid chipping the edges, especially when the process is repeated several times. An awkward and time consuming task is to remove from the edges any residual resin without causing any damage.

Another successful method is to use thermal gradients to bend the two gratings apart. This calls for one blank to be warmed, and, if necessary, cooling the second one. This cannot work, of course, when both blanks are made of low expansion materials.

The third method uses specially designed tooling to carefully force the two blanks apart, with tooling details being carefully guarded secrets [17.7].

17.4 Replication Testing

The standard methods for optical testing of gratings are found in Chapter 11. Special to testing replicas, aside from obvious cosmetics, is to make sure of adequate adhesion of resin to substrate. A simple and rapid test consists of pressing a piece of high adhesion scotch tape onto the grating surface and then pulling it off with a snap.

Abrasion of replicas, common in mirror and lens specifications, is not applicable to the delicate metal surfaces of reflection gratings, nor does it make sense for the more rugged transmission gratings.

17.5 Multiple Replication

As spectrographs increase in size, especially for astronomical use, gratings are needed that exceed significantly what can be mechanically ruled or readily made by photoresist methods. The classical approach has been to mount families of two or four of the largest practical size onto a common base. There remains the choice of carefully cementing them in place when they are properly aligned to each other, or providing for fine adjust mounts that enable this adjustment to be made "on the job". Both have been carried out successfully, but require a lot of skill and lead to rather bulky systems [17.8]. Such grating mosaics require the grating faces to not only lie in nearly the same plane, but the grooves must also be parallel to each other within a few arcseconds. Fortunately the groove sets do not have to be phase matched because the purpose is always to collect more light, not to increase resolution, but neither should they accidentally end up exactly out of phase, as this would defeat the purpose.

A useful, but difficult alternative, is to replicate the same master two or more times onto a large substrate blank. Great care is required to ensure that the resin thickness is identical (to maintain coplanarity and avoid wedging), while at the same time control the parallelism of several sets of grooves. Another challenge is that the large mass of both submaster and replica blanks involved needs to be suported kinematically, yet in such a way as to limit gravitational deflection to $\lambda/4$. Proof that this is possible is provided by the successful double replication of a 200 x 840 mm r-4 echelle grating, as described in [17.9].

17.6 Alternative Replication Methods

While grating replication with high quality wavefront properties is of obvious importance, the necessary care does not lead to the low cost that the term usually conjurs up. What possibilities there are in this direction can be imagined by looking at compact disc recordings, the accuracy of whose surface features is of the same order as that of gratings. At a unit cost of less than 50¢, however, a good level of flatness is neither achievable, nor necessary. Constant distance is achieved by servo-controlling the readout head.

17.6.1 Injection Molding

Injection molding is a classic low cost process, and in principle could produce gratings by inserting a Ni electroform replica derived from a precision master into appropriate molds. No matter how great the care, the accuracy will always be limited (especially for anything larger than a few mm) by the high temperature of the process, and especially by the inability to provide a truly uniform cooling of the mold. However, for simple transmission gratings, for example Fresnel lenses, where quality demands are not too great, the process is entirely feasible.

17.6.2 Embossing

An even faster method of making grating replicas involves embossing a plastic film by passing it over a heated cylinder under some pressure from a smooth back-up roll. The cylinder will typically have a Ni electroform wrapped around it, whose corrugated surface has been derived from a suitable master. Since this is a continuous process the unit cost will be minimal, but quality is limited to student experiments, or more likely decorative devices such as holograms.

17.6.3 Soft Replication

There is one additional approach to replication which has the advantage of requiring no application of heat, but also has accuracy limitations. It is based on making working submasters by pouring a layer of a suitable silicone material onto the master. Being flexible it is very easy to remove from the master and then make additional ones. Replication involves cementing this submaster to a glass blank with a UV setting resin, which can be cured quite rapidly. It is easily peeled off for further cycles. Accuracy is limited by the very flexibility that makes it easy to use. Obviously it is much slower than embossing. Also the blank must be transparent to UV light.

References

17.1 J. U. White and W. Frazer: "Method of making optical elements," U. S. Patent, No.2,464,738 (1949).

17.2 M. Seya and K. Goto: "Production of replica gratings," Science of Light, **5,** p.46-48 (1956).

17.3 G. Hass and W. Erbe: "Method for poducing replica mirrors with high quality surfaces," J. Opt. Soc. Am. **44**, 669-671 (1954).

17.4 G. D. Dew: "On preparing plastic copies of diffraction gratings," J. Sci. Instruments **33**, 348-353 (1956).

17.5 W. Neumann: "Replication technique for aspheric optical surfaces," Zeiss Information, **30**, 33-35 (1990).

17.6 R. R. M. Zwiers and G. C. M. Dortant: "Aspherical lenses produced by a fast high-precision replication process using UV-curable coatings," Appl. Opt. **24**, 4483-4486 (1985).

17.7 I. D. Torbin and A. M. Nizhin: "Use of polymerizing cements for making replicas of optical surfaces," Optical Technology **40**, 192-196 (1973).

17.8 G. A. Brealey, J. M. Fletcher, W. A. Grundman, and E. H. Richardson: "Adjustable Mosaic Grating Mounts," SPIE **240**, 225-231 (1980).

17.9 T. Blasiak, J. Hoose, E. Loewen, T. Sroda, R. Wiley, S. Zhelesyak: "Grand Grating," Photonics Spectra, **29**, no.12, 118-120 (1995).

Additional Reading

J. A. Anderson: "Glass and metallic replicas of gratings," Astroph.Jl., **31**, 171-174 (1910).

P. Assus and A. Glenzlin: "The replication of optical mirrors," J. Optics (Paris) **20**, 219-223 (1989).

A. P. Bradford, W. W. Erbe, and G. Hass: "Two-step method for producing replica mirrors with epoxy resins," J. Opt. Soc. Am. **49**, 990-991 (1959).

H. Dislich and E. Hildebrandt: "Über ein Verfahren zum Herstellen von Kunstoff Beugungsgittern mit behinderter thermische Ausdehning," ("On a process for making plastic diffraction gratings with reduced thermal expansion,") Optik **28**, 126-131 (1968).

M. T. Gale, L. G. Baraldi, and R. E. Kurty: "Replicated microstructures for integrated optics," SPIE **2213**, 2-10 (1994).

E. Heynacher: "Fertigung asphärischer Flächen durch formgebende Bearbeiting und durch Abgiessen," Optik, **45**, 249-267 (1976).

D. F. Horne: *Optical production technology*, pp.167-170, (Adam Hilger, Bristol 1983).

E. G. Loewen: "Replication of mirrors and diffraction gratings," Tutorial T10, SPIE Intern. Conf., Geneva, April 18 (1983).

M. J. Riedl: "Replicated optics - summary and update," SPIE **1168**, 9-18 (1989).

J. W. Strutt (Lord Rayleigh): "Preliminary note on the reproduction of diffraction grating by means of photography," Proc.Royal Soc., **XX**, 414-417 (1872).

J. W. Strutt (Lord Rayleigh): "On the manufacture and theory of diffraction gratings," Phil.Mag. **XLII**, 81-93, 193-205 (1874).

H. M. Weissman: "Epoxy Replication of Optics," Opt. Engineer. **15**, 435-441 (1976).

Subject Index

ISBN 0-8247-9923-2